## Absolute-Value Inequalities

$|x - a| < b$ is equivalent to $-b < x - a < b$

$|x - a| > b$ is equivalent to $x - a > b$ and $x - a < -b$

## The Number $e$

$e = 2.718281828459\ldots$

## Definitions of the Number $e$

$$e = \lim_{x \to \infty} \left(1 + \frac{1}{x}\right)^x$$

$$e = \lim_{h \to 0} (1 + h)^{1/h}$$

## Definition of a Logarithm

$y = \log_b x$ is equivalent to $x = b^y$

## Natural Logarithm

$\log_e x = \ln x$

$y = \ln x$ is equivalent to $x = e^y$

## Laws of Logarithms

$\log_b MN = \log_b M + \log_b N$

$\log_b \dfrac{M}{N} = \log_b M - \log_b N$

$\log_b M^c = c \log_b M$

## Properties of Logarithms

$\log_b b = 1$

$\log_b 1 = 0$

$\log_b b^x = x$

$b^{\log_b x} = x$

## Changing From Base $b$ to Base $e$

$\log_b x = \dfrac{\ln x}{\ln b}$

# Review of Graphs

## To Find Intercepts

$y$-intercepts: Set $x = 0$ in an equation and solve for $y$

$x$-intercepts: Set $y = 0$ in an equation and solve for $x$

## Vertex $(h, k)$ of a Parabola

Complete the square in $x$ for $f(x) = ax^2 + bx + c$ to obtain $f(x) = a(x - h)^2 + k$. Alternatively, compute the coordinates

$$\left(-\frac{b}{2a}, f\left(-\frac{b}{2a}\right)\right).$$

## Even and Odd Functions

Even: $f(-x) = f(x)$;  Symmetry of graph: $y$-axis

Odd:  $f(-x) = -f(x)$;  Symmetry of graph: origin

## Rigid Transformations

Graph of $y = f(x)$ for $c > 0$:

$y = f(x) + c$, shifted up $c$ units

$y = f(x) - c$, shifted down $c$ units

$y = f(x + c)$, shifted left $c$ units

$y = f(x - c)$, shifted right $c$ units

$y = f(-x)$, reflection in $y$-axis

$y = -f(x)$, reflection in $x$-axis

## Asymptotes

If the polynomial functions $P$ and $Q$ have no common factors, then the graph of a rational function

$$f(x) = \frac{P(x)}{Q(x)} = \frac{a_n x^n + \cdots + a_1 x + a_0}{b_m x^m + \cdots + b_1 x + b_0}$$

has a vertical asymptote where $Q(x) = 0$. The graph has the horizontal asymptote

$$y = a_n/b_m \text{ when } n = m,$$

and the horizontal asymptote

$$y = 0 \text{ when } n < m.$$

The graph has no horizontal asymptote when $n > m$. The graph has a slant asymptote when $n = m + 1$.

# Precalculus

## with Calculus Previews

### FOURTH EDITION

Dennis G. Zill
Loyola Marymount University

Jacqueline M. Dewar
Loyola Marymount University

**JONES AND BARTLETT PUBLISHERS**

*Sudbury, Massachusetts*

BOSTON    TORONTO    LONDON    SINGAPORE

*World Headquarters*
Jones and Bartlett Publishers
40 Tall Pine Drive
Sudbury, MA 01776
978-443-5000
info@jbpub.com
www.jbpub.com

Jones and Bartlett Publishers Canada
6339 Ormindale Way
Mississauga, Ontario L5V 1J2
CANADA

Jones and Bartlett Publishers International
Barb House, Barb Mews
London W6 7PA
UK

Jones and Bartlett's books and products are available through most bookstores and online booksellers. To contact Jones and Bartlett Publishers directly, call 800-832-0034, fax 978-443-8000, or visit our website, www.jbpub.com.

Substantial discounts on bulk quantities of Jones and Bartlett's publications are available to corporations, professional associations, and other qualified organizations. For details and specific discount information, contact the special sales department at Jones and Bartlett via the above contact information or send an email to specialsales@jbpub.com.

**Production Credits**
Chief Executive Officer: Clayton Jones
Chief Operating Officer: Don W. Jones, Jr.
President, Higher Education and Professional Publishing: Robert W. Holland, Jr.
V.P., Design and Production: Anne Spencer
V.P., Manufacturing and Inventory Control: Therese Connell
V.P., Sales and Marketing: William J. Kane
Production Director: Amy Rose
Acquisitions Editor: Timothy Anderson
Editorial Assistant: Laura Pagluica
Marketing Manager: Andrea DeFronzo
Composition: WestWords, Inc.
Cover and Interior Design: Anne Spencer
Senior Photo Researcher and Photographer: Kimberly Potvin
Associate Photo Researcher and Photographer: Christine McKeen
Cover Image: © Ingrid E. Stamatson/Shutterstock, Inc.
Printing and Binding: Courier Kendallville
Cover Printing: Courier Kendallville

Library of Congress Cataloging-in-Publication Data
Zill, Dennis G., 1940-
  Precalculus with calculus previews / Dennis G. Zill, Jacqueline M. Dewar. — 4th ed.
     p. cm.
  Includes bibliographical references and index.
  ISBN-13: 978-0-7637-3779-5
  ISBN-10: 0-7637-3779-8
  1. Mathematics. I. Dewar, Jacqueline M. II. Title.
  QA39.3.Z55 2007
  512'.1—dc22

                                    2006016153

6048

Printed in the United States of America
10 09 08 07 06    10 9 8 7 6 5 4 3 2 1

# Contents

## 6 Conic Sections 329

# To the Instructor

Some years ago, we published a short text called *Basic Mathematics for Calculus*. In the preface of the first edition, we stated:

> An instructor is always faced with the dilemma of too much material and too little time. In the vast precalculus market, one can find texts that cover everything from topics in elementary algebra to topics in matrix algebra. . . . We feel there is a great need for a text that quickly gets to the heart of the matter, a text that presents only those topics that will be of direct and immediate use in most calculus courses.

After three editions, *Basic Mathematics for Calculus* faded into obscurity as the original publishing company was merged into a larger company. In the intervening years, we have not changed our opinion. If anything, we feel more strongly that many precalculus texts fall short in thinking about the specific mathematics that are needed in the study of calculus. Because modern texts tend to be almost encyclopedic in length, the instructor can either move through many chapters at a very fast pace or proceed through a reasonable number of chapters leaving much of a very expensive book unused. This is why we decided to reprise our *Basic Mathematics for Calculus*. The present text, with a name change, represents a substantial revision of this earlier work.

## ☐ Philosophy

The following list reflects some of our pedagogical philosophy that underpins this new edition.

- Throughout the revision, we held firmly to our belief in a "no nonsense" approach to precalculus. We have deliberately kept the coverage to a reasonable number of topics. The six chapters that comprise this text can easily be covered in a one-term course. Our style is informal, intuitive, and straightforward—we have avoided the theorem-proof format. We try to talk directly to the student.
- We emphasize the basics, especially algebra. Through the many examples and numerous and varied exercises, we provide opportunities for students to practice operations such as factoring, expanding a power of a binomial, completing the square, synthetic and long division, rationalization, and solving inequalities and equations in situations similar to those they will encounter in calculus. Throughout, we stress the importance of being familiar with key formulas from

algebra, the laws of exponents, the laws of logarithms, and fundamental trigonometric identities.

- The topics presented in this text are those we feel are essential for success in calculus courses. Thus we do not cover topics such as matrices, sequences, induction, or probability. This should allow time for the instructor to work with their students to strengthen their algebraic, logarithmic, and trigonometric skills.

- We pace the introduction of new topics. Students can be overwhelmed when a text presents various types of functions and related functional concepts all in one or two sections. Hence our pace, especially in Chapter 2, *Functions and Graphs*, is more deliberate. For example, we hold off introducing word problems until after essential function concepts are presented. Similarly, the important topic of piecewise-defined functions is delayed and is given its own section.

- Throughout the text we envision the course as a bridge to calculus. In particular, we use some of the terminology of calculus in an informal way to acclimatize the student to these terms. For example, we use the words "continuous function" when describing graphs of polynomial and exponential functions and "discontinuous functions" in the context of rational, piecewise-defined, and logarithmic functions. When the concept of secant lines for the graph of a function is introduced, we use the words "difference quotient" to describe their slopes.

- In calculus it is extremely important to be able to sketch graphs of basic functions and equations quickly and accurately. Therefore we have placed a great emphasis on honing the student's understanding of how concepts such as intercepts, symmetry, rigid and nonrigid transformations, asymptotes, and end behaviors, in conjunction with recognition of the type of function or equation, are valuable aids in sketching its graph by hand. The use of technology is limited to problems of the sort where this mathematical analysis fails. These problems are placed near the end of an exercise set and are clearly marked.

- We firmly believe that a figure should be used to illuminate an example, a discussion, or a problem in the exercises whenever possible. So there are numerous figures in this text—approximately 1600.

- As in previous editions, our approach to trigonometry is through the unit circle.

Here are some items that are new to this edition.

## ☐ Features of This Revision

***An Emphasis on Algebra*** Many times we have seen students in a calculus class perform an operation such as differentiation flawlessly but fail to complete the problem because they had difficulty simplifying the resulting expression or solving a related equation. So as mentioned above, in this revision we have made a pointed effort to reinforce algebraic skills. Marginal side notes and in-text annotations fill in the details of solutions of examples and convey additional information to the reader.

***Translating Words into Functions*** As teachers, we know that the related rate and applied max-min, or optimization, problems can be a discouraging experience for some students of calculus. Typically, correctly interpreting the words of such a problem in order to set up an equation or a function presents the greatest challenge for many students. It follows then that it is appropriate to emphasize such material in a precalculus course.

In Section 2.8, *Translating Words into Functions*, we begin by illustrating how to translate a verbal description into a symbolic representation of a function. We then present actual problems taken from a calculus text and demonstrate how to decode the statement of the problem and transform those words into an objective function. We discuss the importance of drawing pictures, using variables to describe pertinent quantities, identifying a constraint between the variables, using the constraint to eliminate an extra variable, and observing that the domain of the objective function may not be the same as its implicit domain. To ensure that the focus is squarely on the process of fashioning a symbolic function from the words, we have chosen not to discuss how such optimization problems are actually solved.

*Notes from the Classroom*  Selected sections of this text conclude with remarks called *Notes from the Classroom*. These remarks are aimed directly at the student and address a variety of student/textbook/classroom/calculus issues such as alternative terminology, reinforcement of important concepts, what material is or is not recommended for memorization, misinterpretations, common errors, solution procedures, calculators, and advice on the importance of neatness and organization.

*Calculus Previews*  The chapters in this text conclude with a section subtitled *Calculus Previews*. Each of these sections is devoted to a single calculus concept:

- Chapter 1, Section 1.5: *Algebra and Limits*
- Chapter 2, Section 2.9: *The Tangent Line Problem*
- Chapter 3, Section 3.7: *The Area Problem*
- Chapter 4, Section 4.10: *The Limit Concept Revisited*
- Chapter 5, Section 5.4: *The Number e*
- Chapter 6, Section 6.7: *Parametric Equations*

In these sections, the discussion is kept at a level easily within the reach of a precalculus student. The emphasis is *not* on the calculus; the calculus topic provides a framework and motivation for the precalculus mathematics we discuss. The focus in these sections is on the algebraic, logarithmic, and trigonometric manipulations that are necessary for the successful completion of typical calculus problems related to the *Calculus Preview* topic. Consequently the *Calculus Previews* are intended to be taught as part of a regular course in precalculus mathematics. In *Algebra and Limits* we examine the analytical calculation of a limit as $x \to a$, where $a$ represents a real number. The material is presented in such a manner that the instructor has a choice: he/she can choose either to review the important algebra of simplifying fractional expressions using binomial expansions, factoring, common denominators, and rationalizations or to go the extra step and actually calculate a limit. Similarly, we discuss *The Tangent Line Problem* and then examine the four-step calculation of the limit of the difference quotient $\dfrac{f(x + h) - f(x)}{h}$ as $h \to 0$. Once again the stress is on the non-calculus steps, but the instructor can choose to extend the discussion to include the concept of a derivative of a function. In Sections 1.5 and 2.9, we do not go into theoretical aspects of the existence or nonexistence of limits. All limits given in the exercises or illustrated in examples exist. In *The Area Problem*, we discuss the geometry and algebra required to use a limiting process to find the area under a curve. In *The Limit Concept Revisited*, we examine evaluation of some trigonometric limits and the calculation of the difference quotient $\dfrac{f(x + h) - f(x)}{h}$ when $f$ is either the sine or cosine function. In *The Number e*, we use the difference quotient to show the student why the number $e$ is the most natural base for the exponential function in a calculus setting. In *Parametric Equations*, we examine this powerful means of defining a curve in the plane or in three-dimensional space.

***Final Examination***  Following the six chapters of the text, we present a list of 62 questions called the *Final Examination*. This "test" is mostly fill-in-the-blank and true/false questions. It was not our intention to emulate an actual final examination in precalculus; rather our thought was to offer a vehicle for an informal wrap-up of the entire course. We suggest that a part of a class period be devoted to a discussion of these questions to help students prepare for their actual final examination and their subsequent transition to calculus. To facilitate the students' review, the answers of the *Final Examination* are given both in the *Student Resource Manual* as well as in the instructor's *Complete Solutions Manual*. Of course, the instructor is free to utilize this material in whatever manner he or she chooses (including ignoring it completely).

***Student Resource Manual***  We feel that this manual can be of significant help to a student's success in this course as well as in calculus. Unlike the traditional student solutions manual, where a selected subset of the problems are worked out, the *SRM* is divided into five sections:

- ALGEBRA TOPICS • USE OF A CALCULATOR • BASIC SKILLS
- SELECTED SOLUTIONS • ANSWERS TO THE FINAL EXAM

In ALGEBRA TOPICS, selected topics from algebra (such as multiplication of an inequality by an unknown, implicit conditions in a word problem, Pascal's triangle, factoring techniques, the binomial theorem, rationalization of a numerator or a denominator, adding symbolic fractions, complex numbers and their properties, long division of polynomials, synthetic division, and factorial notation) are reviewed because of their relevance to calculus. Since we do not discuss how to use technology within the text proper, we have devoted the section USE OF A CALCULATOR to the review of graphing calculator essentials. In SELECTED SOLUTIONS, a detailed solution of every third problem in the exercise sets is given. ANSWERS TO THE FINAL EXAM is list of answers for all the questions in the *Final Examination*.

***Exercises***  All of the exercises sets have been updated and many new problems have been added. Most of the exercise sets conclude with conceptual problems that are labeled *For Discussion*.

***Four Colors***  This revision also benefits from the use of the four color format. Besides making the text simply more attractive, we feel that the figures now convey more information and are easier to interpret. Also, important concepts in the text stand out more emphatically when highlighted in different colors.

***Figures***  A final word about the numbering of the figures is in order. Because of the great number of figures in this text, we have changed to a double-decimal numeration system. For example, the interpretation of "Figure 1.2.3" is

$$\begin{array}{c} \text{Chapter Section} \\ \downarrow \ \downarrow \\ 1.2.3 \ \leftarrow \text{Third figure in the section} \end{array}$$

We feel that this type of numeration will make it easier to find figures when they are referred to in later sections or chapters. In addition, to better link a figure with the text, the first textual reference to each figure is done in the same font style and color as the figure number. Also, in this revision all the figures have brief explanatory captions.

# □ Supplements

## *For the Instructor*

- *Complete Solutions Manual (CSM)* by Warren S. Wright is available for instructors both in printed form and on CD-Rom. This detailed manual contains detailed solutions of almost every problem in the text.
- *Instructor's Tool Kit (ITK).* This is located on a password-protected website. It offers our precalculus instructors a computerized test bank that allows you to create customized tests and quizzes that can be printed or administered online using a Course Administration tool such as WebCT or BlackBoard. The questions and answers are sorted by chapter and can be easily installed on your computer desktop for accessibility. The ITK also includes PowerPoint® Lecture Slides which feature all labeled figures as they appear in this text. This useful tool allows instructors the ability to easily display and discuss the figures in the classroom. Online registration to access the ITK resources is located at the text's website: http://math.jbpub.com/catalog/0763737798/.

## *For the Student*

- *Student Resource Manual (SRM)* was prepared by Warren S. Wright and Carol Wright. This printed manual can be bundled with the text at a substantial savings compared to buying the text and SRM separately. For a complete description of this effective student tutorial, please turn to page ix of this preface.
- *Precalculus eLearning Center.* This online tutorial learning center can be accessed at any time and at no cost to the student. The resources are tied directly to the text and include: Practice Quizzes, an Online Glossary of Key Terms, a Student Lecture Companion Note Taking Guide, and Animated Flashcards. These resources can be accessed at http://math.jbpub.com/precalculus/elearning.cfm.
- *Maple® 10 Student Edition* (12-month term). Maple® is the ultimate productivity tool for solving mathematical problems and creating interactive technical applications. Intuitive and easy to use, it delivers the most advanced, complete, reliable mathematical capabilities that can only come from a market-leading tool that has been developed and tested over 25 years. Maple® allows you to create rich, executable technical documents that provide both the answer and the thinking behind the analysis. Maple® documents seamlessly combine numeric and symbolic calculations, explorations, mathematical notation, documentation, buttons and sliders, graphics, and animations.

  Maple® 10 Student Edition (12-month term) is only available through J&B when bundled with this text. Please contact your J&B Publisher's representative for details.

**For further details regarding any of the Instructor or Student Supplements, please contact your Jones and Bartlett Publisher's Representative at 1-800-832-0034 or visit http://www.jbpub.com.**

# To the Student

After teaching collegiate mathematics for many years, we have seen almost every type of student, from a budding genius who invented his own calculus, to students who struggled to master the most rudimentary mechanics of the subject. Frequently, the source of difficulty in calculus can be traced to weak algebra skills and an inadequate background in trigonometry. Calculus builds immediately on your prior knowledge and skills and there is much new ground to be covered. Consequently, there is very little time to review precalculus mathematics in the calculus classroom. So those who teach calculus must assume that you can factor, simplify and solve equations, solve inequalities, handle absolute values, use a calculator, apply the laws of exponents, find equations of lines, plot points, sketch basic graphs, and apply important logarithmic and trigonometric identities. The ability to do algebra and trigonometry, work with exponentials and logarithms, and sketch *by hand* basic graphs quickly and accurately are keys to success in a calculus course. This book focuses on the specific mathematical topics and skills we consider essential for calculus.

In this text, we have tried to give you as much help as possible within the confines of the printed page using such features as marginal annotations, annotations within examples, notes of caution, *Notes From the Classroom*, and the *Final Examination*. The many marginal and in-text annotations provide additional information or further explanation of the steps in the solution of an example. The *Student Resource Manual* (described above) was written just for you. It contains review material not found in the text, extra examples, information on calculators, solutions of problems, and answers to the *Final Examination*.

Those of us who teach and write mathematics texts strive to communicate clearly *how* to do mathematics. This text reflects our philosophy that a mathematics text for the beginning college/university level should be readable, straightforward, and loaded with motivation. The principal reason for studying precalculus is to become well-prepared for calculus. To show you how the material covered in this text is essential for success in calculus, we end each chapter with a section called *Calculus Preview*. In each of these previews a calculus topic provides a framework and motivation for precalculus mathematics and shows you how these mathematics are a vital part of the calculus problem.

Finally, we caution you that *learning* mathematics is not like learning how to ride a bicycle—that once learned, the ability sticks for a lifetime. Mathematics is more like learning another language or learning to play a musical instrument; it requires time and effort to memorize basic formulas and to understand when to apply them, and most importantly, it requires a lot of practice to develop and maintain proficiency. Even experienced musicians still practice the fundamental scales. So ultimately, you the student can learn mathematics (that is, make it stick) only through the hard work of doing mathematics.

In conclusion, we wish you the best of luck in this preparatory course and in your subsequent course in calculus.

# Acknowledgments

Compiling a mathematics text, even one of this modest length, is a monumental task. Besides ourselves, many people—some we do not even know—put much time and hard work into this project. We would like to single out the following individuals for special recognition:

- the editorial, production, and marketing staff at Jones and Bartlett Publishers,
- our editor Tim Anderson, production director Amy Rose, text and cover designer Anne Spencer, for their seemingly inexhaustible supply of patience and cooperation as well as for an occasional but necessary prodding,
- George Nichols for his beautiful renditions of our not-so-beautiful originals,
- Scott and Carol Wright for their careful proof reading of the preliminary versions of this text,
- Cathy Herrera, our departmental administrative assistant, who with never-failing good humor contributed in so many little ways to this writing project–from typing portions of the manuscript and for going the extra "mile" (to the campus post office), and
- the instructors, students, and reviewers who have provided us with comments and suggestions over the years.

We give to each a heart-felt *Thank You*!

Even with all this help, the accuracy of every letter, word, symbol, equation, and figure is the responsibility of the authors. We would be very grateful to have any errors or "typos" called to our attention. You can email them directly to our editor at:

tanderson@jbpub.com.

Dennis G. Zill

Jacqueline M. Dewar

$$\int \left(x - \frac{1}{x}\right)^2 dx = $$
$$= \frac{x^3}{3} - 2x + x + $$

## Chapter Outline

# Inequalities, Equations, and Graphs

**1**

**The Real Line**

☐ **Introduction** In calculus you will study quantities described by real numbers. Therefore, we begin with a review of the set of real numbers using the terminology and notation you will encounter in calculus.

Recall that the set $R$ of **real numbers** consists of numbers that are either **rational** or **irrational**. Rational numbers are numbers of the form $a/b$, where $a$ and $b \neq 0$ are integers. For example, $-3$, $-\frac{1}{2}$, $\frac{2}{3}$, $5$, and $\frac{127}{4}$ are rational numbers. Irrational numbers are numbers that are not rational, that is, they are numbers that cannot be expressed as a quotient of integers. For example, $\sqrt{2}$ and $\pi$ are irrational numbers. Every real number can also be written as a decimal. A rational number can be expressed either as a *terminating decimal,* such as $\frac{1}{8} = 0.125$, or a *nonterminating and repeating decimal,* such as $\frac{1}{3} = 0.333\ldots$. Repeating decimals, such as $0.666\ldots$ and $8.545454\ldots$, are often written as $0.\overline{6}$ and $8.\overline{54}$, respectively, where the bar indicates the digit or block of digits that repeat. An irrational number is always a *nonterminating and nonrepeating decimal* such as $\sqrt{2} = 1.41421\ldots$ or $\pi = 3.14159\ldots$. The following chart summarizes the relationship between the principal sets of real numbers.

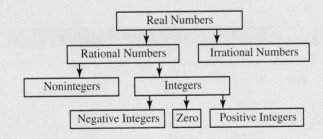

☐ **The Real Line** The set $R$ of real numbers can be put into a one-to-one correspondence with the set of points on a line. As a consequence, we can visualize or represent real numbers as points on a *horizontal line* called the **number line** or **coordinate line**. The point chosen to represent the number 0 is called the **origin**. The direction to the right of 0 is said to be the **positive direction** on the number line; the direction to the left of 0 is the **negative direction**. Real numbers corresponding to points to the right of 0 are called **positive numbers** and numbers corresponding to points to the left of 0 are **negative numbers**. As indicated in **FIGURE 1.1.1**, the number 0 is considered to be neither positive nor negative. From here on, we will not distinguish between a point on the number line and the number that corresponds to this point.

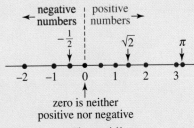

**FIGURE 1.1.1** The real line

3

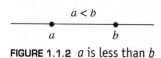

$a < b$

**FIGURE 1.1.2** $a$ is less than $b$

☐ **Inequalities** The number line is useful in demonstrating order relations between two real numbers $a$ and $b$. As shown in **FIGURE 1.1.2**, we say that the number $a$ is **less than** the number $b$, and write $a < b$, whenever the number $a$ lies to the left of the number $b$ on the number line. Equivalently, because the number $b$ lies to the right of $a$ on the number line we say that $b$ is **greater than** $a$ and write $b > a$. For example, $4 < 9$ is the same as $9 > 4$. We also use the notation $a \leq b$ if the number $a$ is either **less than or equal** to the number $b$. Similarly, $b \geq a$ means $b$ is **greater than or equal** to $a$. For example, $2 \leq 5$ since $2 < 5$. Also, $4 \geq 4$ because $4 = 4$. For any two real numbers $a$ and $b$, exactly *one* of the following is true:

$$a < b, \qquad a = b, \qquad \text{or} \qquad a > b.$$

The symbols $<$, $>$, $\geq$, and $\leq$ are called **inequality symbols** and expressions such as $a < b$ or $b \geq a$ are called **inequalities**. The inequality $a > 0$ means the number $a$ lies to the right of the number 0 on the number line, and so $a$ is **positive**. We signify that a number $a$ is **negative** by the inequality $a < 0$. Because the inequality $a \geq 0$ means $a$ is either greater than 0 (positive) or equal to 0 (which is neither positive nor negative), we say that $a$ is **nonnegative**.

☐ **Solving Inequalities** We are interested in solving various kinds of inequalities containing a variable. If a real number $a$ is substituted for the variable $x$ in an inequality such as

$$8x + 4 < 16 + 5x, \tag{1}$$

and if the result is a true statement, then $a$ is said to be a **solution** of the inequality. For example, $-2$ is a solution of (1) because if $x$ is replaced by $-2$, then the resulting inequality $8(-2) + 4 < 16 + 5(-2)$ simplifies to the true statement $-12 < 6$. The word *solve* means that we are to find the set of *all* solutions of an inequality such as (1). This set is called the **solution set** of the inequality. Two inequalities are said to be **equivalent** if they have exactly the same solution set. The representation of the solution set on the number line is the **graph** of the inequality.

We solve an inequality by finding an equivalent inequality with obvious solutions. The following list summarizes three operations that yield equivalent inequalities.

### PROPERTIES OF INEQUALITIES

Suppose $a$ and $b$ are real numbers and $c$ is a nonzero real number. Then the inequality $a < b$ is equivalent to:

(*i*) $a + c < b + c$,
(*ii*) $ac < bc$,     for     $c > 0$,
(*iii*) $ac > bc$,     for     $c < 0$.

Property (*iii*) is frequently forgotten. In words, (*iii*) states that:

> *If an inequality is multiplied by a negative number, then the direction of the resulting inequality is reversed.*

For example, if we multiply the inequality $-2 < 5$ by $-3$ then the *less than* symbol is changed to a *greater than* symbol:

$$-2(-3) > 5(-3) \qquad \text{or} \qquad 6 > -15.$$

CHAPTER 1 INEQUALITIES, EQUATIONS, AND GRAPHS

## EXAMPLE 1      Solving the Inequality (1)

Solve $8x + 4 < 16 + 5x$.

**Solution** We solve the inequality by using the properties of inequalities to obtain a sequence of equivalent inequalities:

$$
\begin{aligned}
8x + 4 &< 16 + 5x \\
8x + 4 - 4 &< 16 + 5x - 4 \quad &\leftarrow \text{by } (i) \\
8x &< 12 + 5x \\
8x - 5x &< 12 + 5x - 5x \quad &\leftarrow \text{by } (i) \\
3x &< 12 \\
(\tfrac{1}{3})3x &< (\tfrac{1}{3})12 \quad &\leftarrow \text{by } (ii) \\
x &< 4.
\end{aligned}
$$

Using set-builder notation, the solution set is $\{x \mid x \text{ real and } x < 4\}$. ∎

☐ **Interval Notation** The solution set in Example 1 is graphed on the number line in **FIGURE 1.1.3** as a colored arrow over the line pointing to the left. In the figure, the right parenthesis at 4 indicates that number 4 is *not* included in the solution set. Because the solution set extends indefinitely to the left—the negative direction—the inequality $x < 4$ can also be written as $-\infty < x < 4$, where $\infty$ is the infinity symbol. In other words, the solution set of the inequality $x < 4$ is

$$\{x \mid x \text{ real and } x < 4\} = \{x \mid -\infty < x < 4\}.$$

Using **interval notation** this set of real numbers is written $(-\infty, 4)$ and is an example of an **unbounded interval**. Table 1.1 summarizes various inequalities and their solution sets, as well as interval notations, names, and graphs. In each of the first four entries

**FIGURE 1.1.3** Solution set in Example 1 in interval notation is $(-\infty, 4)$

## TABLE 1.1     Inequalities and Intervals

| Inequality | Solution Set | Interval Notation | Name | Graph |
|---|---|---|---|---|
| $a < x < b$ | $\{x \mid a < x < b\}$ | $(a, b)$ | Open interval | |
| $a \leq x \leq b$ | $\{x \mid a \leq x \leq b\}$ | $[a, b]$ | Closed interval | |
| $a < x \leq b$ | $\{x \mid a < x \leq b\}$ | $(a, b]$ | Half-open interval | |
| $a \leq x < b$ | $\{x \mid a \leq x < b\}$ | $[a, b)$ | Half-open interval | |
| $a < x$ | $\{x \mid a < x < \infty\}$ | $(a, \infty)$ | | |
| $x < b$ | $\{x \mid -\infty < x < b\}$ | $(-\infty, b)$ | | |
| $x \leq b$ | $\{x \mid -\infty < x \leq b\}$ | $(-\infty, b]$ | Unbounded intervals | |
| $a \leq x$ | $\{x \mid a \leq x < \infty\}$ | $[a, \infty)$ | | |
| $-\infty < x < \infty$ | $\{x \mid -\infty < x < \infty\}$ | $(-\infty, \infty)$ | | |

of the table, the numbers $a$ and $b$ are called the **endpoints** of the interval. As a set, the **open interval**

$$(a, b) = \{x \mid a < x < b\}$$

does not include either endpoint, whereas the **closed interval**

$$[a, b] = \{x \mid a \le x \le b\}$$

includes both endpoints. Note, too, that the graph of the last interval in Table 1.1, which extends indefinitely both to the left and to the right, is the entire real number line. In calculus the interval notation $(-\infty, \infty)$ is generally used to represent the set $R$ of real numbers.

A word of caution is in order as you peruse Table 1.1. The **infinity symbols** $-\infty$ ("minus infinity") and $\infty$ ("infinity") do not represent real numbers and should *never* be manipulated arithmetically like a number. The infinity symbols are merely notational devices: $-\infty$ and $\infty$ are used to indicate unboundedness in the negative direction and in the positive direction, respectively. Thus when using interval notation, the symbols $-\infty$ and $\infty$ can never appear next to a square bracket. For example, the expression $(2, \infty]$ is meaningless.

An inequality of the form $a < x < b$ is sometimes referred to as a **simultaneous inequality** because the number $x$ is *between* the numbers $a$ and $b$. In other words, $x > a$ *and* simultaneously $x < b$. For example, the real numbers that satisfy $2 < x < 5$ is the intersection of the intervals defined by the inequalities $2 < x$ and $x < 5$. Recall that the **intersection** of two sets $A$ and $B$, written $A \cap B$, is the set of elements that are in $A$ *and* in $B$—in other words, the elements that are common to both sets. As illustrated in **FIGURE 1.1.4** by the overlapping arrows extending indefinitely to the right and to the left, the solution set of the inequality $2 < x < 5$ can be written as the intersection $(2, \infty) \cap (-\infty, 5) = (2, 5)$.

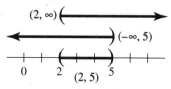

**FIGURE 1.1.4** The numbers in $(2, 5)$ are the numbers common to both $(2, \infty)$ and $(-\infty, 5)$

### EXAMPLE 2　　Solving a Simultaneous Inequality

Solve $-2 \le 1 - 2x < 3$.

**Solution** As previously discussed, one way of proceeding is to solve two inequalities:

$$-2 \le 1 - 2x \qquad \text{and} \qquad 1 - 2x < 3$$

and then take the intersection of the two solution sets. A faster method is to solve both of the inequalities simultaneously in the following manner:

$$-2 \le 1 - 2x < 3$$
$$-1 - 2 \le -1 + 1 - 2x < -1 + 3 \qquad \leftarrow \text{by } (i)$$
$$-3 \le -2x < 2.$$

We isolate the variable $x$ in the middle of the last simultaneous inequality by multiplying by $-\frac{1}{2}$:

$$(-\tfrac{1}{2})(-3) \ge (-\tfrac{1}{2})(-2x) > (-\tfrac{1}{2})2 \qquad \leftarrow \text{by } (iii)$$
$$\tfrac{3}{2} \ge x > -1,$$

where we note that multiplication by the negative number has reversed the direction of the inequalities. To express this inequality in interval notation, we first rewrite it with the leftmost number on the number line on the left side of the inequality: $-1 < x \le \frac{3}{2}$. The solution set of the last inequality is the half-open interval $(-1, \frac{3}{2}]$; the square bracket on the right signifies that $\frac{3}{2}$ is included in the solution set. The graph of this interval is given in **FIGURE 1.1.5**.　■

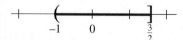

**FIGURE 1.1.5** Solution set in Example 2

☐ **Sign-Chart Method** In Examples 1 and 2 we solved linear inequalities, that is, inequalities containing a single variable $x$ that can be put into the forms $ax + b > 0$, $ax + b \leq 0$, and so on. In the next several examples we illustrate the **sign-chart method** used in calculus for solving nonlinear inequalities. The two properties of real numbers given next are fundamental to constructing a sign chart of an inequality.

---

### SIGN PROPERTIES OF PRODUCTS

(*i*) The product of two real numbers is **positive** if and only if the numbers have the same signs, that is, either $(+)(+)$ or $(-)(-)$.

(*ii*) The product of two real numbers is **negative** if and only if the numbers have opposite signs, that is, $(+)(-)$ or $(-)(+)$.

---

Here are some of the basic steps of the sign-chart method illustrated in the next example.

- Use the properties of inequalities to recast the given inequality into a form where all variables and nonzero constants are on the same side of the inequality symbol and the number 0 is on the other side.
- Then, if possible, factor the expression involving the variables and constants into linear factors $ax + b$.
- Mark the number line at the points where the factors are zero. These points divide the number line into intervals.
- In each of these intervals, determine the sign of each factor and then the sign of the product using (*i*) and (*ii*) of the sign properties of products.

---

### ▌EXAMPLE 3    Solving a Nonlinear Inequality

Solve $x^2 \geq -2x + 15$.

**Solution** We begin by rewriting the inequality with all terms to the left of the inequality symbol and 0 to the right. By (*i*) of the properties of inequalities,

$$x^2 \geq -2x + 15 \qquad \text{is equivalent to} \qquad x^2 + 2x - 15 \geq 0.$$

Factoring, the last expression is the same as $(x + 5)(x - 3) \geq 0$.

Then we indicate on the number line where each factor is 0—in this case, $x = -5$ and $x = 3$. As shown in **FIGURE 1.1.6**, this divides the number line into three disjoint, or nonintersecting, intervals: $(-\infty, -5)$, $(-5, 3)$, and $(3, \infty)$. Note, too, that since the given inequality requires the product to be nonnegative, that is, "greater than or *equal to* 0," the numbers $-5$ and 3 are two solutions. Next, we must determine the signs of the factors $x + 5$ and $x - 3$ on each of the three intervals. We are looking for those intervals on which the two factors are either both positive or both negative, for then their product will be positive. Since the linear factors $x + 5$ and $x - 3$ cannot change signs within these intervals, it suffices to obtain the sign of each factor at just *one* test value chosen from inside each interval. For example, on the interval $(-\infty, -5)$, if we use $x = -10$, then

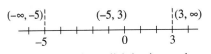

**FIGURE 1.1.6** Three disjoint intervals

◀ See (*i*) in the sign properties of products.

| Interval | $(-\infty, -5)$ |
|---|---|
| Sign of $x + 5$ | $-$ |
| Sign of $x - 3$ | $-$ |
| Sign of $(x + 5)(x - 3)$ | $+$ |

← at $x = -10$, $x + 5 = -10 + 5 < 0$

← at $x = -10$, $x - 3 = -10 - 3 < 0$

← $(-)(-)$ is $(+)$

Continuing in this manner for the remaining two intervals we get the sign chart in **FIGURE 1.1.7.** As can be seen from the third line of this figure, the product $(x + 5)(x - 3)$ is nonnegative on either of the unbounded intervals $(-\infty, -5]$ or $[3, \infty)$.

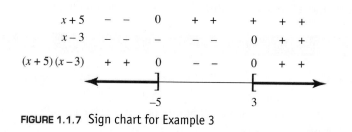

**FIGURE 1.1.7** Sign chart for Example 3

Because the solution set in Example 3 consists of two nonintersecting, or disjoint, intervals it cannot be expressed as a single interval. The best we can do is to write the solution set as the union of the two intervals. Recall that the **union** of two sets $A$ and $B$, written $A \cup B$, is the set of elements that are in either $A$ or in $B$, or in both. Thus the solution set in Example 3 can be written $(-\infty, -5] \cup [3, \infty)$.

### EXAMPLE 4        Solving a Nonlinear Inequality

Solve $(x - 4)^2(x + 8)^3 > 0$.

**Solution** Since the given inequality already has the form appropriate for the sign-chart method (a factored expression to the left of the inequality symbol and 0 to the right), we begin by finding the numbers where each factor is 0, in this case, $x = 4$ and $x = -8$. We place these numbers on the number line and determine three intervals. Then in each interval we consider the signs of the powers of each linear factor. Because of the even power, we see that $(x - 4)^2$ is never negative. However, because of the odd power, $(x + 8)^3$ has the same sign as the factor $x + 8$. Observe that the numbers $x = 4$ and $x = -8$ are not solutions of the inequality because of the "greater than" symbol. Therefore, as we see in **FIGURE 1.1.8**, the solution set is $(-8, 4) \cup (4, \infty)$.

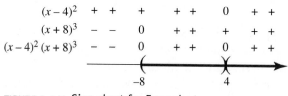

**FIGURE 1.1.8** Sign chart for Example 4

### EXAMPLE 5        Solving a Nonlinear Inequality

Solve $x \leq 3 - \dfrac{6}{x + 2}$.

**Solution** We begin by rewriting the inequality with all variables and nonzero constants to the left and 0 to the right of the inequality sign,

$$x - 3 + \frac{6}{x + 2} \leq 0.$$

Next we put the terms over a common denominator,

One thing we *don't do* is clear the denominator by multiplying the inequality by $x + 2$. See Problem 68 in Exercises 1.1.

$$\frac{(x - 3)(x + 2) + 6}{x + 2} \leq 0 \quad \text{and simplify to} \quad \frac{x(x - 1)}{x + 2} \leq 0. \qquad (2)$$

Now the numbers that make the three linear factors in the last expression equal to 0 are $-2, 0$, and 1. On the number line these three numbers determine four intervals. As a result of the "less than or *equal to* 0," we see that 0 and 1 are members of the solution set. However, $-2$ is excluded from the solution set since substituting this value into the fractional expression results in a zero denominator (making the fraction undefined). As we can see from the sign chart in **FIGURE 1.1.9**, the solution set is $(-\infty, -2) \cup [0, 1]$.

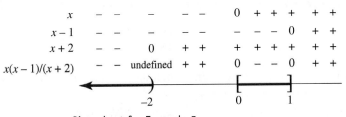

**FIGURE 1.1.9** Sign chart for Example 5

## NOTES FROM THE CLASSROOM

($i$) Terminology used in mathematics often varies from teacher to teacher and from textbook to textbook. For example, inequalities using the symbols $<$ or $>$ are sometimes called *strict* inequalities, whereas inequalities using $\le$ or $\ge$ are called *nonstrict*. As another example, the *positive integers* 1, 2, 3, . . . are often referred to as the *natural numbers*.

($ii$) Suppose the solution set of an inequality consists of the numbers such that $x < -1$ *or* $x > 3$. An answer seen very often on homework, quizzes, and tests is $3 < x < -1$. This is a misunderstanding of the notion of *simultaneity*. The statement $3 < x < -1$ means that $x > 3$ *and* at the same time $x < -1$. If you sketch this on the number line you will see that it is impossible for the same $x$ to satisfy both inequalities. The best we can do in rewriting "$x < -1$ or $x > 3$" is to use the union of intervals $(-\infty, -1) \cup (3, \infty)$.

($iii$) Here is another frequent error: The notation $a < x > b$ is meaningless. If, say, we have $x > -2$ *and* $x > 6$, then only the numbers $x > 6$ satisfy *both* conditions.

($iv$) In the classroom we frequently hear the response "positive" when in reality the student means "nonnegative." Question: $x$ under the square root sign $\sqrt{x}$ must be positive, right? Raise your hand if you agree. Invariably, lots of hands go up. Correct answer: $x$ must be nonnegative, that is, $x \ge 0$. Don't forget that $\sqrt{0} = 0$.

**1.1** | **Exercises** Answers to selected odd-numbered problems begin on page ANS–1.

In Problems 1–6, write the given statement as an inequality.

**1.** $a + 2$ is positive

**2.** $4y$ is negative

**3.** $a + b$ is nonnegative

**4.** $a$ is less than $-3$

**5.** $2b + 4$ is greater than or equal to 100

**6.** $c - 1$ is less than or equal to 5

In Problems 7–14, write the given inequality using interval notation and then graph the interval.

**7.** $x < 0$

**8.** $0 < x < 5$

**9.** $x \geq 5$

**10.** $-1 \leq x$

**11.** $8 < x \leq 10$

**12.** $-5 < x \leq -3$

**13.** $-2 \leq x \leq 4$

**14.** $x > -7$

In Problems 15–18, write the given interval as an inequality.

**15.** $[-7, 9]$

**16.** $[1, 15)$

**17.** $(-\infty, 2)$

**18.** $[-5, \infty)$

In Problems 19–34, solve the given linear inequality. Write the solution set using interval notation. Graph the solution set.

**19.** $x + 3 > -2$

**20.** $3x - 9 < 6$

**21.** $\frac{3}{2}x + 4 \leq 10$

**22.** $5 - \frac{5}{4}x \geq -4$

**23.** $\frac{3}{2} - x > x$

**24.** $-(1 - x) \geq 2x - 1$

**25.** $2 + x \geq 3(x - 1)$

**26.** $-7x + 3 \leq 4 - x$

**27.** $-\frac{20}{3} < \frac{2}{3}x < 4$

**28.** $-3 \leq -x < 2$

**29.** $-7 < x - 2 < 1$

**30.** $3 < x + 4 \leq 10$

**31.** $7 < 3 - \frac{1}{2}x \leq 8$

**32.** $100 + x \leq 41 - 6x \leq 121 + x$

**33.** $-1 \leq \dfrac{x - 4}{4} < \frac{1}{2}$

**34.** $2 \leq \dfrac{4x + 2}{-3} \leq 10$

In Problems 35–58, solve the given nonlinear inequality. Write the solution set using interval notation. Graph the solution set.

**35.** $x^2 - 9 < 0$

**36.** $x^2 \geq 16$

**37.** $x(x - 5) \geq 0$

**38.** $4x^2 + 7x < 0$

**39.** $x^2 - 8x + 12 < 0$

**40.** $(3x + 2)(x - 1) \leq 0$

**41.** $9x \geq 2x^2 - 18$

**42.** $4x^2 > 9x + 9$

**43.** $(x + 1)(x - 2)(x - 4) < 0$

**44.** $(1 - x)(x + \frac{1}{2})(x - 3) \leq 0$

**45.** $(x^2 - 1)(x^2 - 4) \leq 0$

**46.** $(x - 1)^2(x + 3)(x - 5) \geq 0$

**47.** $\dfrac{5}{x + 8} < 0$

**48.** $\dfrac{10}{x^2 + 2} > 0$

**49.** $\dfrac{5}{x} \geq -1$

**50.** $\dfrac{x - 3}{x + 2} < 0$

**51.** $\dfrac{x + 1}{x - 1} + 2 > 0$

**52.** $\dfrac{x - 2}{x + 3} \leq 1$

**53.** $\dfrac{x(x - 1)}{x + 5} \geq 0$

**54.** $\dfrac{(1 + x)(1 - x)}{x} \leq 0$

**55.** $\dfrac{x^2 - 2x + 3}{x + 1} \leq 1$

**56.** $\dfrac{x}{x^2 - 16} > 0$

**57.** $\dfrac{2}{x + 3} - \dfrac{1}{x + 1} < 0$

**58.** $\dfrac{4x + 5}{x^2} \geq \dfrac{4}{x + 5}$

**59.** If 7 times a number is decreased by 6, the result is less than 50. What can be determined about the number?

**60.** The sides of a square are extended to form a rectangle. As shown in **FIGURE 1.1.10**, one side is extended 2 inches and the other side is extended 5 inches. If the area of the resulting rectangle is less than 130 in.$^2$, what are the possible lengths of a side of the original square?

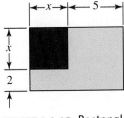

**FIGURE 1.1.10** Rectangle in Problem 60

**61.** A polygon is a closed figure made by joining line segments. For example, a *triangle* is a three-sided polygon. Shown in **FIGURE 1.1.11** is an eight-sided polygon called an *octagon*. A *diagonal* of a polygon is defined to be a line segment that joins any two nonadjacent vertices. The number of diagonals $d$ in a polygon with $n$ sides is given by $d = \frac{1}{2}(n-1)n - n$. For what polygons will the number of diagonals exceed 35?

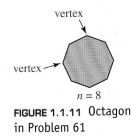

**FIGURE 1.1.11** Octagon in Problem 61

**62.** The total number $N$ of dots in a triangular array with $n$ rows is given by the formula $N = \frac{1}{2}n(n+1)$. See **FIGURE 1.1.12**. How many rows can the array have if the total number of dots is to be less than 5050?

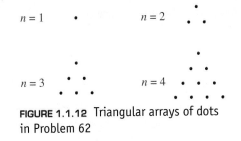

**FIGURE 1.1.12** Triangular arrays of dots in Problem 62

## Miscellaneous Applications

**63. Flower Garden** A rectangular flower bed is to be twice as long as it is wide. If the area enclosed must be greater than 98 m², what can you conclude about the width of the flower bed?

**64. Fever** The relationship between degrees Celsius $T_C$ and degrees Fahrenheit $T_F$ is given by $T_F = \frac{9}{5}T_C + 32$. A person is considered to have a fever if he or she has an oral temperature greater than 98.6°F. What temperatures on the Celsius scale indicate a fever?

Oral thermometer

**65. Parallel Resistors** A 5-ohm resistor and a variable resistor are placed in parallel. The resulting resistance is given by

$R_T = \dfrac{5R}{5+R}$. Determine the values of the variable resistor $R$ for which the resulting resistance $R_T$ will be greater than 2 ohms.

**66. What Goes Up . . .** With the aid of calculus it is easy to show that the height $s$ of a projectile launched straight upward from an initial height $s_0$ with an initial velocity $v_0$ is given by $s = -\frac{1}{2}gt^2 + v_0t + s_0$, where $t$ is in seconds and $g = 32$ ft/s$^2$. If a toy rocket is shot straight upward from ground level, then $s_0 = 0$. If its initial velocity is 72 ft/s, during what time interval will the rocket be more than 80 ft above the ground?

### For Discussion

**67.** Discuss how you might determine the set of numbers for which the given expression is a real number.

  **(a)** $\sqrt{2x - 3}$    **(b)** $\sqrt{4 - 10x}$   **(c)** $\sqrt{x(x - 5)}$   **(d)** $\dfrac{1}{\sqrt{x + 2}}$

  Carry out your ideas.

**68.** In Example 5, explain why one should not multiply the last expression in (2) by $x + 2$.

**1.2**    # Absolute Value

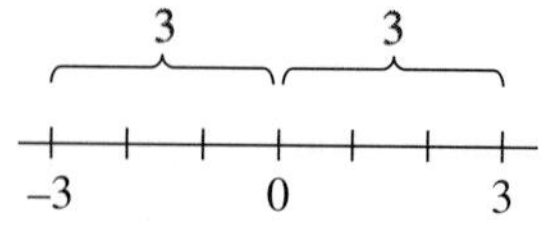

☐ **Introduction** We can use the number line to picture distance. As shown in **FIGURE 1.2.1**, the distance between the number 0 and the number 3 is 3, and the distance between $-3$ and 0 is also 3. In general, for any *positive* real number $x$, the distance between $x$ and 0 is $x$. If $x$ represents a *negative* number, then the distance between $x$ and 0 is $-x$. The concept of distance from a number on the number line to the number 0 is described by the **absolute value** of that number.

**FIGURE 1.2.1** Distance is 3 units

> ### ABSOLUTE VALUE
>
> For any real number $x$, the **absolute value** of $x$, denoted by $|x|$, is
>
> $$|x| = \begin{cases} x, & \text{if } x \geq 0 \\ -x, & \text{if } x < 0. \end{cases} \qquad (1)$$

Be careful. It is a common mistake to think that $-x$ represents a negative quantity simply because of the presence of the minus sign. If a symbol $x$ represents a negative number (that is, $x < 0$), then $-x$ is a positive number. For example, if $x = -10 < 0$, then $|x| = -x = -(-10) = 10$.

As our first example shows, the symbol $x$ in (1) is a placeholder. Other quantities can be placed inside the absolute values symbols $|\ |$.

**EXAMPLE 1**     **Absolute Value**

Write $|x - 5|$ without absolute value symbols.

  **Solution** Wherever the symbol $x$ appears in (1) we replace it by $x - 5$:

$$|x - 5| = \begin{cases} x - 5, & \text{if } x - 5 \geq 0 \\ -(x - 5), & \text{if } x - 5 < 0. \end{cases}$$

Let's consider each part of the foregoing definition separately. First, the inequality $x - 5 \geq 0$ means that $x \geq 5$. Therefore,

$$|x - 5| = x - 5 \quad \text{if} \quad x \geq 5.$$

Check this result (that is, $x - 5$ is nonnegative) by substituting numbers such 5, 8, and 10. Next, $x - 5 < 0$ means that $x < 5$. In this case,

$$\overset{\text{distributive law}}{|x - 5| = \overset{\downarrow}{-(x - 5)} = \overset{\downarrow}{-x + 5}} \qquad \text{if} \qquad x < 5.$$

Again, you should convince yourself that this is correct (that is, $-x + 5$ is positive) by substituting a few numbers, such as 2 and $-3$. ∎

As illustrated in Figure 1.2.1, for any real number $x$ and its negative $-x$, the distance to 0 is the same. That is, $|x| = |-x|$. This is one property in a list of properties of the absolute value that is given next.

## PROPERTIES OF ABSOLUTE VALUES

(i) $|a| = |-a|$
(ii) $|a| = 0$ if and only if $a = 0$
(iii) $|ab| = |a||b|$
(iv) $\left|\dfrac{a}{b}\right| = \dfrac{|a|}{|b|}, \qquad b \neq 0$
(v) $|a + b| \leq |a| + |b|$      (**Triangle inequality**)

For example, by virtue of property (iii) we can rewrite the expression $|-2x|$ as $|-2||x| = 2|x|$.

☐ **Distance Again** If we wish to find the distance between any two numbers on the number line, then all we have to do is subtract the leftmost number from the rightmost number. For example, the distance between 10 and $-2$ is

$$\overset{\text{rightmost number}}{\overset{\downarrow}{10}} - \overset{\text{leftmost number}}{\overset{\downarrow}{(-2)}} = 12.$$

As we saw in the introduction, the distance between $-3$ and 0 is $0 - (-3) = 3$. If an absolute value is used to define the distance, then we do not have to worry about the order of subtraction.

## DISTANCE BETWEEN TWO NUMBERS

If $a$ and $b$ are any two numbers on the number line, the **distance** between $a$ and $b$ is

$$d(a, b) = |b - a|. \tag{2}$$

Using the properties of absolute values,

$$|b - a| = \overset{\text{by property (iii)}}{|(-1)(a - b)|} = |-1|\,|a - b| = |a - b|,$$

and so we have $d(a, b) = d(b, a)$. For example, the distance between $\sqrt{2}$ and 3 is

$$d(\sqrt{2}, 3) = |3 - \sqrt{2}| = 3 - \sqrt{2}$$

because $\qquad 3 > \sqrt{2} \quad$ or $\quad 3 - \sqrt{2} > 0,$

or $\qquad d(3, \sqrt{2}) = |\sqrt{2} - 3| = -(\sqrt{2} - 3) = 3 - \sqrt{2}$

because $\qquad \sqrt{2} < 3 \quad$ or $\quad \sqrt{2} - 3 < 0.$

☐ **Midpoint** Suppose $a$ and $b$ represent two distinct numbers on the number line such that $a < b$. The **midpoint** $m$ of the line segment between the numbers $a$ and $b$ is given by the average of the two endpoints of the interval $[a, b]$, that is

$$m = \frac{a + b}{2}. \qquad (3)$$

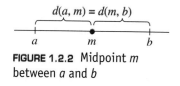

**FIGURE 1.2.2** Midpoint $m$ between $a$ and $b$

As shown in **FIGURE 1.2.2**, (3) is easy to verify by using (2) to show that $d(a, m) = d(m, b)$.

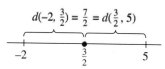

**FIGURE 1.2.3** Midpoint in Example 2

### ■ EXAMPLE 2　　Midpoint

From (3), the midpoint of the line segment joining the numbers $-2$ and $5$ is

$$\frac{(-2) + 5}{2} = \frac{3}{2}.$$

See **FIGURE 1.2.3**. ∎

☐ **Equations** Since ($i$) of the properties of absolute values implies that $|-6| = |6| = 6$, we can conclude that the simple equation $|x| = 6$ has two solutions, either $x = -6$ or $x = 6$. In general, if $a$ is a positive real number, then

$$|x| = a \qquad \text{if and only if} \qquad x = a \qquad \text{or} \qquad x = -a. \qquad (4)$$

### ■ EXAMPLE 3　　An Absolute-Value Equation

Solve **(a)** $|5x - 3| = 8$　**(b)** $|x - 4| = -3$.

**Solution**

**(a)** In (4) the symbol $x$ is a placeholder for any quantity. By replacing $x$ by $5x - 3$, the given equation is equivalent to two equations

$$5x - 3 = 8 \qquad \text{or} \qquad 5x - 3 = -8.$$

We solve each of these. From $5x - 3 = 8$, we obtain

$$5x = 11 \qquad \text{which implies} \qquad x = \frac{11}{5}.$$

From $5x - 3 = -8$, we have

$$5x = -5 \qquad \text{which implies} \qquad x = -1.$$

Therefore, the solutions are $\frac{11}{5}$ and $-1$.

**(b)** Since the absolute value of a real number is always nonnegative, there is no solution to an equation such as $|x - 4| = -3$. ∎

☐ **Inequalities** Many important applications of inequalities involve absolute values. We have just seen that $|x|$ represents the distance along the number line between the number $x$ and the number 0. Thus the inequality $|x| < a$, where $a > 0$, means that the distance between $x$ and 0 is less than $a$. We can see in **FIGURE 1.2.4(a)** that this is the set of real numbers $x$ such that $-a < x < a$. On the other hand, $|x| > a$ means that the distance between $x$ and 0 is greater than $a$. In Figure 1.2.4(b), we see that these are the numbers that satisfy either $x > a$ or $x < -a$. These graphical observations suggest two additional properties of absolute value.

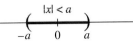

(a) The distance between $x$ and 0 is *less* than $a$

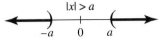

(b) The distance between $x$ and 0 is *greater* than $a$

**FIGURE 1.2.4** Graphical interpretation of properties (*vi*) and (*vii*)

---

### PROPERTIES OF ABSOLUTE VALUES (CONTINUED)

Let $a$ be a positive real number.

(*vi*) $|x| < a$  if and only if  $-a < x < a$.
(*vii*) $|x| > a$  if and only if  $x > a$ or $x < -a$.

---

Properties (*vi*) and (*vii*) also hold with the inequality symbols $<$ and $>$ replaced by $\leq$ and $\geq$, respectively.

| **EXAMPLE 4** | **Two Absolute-Value Inequalities** |

(a) From (*vi*) of the properties of absolute values, the inequality $|x| < 1$ is equivalent to the simultaneous inequality $-1 < x < 1$.

(b) From (*vii*) of the properties of absolute values, the inequality $|x| \geq 5$ is equivalent to two inequalities: $x \geq 5$ or $x \leq -5$. ∎

| **EXAMPLE 5** | **Two Absolute-Value Inequalities** |

Solve (a) $|3x - 7| < 1$    (b) $|2x - 5| \leq 0$.

**Solution**

(a) As in Example 3, the symbol $x$ in the inequality $|x| < a$ is simply a placeholder for other quantities. If we replace $x$ by $3x - 7$ and $a$ by the number 1, then property (*vi*) yields the simultaneous inequality

$$-1 < 3x - 7 < 1$$

which we solve in the usual manner (see Example 2 in Section 1.1):

$$-1 + 7 < 3x - 7 + 7 < 1 + 7$$
$$6 < 3x < 8$$
$$(\tfrac{1}{3})6 < (\tfrac{1}{3})3x < (\tfrac{1}{3})8$$
$$2 < x < \tfrac{8}{3}.$$

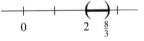

**FIGURE 1.2.5** Solution set in Example 5

The solution set is the open interval $(2, \tfrac{8}{3})$ shown in **FIGURE 1.2.5**.

(b) Since the absolute value of any expression is never negative, the values of $x$ that satisfy the inequality $\leq$ are those for which $|2x - 5| = 0$. By (*ii*) of the properties of absolute values we conclude that $2x - 5 = 0$. Hence the only solution is $\tfrac{5}{2}$. ∎

An inequality such as $|x - b| < a$ can also be interpreted in terms of distance along the number line. Since $|x - b|$ is distance between $x$ and $b$, an inequality such as $|x - b| < a$ is satisfied by all real numbers $x$ whose distance between $x$ and $b$ is less than $a$. This interval is shown in **FIGURE 1.2.6**. Note that when $b = 0$ we get property (*vi*). Similarly, the set of numbers satisfying $|x - b| > a$ are the numbers $x$ whose distance between $x$ and $b$ is greater than $a$.

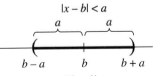

**FIGURE 1.2.6** The distance between $x$ and $b$ is *less* than $a$

## EXAMPLE 6     An Absolute-Value Inequality

Solve $|4 - \frac{1}{2}x| \geq 7$.

**Solution** If we replace $x$ and $a$ in $|x| \geq a$ by $4 - \frac{1}{2}x$ and 7, respectively, then we see from property *(vii)* that $|4 - \frac{1}{2}x| \geq 7$ is equivalent to the two different inequalities

$$4 - \tfrac{1}{2}x \geq 7 \quad \text{or} \quad 4 - \tfrac{1}{2}x \leq -7.$$

We solve each of these inequalities separately. First, we solve

$$4 - \tfrac{1}{2}x \geq 7$$
$$-\tfrac{1}{2}x \geq 3$$
$$x \leq -6. \quad \leftarrow \begin{array}{l}\text{Multiplication by } -2 \text{ reverses}\\ \text{the direction of the inequality}\end{array}$$

In interval notation the solution set of this inequality is $(-\infty, -6]$. Next, we solve

$$4 - \tfrac{1}{2}x \leq -7$$
$$-\tfrac{1}{2}x \leq -11$$
$$(-2)(-\tfrac{1}{2})x \geq (-2)(-11) \quad \leftarrow \begin{array}{l}\text{Multiplication by } -2 \text{ reverses}\\ \text{the direction of the inequality}\end{array}$$
$$x \geq 22.$$

In interval notation the solution set is $[22, \infty)$.

Since the two intervals are disjoint, the solution set is the union of intervals: $(-\infty, -6] \cup [22, \infty)$. The graph of this solution set is shown in **FIGURE 1.2.7**. ∎

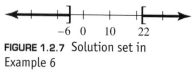

**FIGURE 1.2.7** Solution set in Example 6

Note in Figure 1.2.4(a) that the number 0 is the midpoint of the solution interval for $|x| < a$ and in Figure 1.2.6 that the number $b$ is the midpoint of the solution interval for the inequality $|x - b| < a$. With this in mind, work through the next example.

## EXAMPLE 7     Constructing an Inequality

Find an inequality of the form $|x - b| < a$ for which the open interval $(4, 8)$ is its solution set.

**Solution** The midpoint of the interval $(4, 8)$ is $m = \dfrac{4 + 8}{2} = 6$. The distance between the midpoint $m$ and one of the endpoints of the interval is $d(m, 8) = |8 - 6| = 2$. Therefore the required inequality is $|x - 6| < 2$. ∎

---

## 1.2   Exercises   Answers to selected odd-numbered problems begin on page ANS–1.

In Problems 1–6, write the given quantity without the absolute value symbols.

**1.** $|\pi - 4|$                    **2.** $|\sqrt{5} - 3|$
**3.** $|8 - \sqrt{63}|$               **4.** $|\sqrt{5} - 2.3|$
**5.** $|-6| - |-2|$              **6.** $||-3| - |10||$

In Problems 7–12, write the given expression without the absolute value symbols.

**7.** $|h|$, if $h$ is negative          **8.** $|-h|$, if $h$ is negative
**9.** $|x - 6|$, if $x < 6$            **10.** $|2x - 1|$, if $x \geq \frac{1}{2}$

**11.** $|x - y| - |y - x|$           **12.** $\dfrac{|x - y|}{|y - x|}$, $x \neq y$

In Problems 13–16, write the expression $|x - 2| + |x - 5|$ without the absolute value symbols if $x$ is in the given interval.

**13.** $(-\infty, 1)$      **14.** $(7, \infty)$      **15.** $(3, 4]$      **16.** $[2, 5]$

In Problems 17–20, write the expression $|x + 1| - |x - 3|$ without the absolute value symbols if $x$ is in the given interval.

**17.** $[-1, 3)$      **18.** $(0, 1)$      **19.** $(\pi, \infty)$      **20.** $(-\infty, -5)$

In Problems 21–24, find the distance between the given numbers and find the midpoint of the line segment between them.

**21.** $3, 7$      **22.** $-100, 255$      **23.** $-\frac{3}{2}, \frac{3}{2}$      **24.** $-\frac{1}{4}, \frac{7}{4}$

In Problems 25–28, $m$ is the midpoint of the line segment joining $a$ (the left endpoint) and $b$ (the right endpoint). Use the given conditions to find the indicated quantities.

**25.** $m = 5$, $d(a, m) = 3$; $a$ and $b$      **26.** $m = -1$, $d(m, b) = 2$; $a$ and $b$
**27.** $a = 4$, $d(a, m) = \pi$; $m$ and $b$      **28.** $a = 10$, $d(m, b) = 5$; $m$ and $b$

In Problems 29–34, solve the given equation.

**29.** $|4x - 1| = 2$            **30.** $|5v - 4| = 7$
**31.** $|\frac{1}{4} - \frac{3}{2}y| = 1$          **32.** $|2 - 16t| = 0$
**33.** $\left|\dfrac{x}{x - 1}\right| = 2$        **34.** $\left|\dfrac{x + 1}{x - 2}\right| = 4$

In Problems 35–46, solve the given inequality. Write the solution set using interval notation. Graph the solution set.

**35.** $|-5x| < 4$          **36.** $|3x| > 18$
**37.** $|3 + x| > 7$        **38.** $|x - 4| \le 9$
**39.** $|2x - 7| \le 1$       **40.** $|5 - \frac{1}{3}x| < \frac{1}{2}$
**41.** $|x + \sqrt{2}| \ge 1$       **42.** $|6x + 4| > 4$
**43.** $\left|\dfrac{3x - 1}{-4}\right| < 2$      **44.** $\left|\dfrac{2 - 5x}{3}\right| \ge 5$
**45.** $|x - 5| < 0.01$      **46.** $|x - (-2)| < 0.001$

In Problems 47–50, proceed as in Example 7 and find an inequality $|x - b| < a$ or $|x - b| > a$ for which the given interval is its solution set.

**47.** $(-3, 11)$          **48.** $(1, 2)$
**49.** $(-\infty, 1) \cup (9, \infty)$     **50.** $(-\infty, -3) \cup (13, \infty)$

In Problems 51 and 52, find an inequality whose solution is the set of real numbers $x$ satisfying the given condition. Express each set using interval notation.

**51.** Greater than or equal to 2 units from $-3$
**52.** Less than $\frac{1}{2}$ unit from 3.5

## Miscellaneous Applications

**53. Comparing Ages** Bill and Mary's ages, $A_B$ and $A_M$, differ by at most 3 years. Write this fact as an inequality using absolute value symbols.

**54. Survival** Your score on the first exam is 72%. The midterm grade is the average of the first exam score with the midterm exam score. If the B range is from 80% to 89%, what scores can you obtain on the midterm exam so that your mid-semester grade is B?

**55. Weight of Coffee** The weight $w$ of the coffee in cans filled by a food processing company satisfies

$$\left| \frac{w - 12}{0.05} \right| \leq 1,$$

where $w$ is measured in ounces. Determine the interval in which $w$ lies.

**56. Weight of Cans** A grocery scale is designed to be accurate to within 0.25 oz. If two identical cans of soup placed on the scale have a combined weight of 33.15 oz, what are the largest and smallest possible weights of one of the cans?

## For Discussion

**57.** Discuss how you might solve the following inequalities.

(a) $\left| \dfrac{x + 5}{x - 2} \right| \leq 3$    (b) $|5 - x| = |1 - 3x|$

Carry out your ideas.

**58.** The distance between the number $x$ and 5 is $|x - 5|$.

(a) In words, describe the graphical interpretation of the inequalities $0 < |x - 5|$ and $0 < |x - 5| < 3$.

(b) Solve each inequality in part (a) and write each solution set using interval notation.

**59.** (a) Interpret $|x - 3|$ as distance between the numbers $x$ and 3. Sketch on the number line the set of real numbers that satisfy $2 < |x - 3| < 5$.

(b) Now solve the simultaneous inequality $2 < |x - 3| < 5$ by first solving $|x - 3| < 5$ and then $2 < |x - 3|$. Take the intersection of the two solution sets and compare with your sketch in part (a).

**60.** Here is a statement you may encounter in the beginning of a course in calculus. Express the following statement as best you can in words:

*For every $\epsilon > 0$ there exists a $\delta > 0$ such that $|y - L| < \epsilon$ whenever $0 < |x - a| < \delta$.*

Do not use the symbols $>, <,$ or $|\ |$. The symbols $\epsilon$ and $\delta$ are the Greek letters epsilon and delta and represent real numbers.

---

## 1.3 | The Rectangular Coordinate System

☐ **Introduction** In Section 1.1 we saw that each real number can be associated with exactly one point on the number, or coordinate, line. We now examine a correspondence between points in a plane and ordered pairs of real numbers.

☐ **The Coordinate Plane** A **rectangular coordinate system** is formed by two perpendicular number lines that intersect at the point corresponding to the number 0 on each line. This point of intersection is called the **origin** and is denoted by the symbol $O$. The horizontal and vertical number lines are called the **x-axis** and the **y-axis**, respectively.

These axes divide the plane into four regions, called **quadrants**, which are numbered as shown in **FIGURE 1.3.1(a)**. As we can see in **FIGURE 1.3.1(b)**, the scales on the $x$- and $y$-axes need not be the same. Throughout this text, if tick marks are *not* labeled on the coordinates axes, as in Figure 1.3.1(a), then you may assume that one tick corresponds to one unit. A plane containing a rectangular coordinate system is called an **$xy$-plane** or a **coordinate plane**.

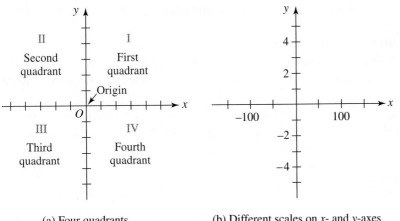

(a) Four quadrants

(b) Different scales on $x$- and $y$-axes

**FIGURE 1.3.1** Coordinate plane

The rectangular coordinate system and the coordinate plane are also called the **Cartesian coordinate system** and the **Cartesian plane** after the famous French mathematician and philosopher René Descartes (1596–1650).

□ **Coordinates of a Point** Let $P$ represent a point in the coordinate plane. We associate an ordered pair of real numbers with $P$ by drawing a vertical line from $P$ to the $x$-axis and a horizontal line from $P$ to the $y$-axis. If the vertical line intersects the $x$-axis at the number $a$ and the horizontal line intersects the $y$-axis at the number $b$, we associate the ordered pair of real numbers $(a, b)$ with the point. Conversely, to each ordered pair $(a, b)$ of real numbers there corresponds a point $P$ in the plane. This point lies at the intersection of the vertical line through $a$ on the $x$-axis and the horizontal line passing through $b$ on the $y$-axis. Hereafter we will refer to an ordered pair as a **point** and denote it by either $P(a, b)$ or $(a, b)$.* The number $a$ is the **$x$-coordinate** of the point and the number $b$ is the **$y$-coordinate** of the point and we say that $P$ has **coordinates** $(a, b)$. For example, the coordinates of the origin are $(0, 0)$. See **FIGURE 1.3.2**.

The algebraic signs of the $x$-coordinate and the $y$-coordinate of any point $(x, y)$ in each of the four quadrants are indicated in **FIGURE 1.3.3**. Points on either of the two axes are not considered to be in any quadrant. Because a point on the $x$-axis has the form $(x, 0)$, an equation that describes the $x$-axis is $y = 0$. Similarly, a point on the $y$-axis has the form $(0, y)$ and so an equation of the $y$-axis is $x = 0$. When we locate a point in the coordinate plane corresponding to an ordered pair of numbers and represent it using a solid dot, we say that we **plot** or **graph** the point.

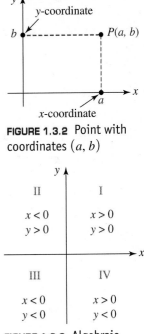

**FIGURE 1.3.2** Point with coordinates $(a, b)$

**FIGURE 1.3.3** Algebraic signs of coordinates in the four quadrants

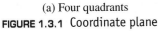

**EXAMPLE 1**    **Plotting Points**

Plot the points $A(1, 2)$, $B(-4, 3)$, $C(-\frac{3}{2}, -2)$, $D(0, 4)$, and $E(3.5, 0)$. Specify the quadrant in which each point lies.

---

*This is the same notation used to denote an open interval. It should be clear from the context of the discussion whether we are considering a point $(a, b)$ or an open interval $(a, b)$.

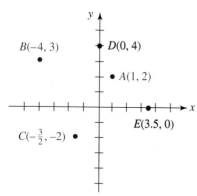

**FIGURE 1.3.4** Plots of five points in Example 1

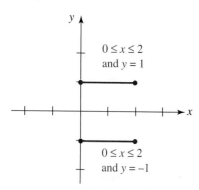

**FIGURE 1.3.5** Set of points in Example 2

**Solution** The five points are plotted in the coordinate plane in **FIGURE 1.3.4**. Point $A$ lies in the first quadrant (quadrant I), $B$ in the second quadrant (quadrant II), and $C$ is in the third quadrant (quadrant III). Points $D$ and $E$, which lie on the $y$- and $x$-axes, respectively, are not in any quadrant. ∎

### ■ EXAMPLE 2     Plotting Points

Sketch the set of points $(x, y)$ in the $xy$-plane whose coordinates satisfy both $0 \le x \le 2$ and $|y| = 1$.

**Solution** First, recall that the absolute-value equation $|y| = 1$ implies that $y = -1$ or $y = 1$. Thus the points that satisfy the given conditions are the points whose coordinates $(x, y)$ *simultaneously* satisfy the conditions: each $x$-coordinate is a number in the closed interval $[0, 2]$ and each $y$-coordinate is either $y = -1$ or $y = 1$. For example, $(1, 1)$, $(\frac{1}{2}, -1)$, $(2, -1)$ are a few of the points that satisfy the two conditions. Graphically, the set of all points satisfying the two conditions are points on the two parallel line segments shown in **FIGURE 1.3.5**. ∎

### ■ EXAMPLE 3     Regions Defined by Inequalities

Sketch the set of points $(x, y)$ in the $xy$-plane whose coordinates satisfy each of the following conditions. **(a)** $xy < 0$     **(b)** $|y| \ge 2$

**Solution**

**(a)** From (*ii*) of the sign properties of products in Section 1.1, we know that a product of two real numbers $x$ and $y$ is negative when one of the numbers is positive and the other is negative. Thus, $xy < 0$ when $x > 0$ and $y < 0$ *or* when $x < 0$ and $y > 0$. We see from Figure 1.3.3 that $xy < 0$ for all points $(x, y)$ in the second and fourth quadrants. Hence we can represent the set of points for which $xy < 0$ by the shaded regions in **FIGURE 1.3.6**. The coordinate axes are shown as dashed lines to indicate that the points on these axes are not included in the solution set.

**(b)** In Section 1.2 we saw that $|y| \ge 2$ means that either $y \ge 2$ or $y \le -2$. Since $x$ is not restricted in any way it can be any real number, and so the points $(x, y)$ for which

$$y \ge 2 \text{ and } -\infty < x < \infty \qquad \text{or} \qquad y \le -2 \text{ and } -\infty < x < \infty$$

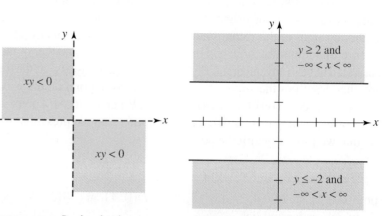

**FIGURE 1.3.6** Region in the $xy$-plane satisfying the condition in (a) of Example 3

**FIGURE 1.3.7** Region in the $xy$-plane satisfying the condition in (b) of Example 3

can be represented by the two shaded regions in **FIGURE 1.3.7**. We use solid lines to represent the boundaries $y = -2$ and $y = 2$ of the region to indicate that the points on these boundaries are included in the solution set.  ■

□ **Distance Formula** Suppose $P_1(x_1, y_1)$ and $P_2(x_2, y_2)$ are two distinct points in the $xy$-plane that are not on a vertical line or on a horizontal line. As a consequence, $P_1$, $P_2$, and $P_3(x_1, y_2)$ are vertices of a right triangle, as shown in **FIGURE 1.3.8**. The length of the side $P_3P_2$ is $|x_2 - x_1|$ and the length of the side $P_1P_3$ is $|y_2 - y_1|$. If we denote the length of $P_1P_2$ by $d$, then

$$d^2 = |x_2 - x_1|^2 + |y_2 - y_1|^2 \tag{1}$$

by the Pythagorean theorem. Since the square of any real number is equal to the square of its absolute value, we can replace the absolute value signs in (1) with parentheses. The distance formula given next follows immediately from (1).

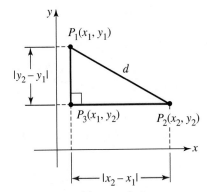

**FIGURE 1.3.8** Distance between points $P_1$ and $P_2$

### DISTANCE FORMULA

The **distance** between any two points $P_1(x_1, y_1)$ and $P_2(x_2, y_2)$ in the $xy$-plane is given by

$$d(P_1, P_2) = \sqrt{(x_2 - x_1)^2 + (y_2 - y_1)^2}. \tag{2}$$

Although we derived this equation for two points not on a vertical or horizontal line, (2) holds in these cases as well. Also, because $(x_2 - x_1)^2 = (x_1 - x_2)^2$, it makes no difference which point is used first in the distance formula, that is, $d(P_1, P_2) = d(P_2, P_1)$.

### EXAMPLE 4　　Distance Between Two Points

Find the distance between the points $A(8, -5)$ and $B(3, 7)$.

　　**Solution** From (2), with $A$ and $B$ playing the parts of $P_1$ and $P_2$:

$$d(A, B) = \sqrt{(3 - 8)^2 + (7 - (-5))^2}$$
$$= \sqrt{(-5)^2 + (12)^2} = \sqrt{169} = 13.$$

The distance $d$ is illustrated in **FIGURE 1.3.9**.  ■

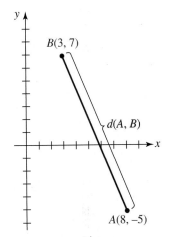

**FIGURE 1.3.9** Distance between two points in Example 4

### EXAMPLE 5　　Three Points Form a Triangle

Determine whether the points $P_1(7, 1)$, $P_2(-4, -1)$, and $P_3(4, 5)$ are the vertices of a right triangle.

　　**Solution** From plane geometry we know that a triangle is a right triangle if and only if the sum of the squares of the lengths of two of its sides is equal to the square of the length of the remaining side. Now, from the distance formula (2), we have

$$d(P_1, P_2) = \sqrt{(-4 - 7)^2 + (-1 - 1)^2}$$
$$= \sqrt{121 + 4} = \sqrt{125},$$
$$d(P_2, P_3) = \sqrt{(4 - (-4))^2 + (5 - (-1))^2}$$
$$= \sqrt{64 + 36} = \sqrt{100} = 10,$$
$$d(P_3, P_1) = \sqrt{(7 - 4)^2 + (1 - 5)^2}$$
$$= \sqrt{9 + 16} = \sqrt{25} = 5.$$

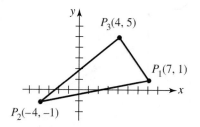

FIGURE 1.3.10 Triangle in Example 5

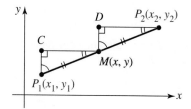

FIGURE 1.3.11 $M$ is the midpoint of the line segment joining $P_1$ and $P_2$

Since

$$[d(P_3, P_1)]^2 + [d(P_2, P_3)]^2 = 25 + 100 = 125 = [d(P_1, P_2)]^2,$$

we conclude that $P_1$, $P_2$, and $P_3$ are the vertices of a right triangle with the right angle at $P_3$. See **FIGURE 1.3.10**. ■

☐ **Midpoint Formula** In Section 1.2 we saw that the midpoint of a line segment between two numbers $a$ and $b$ on the number line is the average, $(a + b)/2$. In the $xy$-plane, each coordinate of the midpoint $M$ of a line segment joining two points $P_1(x_1, y_1)$ and $P_2(x_2, y_2)$ shown in **FIGURE 1.3.11** is the average of the corresponding coordinates of the endpoints of the intervals $[x_1, x_2]$ and $[y_1, y_2]$.

To prove this, we note in Figure 1.3.11 that triangles $P_1CM$ and $MDP_2$ are congruent because corresponding angles are equal and $d(P_1, M) = d(M, P_2)$. Hence, $d(P_1, C) = d(M, D)$, or $y - y_1 = y_2 - y$. Solving the last equation for $y$ gives $y = \dfrac{y_1 + y_2}{2}$. Similarly, $d(C, M) = d(D, P_2)$, so that $x - x_1 = x_2 - x$, and therefore $x = \dfrac{x_1 + x_2}{2}$. We summarize the result.

**MIDPOINT FORMULA**

The coordinates of the **midpoint** of the line segment joining the points $P_1(x_1, y_1)$ and $P_2(x_2, y_2)$ are given by

$$\left( \frac{x_1 + x_2}{2}, \frac{y_1 + y_2}{2} \right). \tag{3}$$

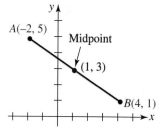

FIGURE 1.3.12 Midpoint of line segment in Example 6

**EXAMPLE 6**     **Midpoint of a Line Segment**

Find the coordinates of the midpoint of the line segment joining $A(-2, 5)$ and $B(4, 1)$.

**Solution** From the midpoint formula (3), the coordinates of the midpoint are given by

$$\left( \frac{-2 + 4}{2}, \frac{5 + 1}{2} \right) \quad \text{or} \quad (1, 3).$$

This point is indicated in color in **FIGURE 1.3.12**. ■

**1.3   Exercises**   Answers to selected odd-numbered problems begin on page ANS–2.

In Problems 1–4, plot the given points.

**1.** $(2, 3)$, $(4, 5)$, $(0, 2)$, $(-1, -3)$      **2.** $(1, 4)$, $(-3, 0)$, $(-4, 2)$, $(-1, -1)$
**3.** $\left(-\frac{1}{2}, -2\right)$, $(0, 0)$, $\left(-1, \frac{4}{3}\right)$, $(3, 3)$      **4.** $(0, 0.8)$, $(-2, 0)$, $(1.2, -1.2)$, $(-2, 2)$

In Problems 5–16, determine the quadrant in which the given point lies if $(a, b)$ is in quadrant I.

**5.** $(-a, b)$      **6.** $(a, -b)$      **7.** $(-a, -b)$      **8.** $(b, a)$
**9.** $(-b, a)$      **10.** $(-b, -a)$      **11.** $(a, a)$      **12.** $(b, -b)$

**13.** $(-a, -a)$     **14.** $(-a, a)$     **15.** $(b, -a)$     **16.** $(-b, b)$

**17.** Plot the points given in Problems 5–16 if $(a, b)$ is the point shown in **FIGURE 1.3.13**.

**18.** Give the coordinates of the points shown in **FIGURE 1.3.14**.

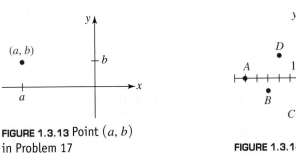

**FIGURE 1.3.13** Point $(a, b)$ in Problem 17

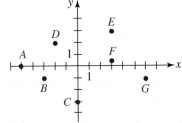

**FIGURE 1.3.14** Points in Problem 18

**19.** The points $(-2, 0)$, $(-2, 6)$, and $(3, 0)$ are vertices of a rectangle. Find the fourth vertex.

**20.** Describe the set of all points $(x, x)$ in the coordinate plane. The set of all points $(x, -x)$.

In Problems 21–26, sketch the set of points $(x, y)$ in the $xy$-plane whose coordinates satisfy the given conditions.

**21.** $xy = 0$

**22.** $xy > 0$

**23.** $|x| \leq 1$ and $|y| \leq 2$

**24.** $x \leq 2$ and $y \geq -1$

**25.** $|x| > 4$

**26.** $|y| \leq 1$

In Problems 27–32, find the distance between the given points.

**27.** $A(1, 2)$, $B(-3, 4)$

**28.** $A(-1, 3)$, $B(5, 0)$

**29.** $A(2, 4)$, $B(-4, -4)$

**30.** $A(-12, -3)$, $B(-5, -7)$

**31.** $A\left(-\frac{3}{2}, 1\right)$, $B\left(\frac{5}{2}, -2\right)$

**32.** $A\left(-\frac{5}{3}, 4\right)$, $B\left(-\frac{2}{3}, -1\right)$

In Problems 33–36, determine whether the points $A$, $B$, and $C$ are vertices of a right triangle.

**33.** $A(8, 1)$, $B(-3, -1)$, $C(10, 5)$

**34.** $A(-2, -1)$, $B(8, 2)$, $C(1, -11)$

**35.** $A(2, 8)$, $B(0, -3)$, $C(6, 5)$

**36.** $A(4, 0)$, $B(1, 1)$, $C(2, 3)$

**37.** Determine whether the points $A(0, 0)$, $B(3, 4)$, and $C(7, 7)$ are vertices of an isosceles triangle.

**38.** Find all points on the $y$-axis that are 5 units from the point $(4, 4)$.

**39.** Consider the line segment joining $A(-1, 2)$ and $B(3, 4)$.
    **(a)** Find an equation that expresses the fact that a point $P(x, y)$ is equidistant from $A$ and from $B$.
    **(b)** Describe geometrically the set of points described by the equation in part (a).

**40.** Use the distance formula to determine whether the points $A(-1, -5)$, $B(2, 4)$, and $C(4, 10)$ lie on a straight line.

**41.** Find all points each with $x$-coordinate 6 such that the distance from each point to $(-1, 2)$ is $\sqrt{85}$.

**42.** Which point, $(1/\sqrt{2}, 1/\sqrt{2})$ or $(0.25, 0.97)$, is closer to the origin?

In Problems 43–48, find the midpoint of the line segment joining the points A and B.

**43.** $A(4, 1), B(-2, 4)$

**44.** $A(\frac{2}{3}, 1), B(\frac{7}{3}, -3)$

**45.** $A(-1, 0), B(-8, 5)$

**46.** $A(\frac{1}{2}, -\frac{3}{2}), B(-\frac{5}{2}, 1)$

**47.** $A(2a, 3b), B(4a, -6b)$

**48.** $A(x, x), B(-x, x + 2)$

In Problems 49–52, find the point B if M is the midpoint of the line segment joining points A and B.

**49.** $A(-2, 1), M(\frac{3}{2}, 0)$

**50.** $A(4, \frac{1}{2}), M(7, -\frac{5}{2})$

**51.** $A(5, 8), M(-1, -1)$

**52.** $A(-10, 2), M(5, 1)$

**53.** Find the distance from the midpoint of the line segment joining $A(-1, 3)$ and $B(3, 5)$ to the midpoint of the line segment joining $C(4, 6)$ and $D(-2, -10)$.

**54.** Find all points on the x-axis that are 3 units from the midpoint of the line segment joining $(5, 2)$ and $(-5, -6)$.

**55.** The x-axis is the perpendicular bisector of the line segment through $A(2, 5)$ and $B(x, y)$. Find x and y.

**56.** Consider the line segment joining the points $A(0, 0)$ and $B(6, 0)$. Find a point $C(x, y)$ in the first quadrant such that A, B, and C are vertices of an equilateral triangle.

**57.** Find the points $P_1(x_1, y_1)$, $P_2(x_2, y_2)$, and $P_3(x_3, y_3)$ on the line segment joining $A(3, 6)$ and $B(5, 8)$ that divide the line segment into four equal parts.

## Miscellaneous Applications

**58. Going to Chicago** Kansas City and Chicago are not directly connected by an interstate highway, but each city is connected to St. Louis and Des Moines. See **FIGURE 1.3.15**. Des Moines is approximately 40 mi east and 180 mi north of Kansas City, St. Louis is approximately 230 mi east and 40 mi south of Kansas City, and Chicago is approximately 360 mi east and 200 mi north of Kansas City. Assume that this part of the Midwest is a flat plane and that the connecting highways are straight lines. Which route from Kansas City to Chicago, through St. Louis or through Des Moines, is shorter?

**FIGURE 1.3.15** Map for Problem 58

## For Discussion

**59.** The points $A(1, 0)$, $B(5, 0)$, $C(4, 6)$, and $D(8, 6)$ are vertices of a parallelogram. Discuss: How can it be shown that the diagonals of the parallelogram bisect each other? Carry out your ideas.

**60.** The points $A(0, 0)$, $B(a, 0)$, and $C(a, b)$ are vertices of a right triangle. Discuss: How can it be shown that the midpoint of the hypotenuse is equidistant from the vertices? Carry out your ideas.

# 1.4 Circles and Graphs

☐ **Introduction** An **equation in two variables**, say $x$ and $y$, is simply a mathematical statement that asserts two quantities involving these variables are equal. In the fields of the physical sciences, engineering, and business, equations are a means of communication. For example, if a physicist wants to tell someone how far a rock dropped from a great height travels in a certain time $t$, he/she will write $s = 16t^2$. A mathematician will look at $s = 16t^2$ and immediately classify it as a certain *type* of equation. The classification of an equation carries with it information about properties shared by all equations of that kind. The remainder of this text is devoted to examining different kinds of equations involving two variables and studying their properties. Here is a sample of some of the equations you will see:

$$x = 1, \quad x^2 + y^2 = 1, \quad y = x^2, \quad y = \sqrt{x},$$
$$y = 5x - 1, \quad y = x^3 - 3x, \quad y = 2^x, \quad y = \ln x, \qquad (1)$$
$$y = \sin x, \quad y^2 = x - 1, \quad \frac{x^2}{4} + \frac{y^2}{9} = 1, \quad \tfrac{1}{2}x^2 - y^2 = 1.$$

A **solution** of an equation in two variables $x$ and $y$ is an ordered pair of numbers $(a, b)$ that yields a true statement when $x = a$ and $y = b$ are substituted into the equation. For example, $(-2, 4)$ is a solution of the equation $y = x^2$ because

$$\overset{\overset{\textstyle y = 4}{\downarrow}}{4} = \overset{\overset{\textstyle x = -2}{\downarrow}}{(-2)^2}$$

is a true statement. We also say that the coordinates $(-2, 4)$ **satisfy** the equation. The set of all solutions of an equation is called its **solution set**. Two equations are said to be **equivalent** if they have the same solution set. For example, we will see in Example 4 of this section that the equation $x^2 + y^2 + 10x - 2y + 17 = 0$ is equivalent to $(x + 5)^2 + (y - 1)^2 = 3^2$.

In the list given in (1), you might object that the first equation $x = 1$ does not involve two variables. It is a matter of interpretation! Because there is no explicit $y$ dependence in the equation, $x = 1$ can be interpreted to mean the set

$$\{(x, y) \mid x = 1, \text{ where } y \text{ is any real number}\}.$$

The solutions of $x = 1$ are then ordered pairs $(1, y)$, where you are free to choose $y$ arbitrarily so long as it is a real number. For example, $(1, 0)$ and $(1, 3)$ are solutions of the equation $x = 1$. The **graph** of an equation is the visual representation in the coordinate plane of the set of points whose coordinates $(a, b)$ satisfy the equation. The graph of $x = 1$ is the vertical line shown in **FIGURE 1.4.1**.

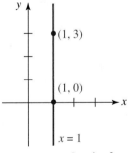

**FIGURE 1.4.1** Graph of equation $x = 1$

☐ **Circles** The distance formula discussed in Section 1.3 can be used to define a set of points in the coordinate plane. One such important set is defined as follows.

---

### CIRCLE

A **circle** is the set of all points $P(x, y)$ in the coordinate plane that are a given fixed distance $r$, called the **radius**, from a given fixed point $C$, called the **center**.

---

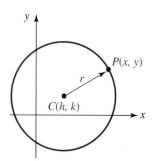

FIGURE 1.4.2 Circle with radius $r$ and center $(h, k)$

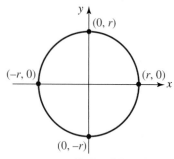

FIGURE 1.4.3 Circle with radius $r$ and center $(0, 0)$

If the center has coordinates $C(h, k)$, then from the preceding definition a point $P(x, y)$ lies on a circle of radius $r$ if and only if

$$d(P, C) = r \qquad \text{or} \qquad \sqrt{(x - h)^2 + (y - k)^2} = r.$$

Since $(x - h)^2 + (y - k)^2$ is always nonnegative, we obtain an equivalent equation when both sides are squared. We conclude that a circle of radius $r$ and center $C(h, k)$ has the equation

$$(x - h)^2 + (y - k)^2 = r^2. \tag{2}$$

In FIGURE 1.4.2 we have sketched a typical graph of an equation of the form given in (2). Equation (2) is called the **standard form** of the equation of a circle. We note that the symbols $h$ and $k$ in (2) represent real numbers and as such can be positive, zero, or negative. When $h = 0$ and $k = 0$, we see that the standard form of the equation of a circle with center at the origin is

$$x^2 + y^2 = r^2. \tag{3}$$

See FIGURE 1.4.3. When $r = 1$ we say that (2) is an equation of a **unit circle**. For example, $x^2 + y^2 = 1$ is an equation of a unit circle centered at the origin.

### EXAMPLE 1          Center and Radius

Find the center and radius of the circle whose equation is

$$(x - 8)^2 + (y + 2)^2 = 49. \tag{4}$$

**Solution**  To obtain the standard form of the equation, we rewrite (4) as

$$(x - 8)^2 + (y - (-2))^2 = 7^2.$$

From this last form we identify $h = 8$, $k = -2$, and $r = 7$. Thus the circle is centered at $(8, -2)$ and has radius 7. ∎

### EXAMPLE 2          Equation of a Circle

Find an equation of the circle with center $C(-5, 4)$ with radius $\sqrt{2}$.
**Solution**  Substituting $h = -5$, $k = 4$, and $r = \sqrt{2}$ in (2), we obtain

$$(x - (-5))^2 + (y - 4)^2 = (\sqrt{2})^2 \qquad \text{or} \qquad (x + 5)^2 + (y - 4)^2 = 2. \quad ∎$$

### EXAMPLE 3          Equation of a Circle

Find an equation of the circle with center $C(4, 3)$ and passing through $P(1, 4)$.
**Solution**  With $h = 4$ and $k = 3$, we have from (2)

$$(x - 4)^2 + (y - 3)^2 = r^2. \tag{5}$$

Since the point $P(1, 4)$ lies on the circle as shown in FIGURE 1.4.4, its coordinates must satisfy equation (5). That is,

$$(1 - 4)^2 + (4 - 3)^2 = r^2 \qquad \text{or} \qquad 10 = r^2.$$

Thus the required equation in standard form is

$$(x - 4)^2 + (y - 3)^2 = 10. \qquad ∎$$

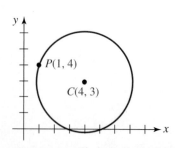

FIGURE 1.4.4 Circle in Example 3

CHAPTER 1 INEQUALITIES, EQUATIONS, AND GRAPHS

☐ **Completing the Square** If the terms $(x - h)^2$ and $(y - k)^2$ are expanded and the like terms grouped together, an equation of a circle in standard form can be written as

$$x^2 + y^2 + ax + by + c = 0. \tag{6}$$

Of course in this last form the center and radius are not apparent. To reverse the process—in other words, to go from (6) to the standard form (2)—we must **complete the square** in both $x$ and $y$. Recall from algebra that adding $(a/2)^2$ to an expression such as $x^2 + ax$ yields $x^2 + ax + (a/2)^2$, which is the perfect square $(x + a/2)^2$. By rearranging the terms in (6),

$$(x^2 + ax \quad) + (y^2 + by \quad) = -c,$$

and then adding $(a/2)^2$ and $(b/2)^2$ to *both* sides of the last equation,

$$\left(x^2 + ax + \left(\frac{a}{2}\right)^2\right) + \left(y^2 + by + \left(\frac{b}{2}\right)^2\right) = \left(\frac{a}{2}\right)^2 + \left(\frac{b}{2}\right)^2 - c,$$

◀ The terms in color added inside the parentheses on the left-hand side are also added to the right-hand side of the equality. This new equation is equivalent to (6).

we obtain the standard form of the equation of a circle:

$$\left(x + \frac{a}{2}\right)^2 + \left(y + \frac{b}{2}\right)^2 = \frac{1}{4}(a^2 + b^2 - 4c).$$

You should *not* memorize the last equation; we strongly recommend that you work through the process of completing the square each time.

### EXAMPLE 4    Completing the Square

Find the center and radius of the circle whose equation is

$$x^2 + y^2 + 10x - 2y + 17 = 0. \tag{7}$$

**Solution** To find the center and radius we rewrite equation (7) in the standard form (2). First, we rearrange the terms,

$$(x^2 + 10x \quad) + (y^2 - 2y \quad) = -17.$$

Then, we complete the square in $x$ and $y$ by adding, in turn, $(10/2)^2$ in the first set of parentheses and $(-2/2)^2$ in the second set of parentheses. Proceed carefully here because we must add these numbers to both sides of the equation:

$$[x^2 + 10x + (\tfrac{10}{2})^2] + [y^2 - 2y + (\tfrac{-2}{2})^2] = -17 + (\tfrac{10}{2})^2 + (\tfrac{-2}{2})^2$$
$$(x^2 + 10x + 25) + (y^2 - 2y + 1) = 9$$
$$(x + 5)^2 + (y - 1)^2 = 3^2.$$

From the last equation we see that the circle is centered at $(-5, 1)$ and has radius 3. See **FIGURE 1.4.5**.

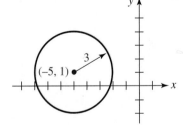

**FIGURE 1.4.5** Circle in Example 4

It is possible that an expression for which we must complete the square has a leading coefficient other than 1. For example,

Note:

$$3x^2 + 3y^2 - 18x + 6y + 2 = 0$$

is an equation of circle. As in Example 4, we start by rearranging the equation:

$$(3x^2 - 18x \quad) + (3y^2 + 6y \quad) = -2.$$

Now, however, we must do one extra step before attempting completion of the square, that is, we must divide both sides of the equation by 3 so that the coefficients of $x^2$ and $y^2$ are each 1:

$$(x^2 - 6x \qquad) + (y^2 + 2y \qquad) = -\tfrac{2}{3}.$$

At this point we can now add the appropriate numbers within each set of parentheses *and* to the right-hand side of the equality. You should verify that the resulting standard form is $(x - 3)^2 + (y + 1)^2 = \tfrac{28}{3}$.

☐ **Semicircles** If we solve (3) for $y$ we get $y^2 = r^2 - x^2$ or $y = \pm\sqrt{r^2 - x^2}$. This last expression is equivalent to two equations, $y = \sqrt{r^2 - x^2}$ and $y = -\sqrt{r^2 - x^2}$. In like manner if we solve (3) for $x$ we obtain $x = \sqrt{r^2 - y^2}$ and $x = -\sqrt{r^2 - y^2}$.

By convention, the symbol $\sqrt{\phantom{x}}$ denotes a nonnegative quantity, thus the $y$-values defined by an equation such as $y = \sqrt{r^2 - x^2}$ are nonnegative. The graphs of the four equations highlighted in color are, in turn, the upper half, lower half, right half, and left half of the circle shown in Figure 1.4.3. Each graph in FIGURE 1.4.6 is called a **semicircle**.

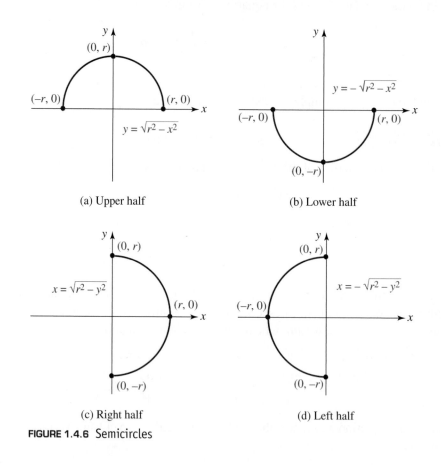

(a) Upper half

(b) Lower half

(c) Right half

(d) Left half

FIGURE 1.4.6 Semicircles

One last point about circles: On occasion we encounter problems where we must sketch the set of points in the $xy$-plane whose coordinates satisfy inequalities such as $x^2 + y^2 < r^2$ or $x^2 + y^2 \geq r^2$. The equation $x^2 + y^2 = r^2$ describes the set of points $(x, y)$ whose distance to the origin $(0, 0)$ is exactly $r$. Therefore the inequality $x^2 + y^2 < r^2$ describes the set of points $(x, y)$ whose distance to the origin is less than $r$. In other words, the points $(x, y)$ whose coordinates satisfy the inequality $x^2 + y^2 < r^2$ are in the *interior* of the circle. Similarly, the points $(x, y)$ whose coordinates satisfy $x^2 + y^2 \geq r^2$ lie either *on* the circle or are *exterior* to it.

☐ **Graphs** It is difficult to read a newspaper, read a science or business text, surf the Internet, or even watch the news on TV without seeing graphical representations of data. It may even be impossible to get past the first page in a mathematics text without seeing some kind of graph. So many diverse quantities are connected by means of equations, and so many questions about the behavior of the quantities linked by the equation can be answered by means of a graph, that the ability to graph equations quickly and accurately—like the ability to do algebra quickly and accurately—is high on the list of skills essential to your success in a course in calculus. For the rest of this section we are going to talk about graphs in general, and more specifically about two important aspects of graphs of equations.

☐ **Intercepts** Locating the points at which the graph of an equation crosses the coordinate axes can be helpful when sketching a graph by hand. The **x-intercepts** of a graph of an equation are the points at which the graph crosses the $x$-axis. Since every point on the $x$-axis has $y$-coordinate 0, the $x$-coordinates of these points (if there are any) can be found from the given equation by setting $y = 0$ and solving for $x$. In turn, the **y-intercepts** of the graph of an equation are the points at which its graph crosses the $y$-axis. The $y$-coordinates of these points can found by setting $x = 0$ in the equation and solving for $y$. See **FIGURE 1.4.7**.

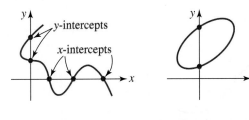
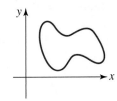

(a) Five intercepts  (b) Two $y$-intercepts  (c) Graph has no intercepts

**FIGURE 1.4.7** Intercepts of a graph

**EXAMPLE 5**                **Intercepts**

Find the intercepts of the graphs of the equations
(a) $x^2 - y^2 = 9$    (b) $y = 2x^2 + 5x - 12$.

**Solution**

(a) To find the $x$-intercepts we set $y = 0$ and solve the resulting equation $x^2 = 9$ for $x$:
$$x^2 - 9 = 0 \quad \text{or} \quad (x + 3)(x - 3) = 0$$
gives $x = -3$ and $x = 3$. The $x$-intercepts of the graph are the points $(-3, 0)$ and $(3, 0)$. To find the $y$-intercepts we set $x = 0$ and solve $-y^2 = 9$ or $y^2 = -9$ for $y$. Because there are no real numbers whose square is negative we conclude the graph of the equation does not cross the $y$-axis.

(b) Setting $y = 0$ yields $2x^2 + 5x - 12 = 0$. This is a quadratic equation and can be solved either by factoring or by the quadratic formula. Factoring gives
$$(x + 4)(2x - 3) = 0$$
and so $x = -4$ and $x = \frac{3}{2}$. The $x$-intercepts of the graph are the points $(-4, 0)$ and $(\frac{3}{2}, 0)$. Now, setting $x = 0$ in the equation $y = 2x^2 + 5x - 12$ immediately gives $y = -12$. The $y$-intercept of the graph is the point $(0, -12)$. ∎

EXAMPLE 6          **Example 4 Revisited**

Let's return to the circle in Example 4 and determine its intercepts from the equation in (7). Setting $y = 0$ in $x^2 + y^2 + 10x - 2y + 17 = 0$ and using the quadratic formula to solve $x^2 + 10x + 17 = 0$ shows the $x$-intercepts of this circle are $(-5 - 2\sqrt{2}, 0)$ and $(-5 + 2\sqrt{2}, 0)$. If we let $x = 0$, then the quadratic formula shows that the roots of the equation $y^2 - 2y + 17 = 0$ are complex numbers. As seen in Figure 1.4.5, the circle does not cross the $y$-axis.                                    ■

☐ **Symmetry**  A graph can also possess symmetry. You may already know that the graph of the equation $y = x^2$ is called a *parabola*. **FIGURE 1.4.8** shows that the graph of $y = x^2$ is symmetric with respect to the $y$-axis since the portion of the graph that lies in the second quadrant is the *mirror image* or *reflection* of that portion of the graph in the first quadrant. In general, a graph is **symmetric with respect to the $y$-axis** if whenever $(x, y)$ is a point on the graph, $(-x, y)$ is also a point on the graph. Note in Figure 1.4.8 that the points $(1, 1)$ and $(2, 4)$ are on the graph. Because the graph possesses $y$-axis symmetry, the points $(-1, 1)$ and $(-2, 4)$ must also be on the graph. A graph is said to be **symmetric with respect to the $x$-axis** if whenever $(x, y)$ is a point on the graph, $(x, -y)$ is also a point on the graph. Finally, a graph is **symmetric with respect to the origin** if whenever $(x, y)$ is on the graph, $(-x, -y)$ is also a point on the graph. **FIGURE 1.4.9** illustrates these three types of symmetries.

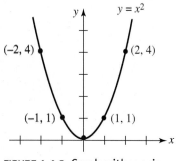

**FIGURE 1.4.8** Graph with $y$-axis symmetry

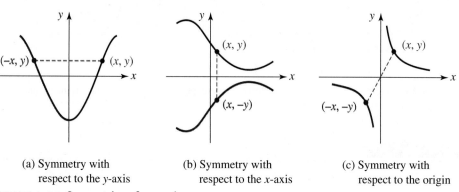

(a) Symmetry with respect to the $y$-axis

(b) Symmetry with respect to the $x$-axis

(c) Symmetry with respect to the origin

**FIGURE 1.4.9** Symmetries of a graph

Observe the graph of the circle given in Figure 1.4.3 possesses all three of these symmetries.

As a practical matter we would like to know whether a graph possesses any symmetry in advance of plotting it. This can be done by applying the following tests to the equation that defines the graph.

### TESTS FOR SYMMETRY

The graph of an equation is symmetric with respect to:

    (*i*) the **$y$-axis** if replacing $x$ by $-x$ results in an equivalent equation;
    (*ii*) the **$x$-axis** if replacing $y$ by $-y$ results in an equivalent equation;
    (*iii*) the **origin** if replacing $x$ and $y$ by $-x$ and $-y$ results in an equivalent equation.

The advantage of using symmetry in graphing should be apparent: If, say, the graph of an equation is symmetric with respect to the $x$-axis, then we need only produce the graph for $y \geq 0$ since points on the graph for $y < 0$ are obtained by taking the mirror images, through the $x$-axis, of the points in the first and second quadrants.

 EXAMPLE 7          **Test for Symmetry**
_____

By replacing $x$ by $-x$ in the equation $y = x^2$ and using $(-x)^2 = x^2$, we see that

$$y = (-x)^2 \quad \text{is equivalent to} \quad y = x^2.$$

This proves what is apparent in Figure 1.4.8: that the graph of $y = x^2$ is symmetric with respect to the $y$-axis.                                                                    ∎

 EXAMPLE 8          **Intercepts and Symmetry**
_____

Determine the intercepts and any symmetry for the graph of

$$x + y^2 = 10. \tag{8}$$

   **Solution** _Intercepts_: Setting $y = 0$ in equation (8) immediately gives $x = 10$. The graph of the equation has a single $x$-intercept, $(10, 0)$. When $x = 0$, we get $y^2 = 10$, which implies that $y = -\sqrt{10}$ or $y = \sqrt{10}$. Thus there are two $y$-intercepts, $(0, -\sqrt{10})$ and $(0, \sqrt{10})$.
   _Symmetry_: If we replace $x$ by $-x$ in the equation $x + y^2 = 10$ we get $-x + y^2 = 10$. This is not equivalent to equation (8). You should also verify that replacing $x$ and $y$ by $-x$ and $-y$ in (8) does not yield an equivalent equation. However, if we replace $y$ by $-y$, we find that

$$x + (-y)^2 = 10 \quad \text{is equivalent to} \quad x + y^2 = 10.$$

Thus, the graph of the equation is symmetric with respect to the $x$-axis.
   _Graph_: In the graph of the equation given in **FIGURE 1.4.10**, the intercepts are indicated and the $x$-axis symmetry should be apparent.                                    ∎

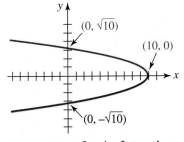

**FIGURE 1.4.10** Graph of equation in Example 8

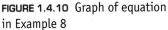

In Problems 1–6, find the center and the radius of the given circle. Sketch its graph.

**1.** $x^2 + y^2 = 5$                        **2.** $x^2 + y^2 = 9$
**3.** $x^2 + (y - 3)^2 = 49$        **4.** $(x + 2)^2 + y^2 = 36$
**5.** $(x - \frac{1}{2})^2 + (y - \frac{3}{2})^2 = 1$        **6.** $(x + 3)^2 + (y - 5)^2 = 25$

In Problems 7–14, complete the square in $x$ and $y$ to find the center and the radius of the given circle.

**7.** $x^2 + y^2 + 8y = 0$                        **8.** $x^2 + y^2 - 6x = 0$
**9.** $x^2 + y^2 + 2x - 4y - 4 = 0$        **10.** $x^2 + y^2 - 18x - 6y - 10 = 0$
**11.** $x^2 + y^2 - 20x + 16y + 128 = 0$        **12.** $x^2 + y^2 + 3x - 16y + 63 = 0$
**13.** $2x^2 + 2y^2 + 4x + 16y + 1 = 0$        **14.** $\frac{1}{2}x^2 + \frac{1}{2}y^2 + \frac{5}{2}x + 10y + 5 = 0$

In Problems 15–24, find an equation of the circle that satisfies the given conditions.

**15.** center $(0, 0)$, radius 1  
**16.** center $(1, -3)$, radius 5  
**17.** center $(0, 2)$, radius $\sqrt{2}$  
**18.** center $(-9, -4)$, radius $\frac{3}{2}$  
**19.** endpoints of a diameter at $(-1, 4)$ and $(3, 8)$  
**20.** endpoints of a diameter at $(4, 2)$ and $(-3, 5)$  
**21.** center $(0, 0)$, graph passes through $(-1, -2)$  
**22.** center $(4, -5)$, graph passes through $(7, -3)$  
**23.** center $(5, 6)$, graph tangent to the $x$-axis  
**24.** center $(-4, 3)$, graph tangent to the $y$-axis  

In Problems 25–28, sketch the semicircle defined by the given equation.

**25.** $y = \sqrt{4 - x^2}$  
**26.** $x = 1 - \sqrt{1 - y^2}$  
**27.** $x = \sqrt{1 - (y - 1)^2}$  
**28.** $y = -\sqrt{9 - (x - 3)^2}$  

**29.** Find an equation for the upper half of the circle $x^2 + (y - 3)^2 = 4$. Repeat for the right half of the circle.

**30.** Find an equation for the lower half of the circle $(x - 5)^2 + (y - 1)^2 = 9$. Repeat for the left half of the circle.

In Problems 31–34, sketch the set of points in the $xy$-plane whose coordinates satisfy the given inequality.

**31.** $x^2 + y^2 \geq 9$  
**32.** $(x - 1)^2 + (y + 5)^2 \leq 25$  
**33.** $1 \leq x^2 + y^2 \leq 4$  
**34.** $x^2 + y^2 > 2y$  

In Problems 35 and 36, find the $x$- and $y$-intercepts of the given circle.

**35.** the circle with center $(3, -6)$ and radius 7  
**36.** the circle $x^2 + y^2 + 5x - 6y = 0$  

In Problems 37–62, find any intercepts of the graph of the given equation. Determine whether the graph of the equation possesses symmetry with respect to the $x$-axis, $y$-axis, or origin. Do not graph.

**37.** $y = -3x$  
**38.** $y - 2x = 0$  
**39.** $-x + 2y = 1$  
**40.** $2x + 3y = 6$  
**41.** $x = y^2$  
**42.** $y = x^3$  
**43.** $y = x^2 - 4$  
**44.** $x = 2y^2 - 4$  
**45.** $y = x^2 - 2x - 2$  
**46.** $y^2 = 16(x + 4)$  
**47.** $y = x(x^2 - 3)$  
**48.** $y = (x - 2)^2(x + 2)^2$  
**49.** $x = -\sqrt{y^2 - 16}$  
**50.** $y^3 - 4x^2 + 8 = 0$  
**51.** $4y^2 - x^2 = 36$  
**52.** $\dfrac{x^2}{25} + \dfrac{y^2}{9} = 1$  
**53.** $y = \dfrac{x^2 - 7}{x^3}$  
**54.** $y = \dfrac{x^2 - 10}{x^2 + 10}$  
**55.** $y = \dfrac{x^2 - x - 20}{x + 6}$  
**56.** $y = \dfrac{(x + 2)(x - 8)}{x + 1}$  
**57.** $y = \sqrt{x} - 3$  
**58.** $y = 2 - \sqrt{x + 5}$  
**59.** $y = |x - 9|$  
**60.** $x = |y| - 4$  
**61.** $|x| + |y| = 4$  
**62.** $x + 3 = |y - 5|$

In Problems 63–66, state all the symmetries of the given graph.

**63.**

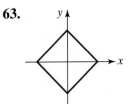

**FIGURE 1.4.11** Graph
for Problem 63

**64.**

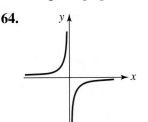

**FIGURE 1.4.12** Graph
for Problem 64

**65.**

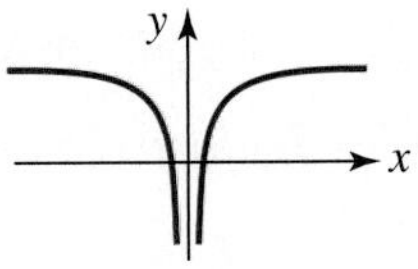

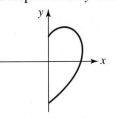

**FIGURE 1.4.13** Graph
for Problem 65

**66.**

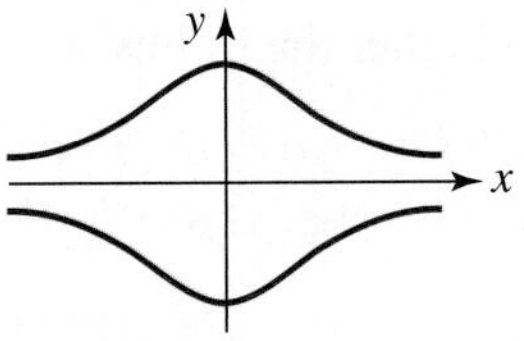

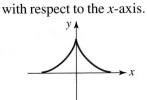

**FIGURE 1.4.14** Graph
for Problem 66

In Problems 67–72, use symmetry to complete the given graph.

**67.** The graph is symmetric
with respect to the *y*-axis.

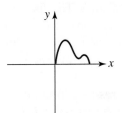

**FIGURE 1.4.15** Graph
for Problem 67

**68.** The graph is symmetric
with respect to the *x*-axis.

**FIGURE 1.4.16** Graph
for Problem 68

**69.** The graph is symmetric
with respect to the origin.

**FIGURE 1.4.17** Graph
for Problem 69

**70.** The graph is symmetric
with respect to the *y*-axis.

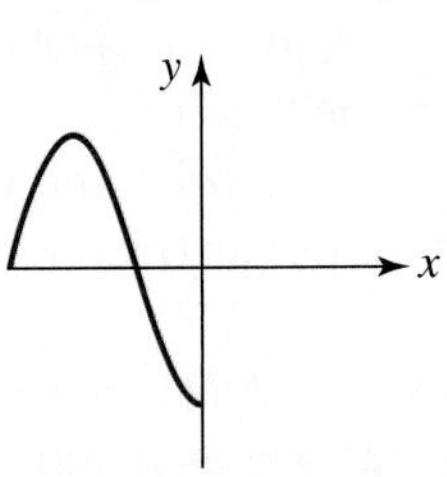

**FIGURE 1.4.18** Graph
for Problem 70

**71.** The graph is symmetric with respect to the $x$- and $y$-axes.

**72.** The graph is symmetric with respect to the origin.

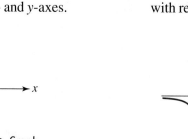

FIGURE 1.4.19 Graph for Problem 71

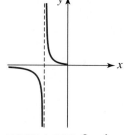

FIGURE 1.4.20 Graph for Problem 72

### For Discussion

**73.** Determine whether the following statement is true or false. Defend your answer.

*If a graph has two of the three symmetries defined on page 30, then the graph must necessarily possess the third symmetry.*

**74. (a)** The radius of the circle in **FIGURE 1.4.21(a)** is $r$. What is its equation in standard form?

**(b)** The center of the circle in **FIGURE 1.4.21(b)** is $(h, k)$. What is its equation in standard form?

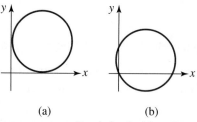

(a)                    (b)

FIGURE 1.4.21 Graph for Problem 74

**75.** Discuss whether the following statement is true or false:

*Every equation of the form $x^2 + y^2 + ax + by + c = 0$ is a circle.*

---

## 1.5 Algebra and Limits

 □ **Introduction** A calculus problem often consists of a sequence of steps, where most of the steps are algebra and only the last few—sometimes just the last step—involve calculus. The discussion that follows focuses on one kind of calculus problem: the computation of a certain type of *limit*. Although we give a brief and intuitive introduction to the notion of a limit, the thrust of the discussion is an overview of the type of algebra frequently encountered in such problems.

In this section we are concerned only with **fractional expressions**. It suffices to think of a fractional expression as a quotient of two *algebraic* expressions.* Roughly, an

---

*At this point we are excluding trigonometric functions, logarithms, and exponentials. See Chapters 4 and 5.

**algebraic expression** is one that is the result of performing a finite number of additions, subtractions, multiplications, divisions, or roots on a collection of variables and real numbers. For example, some algebraic expressions in a single variable $x$ are

$$5x^3 - 3x + 1, \qquad \frac{2x^2 - 18}{x + 3}, \qquad \text{and} \qquad x + \sqrt{x - 5}.$$

An area of algebra that causes difficulties in working calculus problems is the manipulation of fractional expressions.

☐ **Factoring** When the distributive law

$$a(b + c) = ab + ac$$

is read right to left,

$$ab + ac = a(b + c),$$

we say that the expression $ab + ac$ has been **factored**. We will see in Chapter 3 that factoring plays an important role in solving equations as well as in graphing. But in the present context we are concerned only with using factoring to simplify fractional expressions.

The following three factorization formulas are important and are used as a matter of course throughout various fields of mathematics.

---

**FACTORIZATIONS WORTH KNOWING**

Difference of two squares: $a^2 - b^2 = (a - b)(a + b)$      (1)
Difference of two cubes: $a^3 - b^3 = (a - b)(a^2 + ab + b^2)$      (2)
Sum of two cubes: $a^3 + b^3 = (a + b)(a^2 - ab + b^2)$      (3)

---

The symbols $a$ and $b$ in (1)–(3) are placeholders. For example, the expression $x^4 - 16$ is of the form given in (1). With the identifications $a = x^2$ and $b = 4$, we have

$$x^4 - 16 = (x^2)^2 - 4^2 = (x^2 - 4)(x^2 + 4). \qquad (4)$$

Since the factor $x^2 - 4$ is also the difference of two squares, (4) continues as

$$x^4 - 16 = (x^2 - 4)(x^2 + 4) = (x - 2)(x + 2)(x^2 + 4). \qquad (5)$$

The factorization in (5) is as far as we can go using real numbers and integer exponents; the sum of two squares $x^2 + 4$ does not factor using real numbers. As another example, consider the expression $2x^2 - 3$. Since any positive real number can be written as the square of its square root we have $2 = (\sqrt{2})^2$ and $3 = (\sqrt{3})^2$, and so from (1) the expression $2x^2 - 3$ factors in the following manner:

$$2x^2 - 3 = \overset{a}{\overbrace{(\sqrt{2}x)^2}} - \overset{b}{\overbrace{(\sqrt{3})^2}} = (\sqrt{2}x - \sqrt{3})(\sqrt{2}x + \sqrt{3}).$$

We use factoring and the cancellation property to simplify a fractional expression.

◄ Cancellation Property: If $a$, $b$, and $c$ are real numbers, then
$$\frac{ac}{bc} = \frac{a}{b}, \quad c \neq 0.$$

**EXAMPLE 1**      **Factoring and Canceling**

Simplify **(a)** $\dfrac{x^2 - 1}{x - 1}$    **(b)** $\dfrac{x + 3}{x^2 - 4x - 21}$.

**Solution**

**(a)** From (1) we see

$$\frac{x^2 - 1}{x - 1} = \frac{(x - 1)(x + 1)}{x - 1} = x + 1.$$

The cancellation of $x - 1$ in the foregoing expression is only valid for $x \neq 1$. For $x = 1$ we would be dividing by 0.

**(b)** We look for factors $x - a$ and $x - b$ such that

$$x^2 - 4x - 21 = (x - a)(x - b).$$

This implies $ab = -21$ so $a$ and $b$ must be factors of $-21$ whose sum $-(a + b) = -4$. The usual trial and error procedure leads to $a = 7$ and $b = -3$. Therefore,

$$\frac{x + 3}{x^2 - 4x - 21} = \frac{x + 3}{(x + 3)(x - 7)} = \frac{1}{x - 7}, \qquad x \neq -3. \qquad \blacksquare$$

☐ **Binomial Expansion** A two-term algebraic expression $a + b$ is called a **binomial**. You undoubtedly have worked problems where you had to expand powers of binomials such as $(a + b)^2$ and $(a + b)^3$. This occurs so often in mathematics courses that we recommend that you memorize the expansions given in (6) and (7).

---

### BINOMIAL EXPANSIONS WORTH KNOWING

Expansions of $(a + b)^n$ for

$$n = 2: \ (a + b)^2 = a^2 + 2ab + b^2 \qquad\qquad (6)$$
$$n = 3: \ (a + b)^3 = a^3 + 3a^2b + 3ab^2 + b^3 \qquad (7)$$

---

Of course, formulas (6) and (7) work just as well for a binomial in the form of a difference $a - b$. Simply treat $a - b$ as the sum $a + (-b)$ and replace the symbol $b$ in (6) and (7) by $-b$:

$$(a - b)^2 = (a + (-b))^2 = a^2 + 2a(-b) + (-b)^2 = a^2 - 2ab + b^2,$$

and
$$(a - b)^3 = (a + (-b))^3 = a^3 + 3a^2(-b) + 3a(-b)^2 + (-b)^3$$
$$= a^3 - 3a^2b + 3ab^2 - b^3.$$

There are ways of remembering how to obtain the coefficients in the expansion of higher powers such as $(a + b)^4$. **Pascal's triangle** is one such way and is reviewed in the *Student Resource Manual* that accompanies this text.

### EXAMPLE 2     Binomial Expansion

Simplify $\dfrac{(7 + h)^2 - 49}{h}$.

**Solution** We use the expansion of $(a + b)^2$ given in (6) with $a = 7$ and $b = h$:

$$\frac{(7 + h)^2 - 49}{h} = \frac{(7^2 + 2(7)h + h^2) - 49}{h}$$

$$= \frac{49 + 14h + h^2 - 49}{h} \qquad \leftarrow 49 - 49 = 0$$

$$= \frac{h(14 + h)}{h} \quad \leftarrow \text{cancel the } h\text{'s}$$

$$= 14 + h, \quad h \neq 0. \qquad\blacksquare$$

☐ **Addition of Fractional Expressions** Combining two or more fractional expressions, or simplification of a complex fraction where the numerator or denominator is itself a fraction, can be particularly troublesome for some students.

**■ EXAMPLE 3**  **Addition of Fractions**

Write as one fraction $\dfrac{10x}{2x^2 + 3x - 2} - \dfrac{4}{x + 2} + \dfrac{8}{2x - 1}$.

**Solution** Because $2x^2 + 3x - 2 = (2x - 1)(x + 2)$, the least common denominator of the three terms is $(2x - 1)(x + 2)$. Therefore, we multiply the second term by $(2x - 1)/(2x - 1)$ and the third term by $(x - 2)/(x - 2)$:

$$\frac{10x}{(2x - 1)(x + 2)} - \frac{4}{x + 2}\frac{2x - 1}{2x - 1} + \frac{8}{2x - 1}\frac{x + 2}{x + 2}.$$

Adding numerators and simplifying gives

$$\frac{10x - 4(2x - 1) + 8(x + 2)}{(2x - 1)(x + 2)} = \frac{10x - 8x + 4 + 8x + 16}{(2x - 1)(x + 2)}$$

$$= \frac{10x + 20}{(2x - 1)(x + 2)}$$

$$= \frac{10(x + 2)}{(2x - 1)(x + 2)}$$

$$= \frac{10}{2x - 1}.$$

The cancellation of $x + 2$ is permissible provided $x \neq -2$. $\qquad\blacksquare$

We will illustrate the simplification of a complex fraction in Example 9.

☐ **Rationalization** You may have learned **rationalization of a denominator** in a previous mathematics course. Recall that rationalization of a denominator consists of multiplying an expression by a factor equal to 1 with the intent of clearing a radical from a denominator. For example, to rationalize the denominator in $1/\sqrt{2}$ we multiply the fraction by $\sqrt{2}/\sqrt{2}$:

$$\overset{\text{fraction is equal to 1}}{\underset{\downarrow}{\phantom{x}}}$$

$$\frac{1}{\sqrt{2}} = \frac{1}{\sqrt{2}}\frac{\sqrt{2}}{\sqrt{2}} = \frac{\sqrt{2}}{(\sqrt{2})^2} = \frac{\sqrt{2}}{2}.$$

There is no rule in mathematics that says only denominators must be rationalized. There are times in calculus when we are interested in rationalization not only of denominators but numerators as well. The next example uses the factorization of the difference of two squares in a slightly different manner. For $a > 0$ and $b > 0$, we can write $(\sqrt{a})^2 = a$, $(\sqrt{b})^2 = b$, and so we can write $a - b = (\sqrt{a})^2 - (\sqrt{b})^2$. It then follows from (1) that

$$a - b = (\sqrt{a} - \sqrt{b})(\sqrt{a} + \sqrt{b}). \qquad (8)$$

A variation of (8) is $a^2 - b = (a - \sqrt{b})(a + \sqrt{b})$. Thus if a numerator or denominator of a fractional expression contains a binomial term that includes at least one radical, such as

$$a - \sqrt{b}, a + \sqrt{b}, \sqrt{a} - b, \sqrt{a} + b, \sqrt{a} - \sqrt{b}, \text{ or } \sqrt{a} + \sqrt{b},$$

we multiply the numerator and the denominator of the fraction by the corresponding **conjugate factor**

$$a + \sqrt{b}, a - \sqrt{b}, \sqrt{a} + b, \sqrt{a} - b, \sqrt{a} + \sqrt{b}, \text{ or } \sqrt{a} - \sqrt{b}.$$

For example, to rationalize the denominator of $3/(\sqrt{2} - \sqrt{5})$ we use (8) to write

$$\frac{3}{\sqrt{2} - \sqrt{5}} = \frac{3}{\sqrt{2} - \sqrt{5}} \overset{\text{fraction is equal to 1}}{\frac{\sqrt{2} + \sqrt{5}}{\sqrt{2} + \sqrt{5}}} = \frac{3(\sqrt{2} + \sqrt{5})}{(\sqrt{2})^2 - (\sqrt{5})^2}$$

conjugate factor of denominator

$$= \frac{3(\sqrt{2} + \sqrt{5})}{-3} = -(\sqrt{2} + \sqrt{5}).$$

## ▮ EXAMPLE 4       Rationalization of a Numerator

Rationalize the numerator in $\dfrac{\sqrt{4 + x} - 2}{x}$.

**Solution** Think of the numerator as $\sqrt{a} - b$ where $a = 4 + x$ and $b = 2$. Because the conjugate factor of $\sqrt{a} - b$ is $\sqrt{a} + b$, we are able to clear the radical in the numerator by multiplying the numerator and denominator of the given fractional expression by $\sqrt{4 + x} + 2$:

$$\frac{\sqrt{4 + x} - 2}{x} = \frac{\sqrt{4 + x} - 2}{x} \frac{\sqrt{4 + x} + 2}{\sqrt{4 + x} + 2} = \frac{(\sqrt{4 + x})^2 - 2^2}{x(\sqrt{4 + x} + 2)}$$

$$= \frac{4 + x - 4}{x(\sqrt{4 + x} + 2)} = \frac{x}{x(\sqrt{4 + x} + 2)}.$$

After canceling the $x$'s in the numerator and the denominator in the last term the rationalization is complete:

$$\frac{\sqrt{4 + x} - 2}{x} = \frac{1}{\sqrt{4 + x} + 2}, \qquad x \neq 0. \qquad ▪$$

☐ **Limits—The Calculus Connection** Consider the fractional algebraic expression $\dfrac{x^2 - 1}{x - 1}$. Observe that this fraction cannot be evaluated at $x = 1$ because substituting 1 into the expression results in the undefined quantity 0/0. However, the fractional expression can be evaluated at any other real number; in particular, it can be evaluated at numbers that are very *close* to 1. The numerical values of the fractional expression given in the following two tables are easily obtained using the simplification in part (a) of Example 1:

| $x$ | 0.9 | 0.99 | 0.999 |
|---|---|---|---|
| $\dfrac{x^2 - 1}{x - 1}$ | 1.9 | 1.99 | 1.999 |

| $x$ | 1.1 | 1.01 | 1.001 |
|---|---|---|---|
| $\dfrac{x^2 - 1}{x - 1}$ | 2.1 | 2.01 | 2.001 |

(9)

$$\frac{x^2 - 1}{x - 1} = x + 1, \qquad \text{for } x \neq 1.$$

If we let the arrow symbol $\to$ represent the word *approach*, then the symbolism $x \to a^-$ indicates that $x$ approaches the number $a$ from the *left*, that is, through numbers that are less than $a$, and $x \to a^+$ means that $x$ approaches $a$ from the *right*, or through numbers that are greater than $a$. In the left-hand table in (9) we are letting $x \to 1^-$, and in the right-hand table $x \to 1^+$. Each table in (9) shows that the fractional expression $\dfrac{x^2 - 1}{x - 1}$ is close to the number 2 when $x$ is close to 1, that is,

$$\frac{x^2 - 1}{x - 1} \to 2 \text{ as } x \to 1^- \quad \text{and} \quad \frac{x^2 - 1}{x - 1} \to 2 \text{ as } x \to 1^+. \qquad (10)$$

We say that 2 is the **limit** of $\dfrac{x^2 - 1}{x - 1}$ as $x$ approaches 1 and write

$$\lim_{x \to 1} \frac{x^2 - 1}{x - 1} = 2. \qquad (11)$$

Before proceeding any further, we should make it clear that *a limit of an expression need not exist*. In the next two tables, consider $1/x$ as $x$ approaches zero:

| $x \to 0^-$ | $-0.1$ | $-0.01$ | $-0.001$ |
|---|---|---|---|
| $1/x$ | $-10$ | $-100$ | $-1000$ |

| $x \to 0^+$ | 0.1 | 0.01 | 0.001 |
|---|---|---|---|
| $1/x$ | 10 | 100 | 1000 |

As can be seen in the tables, as $x$ gets closer and closer to 0, the values of $1/x$ are becoming larger and larger in absolute value. In other words, $1/x$ is becoming unbounded. In this case we write

$$\frac{1}{x} \to -\infty \text{ as } x \to 0^- \quad \text{and} \quad \frac{1}{x} \to \infty \text{ as } x \to 0^+,$$

where $\infty$ is the infinity symbol. We say that $\lim\limits_{x \to 0} 1/x$ does not exist.

Suppose $f(x)$ denotes an expression involving a single variable $x$ and that the symbols $a$ and $L$ represent real numbers. If, as illustrated in (10),

$$f(x) \to L \text{ as } x \to a^- \quad \text{and} \quad f(x) \to L \text{ as } x \to a^+,$$

then we say that $\lim\limits_{x \to a} f(x)$ **exists** and write

$$\lim_{x \to a} f(x) = L.$$

In calculus, you will not be asked to find a limit by constructing tables of numerical values, although you surely will be asked to construct such tables because they are useful in convincing yourself of either the existence or the nonexistence of a limit. (See Problems 47 and 48 in Exercises 1.5.) Limits are either found or are proved to exist using analytical methods, in many cases using proven laws or properties of limits. Because it is not our goal to delve into theoretical or geometrical interpretations of a limit, and because we want to make the point that the calculus part of *some* problems is often the least significant part of the solution, we will accept three results from calculus without proof: If $a$ and $c$ are real numbers, then

$$\lim_{x \to a} c = c, \qquad \lim_{x \to a} x = a, \qquad \text{and} \qquad \lim_{x \to a} x^n = a^n, \tag{12}$$

where $n$ is a positive integer. For example, (12) allows us to write*

$$\lim_{x \to 3} (5x + 4) = 5(3) + 4 = 19$$

and

$$\lim_{x \to 3} (2x^2 + x + 1) = 2(3)^2 + 3 + 1 = 22.$$

In the preceding line we used $\lim_{x \to 3} x = 3$, $\lim_{x \to 3} x^2 = 9$, $\lim_{x \to 3} 4 = 4$, and $\lim_{x \to 3} 1 = 1$.

The limit concept is the foundation of calculus, and one kind of limit is of particular significance in calculus: the limit of a fractional expression where *both* the numerator and the denominator are approaching 0. Such a limit is said to have the **indeterminate form 0/0**. For example, in view of the results in (12), $\lim_{x \to 1} (x - 1) = 0$ and $\lim_{x \to 1} (x^2 - 1) = 0$.

Therefore $\lim_{x \to 1} \dfrac{x^2 - 1}{x - 1}$ has the indeterminate form 0/0. Of course, not all limit problems have this indeterminate form, but because of their importance (see Section 2.9), the limits in the remaining five examples, as well as *all* the limits in Exercises 1.5, have the form 0/0. Moreover, for simplicity we will only consider limits that actually exist.

We now show you how to find $\lim_{x \to 1} \dfrac{x^2 - 1}{x - 1}$ without the help of numerical tables:

algebra from Example 1(a)<br/>↓

$$\lim_{x \to 1} \frac{x^2 - 1}{x - 1} = \lim_{x \to 1} (x + 1) = 1 + 1 = 2.$$

Done! The intermediate steps were all algebra performed to rewrite the expression in a more tractable form, a form where the actual limit can be computed with minimal effort.

### ▮EXAMPLE 5　　　　Example 1 Revisited

Find $\lim_{x \to -3} \dfrac{x + 3}{x^2 - 4x - 21}$.

**Solution** This is the fractional expression in part (b) of Example 1. Observe that as $x \to -3$, the given limit has the indeterminate form 0/0. Now, using the algebraic simplification of this expression done in Example 1, we find that

algebra from Example 1(b)<br/>↓

$$\lim_{x \to -3} \frac{x + 3}{x^2 - 4x - 21} = \lim_{x \to -3} \frac{1}{x - 7} = \frac{1}{-10} = -\frac{1}{10}. \qquad ▪$$

### ▮EXAMPLE 6　　　　Example 2 Revisited

Find $\lim_{h \to 0} \dfrac{(7 + h)^2 - 49}{h}$.

---

*We are actually using several other properties of limits here. However, we do not feel this is the place to discuss all the properties of the limit concept.

**Solution** Using the algebra from Example 2,

$$\lim_{h \to 0} \frac{(7 + h)^2 - 49}{h} = \lim_{h \to 0} (14 + h) = 14.$$ ∎

**EXAMPLE 7**              **Example 3 Revisited**

Find $\displaystyle\lim_{x \to -2} \left[ \frac{10x}{2x^2 + 3x - 2} - \frac{4}{x + 2} + \frac{8}{2x - 1} \right]$.

**Solution** Were this a calculus course, you should observe that the first and second terms are of the form $1/0$ as $x \to -2$. You may think that this is the situation $\infty - \infty$, and so gives 0. No. Remember we never treat $\infty$ as we would a number. The observation that the given algebraic expression contains these undefined quantities should trigger the idea that combining the fractions into *one* fractional expression would be a way to proceed. After carrying out the algebra, as done in Example 3, you would then finish the problem as follows:

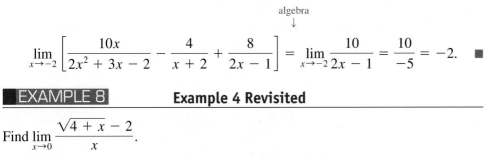

algebra
↓

$$\lim_{x \to -2} \left[ \frac{10x}{2x^2 + 3x - 2} - \frac{4}{x + 2} + \frac{8}{2x - 1} \right] = \lim_{x \to -2} \frac{10}{2x - 1} = \frac{10}{-5} = -2.$$ ∎

**EXAMPLE 8**              **Example 4 Revisited**

Find $\displaystyle\lim_{x \to 0} \frac{\sqrt{4 + x} - 2}{x}$.

**Solution** Using the algebra from Example 4, we find

algebra
↓

$$\lim_{x \to 0} \frac{\sqrt{4 + x} - 2}{x} = \lim_{x \to 0} \frac{1}{\sqrt{4 + x} + 2} = \frac{1}{\sqrt{4} + 2} = \frac{1}{2 + 2} = \frac{1}{4}.$$ ∎

When finding the value of a limit, the algebra can be done as a side problem (as we have done in Examples 1–4), and then making use of your work, completing the problem as we have illustrated in Examples 5–8. In our last example, we combine the algebra with the computing the limit. We recommend that you work through this example rather than just read it.

**EXAMPLE 9**              **Limit of a Complex Fraction**

Find $\displaystyle\lim_{x \to 0} \frac{\dfrac{1}{(2 + x)^3} - \dfrac{1}{8}}{x}$.

**Solution** The given expression is an example of a complex fraction, that is, a quotient where either the numerator or denominator is a fractional expression. We begin by finding a common denominator in the numerator:

$$\lim_{x \to 0} \frac{\dfrac{1}{(2 + x)^3} - \dfrac{1}{8}}{x} = \lim_{x \to 0} \frac{\dfrac{1}{(2 + x)^3} \dfrac{8}{8} - \dfrac{1}{8} \dfrac{(2 + x)^3}{(2 + x)^3}}{x}$$

$$= \lim_{x \to 0} \frac{\dfrac{8 - (2 + x)^3}{8(2 + x)^3}}{x}$$

To continue we use the expansion of $(a + b)^3$ given in (7) with $a = 2$ and $b = x$:

$$\lim_{x \to 0} \frac{\dfrac{1}{(2+x)^3} - \dfrac{1}{8}}{x} = \lim_{x \to 0} \frac{\dfrac{8 - (2^3 + 3x(2)^2 + 3x^2(2) + x^3)}{8(2+x)^3}}{x}$$

$$= \lim_{x \to 0} \frac{\dfrac{8 - 8 - 12x - 6x^2 - x^3}{8(2+x)^3}}{x} \qquad \leftarrow 8 - 8 = 0$$

$$= \lim_{x \to 0} \frac{\dfrac{-12x - 6x^2 - x^3}{8(2+x)^3}}{x}.$$

Because the $x$ in the denominator of the last complex fraction is equivalent to the fraction $x/1$ we invert and multiply:

Recall that division of fractions is converted into multiplication of fractions:

$$\frac{\dfrac{a}{b}}{\dfrac{c}{d}} = \frac{a}{b} \times \frac{d}{c} = \frac{ad}{bc}, \quad bc \neq 0$$

$$\lim_{x \to 0} \frac{\dfrac{1}{(2+x)^3} - \dfrac{1}{8}}{x} = \lim_{x \to 0} \frac{\dfrac{-12x - 6x^2 - x^3}{8(2+x)^3}}{\dfrac{x}{1}}$$

$$= \lim_{x \to 0} \frac{-12x - 6x^2 - x^3}{8(2+x)^3} \frac{1}{x}$$

$$= \lim_{x \to 0} \frac{x(-12 - 6x - x^2)}{8(2+x)^3} \frac{1}{x} \qquad \leftarrow \begin{cases} \text{Factor } x \text{ from numerator} \\ \text{and cancel } x\text{'s} \end{cases}$$

$$= \lim_{x \to 0} \frac{-12 - 6x - x^2}{8(2+x)^3}.$$

Finally, we have

$$\lim_{x \to 0} \frac{\dfrac{1}{(2+x)^3} - \dfrac{1}{8}}{x} = \lim_{x \to 0} \frac{-12 - 6x - x^2}{8(2+x)^3} = \frac{-12}{8 \cdot 2^3} = -\frac{3}{16},$$

since $\lim\limits_{x \to 0} x = 0$ and $\lim\limits_{x \to 0} x^2 = 0$ by (12). $\blacksquare$

## NOTES FROM THE CLASSROOM

····················································

($i$) On tests we see students carrying out the expansion of $(a + b)^3$ by brute force, multiplying out $(a + b)(a + b)(a + b)$. This procedure is not recommended; it is slow and you are prone to errors. Instead, you should memorize (6) and (7).

($ii$) In *any* mathematics course—not just calculus—do not erase or leave out important steps of your work. Most mathematics instructors want to see all work. Presenting that work in a neat and orderly fashion is also to your advantage. Finally, in the case of a limit problem such as Example 9, be sure to write down the symbol $\lim\limits_{x \to a}$ at each step. For example, we frequently see *incorrect* statements like this:

$$\lim_{x \to 1} \frac{x^2 - 1}{x - 1} = \frac{1}{x + 1} = \frac{1}{2}$$

CHAPTER 1 INEQUALITIES, EQUATIONS, AND GRAPHS

on students' papers. The *correct* version of the preceding line is

$$\lim_{x \to 1} \frac{x^2 - 1}{x - 1} = \lim_{x \to 1} \frac{1}{x + 1} = \frac{1}{2}.$$

**1.5** Exercises Answers to selected odd-numbered problems begin on page ANS–3.

In Problems 1–12, use factorization to simplify the given expression in part (a). Then, if instructed, find the indicated limit in part (b).

1. (a) $\dfrac{x^2 - 25}{x - 5}$      (b) $\displaystyle\lim_{x \to 5} \dfrac{x^2 - 25}{x - 5}$

2. (a) $\dfrac{y - 3}{y^2 - 9}$      (b) $\displaystyle\lim_{y \to 3} \dfrac{y - 3}{y^2 - 9}$

3. (a) $\dfrac{x^2 - 7x + 6}{x - 1}$      (b) $\displaystyle\lim_{x \to 1} \dfrac{x^2 - 7x + 6}{x - 1}$

4. (a) $\dfrac{2x + 10}{x^2 + 7x + 10}$      (b) $\displaystyle\lim_{x \to -5} \dfrac{2x + 10}{x^2 + 7x + 10}$

5. (a) $\dfrac{x^2 + x - 6}{x^2 - 5x + 6}$      (b) $\displaystyle\lim_{x \to 2} \dfrac{x^2 + x - 6}{x^2 - 5x + 6}$

6. (a) $\dfrac{x^2 - 8x}{x^2 - 6x - 16}$      (b) $\displaystyle\lim_{x \to 8} \dfrac{x^2 - 8x}{x^2 - 6x - 16}$

7. (a) $\dfrac{x^3 - 1}{x - 1}$      (b) $\displaystyle\lim_{x \to 1} \dfrac{x^3 - 1}{x - 1}$

8. (a) $\dfrac{x^2 - 4}{x^3 + 8}$      (b) $\displaystyle\lim_{x \to -2} \dfrac{x^2 - 4}{x^3 + 8}$

9. (a) $\dfrac{x^3 - 1}{x^2 + 3x - 4}$      (b) $\displaystyle\lim_{x \to 1} \dfrac{x^3 - 1}{x^2 + 3x - 4}$

10. (a) $\dfrac{x^5 + 2x^4 + x^3}{x^4 - 2x^2 + 1}$      (b) $\displaystyle\lim_{x \to -1} \dfrac{x^5 + 2x^4 + x^3}{x^4 - 2x^2 + 1}$

11. (a) $\dfrac{x^3 + 3x^2 + 3x + 1}{x^4 + x^3 + x + 1}$      (b) $\displaystyle\lim_{x \to -1} \dfrac{x^3 + 3x^2 + 3x + 1}{x^4 + x^3 + x + 1}$

12. (a) $\dfrac{x^4 - 5x^3 + 4x - 20}{x^4 - 5x^3 + x - 5}$      (b) $\displaystyle\lim_{x \to 5} \dfrac{x^4 - 5x^3 + 4x - 20}{x^4 - 5x^3 + x - 5}$

In Problems 13–20, use binomial expansion to simplify the given expression in part (a). Then, if instructed, find the indicated limit in part (b).

13. (a) $\dfrac{(2 + h)^2 - 4}{h}$      (b) $\displaystyle\lim_{h \to 0} \dfrac{(2 + h)^2 - 4}{h}$

14. (a) $\dfrac{5 - 5(h + 1)^2}{h}$      (b) $\displaystyle\lim_{h \to 0} \dfrac{5 - 5(h + 1)^2}{h}$

15. (a) $\dfrac{(2x + 1)^2 - 9}{x - 1}$      (b) $\displaystyle\lim_{x \to 1} \dfrac{(2x + 1)^2 - 9}{x - 1}$

**16. (a)** $\dfrac{2(x - 1)^2 - 4(x - 1) - 6}{x}$      **(b)** $\displaystyle\lim_{x \to 0} \dfrac{2(x - 1)^2 - 4(x - 1) - 6}{x}$

**17. (a)** $\dfrac{(1 + x)^3 - 1}{x}$      **(b)** $\displaystyle\lim_{x \to 0} \dfrac{(1 + x)^3 - 1}{x}$

**18. (a)** $\dfrac{(x + 1)^3 + (x - 1)^3}{x}$      **(b)** $\displaystyle\lim_{x \to 0} \dfrac{(x + 1)^3 + (x - 1)^3}{x}$

**19. (a)** $\dfrac{2(h + 1)^3 - 5(h + 1)^2 + 3}{h}$      **(b)** $\displaystyle\lim_{h \to 0} \dfrac{2(h + 1)^3 - 5(h + 1)^2 + 3}{h}$

**20. (a)** $\dfrac{(x + 2)^4 - 16}{x}$      **(b)** $\displaystyle\lim_{x \to 0} \dfrac{(x + 2)^4 - 16}{x}$

In Problems 21–26, use addition of algebraic fractions to simplify the given expression in part (a). Then, if instructed, find the indicated limit in part (b).

**21. (a)** $\dfrac{1}{x - 2} - \dfrac{6}{x^2 + 2x - 8}$      **(b)** $\displaystyle\lim_{x \to 2} \left[ \dfrac{1}{x - 2} - \dfrac{6}{x^2 + 2x - 8} \right]$

**22. (a)** $\dfrac{x^2 + 3x - 1}{x} + \dfrac{1}{x}$      **(b)** $\displaystyle\lim_{x \to 0} \left[ \dfrac{x^2 + 3x - 1}{x} + \dfrac{1}{x} \right]$

**23. (a)** $\dfrac{1}{x - 10} - \dfrac{20}{x^2 - 100}$      **(b)** $\displaystyle\lim_{x \to 10} \left[ \dfrac{1}{x - 10} - \dfrac{20}{x^2 - 100} \right]$

**24. (a)** $\dfrac{1}{x} \left[ \dfrac{1}{9} - \dfrac{1}{x + 9} \right]$      **(b)** $\displaystyle\lim_{x \to 0} \dfrac{1}{x} \left[ \dfrac{1}{9} - \dfrac{1}{x + 9} \right]$

**25. (a)** $\dfrac{\dfrac{1}{(2 + h)^2} - \dfrac{1}{4}}{h}$      **(b)** $\displaystyle\lim_{h \to 0} \dfrac{\dfrac{1}{(2 + h)^2} - \dfrac{1}{4}}{h}$

**26. (a)** $\dfrac{1}{t - 1} \left[ \dfrac{1}{(t + 3)^2} - \dfrac{1}{16} \right]$      **(b)** $\displaystyle\lim_{t \to 1} \dfrac{1}{t - 1} \left[ \dfrac{1}{(t + 3)^2} - \dfrac{1}{16} \right]$

In Problems 27–34, use rationalization to simplify the given expression in part (a). Then, if instructed, find the indicated limit in part (b).

**27. (a)** $\dfrac{\sqrt{x} - 3}{x - 9}$      **(b)** $\displaystyle\lim_{x \to 9} \dfrac{\sqrt{x} - 3}{x - 9}$

**28. (a)** $\dfrac{\dfrac{1}{\sqrt{x}} - \dfrac{1}{\sqrt{2}}}{x - 2}$      **(b)** $\displaystyle\lim_{x \to 2} \dfrac{\dfrac{1}{\sqrt{x}} - \dfrac{1}{\sqrt{2}}}{x - 2}$

**29. (a)** $\dfrac{x}{\sqrt{7 + x} - \sqrt{7}}$      **(b)** $\displaystyle\lim_{x \to 0} \dfrac{x}{\sqrt{7 + x} - \sqrt{7}}$

**30. (a)** $\dfrac{\sqrt{u + 4} - 3}{u - 5}$      **(b)** $\displaystyle\lim_{u \to 5} \dfrac{\sqrt{u + 4} - 3}{u - 5}$

**31. (a)** $\dfrac{25 - t}{5 - \sqrt{t}}$      **(b)** $\displaystyle\lim_{t \to 25} \dfrac{25 - t}{5 - \sqrt{t}}$

**32. (a)** $\dfrac{1}{h} \left[ 1 - \dfrac{1}{\sqrt{1 + h}} \right]$      **(b)** $\displaystyle\lim_{h \to 0} \dfrac{1}{h} \left[ 1 - \dfrac{1}{\sqrt{1 + h}} \right]$

**33. (a)** $\dfrac{4y^2}{\sqrt{y^2 + y + 1} - \sqrt{y + 1}}$      **(b)** $\displaystyle\lim_{y \to 0} \dfrac{4y^2}{\sqrt{y^2 + y + 1} - \sqrt{y + 1}}$

**34. (a)** $\dfrac{9t^2}{t + 2 - 2\sqrt{t + 1}}$      **(b)** $\displaystyle\lim_{t \to 0} \dfrac{9t^2}{t + 2 - 2\sqrt{t + 1}}$

## Miscellaneous Calculus-Related Problems

In Problems 35–40, the given algebraic expression is an unsimplified answer to a calculus problem. Simplify the expression.

**35.** $\dfrac{x + \dfrac{1}{x} - a - \dfrac{1}{a}}{x - a}$

**36.** $\dfrac{\dfrac{3}{(x + 1)^2} - \dfrac{3}{(a + 1)^2}}{x - a}$

**37.** $(3x^2 + 4x - 1)(4)(2x - 3)^3(2) + (2x - 3)^4(6x + 4)$

**38.** $(12x - 1)^{1/3}(2)(x^2 - 1)(2x) + (x^2 - 1)^2(\frac{1}{3})(12x - 1)^{-2/3}(12)$

**39.** $\dfrac{2x(-4x + 6)^{1/2} - x^2(\frac{1}{2})(-4x + 6)^{-1/2}(-4)}{[(-4x + 6)^{1/2}]^2}$

**40.** $\dfrac{1}{2}\left(\dfrac{2x - 1}{4x + 1}\right)^{-\frac{1}{2}} \cdot \dfrac{(4x + 1)2 - (2x - 1)4}{(4x + 1)^2}$

In Problems 41–46, the given equation is a partial answer to a calculus problem. Solve the equation for the symbol $y'$

**41.** $3y^2y' - y - xy' = x$

**42.** $y' = 2(x - y)(1 - y')$

**43.** $2yy' + 2x = y'$

**44.** $2xy^2 + x^2(2y)y' - 2 = -3y'$

**45.** $\dfrac{(x - y)(1 + y') - (x + y)(1 - y')}{(x - y)^2} = 1$

**46.** $\dfrac{1}{1 + x^2y^2}(xy' + y) = 2xyy' + y^2$

## Calculator/Computer Problems

In Problems 47 and 48, use a calculator or computer to estimate the given limit by completing each table. Round the entries in each table to eight decimal places.

**47.** $\displaystyle\lim_{x \to 1} \dfrac{x^3 - 1}{\sqrt[3]{x} - 1}$;

| $x \to 0^+$ | 1.1 | 1.01 | 1.001 | 1.0001 | 1.00001 |
| --- | --- | --- | --- | --- | --- |
| $\dfrac{x^3 - 1}{\sqrt[3]{x} - 1}$ | | | | | |

| $x \to 0^-$ | 0.9 | 0.99 | 0.999 | 0.9999 | 0.99999 |
| --- | --- | --- | --- | --- | --- |
| $\dfrac{x^3 - 1}{\sqrt[3]{x} - 1}$ | | | | | |

**48.** $\displaystyle\lim_{x \to 0} (1 + x)^{1/x}$;

| $x \to 0^+$ | 0.1 | 0.01 | 0.001 | 0.0001 | 0.00001 |
| --- | --- | --- | --- | --- | --- |
| $(1 + x)^{1/x}$ | | | | | |

| $x \to 0^-$ | −0.1 | −0.01 | −0.001 | −0.0001 | −0.00001 |
| --- | --- | --- | --- | --- | --- |
| $(1 + x)^{1/x}$ | | | | | |

## For Discussion

In Problems 49 and 50, discuss what algebra is necessary to evaluate the given limit. Carry out your ideas.

**49.** $\displaystyle\lim_{x \to 1} \dfrac{x - 1}{x^8 - 1}$

**50.** $\displaystyle\lim_{x \to 0} \dfrac{\sqrt[3]{x + 27} - 3}{x}$

In Problems 1–20, fill in the blanks.

1. An inequality with $(-\infty, 9]$ as its solution set is _____.
2. The solution set of the inequality $-3 < x \le 8$ as an interval is _____.
3. If the point $(a, b)$ lies in quadrant IV, then $(b, a)$ lies in quadrant _____.
4. The point $(x, -3x)$ in the second quadrant that is 5 units from $(2, -1)$ is _____.
5. If the graph of an equation contains the point $(2, 3)$ and is symmetric with respect to the $x$-axis, then the graph also contains the point _____.
6. If the graph of an equation contains the point $(-1, 6)$ and is symmetric with respect to the origin, then the graph also contains the point _____.
7. An equation of a circle with center $(-2, -5)$ and radius 6 is _____.
8. If $|2 - x| = 15$, then $x =$ _____.
9. The distance from the midpoint of the line segment joining $(4, -6)$ and $(-2, 0)$ to the origin is _____.
10. The graph of $y = 2|x| - 5$ is symmetric with respect to _____.
11. The intercepts of the graph of $y = 2|x| - 5$ are _____.
12. The circle $x^2 - 16x + y^2 = 0$ is symmetric with respect to _____.
13. The center and radius of the circle $x^2 - 16x + y^2 = 0$ are _____.
14. The intercepts of the circle $(x - 1)^2 + (y - 2)^2 = 10$ are _____.
15. Two points on the circle $x^2 + y^2 = 25$ with the same $x$-coordinate $-3$ are _____.
16. The graph of $y = -\sqrt{100 - x^2}$ is a _____.
17. The inequality _____ describes the set of points in the $xy$-plane outside the circle $x^2 + y^2 = 36$.
18. If $(a, a + \sqrt{3})$ lies on the graph of $y = 2x$, then $a =$ _____.
19. The set of real numbers $x$ whose distance between $x$ and $\sqrt{2}$ is greater than 3 is defined by the absolute-value inequality _____.
20. A point $(x, y)$ in the $xy$-plane whose coordinates satisfy $xy < 0$ lies in quadrant(s) _____ or _____.

In Problems 21–40, answer true or false.

21. The word *nonnegative* means the same as the word *positive*. _____
22. The number 0 is neither positive nor negative. _____
23. $-3$ is not greater than $-1$. _____
24. If $a < b$, then $b - a$ is a positive number. _____
25. If $a < b$, then $a^2 < b^2$. _____
26. For any real number $a$, $-a \le a$. _____
27. If $a < 0$, then $\dfrac{a}{-a} < 0$. _____
28. If $a^2 < a$, then $a < 1$. _____
29. If $x$ is a negative number, then $-x$ is a positive number. _____
30. The solution set of $|4x - 6| \ge -1$ is $(-\infty, \infty)$. _____
31. $|-3t + 6| = 3|t - 2|$ _____
32. The point $(5, 0)$ is in quadrant I. _____
33. The point $(-3, 7)$ is in quadrant III. _____
34. The distance between the points $(0, 0)$ and $(3, 6)$ is 9. _____
35. To find $y$-intercepts of the graph of an equation we let $x = 0$ and solve for $y$. _____
36. There is no point on the circle $x^2 + y^2 - 10x + 22 = 0$ with $x$-coordinate 2.

_____

**37.** A circle whose equation can be put into the form $x^2 + y^2 + ax + by = 0$ must pass through the origin. _____

**38.** The points $(0, 0)$, $(a, 0)$, $a > 0$, and $(0, b)$, $b < 0$, are vertices of a right triangle. _____

**39.** The graph of the equation $x^2 y + 4y = x$ is symmetric with respect to the origin.

_____

**40.** The inequality $\dfrac{100}{x^2 + 64} \le 0$ has no solution. _____

In Problems 41–44, assume that $0 < a < b$. Compare the given expressions using inequality symbols.

**41.** $a^2$ and $ab$  **42.** $-a$ and $-b$  **43.** $a$ and $a + b$  **44.** $\dfrac{1}{a}$ and $\dfrac{1}{a + b}$

In Problems 45–50, fill in the blank with either an appropriate inequality symbol or a number.

**45.** If $x - 10 > 5$, then $x +$ _____ $> 25$.
**46.** If $x - 2 \le 7$, then $x$ _____ $9$.
**47.** If $-\frac{1}{3}x \ge 4$, then $x$ _____ $-12$.
**48.** If $3x - 6 \le 4x - 4$, then $x$ _____ $2$.
**49.** If $-2 \le 1 - x \le 5$, then _____ $\le x \le$ _____.
**50.** If $-3 < x < 9$, then _____ $< -2x <$ _____.
**51.** On the number line, $m = 5$ is the midpoint of the line segment joining the number $a$ (left endpoint) and the number $b$ (right endpoint). Use the fact that $d(a, b) = 2$ to find $a$ and $b$.
**52.** In the $xy$-plane, find an equation that describes the set of points $(x, y)$ that are equidistant from $(0, 5)$ and $(x, -5)$.

In Problems 53–66, solve the given inequality. Write the solution set using interval notation.

**53.** $2x - 5 \ge 6x + 7$

**54.** $\frac{1}{4}x - 3 < \frac{1}{2}x + 1$

**55.** $-4 < x - 8 < 4$

**56.** $7 \le 3 - 2x < 11$

**57.** $|x| > 10$

**58.** $|-6x| \le 42$

**59.** $|3x - 4| < 5$

**60.** $|5 - 2x| \ge 7$

**61.** $3x \ge 2x^2 - 5$

**62.** $x^2 > 6x - 9$

**63.** $x^3 > x$

**64.** $(x^2 - x)(x^2 + x) \le 0$

**65.** $\dfrac{1}{x} + x > 2$

**66.** $\dfrac{2x - 6}{x - 1} \ge 1$

In Problems 67–70, simplify the given expression in part (a). Then, if instructed, find the indicated limit in part (b).

**67. (a)** $\dfrac{2x - 1}{4x^2 - 1}$  **(b)** $\lim\limits_{x \to \frac{1}{2}} \dfrac{2x - 1}{4x^2 - 1}$

**68. (a)** $\dfrac{x^2 - 6x + 5}{x - 5}$  **(b)** $\lim\limits_{x \to 5} \dfrac{x^2 - 6x + 5}{x - 5}$

**69. (a)** $\dfrac{x^2 - 16}{\sqrt{x} - 2}$  **(b)** $\lim\limits_{x \to 4} \dfrac{x^2 - 16}{\sqrt{x} - 2}$

**70. (a)** $\dfrac{1}{h}\left(\dfrac{1}{3 + h} - \dfrac{1}{3}\right)$  **(b)** $\lim\limits_{h \to 0} \dfrac{1}{h}\left(\dfrac{1}{3 + h} - \dfrac{1}{3}\right)$

## Chapter Outline

# Functions

**2**

**Functions and Graphs**

☐ **Introduction** Using the objects and the persons around us, it is easy to make up a rule of correspondence that associates, or pairs, the members, or elements, of one set with the members of another set. For example, to each social security number there is a person, to each car registered in the State of California there is a license plate number, to each book there corresponds at least one author, to each state there is a governor, and so on. A natural correspondence occurs between a set of 20 students and a set of, say, 25 desks in a classroom when each student selects and sits in a different desk. In mathematics we are interested in a special type of correspondence, a *single-valued correspondence*, called a function.

> **FUNCTION**
>
> A **function** from a set $X$ to a set $Y$ is a rule of correspondence that assigns to each element $x$ in $X$ exactly one element $y$ in $Y$.

In the student/desk correspondence above suppose the set of 20 students is the set $X$ and the set of 25 desks is the set $Y$. This correspondence is a function from the set $X$ to the set $Y$ provided no student sits in two desks at the same time.

☐ **Terminology** A function is usually denoted by a letter such as $f$, $g$, or $h$. We can then represent a function $f$ from a set $X$ to a set $Y$ by the notation $f : X \rightarrow Y$. The set $X$ is called the **domain** of $f$. The set of corresponding elements $y$ in the set $Y$ is called the **range** of the function. For our student/desk function, the set of students is the domain and the set of 20 desks actually occupied by the students constitutes the range. Notice that the range of $f$ need not be the entire set $Y$. The unique element $y$ in the range that corresponds to a selected element $x$ in the domain $X$ is called the **value** of the function at $x$, or the **image** of $x$, and is written $f(x)$. The latter symbol is read "$f$ of $x$" or "$f$ at $x$," and we write $y = f(x)$.* See **FIGURE 2.1.1**. Since the value of $y$ depends on the choice of $x$, $y$ is called the **dependent variable**; $x$ is called the **independent variable**. Unless otherwise stated, we will assume hereafter that the sets $X$ and $Y$ consist of real numbers.

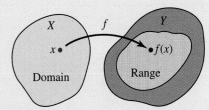

**FIGURE 2.1.1** Domain and range of a function $f$

---

*Many instructors like to call $x$ the *input* of the function and $f(x)$ the *output*.

## EXAMPLE 1　　The Squaring Function

The rule for squaring a real number is given by the equation $y = x^2$ or $f(x) = x^2$. The values of $f$ at $x = -5$ and $x = \sqrt{7}$ are obtained by replacing $x$, in turn, by the numbers $-5$ and $\sqrt{7}$:

$$f(-5) = (-5)^2 = 25 \quad \text{and} \quad f(\sqrt{7}) = (\sqrt{7})^2 = 7.\qquad\blacksquare$$

Occasionally for emphasis we will write a function using parentheses in place of the symbol $x$. For example, we can write the squaring function $f(x) = x^2$ as

$$f(\ ) = (\ )^2. \tag{1}$$

This illustrates the fact that $x$ is a *placeholder* for any number in the domain of the function $y = f(x)$. Thus, if we wish to evaluate (1) at, say, $3 + h$, where $h$ represents a real number, we put $3 + h$ into the parentheses and carry out the appropriate algebra:

See (6) of Section 1.5. ▶

$$f(3 + h) = (3 + h)^2 = 9 + 6h + h^2.$$

If a function $f$ is defined by means of a formula or an equation, then typically the domain of $y = f(x)$ is not expressly stated. We will see that we can usually deduce the domain of $y = f(x)$ either from the structure of the equation or from the context of the problem.

## EXAMPLE 2　　Domain and Range

In Example 1, since any real number $x$ can be squared and the result $x^2$ is another real number, $f(x) = x^2$ is a function from $R$ to $R$, that is, $f:R \rightarrow R$. In other words, the domain of $f$ is the set $R$ of real numbers. Using interval notation, we also write the domain as $(-\infty, \infty)$. The range of $f$ is the set of nonnegative real numbers or $[0, \infty)$; this follows from the fact that $x^2 \geq 0$ for every real number $x$.　　$\blacksquare$

☐ **Domain of a Function** As mentioned earlier, the domain of a function $y = f(x)$ that is defined by a formula is usually not specified. Unless stated or implied to the contrary, it is understood that:

> *The domain of a function $f$ is the largest subset of the set of real numbers for which $f(x)$ is a real number.*

This set is sometimes referred to as the **implicit domain** of the function. For example, we cannot compute $f(0)$ for the reciprocal function $f(x) = 1/x$ since $1/0$ is not a real number. In this case we say that $f$ is **undefined** at $x = 0$. Since every nonzero real number has a reciprocal, the domain of $f(x) = 1/x$ is the set of real numbers except $0$. By the same reasoning, the function $g(x) = 1/(x^2 - 4)$ is not defined at either $x = -2$ or $x = 2$, and so its domain is the set of real numbers with $-2$ and $2$ excluded. The square root function $h(x) = \sqrt{x}$ is not defined at $x = -1$ because $\sqrt{-1}$ is not a real number. In order for $h(x) = \sqrt{x}$ to be defined in the real number system we must require the **radicand**, in this case simply $x$, to be nonnegative. From the inequality $x \geq 0$ we see that the domain of the function $h$ is the interval $[0, \infty)$.

## EXAMPLE 3　　Domain and Range

Determine the domain and range of $f(x) = 4 + \sqrt{x - 3}$.

**Solution** The radicand $x - 3$ must be nonnegative. By solving the inequality $x - 3 \geq 0$ we get $x \geq 3$, and so the domain of $f$ is $[3, \infty)$. Now, since the symbol $\sqrt{\ }$ denotes the nonnegative square root of a number, $\sqrt{x - 3} \geq 0$ for $x \geq 3$ and consequently $4 + \sqrt{x - 3} \geq 4$. The smallest value of $f(x)$ occurs at $x = 3$ and is $f(3) = 4 + \sqrt{0} = 4$. Moreover, because $x - 3$ and $\sqrt{x - 3}$ increase as $x$

takes on increasing larger values, we conclude that $y \geq 4$. Consequently the range of $f$ is $[4, \infty)$. ∎

### EXAMPLE 4    Domain of $f$

Determine the domain of $f(x) = \sqrt{x^2 + 2x - 15}$.

**Solution** As in Example 3, the expression under the radical symbol—the radicand—must be nonnegative, that is, the domain of $f$ is the set of real numbers $x$ for which $x^2 + 2x - 15 \geq 0$ or $(x - 3)(x + 5) \geq 0$. We have already solved the last inequality by means of a sign chart in Example 3 of Section 1.1. The solution set of the inequality $(-\infty, -5] \cup [3, \infty)$ is also the domain of $f$. ∎

### EXAMPLE 5    Domains of Two Functions

Determine the domain of

**(a)** $g(x) = \dfrac{1}{\sqrt{x^2 + 2x - 15}}$ and

**(b)** $h(x) = \dfrac{5x}{x^2 - 3x - 4}$.

**Solution** A function that is given by a fractional expression is not defined at the $x$-values for which its denominator is equal to 0.

**(a)** The expression under the radical is the same as in Example 4. Since $x^2 + 2x - 15$ is in the denominator we must have $x^2 + 2x - 15 \neq 0$. This excludes $x = -5$ and $x = 3$. In addition, since $x^2 + 2x - 15$ appears under a radical, we must have $x^2 + 2x - 15 > 0$ for all other values of $x$. Thus the domain of the function $g$ is the union of two open intervals $(-\infty, -5) \cup (3, \infty)$.

**(b)** Since the denominator of $h(x)$ factors,

$$x^2 - 3x - 4 = (x + 1)(x - 4)$$

we see that $(x + 1)(x - 4) = 0$ for $x = -1$ and $x = 4$. In contrast to the function in part (a), these are the *only* numbers for which $h$ is not defined. Hence, the domain of the function $h$ is the set of real numbers with $x = -1$ and $x = 4$ excluded. ∎

Using interval notation, the domain of $h$ in part (b) of Example 5 can be written as

$$(-\infty, -1) \cup (-1, 4) \cup (4, \infty).$$

As an alternative to this ungainly union of disjoint intervals, this domain can also be written using set-builder notation as $\{x \mid x \neq -1 \text{ and } x \neq 4\}$.

☐ **Graphs** A function is often used to describe phenomena in fields such as science, engineering, and business. In order to interpret and utilize data, it is useful to display this data in the form of a graph. The graph of a function $f$ is the graph of the set of ordered pairs $(x, f(x))$, where $x$ is in the domain of $f$. In the $xy$-plane an ordered pair $(x, f(x))$ is a point, so that the graph of a function is a set of points. If a function is defined by an equation $y = f(x)$, then the graph of $f$ is the graph of the equation. To obtain points on the graph of an equation $y = f(x)$, we judiciously choose numbers $x_1, x_2, x_3, \ldots$ in its domain, compute $f(x_1), f(x_2), f(x_3), \ldots$, plot the corresponding points $(x_1, f(x_1)), (x_2, f(x_2)), (x_3, f(x_3)), \ldots$, and then connect these points with a curve. See **FIGURE 2.1.2**. Keep in mind that:

- a value of $x$ is a directed distance from the $y$-axis, and
- a function value $f(x)$ is a directed distance from the $x$-axis.

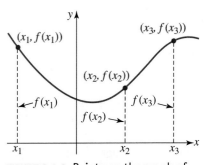

**FIGURE 2.1.2** Points on the graph of an equation $y = f(x)$

A word about the figures in this text is in order. With a few exceptions, it is usually impossible to display the complete graph of a function, and so we often display only the more important features of the graph. In Figure 2.1.3(a), notice that the graph goes down on its left and right sides. Unless indicated to the contrary, we may assume that there are no major surprises beyond what we have shown and the graph simply continues in the manner indicated. The graph in Figure 2.1.3(a) indicates the so-called **end behavior** or **global behavior** of the function: for a point $(x, y)$ on the graph, $y \to -\infty$ as $x \to -\infty$ and $y \to -\infty$ as $x \to \infty$. (More will be said about this concept of global behavior in Chapter 3.) If a graph terminates at either its right or left end, we will indicate this by a dot when clarity demands it. See Figure 2.1.4. We will use a solid dot to represent the fact that the end point is included on the graph and an open dot to signify that the end point is not included on the graph.

☐ **Vertical Line Test** From the definition of a function we know that for each $x$ in the domain of $f$ there corresponds only one value $f(x)$ in the range. This means a vertical line that intersects the graph of a function $y = f(x)$ (this is equivalent to choosing an $x$) can do so in at most one point. Conversely, if *every* vertical line that intersects a graph of an equation does so in at most one point, then the graph is the graph of a function. The last statement is called the **vertical line test** for a function. See FIGURE 2.1.3(a). On the other hand, if *some* vertical line intersects a graph of an equation more than once, then the graph is not that of a function. See Figures 2.1.3(b) and 2.1.3(c). When a vertical line intersects a graph in several points, the same number $x$ corresponds to different values of $y$ in contradiction to the definition of a function.

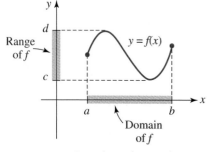

FIGURE 2.1.4 Domain and range interpreted graphically

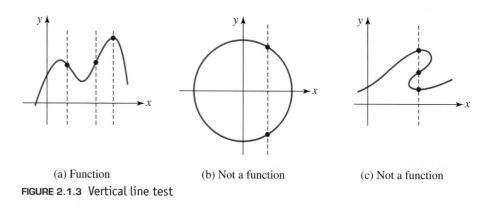

(a) Function    (b) Not a function    (c) Not a function

FIGURE 2.1.3 Vertical line test

If you have an accurate graph of a function $y = f(x)$ it is often possible to *see* the domain and range of $f$. In FIGURE 2.1.4 assume that the colored curve is the entire, or complete, graph of some function $f$. The domain of $f$ then is the interval $[a, b]$ on the $x$-axis and the range is the interval $[c, d]$ on the $y$-axis.

### EXAMPLE 6     Example 3 Revisited

From the graph of $f(x) = 4 + \sqrt{x - 3}$ given in FIGURE 2.1.5, we can see that the domain and range of $f$ are, respectively, $[3, \infty)$ and $[4, \infty)$. This agrees with the results in Example 3. ∎

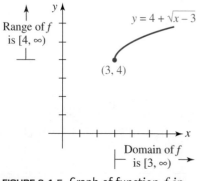

FIGURE 2.1.5 Graph of function $f$ in Example 6

As shown in Figure 2.1.3(b), a circle is not the graph of a function. Actually, an equation such as $x^2 + y^2 = 9$ defines (at least) two functions of $x$. If we solve this equation for $y$ in terms of $x$ we get $y = \pm\sqrt{9 - x^2}$. Because of the single-valued convention of the $\sqrt{\phantom{x}}$ sign, both equations $y = \sqrt{9 - x^2}$ and $y = -\sqrt{9 - x^2}$ define functions. As we saw in Section 1.4, the first equation defines an *upper semicircle* and the second defines a *lower semicircle*. From the graphs shown in FIGURE 2.1.6, the domain of $y = \sqrt{9 - x^2}$ is $[-3, 3]$ and the range is $[0, 3]$; the domain and range of $y = -\sqrt{9 - x^2}$ are $[-3, 3]$ and $[-3, 0]$, respectively.

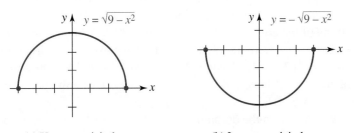

(a) Upper semicircle          (b) Lower semicircle

FIGURE 2.1.6 These semicircles are graphs of functions

☐ **Intercepts** To graph a function defined by an equation $y = f(x)$, it is usually a good idea to first determine whether the graph of $f$ has any intercepts. Recall that all points on the $y$-axis are of the form $(0, y)$. Thus, if 0 is the domain of a function $f$, the **y-intercept** is the point on the $y$-axis whose $y$-coordinate is $f(0)$, in other words, $(0, f(0))$. See FIGURE 2.1.7(a). Similarly, all points on the $x$-axis have the form $(x, 0)$. This means that to find the $x$-intercepts of the graph of $y = f(x)$, we determine the values of $x$ which make $y = 0$. That is, we must solve the equation $f(x) = 0$ for $x$. A number $c$ for which

$$f(c) = 0$$

is referred to as either a **zero** of the function $f$ or a **root** (or **solution**) of the equation $f(x) = 0$. The *real* zeros of a function $f$ are the $x$-coordinates of the $x$-intercepts of the graph of $f$. In Figure 2.1.7(b), we have illustrated a function that has three zeros $x_1, x_2$, and $x_3$ because $f(x_1) = 0$, $f(x_2) = 0$, and $f(x_3) = 0$. The corresponding three $x$-intercepts are the points $(x_1, 0)$, $(x_2, 0)$, and $(x_3, 0)$. Of course, the graph of the function may have no intercepts. This is illustrated in Figure 2.1.5.

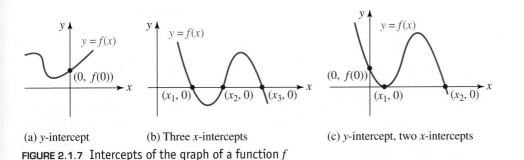

(a) $y$-intercept          (b) Three $x$-intercepts          (c) $y$-intercept, two $x$-intercepts

FIGURE 2.1.7 Intercepts of the graph of a function $f$

A graph does not necessarily have to *cross* a coordinate axis at an intercept; a graph could simply be tangent to, or *touch*, an axis. In Figure 2.1.7(c) the graph of $y = f(x)$ is tangent to the $x$-axis at $(x_1, 0)$. Also, the graph of a function $f$ can have at most one $y$-intercept since, if 0 is the domain of $f$, there can correspond only one $y$-value, namely, $y = f(0)$.

◀ More will be said about this in Chapter 3.

Find, if possible, the $x$- and $y$-intercepts of the given function.

**(a)** $f(x) = x^2 + 2x - 2$    **(b)** $f(x) = \dfrac{x^2 - 2x - 3}{x}$

**Solution**

**(a)** Since 0 is in the domain of $f$, $f(0) = -2$ is the $y$-coordinate of the $y$-intercept of the graph of $f$. The $y$-intercept is the point $(0, -2)$. To obtain the $x$-intercepts we must determine whether $f$ has any real zeros, that is, real solutions of the equation $f(x) = 0$. Since the left-hand side of the equation $x^2 + 2x - 2 = 0$ has no obvious factors, we use the quadratic formula to obtain $x = \frac{1}{2}(2 \pm \sqrt{12})$. Since $\sqrt{12} = \sqrt{4 \cdot 3} = 2\sqrt{3}$ the zeros of $f$ are the numbers $1 - \sqrt{3}$ and $1 + \sqrt{3}$. The $x$-intercepts are the points $(1 - \sqrt{3}, 0)$ and $(1 + \sqrt{3}, 0)$.

**(b)** Because 0 is not in the domain of $f$ ($f(0) = -3/0$ is not defined), the graph of $f$ possesses no $y$-intercept. Now since $f$ is a fractional expression, the only way we can have $f(x) = 0$ is to have the numerator equal zero. Factoring the left-hand side of $x^2 - 2x - 3 = 0$ gives $(x + 1)(x - 3) = 0$. Therefore the numbers $-1$ and 3 are the zeros of $f$. The $x$-intercepts are the points $(-1, 0)$ and $(3, 0)$.

☐ **Approximating Zeros**  Even when it is obvious that the graph of a function $y = f(x)$ possesses $x$-intercepts it is not always a straightforward matter to solve the equation $f(x) = 0$. In fact, it is *impossible* to solve some equations exactly; some times the best we can do is to **approximate** the zeros of the function. One way of doing this is to obtain a very accurate graph of $f$.

With the aid of a graphing utility the graph of the function $f(x) = x^3 - x + 4$ is given in **FIGURE 2.1.8**. From $f(0) = 4$ we see that the $y$-intercept is $(0, 4)$. As we see in the figure, there appears to be only one $x$-intercept with $x$-coordinate close to $-1.7$ or $-1.8$. But there is no convenient way of finding the roots of the equation $x^3 - x + 4 = 0$. We can however approximate the real root of this equation with the aid of the *find root* feature of either a graphing calculator or computer algebra system. We find that $x \approx -1.796$ and so the approximate $x$-intercept is $(-1.796, 0)$. As a check, note that the function value

$$f(-1.796) = (-1.796)^3 - (-1.796) + 4 \approx 0.0028$$

is nearly 0.

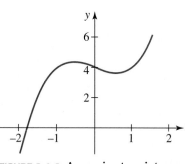

**FIGURE 2.1.8** Approximate $x$-intercept in Example 8

# NOTES FROM THE CLASSROOM

When sketching the graph of a function, you should never resort to plotting a lot of points by hand. That is something a graphing calculator or a computer algebra system does so well. On the other hand, you should not become dependent on a calculator to obtain a graph. Believe it or not, there are precalculus and calculus instructors who do not allow the use of graphing calculators on quizzes or tests. Usually there is no objection to your using calculators or computers as an aid in checking homework problems, but in the classroom instructors want to see the product of your own mind, namely, the ability to analyze. So you are strongly encouraged to develop your graphing skills to the point where you are able to quickly sketch by hand the graph of a function from a basic familiarity of types of functions and by plotting a minimum of well-chosen points.

In Problems 1–6, find the indicated function values.

**1.** If $f(x) = x^2 - 1$; $f(-5), f(-\sqrt{3}), f(3)$, and $f(6)$
**2.** If $f(x) = -2x^2 + x$; $f(-5), f(-\frac{1}{2}), f(2)$, and $f(7)$
**3.** If $f(x) = \sqrt{x + 1}$; $f(-1), f(0), f(3)$, and $f(5)$
**4.** If $f(x) = \sqrt{2x + 4}$; $f(-\frac{1}{2}), f(\frac{1}{2}), f(\frac{5}{2})$, and $f(4)$

**5.** If $f(x) = \dfrac{3x}{x^2 + 1}$; $f(-1), f(0), f(1)$, and $f(\sqrt{2})$

**6.** If $f(x) = \dfrac{x^2}{x^3 - 2}$; $f(-\sqrt{2}), f(-1), f(0)$, and $f(\frac{1}{2})$

In Problems 7 and 8, find

$$f(x), f(2a), f(a^2), f(-5x), f(2a + 1), f(x + h)$$

for the given function $f$ and simplify as much as possible.

**7.** $f(\ ) = -2(\ )^2 + 3(\ )$    **8.** $f(\ ) = (\ )^3 - 2(\ )^2 + 20$

**9.** For what values of $x$ is $f(x) = 6x^2 - 1$ equal to 23?
**10.** For what values of $x$ is $f(x) = \sqrt{x - 4}$ equal to 4?

In Problems 11–20, find the domain of the given function $f$.

**11.** $f(x) = \sqrt{4x - 2}$    **12.** $f(x) = \sqrt{15 - 5x}$

**13.** $f(x) = \dfrac{10}{\sqrt{1 - x}}$    **14.** $f(x) = \dfrac{2x}{\sqrt{3x - 1}}$

**15.** $f(x) = \dfrac{2x - 5}{x(x - 3)}$    **16.** $f(x) = \dfrac{x}{x^2 - 1}$

**17.** $f(x) = \dfrac{1}{x^2 - 10x + 25}$    **18.** $f(x) = \dfrac{x + 1}{x^2 - 4x - 12}$

**19.** $f(x) = \dfrac{x}{x^2 - x + 1}$    **20.** $f(x) = \dfrac{x^2 - 9}{x^2 - 2x - 1}$

In Problems 21–26, use the sign-chart method to find the domain of the given function $f$.

**21.** $f(x) = \sqrt{25 - x^2}$    **22.** $f(x) = \sqrt{x(4 - x)}$
**23.** $f(x) = \sqrt{x^2 - 5x}$    **24.** $f(x) = \sqrt{x^2 - 3x - 10}$

**25.** $f(x) = \sqrt{\dfrac{3 - x}{x + 2}}$    **26.** $f(x) = \sqrt{\dfrac{5 - x}{x}}$

In Problems 27–30, determine whether the graph in the figure is the graph of a function.

**27.**

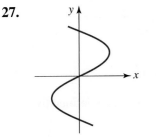

FIGURE 2.1.9 Graph for Problem 27

**28.**

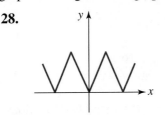

FIGURE 2.1.10 Graph for Problem 28

**29.**

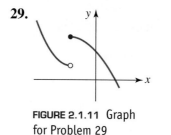

FIGURE 2.1.11 Graph
for Problem 29

**30.**

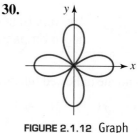

FIGURE 2.1.12 Graph
for Problem 30

In Problems 31–34, use the graph of the function $f$ given in the figure to find its domain and range.

**31.**

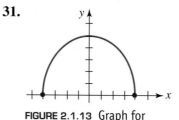

FIGURE 2.1.13 Graph for
Problem 31

**32.**

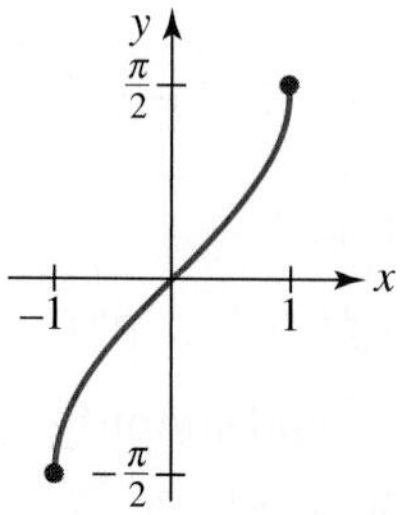

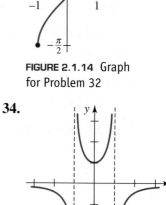

FIGURE 2.1.14 Graph
for Problem 32

**33.**

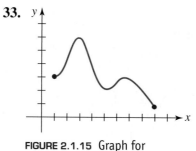

FIGURE 2.1.15 Graph for
Problem 33

**34.**

FIGURE 2.1.16 Graph for
Problem 34

In Problems 35–42, find the zeros of the given function $f$.

**35.** $f(x) = 5x + 6$
**36.** $f(x) = -2x + 9$
**37.** $f(x) = x^2 - 5x + 6$
**38.** $f(x) = x^2 - 2x - 1$
**39.** $f(x) = x(3x - 1)(x + 9)$
**40.** $f(x) = x^3 - x^2 - 2x$
**41.** $f(x) = x^4 - 1$
**42.** $f(x) = 2 - \sqrt{4 - x^2}$

In Problems 43–50, find the $x$- and $y$-intercepts, if any, of the graph of the given function $f$. Do not graph.

**43.** $f(x) = \frac{1}{2}x - 4$
**44.** $f(x) = x^2 - 6x + 5$
**45.** $f(x) = 4(x - 2)^2 - 1$
**46.** $f(x) = (2x - 3)(x^2 + 8x + 16)$
**47.** $f(x) = \dfrac{x^2 + 4}{x^2 - 16}$
**48.** $f(x) = \dfrac{x(x + 1)(x - 6)}{x + 8}$
**49.** $f(x) = \frac{3}{2}\sqrt{4 - x^2}$
**50.** $f(x) = \frac{1}{2}\sqrt{x^2 - 2x - 3}$

In Problems 51 and 52, find two functions $y = f_1(x)$ and $y = f_2(x)$ defined by the given equation. Find the domain of the functions $f_1$ and $f_2$.

**51.** $x = y^2 - 5$
**52.** $x^2 - 4y^2 = 16$

In Problems 53 and 54, use the graph of the function $f$ given in the figure to estimate the values of $f(-3), f(-2), f(-1), f(1), f(2),$ and $f(3)$. Estimate the $y$-intercept.

**53.**

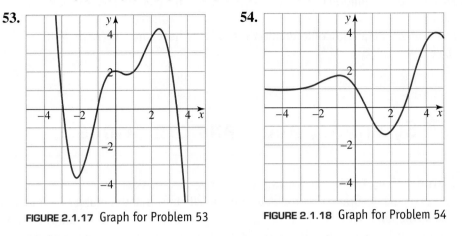

FIGURE 2.1.17 Graph for Problem 53

FIGURE 2.1.18 Graph for Problem 54

In Problems 55 and 56, use the graph of the function $f$ given in the figure to estimate the values of $f(-2), f(-1.5), f(0.5), f(1), f(2),$ and $f(3.2)$. Estimate the $x$-intercepts.

**55.**

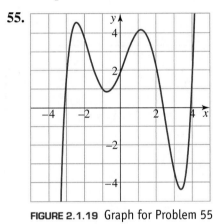

**56.**

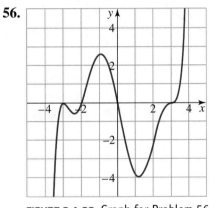

FIGURE 2.1.19 Graph for Problem 55

FIGURE 2.1.20 Graph for Problem 56

## Miscellaneous Calculus-Related Problems

**57.** In calculus some of the functions that you will encounter have as their domain the set of positive integers $n$. The **factorial function** $f(n) = n!$ is defined as the product of the first $n$ positive integers, that is,

$$f(n) = n! = 1 \cdot 2 \cdot 3 \cdots (n-1) \cdot n.$$

**(a)** Evaluate $f(2), f(3), f(5),$ and $f(7)$.
**(b)** Show that $f(n+1) = f(n) \cdot (n+1)$.
**(c)** Simplify $f(n+2)/f(n)$.

**58.** Another function of a positive integer $n$ gives the sum of the first $n$ squared positive integers:

$$S(n) = \tfrac{1}{6}n(n+1)(2n+1) = 1^2 + 2^2 + \cdots + n^2.$$

**(a)** Find the value of the sum $1^2 + 2^2 + \cdots + 99^2 + 100^2$.
**(b)** Find $n$ such that $300 < S(n) < 400$. [*Hint*: Use a calculator.]

**59.** Determine an equation of a function $y = f(x)$ whose domain is **(a)** $[3, \infty)$, **(b)** $(3, \infty)$.

**60.** Determine an equation of a function $y = f(x)$ whose range is **(a)** $[3, \infty)$, **(b)** $(3, \infty)$.

## 2.2 Symmetry and Transformations

☐ **Introduction** In this section we discuss two aids in sketching graphs of functions quickly and accurately. If you determine in advance that the graph of a function possesses *symmetry*, then you can cut your work in half. In addition, sketching a graph of a complicated-looking function is expedited if you recognize that the required graph is actually a *transformation* of the graph of a simpler function. This latter graphing aid is based on your prior knowledge of the graphs of some basic functions.

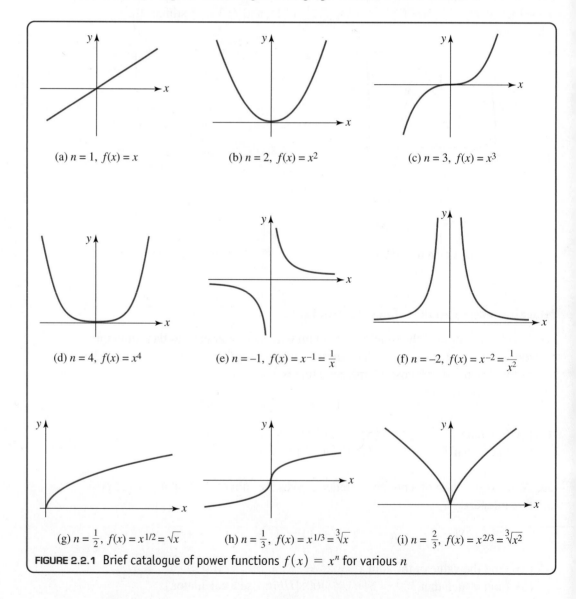

(a) $n = 1$, $f(x) = x$

(b) $n = 2$, $f(x) = x^2$

(c) $n = 3$, $f(x) = x^3$

(d) $n = 4$, $f(x) = x^4$

(e) $n = -1$, $f(x) = x^{-1} = \dfrac{1}{x}$

(f) $n = -2$, $f(x) = x^{-2} = \dfrac{1}{x^2}$

(g) $n = \dfrac{1}{2}$, $f(x) = x^{1/2} = \sqrt{x}$

(h) $n = \dfrac{1}{3}$, $f(x) = x^{1/3} = \sqrt[3]{x}$

(i) $n = \dfrac{2}{3}$, $f(x) = x^{2/3} = \sqrt[3]{x^2}$

**FIGURE 2.2.1** Brief catalogue of power functions $f(x) = x^n$ for various $n$

□ **Power Functions** A function of the form

$$f(x) = x^n,$$

where $n$ represents a real number, is called a **power function**. The domain of a power function depends on the power $n$. For example, we have already seen in Section 2.1 for $n = 2$, $n = \frac{1}{2}$, and $n = -1$, respectively, that:

- the domain of $f(x) = x^2$ is the set $R$ of real numbers or $(-\infty, \infty)$,
- the domain of $f(x) = x^{1/2} = \sqrt{x}$ is $[0, \infty)$,
- the domain of $f(x) = x^{-1} = \dfrac{1}{x}$ is the set $R$ of real numbers except $x = 0$.

Simple power functions, or modified versions of these functions, occur so often in problems in calculus that you do not want to spend valuable time plotting their graphs. We suggest that you know (memorize) the short catalogue of graphs of power functions given in **FIGURE 2.2.1** on the previous page. You might already know that the graph in part (a) of that figure is a **line** and the graph in part (b) is called a **parabola**.

□ **Symmetry** In Section 1.4 we discussed symmetry of a graph with respect to the $y$-axis, the $x$-axis, and the origin. Of those three types of symmetries, the graph of a function can be symmetric with respect to the $y$-axis or with respect to the origin, but the graph of a nonzero function *cannot* be symmetric with respect to the $x$-axis. See Problem 43 in Exercises 2.2. If the graph of a function is symmetric with respect to the $y$-axis, we say that $f$ is an **even function**. A function whose graph is symmetric with respect to the origin is said to be an **odd function**. For functions, the following two tests for symmetry are equivalent to tests (*i*) and (*iii*), respectively, on page 30. See **FIGURES 2.2.2** and **2.2.3**. The function whose graph is given in **FIGURE 2.2.4** is neither even nor odd.

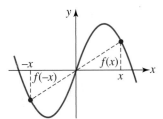

**FIGURE 2.2.2** Even function; graph has $y$-axis symmetry

◀ Can you explain *why* the graph of a function cannot have symmetry with respect to the $x$-axis?

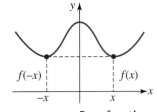

**FIGURE 2.2.3** Odd function; graph has origin symmetry

---

### TESTS FOR SYMMETRY

The graph of a function $f$ with domain $X$ is symmetric with respect to:

(*i*) the $y$-axis if $f(-x) = f(x)$ for every $x$ in $X$, or     (1)
(*ii*) the origin if $f(-x) = -f(x)$ for every $x$ in $X$.     (2)

---

The graphical interpretation of these tests is illustrated in Figures 2.2.2 and 2.2.3. In Figure 2.2.2, observe that if $f$ is an even function and

$$\underset{\downarrow}{f(x)} \qquad\qquad\qquad \underset{\downarrow}{f(-x)}$$
$$(x, y) \text{ is a point on its graph, then necessarily } (-x, y)$$

is also on its graph. Similarly we see in Figure 2.2.3 that if $f$ is an odd function and

$$\underset{\downarrow}{f(x)} \qquad\qquad\qquad \underset{\downarrow}{f(-x) = -f(x)}$$
$$(x, y) \text{ is a point on its graph, then necessarily } (-x, -y)$$

is on its graph.

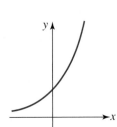

**FIGURE 2.2.4** Function is neither even nor odd; no $y$-axis or origin symmetry

**(a)** $f(x) = x^3$ is an odd function since by (2),

$$f(-x) = (-x)^3 = (-1)^3 x^3 = -x^3 = -f(x).$$

Inspection of Figure 2.2.1(c) shows that the graph of $f$ is symmetric with respect to the origin. For example, since $f(1) = 1, (1, 1)$ is a point on the graph of $y = x^3$. Because $f$ is an odd function, $f(-1) = -f(1)$ implies $(-1, -1)$ is on the same graph.

**(b)** $f(x) = x^{2/3}$ is an even function since by (1) and the laws of exponents

cube root of $-1$ is $-1$
↓

$$f(-x) = (-x)^{2/3} = (-1)^{2/3} x^{2/3} = (\sqrt[3]{-1})^2 x^{2/3} = (-1)^2 x^{2/3} = x^{2/3} = f(x).$$

In Figure 2.2.1(i), we see that the graph of $f$ is symmetric with respect to the $y$-axis. For example, since $f(8) = 8^{2/3} = 2, (8, 2)$ is a point on the graph of $y = x^{2/3}$. Because $f$ is an even function, $f(-8) = f(8)$ implies $(-8, 2)$ is on the same graph.

**(c)** $f(x) = x^3 + 1$ is neither even nor odd. From

$$f(-x) = (-x)^3 + 1 = -x^3 + 1$$

we see that $f(-x) \neq f(x)$, and $f(-x) \neq -f(x).$     ■

The graphs in Figure 2.2.1, with part (g) the only exception, possess either $y$-axis or origin symmetry. The functions in Figures 2.2.1(b), (d), (f), and (i) are even, whereas the functions in Figures 2.2.1(a), (c), (e), and (h) are odd.

Often we can sketch the graph of a function by applying a certain transformation to the graph of a simpler function (such as those given in Figure 2.2.1). We are going to consider two kinds of graphical transformations, rigid and nonrigid.

☐ **Rigid Transformations** A **rigid transformation** of a graph is one that changes only the *position* of the graph in the $xy$-plane but not its shape. We have already examined this concept briefly in the discussion of the circle in Section 1.3. For example, the circle $(x - 2)^2 + (y - 3)^2 = 1$ with center $(2, 3)$ and radius $r = 1$ has *exactly* the same shape as the circle $x^2 + y^2 = 1$ with center at the origin. We can think of the graph of $(x - 2)^2 + (y - 3)^2 = 1$ as the graph of $x^2 + y^2 = 1$ shifted horizontally two units to the right followed by an upward vertical shift of three units. For the graph of a function $y = f(x)$ we examine four kinds of shifts or translations.

---

## VERTICAL AND HORIZONTAL SHIFTS

Suppose $y = f(x)$ is a function and $c$ is a positive constant. Then the graph of

    (*i*) $y = f(x) + c$ is the graph of $f$ shifted vertically **up** $c$ units,
    (*ii*) $y = f(x) - c$ is the graph of $f$ shifted vertically **down** $c$ units,
    (*iii*) $y = f(x + c)$ is the graph of $f$ shifted horizontally to the **left** $c$ units,
    (*iv*) $y = f(x - c)$ is the graph of $f$ shifted horizontally to the **right** $c$ units.

---

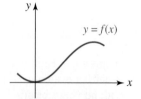

**FIGURE 2.2.5** Graph of $y = f(x)$

Consider the graph of a function $y = f(x)$ given in **FIGURE 2.2.5**. The shifts of this graph described in (*i*)–(*iv*) are the graphs in red in parts (a)–(d) of Figure 2.2.6. If

$(x, y)$ is a point on the graph of $y = f(x)$ and the graph of $f$ is shifted, say, upward by $c > 0$ units, then $(x, y + c)$ is a point on the new graph. In general, the $x$-coordinates do not change as a result of a vertical shift. See **FIGURES 2.2.6(a)** and 2.2.6(b). Similarly, in a horizontal shift the $y$-coordinates of points on the shifted graph are the same as on the original graph. See Figures 2.2.6(c) and 2.2.6(d).

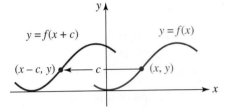

(a) Vertical shift up

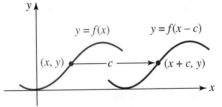

(b) Vertical shift down

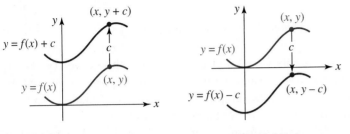

(c) Horizontal shift left          (d) Horizontal shift right

**FIGURE 2.2.6** Vertical and horizontal shifts of the graph of $y = f(x)$ by an amount $c > 0$.

### EXAMPLE 2          Vertical and Horizontal Shifts

The graphs of $y = x^2 + 1$, $y = x^2 - 1$, $y = (x + 1)^2$, and $y = (x - 1)^2$ are obtained from the graph of $f(x) = x^2$ in **FIGURE 2.2.7(a)** by shifting this graph, in turn, 1 unit up (Figure 2.2.7(b)), 1 unit down (Figure 2.2.7(c)), 1 unit to the left (Figure 2.2.7(d)), and 1 unit to the right (Figure 2.2.7(e)).

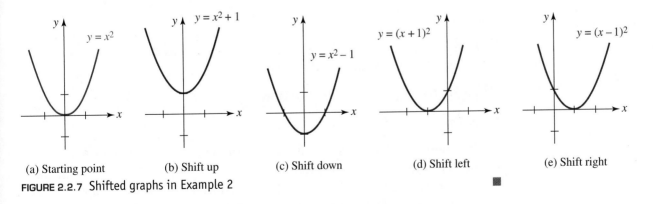

(a) Starting point    (b) Shift up    (c) Shift down    (d) Shift left    (e) Shift right

**FIGURE 2.2.7** Shifted graphs in Example 2

☐ **Combining Shifts** In general, the graph of a function

$$y = f(x \pm c_1) \pm c_2, \tag{3}$$

where $c_1$ and $c_2$ are positive constants, combines a horizontal shift (left or right) with a vertical shift (up or down). For example, the graph of $y = f(x - c_1) + c_2$ is the graph of $y = f(x)$ shifted $c_1$ units to the right and then $c_2$ units up.

◀ The order in which the shifts are done is irrelevant. We could do the upward shift first followed by the shift to the right.

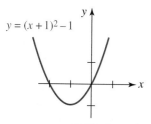

$y = (x+1)^2 - 1$

**FIGURE 2.2.8** Shifted graph in Example 3

Graph $y = (x + 1)^2 - 1$.

**Solution** From the preceding paragraph we identify in (3) the form $y = f(x + c_1) - c_2$ with $c_1 = 1$ and $c_2 = 1$. Thus, the graph of $y = (x + 1)^2 - 1$ is the graph of $f(x) = x^2$ shifted 1 unit to the left followed by a downward shift of 1 unit. The graph is given in **FIGURE 2.2.8**. ∎

From the graph in Figure 2.2.8 we see immediately that the range of the function $y = (x + 1)^2 - 1 = x^2 + 2x$ is the interval $[-1, \infty)$ on the $y$-axis. Note also that the graph has $x$-intercepts $(0, 0)$ and $(-2, 0)$; you should verify this by solving $x^2 + 2x = 0$. Also, if you reexamine Figure 2.1.5 in Section 2.1 you will see that the graph of $y = 4 + \sqrt{x - 3}$ is the graph of the square root function $f(x) = \sqrt{x}$ (Figure 2.2.1(g)) shifted 3 units to the right and then 4 units up.

Another way of rigidly transforming a graph of a function is by a **reflection** in a coordinate axis.

## REFLECTIONS

Suppose $y = f(x)$ is a function. Then the graph of

    (*i*) $y = -f(x)$ is the graph of $f$ reflected in the **$x$-axis**,
    (*ii*) $y = f(-x)$ is the graph of $f$ reflected in the **$y$-axis**.

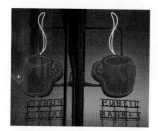

Reflection or mirror image

In part (a) of **FIGURE 2.2.9** we have reproduced the graph of a function $y = f(x)$ given in Figure 2.2.5. The reflections of this graph described in (*i*)–(*ii*) are illustrated in Figures 2.2.9(b) and 2.2.9(c). If $(x, y)$ denotes a point on the graph of $y = f(x)$, then the point $(x, -y)$ is on the graph of $y = -f(x)$, and $(-x, y)$ is on the graph of $y = f(-x)$. Each of these reflections is a mirror image of the graph of $y = f(x)$ in the respective coordinate axis.

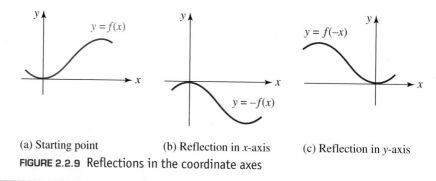

(a) Starting point      (b) Reflection in $x$-axis      (c) Reflection in $y$-axis
**FIGURE 2.2.9** Reflections in the coordinate axes

EXAMPLE 4      **Reflections**

Graph **(a)** $y = -\sqrt{x}$    **(b)** $y = \sqrt{-x}$

**Solution** The starting point is the graph of $f(x) = \sqrt{x}$ given in **FIGURE 2.2.10(a)**.

**(a)** The graph of $y = -\sqrt{x}$ is the reflection of the graph of $f(x) = \sqrt{x}$ in the $x$-axis. Observe in Figure 2.2.10(b) that since $(1, 1)$ is on the graph of $f$, the point $(1, -1)$ is on the graph of $y = -\sqrt{x}$.

**(b)** The graph of $y = \sqrt{-x}$ is the reflection of the graph of $f(x) = \sqrt{x}$ in the $y$-axis. Observe in Figure 2.2.10(c) that since $(1, 1)$ is on the graph of $f$ the point $(-1, 1)$ is on the graph of $y = \sqrt{-x}$. The function $y = \sqrt{-x}$ looks a little strange, but bear in mind that its domain is determined by the requirement that $-x \geq 0$, or equivalently $x \leq 0$, and so the reflected graph is defined on the interval $(-\infty, 0]$.

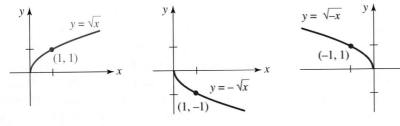

| (a) Starting point | (b) Reflection in *x*-axis | (c) Reflection in *y*-axis |

**FIGURE 2.2.10** Graphs in Example 4

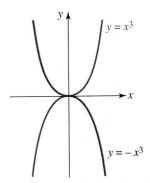

**FIGURE 2.2.11** Reflection of an odd function in *y*-axis

If a function $f$ is even, then $f(-x) = f(x)$ shows that a reflection in the *y*-axis would give precisely the same graph. If a function is odd, then from $f(-x) = -f(x)$ we see that a reflection of the graph of $f$ in the *y*-axis is identical to the graph of $f$ reflected in the *x*-axis. In **FIGURE 2.2.11** the blue curve is the graph of the odd function $f(x) = x^3$; the red curve is the graph of $y = f(-x) = (-x)^3 = -x^3$. Notice that if the blue curve is reflected in either the *y*-axis or the *x*-axis, we get the red curve.

☐ **Nonrigid Transformations** If a function $f$ is multiplied by a constant $c > 0$ the shape of the graph is changed but retains, *roughly*, its original shape. The graph of $y = cf(x)$ is the graph of $y = f(x)$ distorted vertically; the graph of $f$ is either stretched (or elongated) vertically or is compressed (or flattened) vertically depending on the value of $c$. Stretching or compressing a graph are examples of **nonrigid transformations**.

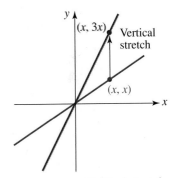

**FIGURE 2.2.12** Vertical stretch of the graph of $f(x) = x$

> ## VERTICAL STRETCHES AND COMPRESSIONS
>
> Suppose $y = f(x)$ is a function and $c$ a positive constant. Then the graph of $y = cf(x)$ is the graph of $f$
>
>     (*i*) vertically stretched by a factor of $c$ units if $c > 1$,
>     (*ii*) vertically compressed by a factor of $c$ units if $0 < c < 1$.

If $(x, y)$ represents a point on the graph of $f$, then the point $(x, cy)$ is on the graph of $cf$. The graphs of $y = x$ and $y = 3x$ are compared in **FIGURE 2.2.12**; the *y*-coordinate of a point on the graph of $y = 3x$ is 3 times as large as the *y*-coordinate of the point with the same *x*-coordinate on the graph of $y = x$. The comparison of the graphs of $y = 10x^2$ (blue graph) and $y = \frac{1}{10}x^2$ (red graph) in **FIGURE 2.1.13** is a little more dramatic; the graph of $y = \frac{1}{10}x^2$ exhibits considerable vertical flattening, especially in a neighborhood of the origin. Note that $c$ is positive in this discussion. To sketch the graph of $y = -10x^2$ we think of it as $y = -(10x^2)$, which means we first stretch the graph of $y = x^2$ vertically by a factor of 10 units and then reflect that graph in the *x*-axis.

The next example illustrates shifting, reflecting, and stretching of a graph.

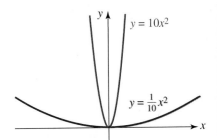

**FIGURE 2.2.13** Vertical stretch (blue) and vertical compression (red) of the graph of $f(x) = x^2$

| **EXAMPLE 5** | **Combining Transformations** |

Graph $y = 2 - 2\sqrt{x - 3}$.

    **Solution** You should recognize that the given function consists of four transformations of the basic function $f(x) = \sqrt{x}$:

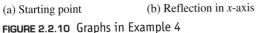

$$y = 2 - 2\sqrt{x - 3}$$

vertical shift up     horizontal shift to right

reflection in *x*-axis    vertical stretch

We start with the graph of $f(x) = \sqrt{x}$ in **FIGURE 2.2.14(a)**. Then stretch this graph vertically by a factor of 2 to obtain $y = 2\sqrt{x}$ in Figure 2.2.14(b). Reflect this second graph in the $x$-axis to obtain $y = -2\sqrt{x}$ in Figure 2.2.14(c). Shift this third graph 3 units to the right to obtain $y = -2\sqrt{x-3}$ in Figure 2.2.14(d). Finally, shift the fourth graph upward 2 units to obtain $y = 2 - 2\sqrt{x-3}$ in Figure 2.2.14(e). Note that the point $(0, 0)$ on the graph of $f(x) = \sqrt{x}$ remains fixed in the vertical stretch and the reflection in the $x$-axis, but under the first (horizontal) shift $(0, 0)$ moves to $(3, 0)$ and under the second (vertical) shift $(3, 0)$ moves to $(3, 2)$.

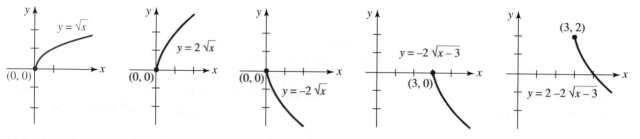

(a) Starting point      (b) Vertical stretch      (c) Reflection in $x$-axis      (d) Shift right      (d) Shift up

**FIGURE 2.2.14** Graph of function in Example 5

---

**2.2**    **Exercises**   Answers to selected odd-numbered problems begin on page ANS-3.

In Problems 1–10, use (1) and (2) to determine whether the given function $y = f(x)$ is even, odd, or neither even nor odd. Do not graph.

**1.** $f(x) = 4 - x^2$            **2.** $f(x) = x^2 + 2x$

**3.** $f(x) = x^3 - x + 4$        **4.** $f(x) = x^5 + x^3 + x$

**5.** $f(x) = 3x - \dfrac{1}{x}$          **6.** $f(x) = \dfrac{x}{x^2 + 1}$

**7.** $f(x) = 1 - \sqrt{1 - x^2}$      **8.** $f(x) = \sqrt[3]{x^3 + x}$

**9.** $f(x) = |x^3|$                **10.** $f(x) = x|x|$

In Problems 11–14, classify the function $y = f(x)$ whose graph is given as even, odd, or neither even nor odd.

**11.**

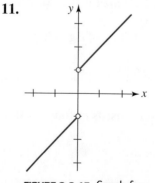

**FIGURE 2.2.15** Graph for Problem 11

**12.**

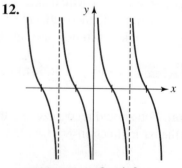

**FIGURE 2.2.16** Graph for Problem 12

**13.**

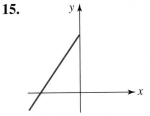

**FIGURE 2.2.17** Graph for Problem 13

**14.**

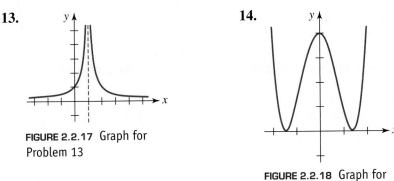

**FIGURE 2.2.18** Graph for Problem 14

In Problems 15–18, complete the graph of the given function $y = f(x)$ if **(a)** $f$ is an even function and **(b)** $f$ is an odd function.

**15.**

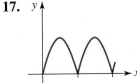

**FIGURE 2.2.19** Graph for Problem 15

**16.**

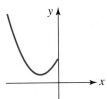

**FIGURE 2.2.20** Graph for Problem 16

**17.**

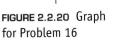

**FIGURE 2.2.21** Graph for Problem 17

**18.**

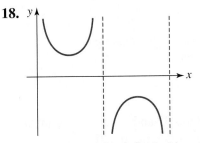

**FIGURE 2.2.22** Graph for Problem 18

In Problems 19 and 20, suppose that $f(-2) = 4$ and $f(3) = 7$. Determine $f(2)$ and $f(-3)$.

**19.** If $f$ is an even function.　　　　**20.** If $f$ is an odd function.

In Problems 21 and 22, suppose that $g(-1) = -5$ and $g(4) = 8$. Determine $g(1)$ and $g(-4)$.

**21.** If $g$ is an odd function.　　　　**22.** If $g$ is an even function.

In Problems 23–32, the points $(-2, 1)$ and $(3, -4)$ are on the graph of the function $y = f(x)$. Find the corresponding points on the graph obtained by the given transformations.

**23.** the graph of $f$ shifted up 2 units
**24.** the graph of $f$ shifted down 5 units
**25.** the graph of $f$ shifted to the left 6 units
**26.** the graph of $f$ shifted to the right 1 unit
**27.** the graph of $f$ shifted up 1 unit and to the left 4 units
**28.** the graph of $f$ shifted down 3 units and to the right 5 units

**29.** the graph of $f$ reflected in the $y$-axis

**30.** the graph of $f$ reflected in the $x$-axis

**31.** the graph of $f$ stretched vertically by a factor of 15 units

**32.** the graph of $f$ compressed vertically by a factor of $\frac{1}{4}$ unit, then reflected in the $x$-axis

In Problems 33–36, use the graph of the function $y = f(x)$ given in the figure to graph the following functions.

**(a)** $y = f(x) + 2$             **(b)** $y = f(x) - 2$
**(c)** $y = f(x + 2)$          **(d)** $y = f(x - 5)$
**(e)** $y = -f(x)$            **(f)** $y = f(-x)$

**33.**

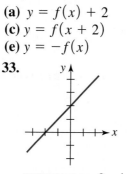

FIGURE 2.2.23 Graph for Problem 33

**34.**

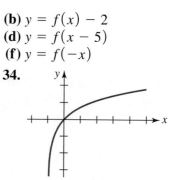

FIGURE 2.2.24 Graph for Problem 34

**35.**

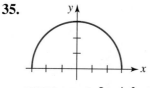

FIGURE 2.2.25 Graph for Problem 35

**36.**

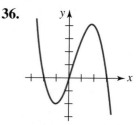

FIGURE 2.2.26 Graph for Problem 36

In Problems 37 and 38, use the graph of the function $y = f(x)$ given in the figure to graph the following functions.

**(a)** $y = f(x) + 1$           **(b)** $y = f(x) - 1$
**(c)** $y = f(x + \pi)$         **(d)** $y = f\left(x - \dfrac{\pi}{2}\right)$
**(e)** $y = -f(x)$           **(f)** $y = f(-x)$
**(g)** $y = 3f(x)$           **(h)** $y = -\frac{1}{2}f(x)$

**37.**

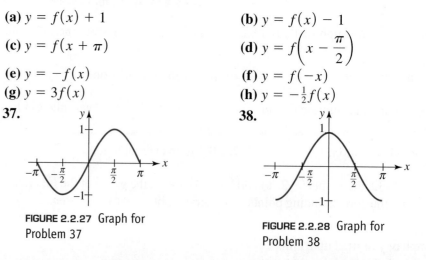

FIGURE 2.2.27 Graph for Problem 37

**38.**

FIGURE 2.2.28 Graph for Problem 38

In Problems 39–42, find the equation of the final graph after the given transformations are applied to the graph of $y = f(x)$.

**39.** the graph of $f(x) = x^3$ shifted up 5 units and right 1 unit

**40.** the graph of $f(x) = x^{2/3}$ stretched vertically by a factor of 3 units, then shifted right 2 units

**41.** the graph of $f(x) = x^4$ reflected in the *x*-axis, then shifted left 7 units

**42.** the graph of $f(x) = \dfrac{1}{x}$ reflected in the *y*-axis, then shifted left 5 units and down 10 units

## For Discussion

**43.** Explain why the graph of a nonzero function cannot be symmetric with respect to the *x*-axis.

**44.** What points, if any, on the graph of $y = f(x)$ remain fixed, that is, the same on the resulting graph after a vertical stretch or compression? After a reflection in the *x*-axis? After a reflection in the *y*-axis?

**45.** Discuss the relationship between the graphs of $y = f(x)$ and $y = f(|x|)$.

**46.** Discuss the relationship between the graphs of $y = f(x)$ and $y = f(cx)$, where $c > 0$ is a constant. Consider two cases: $0 < c < 1$ and $c > 1$.

**47.** Review the graphs of $y = x$ and $y = \dfrac{1}{x}$ in Figure 2.2.1. Then discuss how to obtain the graph of the reciprocal $y = \dfrac{1}{f(x)}$ from the graph of $y = f(x)$. Sketch the graph of $y = \dfrac{1}{f(x)}$ for the function $f$ whose graph is given in Figure 2.2.26.

## 2.3 Linear Functions

☐ **Introduction**  The notion of a line plays an important role in the study of differential calculus. There are three types of lines in the *xy*- or Cartesian plane: horizontal lines, vertical lines, and slant or oblique lines. We will see in this section that an equation of each of these lines stems from a **linear equation**

$$Ax + By + C = 0, \tag{1}$$

where *A*, *B*, and *C* are real constants. The characteristic that gives (1) its name *linear* is that the variables *x* and *y* appear only to the first power. We will refer back to (1) when we review lines and their equations, but let's note the cases of special interest:

$$A = 0, B \neq 0, \text{ gives } y = -\frac{C}{B}, \tag{2}$$

$$A \neq 0, B = 0, \text{ gives } x = -\frac{C}{A}, \tag{3}$$

$$A \neq 0, B \neq 0, \text{ gives } y = -\frac{A}{B}x - \frac{C}{B}. \tag{4}$$

The first and the third of these three equations define functions. By relabeling $-C/B$ in (2) as *b* we get a constant function.

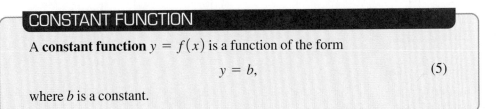

**CONSTANT FUNCTION**

A **constant function** $y = f(x)$ is a function of the form

$$y = b, \tag{5}$$

where *b* is a constant.

The **domain** of a constant function is the set of real numbers $(-\infty, \infty)$. In the definition of a function we are pairing each real number $x$ with the same value of $y$, that is, $(x, b)$. In our student/desk example of a function in Section 2.2 this is equivalent to having all the students in a classroom sit in one desk. On the other hand, the equation in (3) does not define a function. We cannot have one student (the fixed value of $x$) sit in all the desks in a classroom.

By relabeling $-A/B$ and $-C/B$ in (4) as $a$ and $b$, respectively, we get the form of a linear function.

> ## LINEAR FUNCTION
>
> A **linear function** $y = f(x)$ is a function of the form
> $$f(x) = ax + b, \qquad (6)$$
> where $a \neq 0$ and $b$ are constants.

The **domain** of a linear function is the set of real numbers $(-\infty, \infty)$.

☐ **Graphs** Since the graphs of constant and linear functions are straight lines, it is appropriate that we will review equations of all lines. We begin with the recollection from plane geometry that through any two distinct points $(x_1, y_1)$ and $(x_2, y_2)$ in the plane there passes only one line $L$. If $x_1 \neq x_2$, then the number

$$m = \frac{y_2 - y_1}{x_2 - x_1} \qquad (7)$$

is called the **slope** of the line determined by these two points. It is customary to call $y_2 - y_1$ the **change in $y$** or the **rise** of the line; $x_2 - x_1$ is the **change in $x$** or the **run** of the line. Therefore (7) is

$$m = \frac{\text{rise}}{\text{run}}.$$

See **FIGURE 2.3.1(a)**. Any pair of distinct points on a line will determine the same slope. To see why this is so, consider the two similar right triangles in Figure 2.3.1(b). Since we know that the ratios of corresponding sides in similar triangles are equal we have

$$\frac{y_2 - y_1}{x_2 - x_1} = \frac{y_4 - y_3}{x_4 - x_3}.$$

Hence the slope of a line is independent of the choice of points on the line.

In **FIGURE 2.3.2** we compare the graphs of lines with positive, negative, zero, and undefined slopes. In Figure 2.3.2(a) we see, reading the graph from left to right, that a line with positive slope $(m > 0)$ rises as $x$ increases. Figure 2.3.2(b) shows that a line

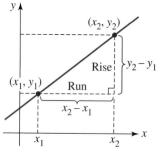

(a) Rise and run

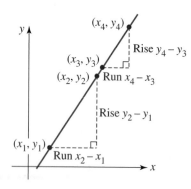

(b) Similar triangles

**FIGURE 2.3.1** Slope of a line

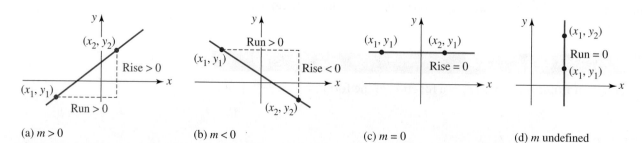

(a) $m > 0$      (b) $m < 0$      (c) $m = 0$      (d) $m$ undefined

**FIGURE 2.3.2** Lines with slope (a)–(c); line with no slope (d)

with negative slope $(m < 0)$ falls as $x$ increases. If $(x_1, y_1)$ and $(x_2, y_2)$ are points on a horizontal line, then $y_1 = y_2$ and so its rise is $y_2 - y_1 = 0$. Hence from (7) the slope is zero $(m = 0)$. See Figure 2.3.2(c). If $(x_1, y_1)$ and $(x_2, y_2)$ are points on a vertical line, then $x_1 = x_2$ and so its run is $x_2 - x_1 = 0$. In this case we say that the slope of the line is **undefined** or that the line has no slope. See Figure 2.3.2(d).

☐ **Point-Slope Equation** We are now in a position to find an equation of a line $L$. To begin, suppose $L$ has slope $m$ and that $(x_1, y_1)$ is on the line. If $(x, y)$ represents any other point on $L$, then (7) gives

$$m = \frac{y - y_1}{x - x_1}.$$

Multiplying both sides of the last equality by $x - x_1$ gives an important equation.

> ## POINT-SLOPE EQUATION OF A LINE
>
> The **point-slope equation** of the line through $(x_1, y_1)$ with slope $m$ is
>
> $$y - y_1 = m(x - x_1). \tag{8}$$

### EXAMPLE 1     Point-Slope Equation

Find an equation of the line with slope 6 and passing through $\left(-\frac{1}{2}, 2\right)$.
   **Solution** Letting $m = 6$, $x_1 = -\frac{1}{2}$, and $y_1 = 2$ we obtain from (8)

$$y - 2 = 6\left[x - \left(-\tfrac{1}{2}\right)\right].$$

Simplifying gives

$$y - 2 = 6\left(x + \tfrac{1}{2}\right) \quad \text{or} \quad y = 6x + 5. \qquad \blacksquare$$

### EXAMPLE 2     Point-Slope Equation

Find an equation of the line passing through the points $(4, 3)$ and $(-2, 5)$.
   **Solution** First we compute the slope of the line through the points. From (7),

$$m = \frac{5 - 3}{-2 - 4} = \frac{2}{-6} = -\frac{1}{3}.$$

The point-slope equation (8) then gives

the distributive law
$\downarrow \qquad \downarrow$

$$y - 3 = -\tfrac{1}{3}(x - 4) \qquad \text{or} \qquad y = -\tfrac{1}{3}x + \tfrac{13}{3}. \qquad \blacksquare$$

◀ The distributive law $a(b + c) = ab + ac$ is the source of many errors on students' papers. A common error goes something like this:

$$-(2x - 3) = -2x - 3.$$

The correct result is:

$$\begin{aligned} -(2x - 3) &= (-1)(2x - 3) \\ &= (-1)2x - (-1)3 \\ &= -2x + 3. \end{aligned}$$

☐ **Slope-Intercept Equation** Any line with slope (that is, any line that is not vertical) must cross the $y$-axis. If this $y$-intercept is $(0, b)$, then with $x_1 = 0$, $y_1 = b$, the point-slope form (8) gives $y - b = m(x - 0)$. The last equation simplifies to the next result.

> ## SLOPE-INTERCEPT EQUATION OF A LINE
>
> The **slope-intercept equation** of the line with slope $m$ and $y$-intercept $(0, b)$ is
>
> $$y = mx + b. \tag{9}$$

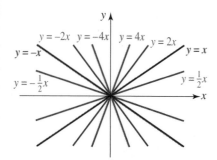

FIGURE 2.3.3 Lines through the origin are $y = mx$

For $m \neq 0$, (8) and (9) gives us the form of the linear function in (6). The coefficient $a$ in (6) is, of course, the slope $m$ of the line. When $b = 0$ in (9), the equation $y = mx$ represents a family of lines that pass through the origin $(0, 0)$. In **FIGURE 2.3.3** we have drawn a few of the members of that family.

### EXAMPLE 3　Example 2 Revisited

We can also use the slope-intercept form (9) to obtain the equation of the line through two points in Example 2. As in that example, we start by finding the slope $m = -\frac{1}{3}$. The equation of the line is then $y = -\frac{1}{3}x + b$. By substituting the coordinates of either point $(4, 3)$ or $(-2, 5)$ into the last equation enables us to determine $b$. If we use $x = 4$ and $y = 3$, then $3 = -\frac{1}{3} \cdot 4 + b$ and so $b = 3 + \frac{4}{3} = \frac{13}{3}$. The equation of the line is $y = -\frac{1}{3}x + \frac{13}{3}$. ∎

☐ **Horizontal and Vertical Lines** We saw in Figure 2.3.2(c) that a horizontal line has slope $m = 0$. An equation of a horizontal line passing through a point $(a, b)$ can be obtained from (8), that is, $y - b = 0(x - a)$. The **equation of a horizontal line** is then

$$y = b. \tag{10}$$

We have already seen this in (5) and in (2) where $-C/B$ played the part of the symbol $b$. A vertical line through $(a, b)$ has undefined slope and all points on the line have the same $x$-coordinate. The **equation of a vertical line** is then

$$x = a. \tag{11}$$

Equation (11) is (3) with $-C/A$ replaced by the symbol $a$.

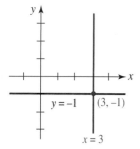

FIGURE 2.3.4 Horizontal and vertical lines in Example 4

### EXAMPLE 4　Vertical and Horizontal Lines

Find equations for the vertical and horizontal lines through $(3, -1)$. Graph these lines.

**Solution** Any point on the vertical line through $(3, -1)$ has $x$-coordinate 3. The equation of this line is then $x = 3$. Similarly, any point on the horizontal line through $(3, -1)$ has $y$-coordinate $-1$. The equation of this line is $y = -1$. Both lines are graphed in **FIGURE 2.3.4**. Don't forget, only $y = -1$ is a function. ∎

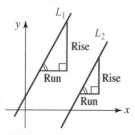

Parallel lines

☐ **Parallel and Perpendicular Lines** Suppose $L_1$ and $L_2$ are two distinct lines with slope. This assumption means that both $L_1$ and $L_2$ are nonvertical lines. Then necessarily $L_1$ and $L_2$ are either parallel or they intersect. If the lines intersect at a right angle they are said to be perpendicular. We can determine whether two lines are parallel or are perpendicular by examining their slopes.

FIGURE 2.3.5 Parallel lines

> ### SLOPES OF PARALLEL AND PERPENDICULAR LINES
>
> If $L_1$ and $L_2$ are lines with slopes $m_1$ and $m_2$, respectively, then
>
> - $L_1$ is **parallel** to $L_2$ if and only if $m_1 = m_2$, and $\qquad$ (12)
> - $L_1$ is **perpendicular** to $L_2$ if and only if $m_1 m_2 = -1$. $\qquad$ (13)

There are several ways of proving these theorems. The proof of (12) can be obtained using similar right triangles, as in **FIGURE 2.3.5**, and the fact that the ratios of corresponding sides in such triangles are equal. We leave the justification of (13) as an exercise. See Problem 62 in Exercises 2.3. Note that the condition $m_1 m_2 = -1$ implies that $m_2 = -1/m_1$, that is, the slopes are negative reciprocals of each other. A horizontal line $y = b$ and a vertical line $x = a$ are perpendicular, but the latter is a line with no slope.

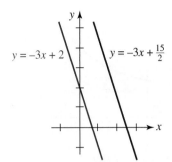

FIGURE 2.3.6 Parallel lines in Example 5

### ▮ EXAMPLE 5　　　Parallel Lines

The linear equations $3x + y = 2$ and $6x + 2y = 15$ can be rewritten in the slope-intercept forms

$$y = -3x + 2 \quad \text{and} \quad y = -3x + \tfrac{15}{2},$$

respectively. As noted in color in the preceding line the slope of each line is $-3$. Therefore the lines are parallel. The graphs of these equations are shown in FIGURE 2.3.6.　▮

### ▮ EXAMPLE 6　　　Perpendicular Lines

Find an equation of the line through $(0, -3)$ that is perpendicular to the graph of $4x - 3y + 6 = 0$.

**Solution** We express the given linear equation in slope-intercept form:

$$4x - 3y + 6 = 0 \quad \text{implies} \quad 3y = 4x + 6.$$

Dividing by 3 gives $y = \tfrac{4}{3}x + 2$. This line, whose graph is given in blue in Figure 2.3.7, has slope $\tfrac{4}{3}$. The slope of any line perpendicular to it is the negative reciprocal of $\tfrac{4}{3}$, namely, $-\tfrac{3}{4}$. Since $(0, -3)$ is the $y$-intercept of the required line, it follows from (9) that its equation is $y = -\tfrac{3}{4}x - 3$. The graph of the last equation is the red line in FIGURE 2.3.7.　▮

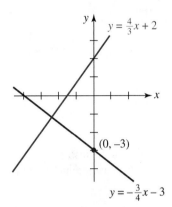

FIGURE 2.3.7 Perpendicular lines in Example 6

☐ **Graphs** As mentioned in the earlier sections of this chapter, when graphing an equation it is always a good habit to try to find $x$- and $y$-intercepts of its graph. Except in the cases of horizontal and vertical lines, and lines through the origin, a line will have distinct $x$- and $y$-intercepts. Of course, that is all we need to draw a line: two points.

### ▮ EXAMPLE 7　　　Graph of a Linear Equation

Graph the linear equation $3x - 2y + 8 = 0$.

**Solution** There is no need to rewrite the linear equation in the form $y = mx + b$. We simply find the intercepts.

$y$-*intercept*: Setting $x = 0$ gives $-2y + 8 = 0$ or $y = 4$. The $y$-intercept is $(0, 4)$.
$x$-*intercept*: Setting $y = 0$ gives $3x + 8 = 0$ or $x = -\tfrac{8}{3}$. The $x$-intercept is $\left(-\tfrac{8}{3}, 0\right)$.

As shown in FIGURE 2.3.8, the line is drawn through the two intercepts $(0, 4)$ and $\left(-\tfrac{8}{3}, 0\right)$.　▮

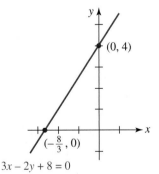

FIGURE 2.3.8 Graph of equation in Example 7

☐ **Increasing-Decreasing Functions** We have just seen in Figures 2.3.2(a) and 2.3.2(b) that if $a > 0$ (which, as we have just seen plays the part of $m$) the values of a linear function $f(x) = ax + b$ increase as $x$-increases, whereas for $a < 0$, the values $f(x)$ decrease as $x$ increases. The notions of increasing and decreasing can be extended to *any* function. The ability to determine intervals over which a function $f$ is either increasing or decreasing plays an important role in applications of calculus.

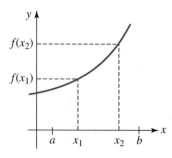

$f(x_2)$

$f(x_1)$

$a \quad x_1 \quad x_2 \quad b$

(a) $f(x_1) < f(x_2)$

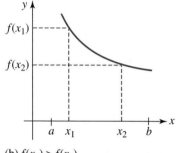

$f(x_1)$

$f(x_2)$

$a \quad x_1 \quad x_2 \quad b$

(b) $f(x_1) > f(x_2)$

**FIGURE 2.3.9** Increasing function in (a); decreasing function in (b)

## INCREASING/DECREASING

Suppose $y = f(x)$ is a function defined on an interval, and $x_1$ and $x_2$ are any two numbers in the interval such that $x_1 < x_2$. Then the function $f$ is

- **increasing** on the interval if $f(x_1) < f(x_2)$, (14)
- **decreasing** on the interval if $f(x_1) > f(x_2)$. (15)

In **FIGURE 2.3.9(a)** the function $f$ is increasing on the interval $[a, b]$, whereas $f$ is decreasing on $[a, b]$ in Figure 2.3.9(b). A linear function $f(x) = ax + b$ increases on the interval $(-\infty, \infty)$ for $a > 0$ and decreases on the interval $(-\infty, \infty)$ for $a < 0$.

☐ **Points of Intersection** We are often interested in finding the points where the graphs of two functions intersect. The $x$-intercepts of the graph of a function $f$ can be interpreted as the points where the graph of $f$ intersects the graph of the constant function $y = 0$. In general, at a point $P$ of intersection of the graphs of two functions $f$ and $g$, the coordinates $(x, y)$ of $P$ must satisfy both equations $y = f(x)$ and $y = g(x)$, and so $f(x) = g(x)$.

### EXAMPLE 8    Intersecting Lines

Find the point where the two lines in Figure 2.3.7 intersect.

**Solution** We equate $y = \frac{4}{3}x + 2$ and $y = -\frac{3}{4}x - 3$ and solve for $x$:

$$\frac{4}{3}x + 2 = -\frac{3}{4}x - 3$$
$$\left(\frac{4}{3} + \frac{3}{4}\right)x = -5$$
$$\frac{25}{12}x = -5$$
$$x = -\frac{12}{5}.$$

By substituting $x = -\frac{12}{5}$ into either equation we find that $y = -\frac{6}{5}$. The point of intersection of the lines is $\left(-\frac{12}{5}, -\frac{6}{5}\right)$.   ∎

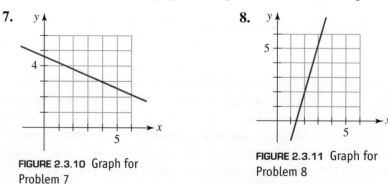

**2.3    Exercises**    Answers to selected odd-numbered problems begin on page ANS-4.

In Problems 1–6, find the slope of the line through the given points. Graph the line through the points.

**1.** $(3, -7), (1, 0)$        **2.** $(-4, -1), (1, -1)$
**3.** $(5, 2), (4, -3)$        **4.** $(1, 4), (6, -2)$
**5.** $(-1, 2), (3, -2)$        **6.** $\left(8, -\frac{1}{2}\right), \left(2, \frac{5}{2}\right)$

In Problems 7 and 8, use the graph of the given to estimate its slope.

**7.**

**FIGURE 2.3.10** Graph for Problem 7

**8.**

**FIGURE 2.3.11** Graph for Problem 8

CHAPTER 2 FUNCTIONS

In Problems 9–16, find the slope and the $x$- and $y$-intercepts of the given line. Graph the line.

**9.** $3x - 4y + 12 = 0$
**10.** $\frac{1}{2}x - 3y = 3$
**11.** $2x - 3y = 9$
**12.** $-4x - 2y + 6 = 0$
**13.** $2x + 5y - 8 = 0$
**14.** $\dfrac{y}{2} - \dfrac{x}{10} - 1 = 0$
**15.** $y + \frac{2}{3}x = 1$
**16.** $y = 2x + 6$

In Problems 17–22, find an equation of the line through $(1, 2)$ with the indicated slope.

**17.** $\frac{2}{3}$
**18.** $\frac{1}{10}$
**19.** $0$
**20.** $-2$
**21.** $-1$
**22.** undefined

In Problems 23–36, find an equation of the line that satisfies the given conditions.

**23.** through $(2, 3)$ and $(6, -5)$
**24.** through $(5, -6)$ and $(4, 0)$
**25.** through $(8, 1)$ and $(-3, 1)$
**26.** through $(2, 2)$ and $(-2, -2)$
**27.** through $(-2, 0)$ and $(-2, 6)$
**28.** through $(0, 0)$ and $(a, b)$
**29.** through $(-2, 4)$ parallel to $3x + y - 5 = 0$
**30.** through $(1, -3)$ parallel to $2x - 5y + 4 = 0$
**31.** through $(5, -7)$ parallel to the $y$-axis
**32.** through the origin parallel to the line through $(1, 0)$ and $(-2, 6)$
**33.** through $(2, 3)$ perpendicular to $x - 4y + 1 = 0$
**34.** through $(0, -2)$ perpendicular to $3x + 4y + 5 = 0$
**35.** through $(-5, -4)$ perpendicular to the line through $(1, 1)$ and $(3, 11)$
**36.** through the origin perpendicular to every line with slope 2

In Problems 37–40, determine which of the given lines are parallel to each other and which are perpendicular to each other.

**37.(a)** $3x - 5y + 9 = 0$
**(b)** $5x = -3y$
**(c)** $-3x + 5y = 2$
**(d)** $3x + 5y + 4 = 0$
**(e)** $-5x - 3y + 8 = 0$
**(f)** $5x - 3y - 2 = 0$
**38.(a)** $2x + 4y + 3 = 0$
**(b)** $2x - y = 2$
**(c)** $x + 9 = 0$
**(d)** $x = 4$
**(e)** $y - 6 = 0$
**(f)** $-x - 2y + 6 = 0$
**39.(a)** $3x - y - 1 = 0$
**(b)** $x - 3y + 9 = 0$
**(c)** $3x + y = 0$
**(d)** $x + 3y = 1$
**(e)** $6x - 3y + 10 = 0$
**(f)** $x + 2y = -8$
**40.(a)** $y + 5 = 0$
**(b)** $x = 7$
**(c)** $4x + 6y = 3$
**(d)** $12x - 9y + 7 = 0$
**(e)** $2x - 3y - 2 = 0$
**(f)** $3x + 4y - 11 = 0$

In Problems 41 and 42, find a linear function (6) that satisfies both of the given conditions.

**41.** $f(-1) = 5, f(1) = 6$
**42.** $f(-1) = 1 + f(2), f(3) = 4f(1)$

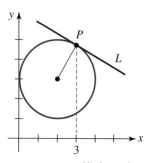

**FIGURE 2.3.12** Graphs for Problem 49

**FIGURE 2.3.13** Circle and tangent line in Problem 50

In Problems 43–46, find the point of intersection of the graphs of the given linear functions. Sketch both lines.

**43.** $f(x) = -2x + 1, g(x) = 4x + 6$  **44.** $f(x) = 2x + 5, g(x) = \frac{3}{2}x + 5$
**45.** $f(x) = 4x + 7, g(x) = \frac{1}{3}x + \frac{10}{3}$  **46.** $f(x) = 2x - 10, g(x) = -3x$

In Problems 47 and 48, for the given linear function compute the quotient

$$\frac{f(x + h) - f(x)}{h},$$

where h is a constant.

**47.** $f(x) = -9x + 12$  **48.** $f(x) = \frac{4}{3}x - 5$
**49.** Find an equation of the line $L$ shown in **FIGURE 2.3.12** if an equation of the blue curve is $y = x^2 + 1$.
**50.** A tangent line $L$ to a circle at a point $P$ on the circle is perpendicular to the line through $P$ and the center of the circle. Find an equation of the line $L$ shown in **FIGURE 2.3.13**.

## Miscellaneous Applications

**51. Thermometers** The functional relationship between degrees Celsius $T_C$ and degrees Fahrenheit $T_F$ is linear.
   **(a)** Express $T_F$ as a function of $T_C$ if $(0°C, 32°F)$ and $(60°C, 140°F)$ are on the graph of $T_F$.
   **(b)** Show that $100°C$ is equivalent to the Fahrenheit boiling point $212°F$. See **FIGURE 2.3.14**.
**52. Thermometers—Continued** The functional relationship between degrees Celsius $T_C$ and degrees Kelvin $T_K$ is linear.
   **(a)** Express $T_K$ as a function of $T_C$ if $(0°C, 273°K)$ and $(27°C, 300°K)$ are on the graph of $T_K$.
   **(b)** Express the boiling point $100°C$ in degrees Kelvin. See Figure 2.3.14.
   **(c)** Absolute zero is defined to be $0°K$. What is $0°K$ in degrees Celsius?
   **(d)** Express $T_K$ as a linear function of $T_F$.
   **(e)** What is $0°K$ in degrees Fahrenheit?

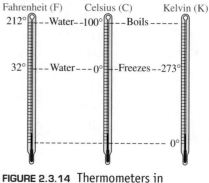

**FIGURE 2.3.14** Thermometers in Problems 51 and 52

**53. Simple Interest** In simple interest, the amount $A$ accrued over time is the linear function $A = P + Prt$, where $P$ is the principal, $t$ is measured in years, and $r$ is the annual interest rate (expressed as a decimal). Compute $A$ after 20 years if the principal is $P = 1000$ and the annual interest rate is 3.4%. At what time is $A = 2200$?

54. **Linear Depreciation** Straight line, or linear, depreciation consists of an item losing all its initial worth of $A$ dollars over a period of $n$ years by an amount $A/n$ each year. If an item costing $20,000 when new is depreciated linearly over 25 years, determine a linear function giving its value $V$ after $x$ years, where $0 \le x \le 25$. What is the value of the item after 10 years?

## For Discussion

55. Consider the linear function $f(x) = \frac{5}{2}x - 4$. If $x$ is changed by one unit, how many units will $y$ change? If $x$ is changed by 2 units? If $x$ is changed by $n$ ($n$ a positive integer) units?

56. Consider the interval $[x_1, x_2]$ and the linear function $f(x) = ax + b$, $a \ne 0$. Show that

$$f\left(\frac{x_1 + x_2}{2}\right) = \frac{f(x_1) + f(x_2)}{2}$$

and interpret this result geometrically for $a > 0$.

57. How would you find an equation of the line that is the perpendicular bisector of the line segment through $\left(\frac{1}{2}, 10\right)$ and $\left(\frac{3}{2}, 4\right)$?

58. Using only the concepts of this section, how would you prove or disprove that the triangle with vertices $(2, 3)$, $(-1, -3)$, and $(4, 2)$ is a right triangle?

59. Using only the concepts of this section, how would you prove or disprove that the quadrilateral with vertices $(0, 4)$, $(-1, 3)$, $(-2, 8)$, and $(-3, 7)$ is a parallelogram?

60. If $C$ is an arbitrary real constant, an equation such as $2x - 3y = C$ is said to define a **family of lines**. Choose four different values of $C$ and plot the corresponding lines on the same coordinate axes. What is true about the lines that are members of this family?

61. Find the equations of the lines through $(0, 4)$ that are tangent to the circle $x^2 + y^2 = 4$.

62. To prove (13) you have to prove two things, the "only if" and the "if" parts of the theorem.

    (a) In **FIGURE 2.3.15**, without loss of generality, we have assumed that two perpendicular lines, $y = m_1x$, $m_1 > 0$ and $y = m_2x$, $m_2 < 0$, intersect at the origin. Use the information in the figure to prove the "only if" part:

    *If $L_1$ and $L_2$ are perpendicular lines with slopes $m_1$ and $m_2$, then $m_1 m_2 = -1$.*

    (b) Reverse your argument in part (a) to prove the "if" part:

    *If $L_1$ and $L_2$ are lines with slopes $m_1$ and $m_2$ such that $m_1 m_2 = -1$, then $L_1$ and $L_2$ are perpendicular.*

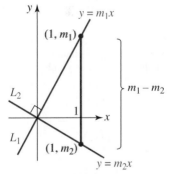

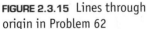

**FIGURE 2.3.15** Lines through origin in Problem 62

## 2.4 Quadratic Functions

☐ **Introduction** The squaring function $y = x^2$ that played an important role in Section 2.2 is a member of a family of functions called **quadratic functions**.

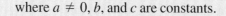

QUADRATIC FUNCTION

A **quadratic function** $y = f(x)$ is a function of the form

$$f(x) = ax^2 + bx + c, \qquad (1)$$

where $a \neq 0, b$, and $c$ are constants.

The **domain** of a quadratic function $f$ is the set of real numbers $(-\infty, \infty)$.

☐ **Graphs** The graph of any quadratic function is called a **parabola**. The graph of a quadratic function has the same basic shape of the squaring function $y = x^2$ shown in **FIGURE 2.4.1**. In the examples that follow we will see that the graphs of quadratic functions (1) are simply transformations of the graph of $y = x^2$:

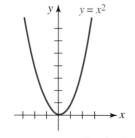

**FIGURE 2.4.1** Graph of simplest parabola

- The graph of $f(x) = ax^2, a > 0$, is the graph of $y = x^2$ **stretched** vertically when $a > 1$, and **compressed** vertically when $0 < a < 1$.
- The graph of $f(x) = ax^2, a < 0$, is the graph of $y = ax^2, a > 0$, **reflected** in the x-axis.
- The graph of $f(x) = ax^2 + bx + c, b \neq 0$, is the graph of $y = ax^2$ **shifted** horizontally or vertically.

From the first two items in the bulleted list, we conclude that the graph of a quadratic function opens upward (as in Figure 2.4.1) if $a > 0$ and opens downward if $a < 0$.

**EXAMPLE 1**     Stretch, Compression, and Reflection

(a) The graphs of $y = 4x^2$ and $y = \frac{1}{10}x^2$ are, respectively, a vertical stretch and a vertical compression of the graph of $y = x^2$. The graphs of these functions are shown in **FIGURE 2.4.2(a)**; the graph of $y = 4x^2$ is shown in red, the graph of $y = \frac{1}{10}x^2$ is green, and the graph of $y = x^2$ is blue.

(b) The graphs of $y = -4x^2$, $y = -\frac{1}{10}x^2$, $y = -x^2$ are obtained from the graphs of the functions in part (a) by reflecting their graphs in the x-axis. See Figure 2.4.2(b). ∎

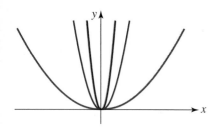

(a) Red graph is a vertical stretch of blue graph; green graph is a vertical compression of blue graph

☐ **Vertex and Axis** If the graph of a quadratic function opens upward $a > 0$ (or downward $a < 0$), the lowest (highest) point $(h, k)$ on the parabola is called its **vertex**. All parabolas are symmetric with respect to a vertical line through the vertex $(h, k)$. The line $x = h$ is called the **axis of symmetry** or simply the **axis** of the parabola. See **FIGURE 2.4.3**.

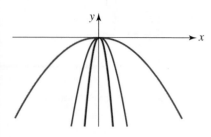

(b) Reflections in x-axis

**FIGURE 2.4.2** Graphs of quadratic functions in Example 1

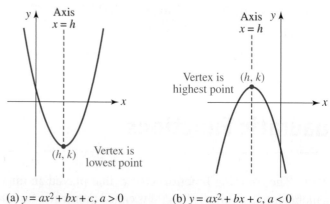

(a) $y = ax^2 + bx + c, a > 0$          (b) $y = ax^2 + bx + c, a < 0$

**FIGURE 2.4.3** Vertex and axis of a parabola

☐ **Standard Form**  The vertex of a parabola can be determined by recasting the equation $f(x) = ax^2 + bx + c$ into the **standard form**

$$f(x) = a(x - h)^2 + k. \tag{2}$$

◀ See Section 1.4.

The form (2) is obtained from the equation (1) by completing the square in $x$. Completing the square in (2) starts with factoring the number $a$ from all terms involving the variable $x$:

$$f(x) = ax^2 + bx + c$$
$$= a\left(x^2 + \frac{b}{a}x\right) + c.$$

Within the parentheses we add and subtract the square of one-half the coefficient of $x$:

$$\text{square } \tfrac{b}{2a}$$
$$\downarrow$$

$$f(x) = a\left(x^2 + \frac{b}{a}x + \frac{b^2}{4a^2} - \frac{b^2}{4a^2}\right) + c \quad \leftarrow \text{ terms in color add to 0}$$
$$= a\left(x^2 + \frac{b}{a}x + \frac{b^2}{4a^2}\right) - \frac{b^2}{4a} + c \quad \leftarrow \text{ note that } a \cdot \left(-\frac{b^2}{4a^2}\right) = -\frac{b^2}{4a} \tag{3}$$
$$= a\left(x + \frac{b}{2a}\right)^2 + \frac{4ac - b^2}{4a}$$

The last expression is equation (2) with the identifications $h = -b/2a$ and $k = (4ac - b^2)/4a$. If $a > 0$, then necessarily $a(x - h)^2 \geq 0$. Hence $f(x)$ in (2) is a minimum when $(x - h)^2 = 0$, that is, for $x = h$. A similar argument shows that if $a < 0$ in (2), $f(x)$ is a maximum value for $x = h$. Thus $(h, k)$ is the vertex of the parabola. The equation of the axis of the parabola is $x = h$ or $x = -b/2a$.

If $a > 0$, then the function $f$ in (2) is decreasing on the interval $(-\infty, h]$ and increasing on the interval $[h, \infty)$. If $a < 0$, we have just the opposite, that is, $f$ is increasing on $(-\infty, h]$ followed by decreasing on $[h, \infty)$.

We strongly suggest that you *do not memorize* the result in the last line of (3), but practice completing the square each time. However, if memorization is permitted by your instructor to save time, then the vertex can be found by computing the coordinates of the point

$$\left(-\frac{b}{2a}, f\left(-\frac{b}{2a}\right)\right). \tag{4}$$

☐ **Intercepts**  The graph of (1) always has a **y-intercept** since $f(0) = c$, and so the $y$-intercept is $(0, c)$. To determine whether the graph has $x$-intercepts we must solve the equation $f(x) = 0$. The last equation can be solved either by factoring or by using the quadratic formula. Recall that a quadratic equation $ax^2 + bx + c = 0$, $a \neq 0$, has the solutions

$$x_1 = \frac{-b - \sqrt{b^2 - 4ac}}{2a}, \qquad x_2 = \frac{-b + \sqrt{b^2 - 4ac}}{2a}.$$

We distinguish three cases according to the algebraic sign of the discriminant $b^2 - 4ac$.

- If $b^2 - 4ac > 0$, then there are two distinct real solutions $x_1$ and $x_2$. The parabola crosses the $x$-axis at $(x_1, 0)$ and $(x_2, 0)$.

2.4 Quadratic Functions

- If $b^2 - 4ac = 0$, then there is a single real solution $x_1$. The vertex of the parabola is located on the $x$-axis at $(x_1, 0)$. The parabola is tangent to, or touches, the $x$-axis at this point.
- If $b^2 - 4ac < 0$, then there are no real solutions. The parabola does not cross the $x$-axis.

As the next example shows, a reasonable sketch of a parabola can be obtained by plotting the intercepts and the vertex.

### ■ EXAMPLE 2        Graph Using Intercepts and Vertex

Graph $f(x) = x^2 - 2x - 3$.

   **Solution** Since $a = 1 > 0$ we know that the parabola will open upward. From $f(0) = -3$ we get the $y$-intercept $(0, -3)$. To see whether there are any $x$-intercepts we solve $x^2 - 2x - 3 = 0$. By factoring

$$(x + 1)(x - 3) = 0,$$

we find the solutions $x = -1$ and $x = 3$. The $x$-intercepts are $(-1, 0)$ and $(3, 0)$. To locate the vertex we complete the square:

$$f(x) = (x^2 - 2x + 1) - 1 - 3 = (x^2 - 2x + 1) - 4.$$

Thus the standard form is $f(x) = (x - 1)^2 - 4$. With the identifications $h = 1$ and $k = -4$, we conclude that the vertex is $(1, -4)$. Using this information we draw a parabola through these four points as shown in **FIGURE 2.4.4**.

   One last observation. By finding the vertex we automatically determine the range of a quadratic function. In our current example, $y = -4$ is the smallest number in the range of $f$ and so the range of $f$ is the interval $[-4, \infty)$ on the $y$-axis.   ■

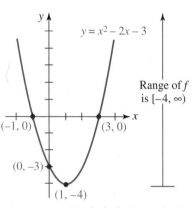

FIGURE 2.4.4 Parabola in Example 2

### ■ EXAMPLE 3        Vertex Is the $x$-intercept

Graph $f(x) = -4x^2 + 12x - 9$.

   **Solution** The graph of this quadratic function is a parabola that opens downward because $a = -4 < 0$. To complete the square we start by factoring $-4$ from the two $x$-terms:

$$\begin{aligned}
f(x) &= -4x^2 + 12x - 9 \\
&= -4(x^2 - 3x) - 9 \\
&= -4\left(x^2 - 3x + \frac{9}{4} - \frac{9}{4}\right) - 9 \\
&= -4\left(x^2 - 3x + \frac{9}{4}\right) - 9 + 9 \\
&= -4\left(x^2 - 3x + \frac{9}{4}\right).
\end{aligned}$$

Thus the standard form is $f(x) = -4(x - \frac{3}{2})^2$. With $h = \frac{3}{2}$ and $k = 0$ we see that the vertex is $(\frac{3}{2}, 0)$. The $y$-intercept is $(0, f(0)) = (0, -9)$. Solving $-4x^2 + 12x - 9 = 0$, we find that there is only one $x$-intercept, namely, $(\frac{3}{2}, 0)$. Of course, this was to be expected because the vertex $(\frac{3}{2}, 0)$ is on the $x$-axis. As shown in **FIGURE 2.4.5** a rough sketch can be obtained from these two points alone. The parabola is tangent to the $x$-axis at $(\frac{3}{2}, 0)$.   ■

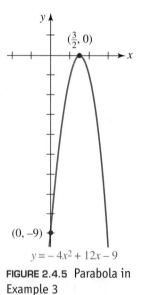

FIGURE 2.4.5 Parabola in Example 3

### ■ EXAMPLE 4        Using (4) to Find the Vertex

Graph $f(x) = x^2 + 2x + 4$.

CHAPTER 2 FUNCTIONS

**Solution** The graph is a parabola that opens upward because $a = 1 > 0$. For the sake of illustration we will use (4) this time to find the vertex. With $b = 2$, $-b/2a = -2/2 = -1$ and

$$f(-1) = (-1)^2 + 2(-1) + 4 = 3,$$

the vertex is $(-1, f(-1)) = (-1, 3)$. Now the $y$-intercept is $(0, f(0)) = (0, 4)$ but the quadratic formula shows that the equation $f(x) = 0$ or $x^2 + 2x + 4 = 0$ has no real solutions. Therefore the graph has no $x$-intercepts. Since the vertex is above the $x$-axis and the parabola opens upward, the graph must lie entirely above the $x$-axis. See **FIGURE 2.4.6**. ∎

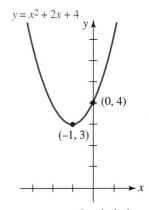

$y = x^2 + 2x + 4$

$(0, 4)$

$(-1, 3)$

**FIGURE 2.4.6** Parabola in Example 4

☐ **Graphs by Transformations** The standard form (2) clearly describes how the graph of any quadratic function is constructed from the graph of $y = x^2$ starting with a non-rigid transformation followed by two rigid transformations:

- $y = ax^2$ is the graph of $y = x^2$ stretched or compressed vertically.
- $y = a(x - h)^2$ is the graph of $y = ax^2$ shifted $|h|$ units horizontally.
- $y = a(x - h)^2 + k$ is the graph of $y = a(x - h)^2$ shifted $|k|$ units vertically.

**FIGURE 2.4.7** illustrates the horizontal and vertical shifting in the case where $a > 0$, $h > 0$, and $k > 0$.

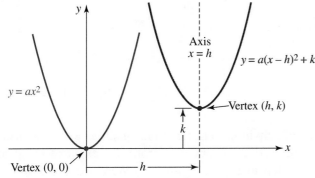

Axis $x = h$

$y = a(x - h)^2 + k$

$y = ax^2$

Vertex $(h, k)$

$k$

Vertex $(0, 0)$

$h$

**FIGURE 2.4.7** The red graph is obtained by shifting the blue graph $h$ units to the right and $k$ units upward.

### ▆ EXAMPLE 5    Horizontally Shifted Graphs

Compare the graphs of **(a)** $y = (x - 2)^2$ and **(b)** $y = (x + 3)^2$.

**Solution** The blue dashed graph in Figure 2.4.8 is the graph of $y = x^2$. Matching the given functions with (2) shows in each case that $a = 1$ and $k = 0$. This means that neither graph undergoes a vertical stretch or compression, and neither graph is shifted vertically.

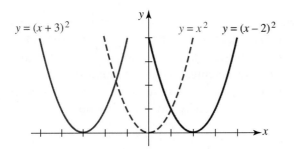

$y = (x + 3)^2$     $y = x^2$     $y = (x - 2)^2$

**FIGURE 2.4.8** Shifted graphs in Example 5

**(a)** With the identification $h = 2$, the graph of $y = (x - 2)^2$ is the graph of $y = x^2$ shifted horizontally 2 units to the right. The vertex $(0, 0)$ for $y = x^2$ becomes the vertex $(2, 0)$ for $y = (x - 2)^2$. See the red graph in **FIGURE 2.4.8**.

**(b)** With the identification $h = -3$, the graph of $y = (x + 3)^2$ is the graph of $y = x^2$ shifted horizontally $|-3| = 3$ units to the left. The vertex $(0, 0)$ for $y = x^2$ becomes the vertex $(-3, 0)$ for $y = (x + 3)^2$. See the green graph in Figure 2.4.8. ∎

### ■ EXAMPLE 6    Shifted Graph

Graph $y = 2(x - 1)^2 - 6$.

**Solution** The graph is the graph of $y = x^2$ stretched vertically upward, followed by a horizontal shift to the right of 1 unit, followed by a vertical shift downward of 6 units. In **FIGURE 2.4.9** on the next page, you should note how the vertex $(0, 0)$ on the graph of $y = x^2$ is moved to $(1, -6)$ on the graph of $y = 2(x - 1)^2 - 6$ as a result of these transformations. You should also follow how the point $(1, 1)$ shown in Figure 2.4.9(a) ends up as $(2, -4)$ in Figure 2.4.9(d). ∎

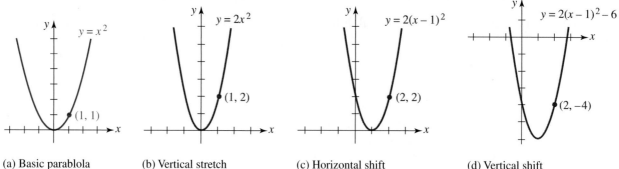

(a) Basic parablola     (b) Vertical stretch     (c) Horizontal shift     (d) Vertical shift

**FIGURE 2.4.9** Graphs in Example 6

Graphs can be of help in solving certain inequalities when a sign chart is not useful because the quadratic does factor conveniently. For example, the quadratic function in Example 6 is equivalent to $y = 2x^2 - 4x - 4$. Were we required to solve the inequality $2x^2 - 4x - 4 \geq 0$, we see in Figure 2.4.9(d) that $y \geq 0$ to the left of the $x$-intercept on the negative $x$-axis and to the right of the $x$-intercept on the positive $x$-axis. The $x$-coordinates of these intercepts, obtained by solving $2x^2 - 4x - 4 = 0$ by the quadratic formula, are $1 - \sqrt{3}$ and $1 + \sqrt{3}$. Thus the solution of $2x^2 - 4x - 4 \geq 0$ is $(-\infty, 1 - \sqrt{3}] \cup [1 + \sqrt{3}, \infty)$.

□ **Freely Falling Object** Suppose an object, such as a ball, is either thrown straight upward (downward) or simply dropped from an initial height $s_0$. Then if the positive direction is taken to be upward, the height $s(t)$ of the object above ground is given by the quadratic function

$$s(t) = \tfrac{1}{2}gt^2 + v_0 t + s_0, \tag{5}$$

where $g$ is the acceleration due to gravity $(-32 \text{ ft/s}^2 \text{ or } -9.8 \text{ m/s}^2)$, $v_0$ is the initial velocity imparted to the object, and $t$ is time measured in seconds. See **FIGURE 2.4.10**. If the object is dropped, then $v_0 = 0$. An assumption in the derivation of (5), a straightforward exercise in integral calculus, is that the motion takes place close to the surface of Earth and so the retarding effects of air resistance is ignored. Also, the velocity of the object while it is in the air is given by the linear function

$$v(t) = gt + v_0. \tag{6}$$

See Problems 47–50 in Exercises 2.4.

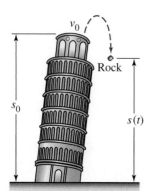

**FIGURE 2.4.10** Rock thrown upward from an initial height $s_0$

In Problems 1–6, sketch the graph of the given function $f$.

**1.** $f(x) = 2x^2$
**2.** $f(x) = -2x^2$
**3.** $f(x) = 2x^2 - 2$
**4.** $f(x) = 2x^2 + 5$
**5.** $f(x) = -2x^2 + 1$
**6.** $f(x) = -2x^2 - 3$

In Problems 7–18, consider the quadratic function $f$.
**(a)** Find all intercepts of the graph of $f$.
**(b)** Express the function $f$ in standard form.
**(c)** Find the vertex and axis of symmetry.
**(d)** Sketch the graph of $f$.

**7.** $f(x) = x(x + 5)$
**8.** $f(x) = -x^2 + 4x$
**9.** $f(x) = (3 - x)(x + 1)$
**10.** $f(x) = (x - 2)(x - 6)$
**11.** $f(x) = x^2 - 3x + 2$
**12.** $f(x) = -x^2 + 6x - 5$
**13.** $f(x) = 4x^2 - 4x + 3$
**14.** $f(x) = -x^2 + 6x - 10$
**15.** $f(x) = -\frac{1}{2}x^2 + x + 1$
**16.** $f(x) = x^2 - 2x - 7$
**17.** $f(x) = x^2 - 10x + 25$
**18.** $f(x) = -x^2 + 6x - 9$

In Problems 19 and 20, find the maximum or the minimum value of the function $f$. Give the range of the function $f$.

**19.** $f(x) = 3x^2 - 8x + 1$
**20.** $f(x) = -2x^2 - 6x + 3$

In Problems 21–24, find the largest interval on which the function $f$ is increasing and the largest interval on which $f$ is decreasing.

**21.** $f(x) = \frac{1}{3}x^2 - 25$
**22.** $f(x) = -(x + 10)^2$
**23.** $f(x) = -2x^2 - 12x$
**24.** $f(x) = x^2 + 8x - 1$

In Problems 25–30, describe in words how the graph of the given function can be obtained from the graph of $y = x^2$ by rigid or nonrigid transformations.

**25.** $f(x) = (x - 10)^2$
**26.** $f(x) = (x + 6)^2$
**27.** $f(x) = -\frac{1}{3}(x + 4)^2 + 9$
**28.** $f(x) = 10(x - 2)^2 - 1$
**29.** $f(x) = (-x - 6)^2 - 4$
**30.** $f(x) = -(1 - x)^2 + 1$

In Problems 31–36, the given graph is the graph of $y = x^2$ shifted/reflected in the $xy$-plane. Write an equation of the graph.

**31.**

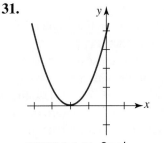

FIGURE 2.4.11 Graph for Problem 31

**32.**

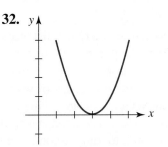

FIGURE 2.4.12 Graph for Problem 32

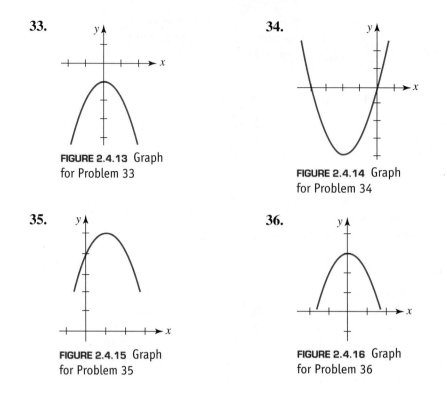

**33.**

FIGURE 2.4.13 Graph for Problem 33

**34.**

FIGURE 2.4.14 Graph for Problem 34

**35.**

FIGURE 2.4.15 Graph for Problem 35

**36.**

FIGURE 2.4.16 Graph for Problem 36

In Problems 37 and 38, find a quadratic function $f(x) = ax^2 + bx + c$ that satisfies the given conditions.

**37.** $f$ has the values $f(0) = 5$, $f(1) = 10$, $f(-1) = 4$
**38.** graph passes through $(2, -1)$, zeros of $f$ are 1 and 3

In Problems 39 and 40, find a quadratic function in standard form $f(x) = a(x - h)^2 + k$ that satisfies the given conditions.

**39.** the vertex of the graph of $f$ is $(1, 2)$, graph passes through $(2, 6)$
**40.** the maximum value of $f$ is 10, axis of symmetry is $x = -1$, and $y$-intercept is $(0, 8)$

In Problems 41–44, sketch the region in the $xy$-plane that is bounded between the graphs of the given functions. Find the points of intersection of the graphs.

**41.** $y = -x + 4$,  $y = x^2 + 2x$
**42.** $y = 2x - 2$,  $y = 1 - x^2$
**43.** $y = x^2 + 2x + 2$,  $y = -x^2 - 2x + 2$
**44.** $y = x^2 - 6x + 1$,  $y = -x^2 + 2x + 1$

**45. (a)** Express the square of the distance $d$ from the point $(x, y)$ on the graph of $y = 2x$ to the point $(5, 0)$ shown in **FIGURE 2.4.17** as a function of $x$.
   **(b)** Use the function in part (a) to find the point $(x, y)$ that is closest to $(5, 0)$.
**46.** As shown in **FIGURE 2.4.18**, an arrow that is shot at a 45° angle with the horizontal travels along a parabolic arc defined by the equation $y = ax^2 + x + c$. Use the fact that the arrow is launched at a vertical height of 6 ft and travels a horizontal distance of 200 ft to find the coefficients $a$ and $c$. What is the maximum height attained by the arrow?

$y = 2x$

$(x, y)$

$d$

$(5, 0)$

FIGURE 2.4.17 Distance in Problem 45

CHAPTER 2 FUNCTIONS

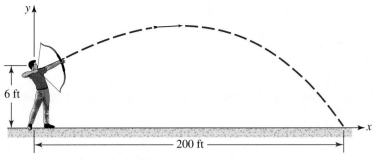

FIGURE 2.4.18 Arrow in Problem 46

**47.** An arrow is shot vertically upward with an initial velocity of 64 ft/s from a point 6 ft above the ground. See **FIGURE 2.4.19**.
   **(a)** Find the height $s(t)$ and the velocity $v(t)$ of the arrow at time $t \geq 0$.
   **(b)** What is the maximum height attained by the arrow? What is the velocity of the arrow at the time the arrow attains its maximum height?
   **(c)** At what time does the arrow fall back to the 6-ft level? What is its velocity at this time?
**48.** The height above ground of a toy rocket launched upward from the top of a building is given by $s(t) = -16t^2 + 96t + 256$.
   **(a)** What is the height of the building?
   **(b)** What is the maximum height attained by the rocket?
   **(c)** Find the time when the rocket strikes the ground.
**49.** A ball is dropped from the roof of a building that is 122.5 meters above ground level.
   **(a)** What is the height and velocity of the ball at $t = 1$ s?
   **(b)** At what time does the ball hit the ground?
   **(c)** What is the impact velocity of the ball when it hits the ground?
**50.** A few years ago a newspaper in the Midwest reported that an escape artist was planning to jump off a bridge into the Mississippi River wearing 70 lb of chains and manacles. The newspaper article stated that the height of the bridge was 48 ft and predicted that the escape artist's impact velocity on hitting the water would be 85 mph. Assuming that he simply dropped from the bridge, then his height (in feet) and velocity (in feet/second) $t$ seconds after jumping off the bridge are given by the functions $s(t) = -16t^2 + 48$ and $v(t) = -32t$, respectively. Determine whether the newspaper's estimate of his impact velocity was accurate.

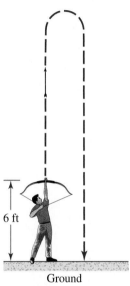

FIGURE 2.4.19 Arrow in Problem 47

## Miscellaneous Applications

**51. Spread of a Disease** One model for the spread a flu virus assumes that within a population of $P$ persons the rate at which a disease spreads is jointly proportional to the number $D$ of persons already carrying the disease and the number $P - D$ of persons not yet infected. Mathematically, the model is given by the quadratic function

$$R(D) = kD(P - D),$$

where $R(D)$ is the rate of spread of the flu virus (in cases per day) and $k > 0$ is a constant of proportionality.
   **(a)** Show that if the population $P$ is a constant, then the disease spreads most rapidly when exactly one-half the population is carrying the flu.

Spreading a virus

**(b)** Suppose that in a town of 10,000 persons, 125 are sick on Sunday, and 37 new cases occur on Monday. Estimate the constant $k$.

**(c)** Use the result of part (b) to estimate the number of new cases on Tuesday. [*Hint*: The number of persons carrying the flu on Monday is $162 = 125 + 37$.]

**(d)** Estimate the number of new cases on Wednesday, Thursday, Friday, and Saturday.

### For Discussion

**52.** In Problems 48 and 50, what is the domain of the function $s(t)$? [*Hint*: It is *not* $(-\infty, \infty)$.]

**53.** On the Moon the acceleration due to gravity is one-sixth the acceleration due to gravity on Earth. If a ball is tossed vertically upward from the surface of the Moon, would it attain a maximum height six times that on Earth when the same initial velocity is used? Defend your answer.

**54.** Suppose the quadratic function $f(x) = ax^2 + bx + c$ has two distinct real zeros. How would you prove that the $x$-coordinate of the vertex is the midpoint of the line segment between the $x$-coordinates of the intercepts? Carry out your ideas.

## 2.5 Piecewise-Defined Functions

☐ **Introduction** A function $f$ may involve two or more expressions or formulas, with each formula defined on different parts of the domain of $f$. A function defined in this manner is called a **piecewise-defined function**. For example,

$$f(x) = \begin{cases} x^2, & x < 0 \\ x + 1, & x \geq 0 \end{cases}$$

is not two functions, but a single function in which the rule of correspondence is given in two pieces. In this case, one piece is used for the negative real numbers $(x < 0)$ and the other part on the nonnegative numbers $(x \geq 0)$; the domain of $f$ is the union of the intervals $(-\infty, 0) \cup [0, \infty) = (-\infty, \infty)$. For example, since $-4 < 0$, the rule indicates that we square the number:

$$f(-4) = (-4)^2 = 16;$$

on the other hand, since $6 \geq 0$ we add 1 to the number:

$$f(6) = 6 + 1 = 7.$$

The USPS first-class mailing rates for a letter, a card, or a package provides a real-world illustration of a piecewise-defined function. As of this writing, the postage for sending a letter in a standard-size envelope by first-class mail depends on its weight in ounces:

$$\text{Postage} = \begin{cases} \$0.39, & 0 < \text{weight} \leq 1 \text{ ounce} \\ \$0.63, & 1 < \text{weight} \leq 2 \text{ ounces} \\ \$0.87, & 2 < \text{weight} \leq 3 \text{ ounces,} \\ \vdots \\ \$3.27, & 12 < \text{weight} \leq 13 \text{ ounces.} \end{cases} \qquad (1)$$

The rule in (1) is a function $P$ consisting of 13 pieces (letters over 13 ounces are sent priority mail). A value $P(w)$ is one of thirteen constants; the constant changes depending on the weight $w$ (in ounces) of the letter. For example,

$$P(0.5) = \$0.39, P(1.7) = \$0.63, P(2.2) = \$.87, P(2.9) = \$.87,$$
$$\text{and } P(12.1) = \$3.27.$$

The domain of the function $P$ is the union of the thirteen intervals:

$$(0, 1] \cup (1, 2] \cup (2, 3] \cup \cdots \cup (12, 13] = (0, 13].$$

EXAMPLE 1            **Graph of a Piecewise-Defined Function**

Graph the piecewise-defined function

$$f(x) = \begin{cases} -1, & x < 0, \\ 0, & x = 0, \\ x + 1, & x > 0. \end{cases} \qquad (2)$$

**Solution** Although the domain of $f$ consists of all real numbers $(-\infty, \infty)$, each piece of the function is defined on a different part of this domain. We draw

the horizontal line $y = -1$ for $x < 0$,
the point $(0, 0)$ for $x = 0$, and
the line $y = x + 1$ for $x > 0$.

The graph is given in **FIGURE 2.5.1**.                                    ■

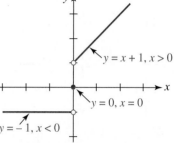

**FIGURE 2.5.1** Graph of piecewise-defined function in Example 1

The solid dot at the origin in Figure 2.5.1 indicates that the function in (2) is defined at $x = 0$ only by $f(0) = 0$; the open dots indicate that the formulas corresponding to $x < 0$ and to $x > 0$ do not define $f$ at $x = 0$. Since we are making up a function, consider the definition:

$$g(x) = \begin{cases} -1, & x \le 0, \\ x + 1, & x > 0. \end{cases} \qquad (3)$$

The graph of $g$ shown in **FIGURE 2.5.2** is very similar to the graph of (2), but (2) and (3) are not the same function since $f(0) = 0$ but $g(0) = -1$.

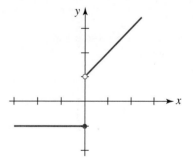

**FIGURE 2.5.2** Graph of function $g$ defined in (3)

☐ **Greatest Integer Function** We consider next a piecewise-defined function that is similar to the "postage stamp" function (1) in that both are examples of *step functions*; each function is constant on an interval and then jumps to another constant value on the next abutting interval. This new function, which has many notations, will be denoted here by $f(x) = [\![x]\!]$, and is defined by the rule

$$[\![x]\!] = n, \quad \text{where } n \text{ is an integer satisfying } n \le x < n + 1. \qquad (4)$$

The function $f$ is called the **greatest integer function** because (4), translated into words, means that:

*$f(x)$ is the greatest integer $n$ that is less than or equal to x.*

For example,

$f(6) = 6$ since $6 \le x = 6$,      $f(-1.5) = -2$ since $-2 \le x = -1.5$,
$f(0.4) = 0$ since $0 \le x = 0.4$,      $f(7.6) = 7$ since $7 \le x = 7.6$,
$f(\pi) = 3$ since $3 \le x = \pi$,      $f(-\sqrt{2}) = -2$ since $-2 \le x = -\sqrt{2}$,

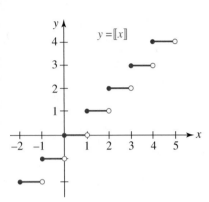

FIGURE 2.5.3 Greatest integer function

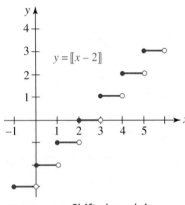

FIGURE 2.5.4 Shifted graph in Example 2

and so on. The domain of $f$ is the set of real numbers and consists of the union of an infinite number of disjoint intervals, in other words, $f(x) = [\![x]\!]$ is a piecewise-defined function given by

$$f(x) = [\![x]\!] = \begin{cases} \vdots \\ -2, & -2 \le x < -1 \\ -1, & -1 \le x < 0 \\ 0, & 0 \le x < 1 \\ 1, & 1 \le x < 2 \\ 2, & 2 \le x < 3 \\ \vdots \end{cases} \quad (5)$$

The range of $f$ is the set of integers. A portion of the graph of $f$ is given on the closed interval $[-2, 5]$ in FIGURE 2.5.3.

In computer science the greatest integer function is known as the **floor function** and is denoted by $f(x) = \lfloor x \rfloor$. See Problems 47, 48, and 53 in Exercises 2.5.

### EXAMPLE 2      Shifted Graph

Graph $y = [\![x - 2]\!]$.

**Solution** The function is $y = f(x - 2)$, where $f(x) = [\![x]\!]$. Thus the graph in Figure 2.5.3 is shifted horizontally 2 units to the right. Note in Figure 2.5.3 that if $n$ is an integer, then $f(n) = [\![n]\!] = n$. But in FIGURE 2.5.4, for $x = n$, $y = n - 2$. ∎

☐ **Continuous Functions** The graph of a **continuous function** has no holes, finite gaps, or infinite breaks. While the formal definition of continuity of a function is an important topic of discussion in calculus, in this course it suffices to think in informal terms. A continuous function is often characterized by saying that its graph can be drawn "without lifting pencil from paper." Parts (a)–(c) of Figure 2.5.5 illustrate functions that are *not* continuous, or **discontinuous**, at $x = 2$. The function

$$f(x) = \frac{x^2 - 4}{x - 2} = x + 2, \quad x \ne 2,$$

in FIGURE 2.5.5(a) has a hole in its graph (there is no point $(2, f(2))$); the function $f(x) = \dfrac{|x - 2|}{x - 2}$ in Figure 2.5.5(b) has a finite gap or jump in its graph at $x = 2$; the function $f(x) = \dfrac{1}{x - 2}$ in Figure 2.5.5(c) has an infinite break in its graph at $x = 2$. The function $f(x) = x^3 - 3x + 2$ is continuous; its graph given in Figure 2.5.5(d) has no holes, gaps, or infinite breaks.

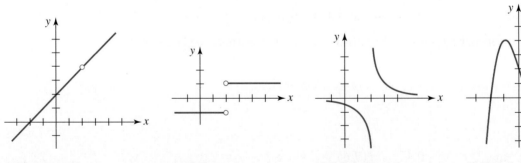

(a) Hole in graph      (b) Finite gap in graph      (c) Infinite break in graph      (d) No holes, gaps, or breaks

FIGURE 2.5.5 Discontinuous functions (a)–(c); continuous function (d)

You should be aware that constant functions, linear functions, and quadratic functions are continuous. Piecewise-defined functions can be continuous or discontinuous. The functions given in (2), (3), and (4) are discontinuous.

☐ **Absolute-Value Function** The function $y = |x|$, called the **absolute-value function**, appears frequently in the study of calculus. To obtain the graph, we graph its two pieces consisting of perpendicular half lines:

$$y = |x| = \begin{cases} -x, & \text{if } x < 0 \\ x, & \text{if } x \geq 0. \end{cases} \qquad (6)$$

See **FIGURE 2.5.6(a)**. Since $y \geq 0$ for all $x$, another way of graphing (6) is simply to sketch the line $y = x$ and then reflect in the $x$-axis that portion of the line that is below the $x$-axis. See Figure 2.5.6(b). The domain of (6) is the set of real numbers $(-\infty, \infty)$, and as is seen in Figure 2.5.6(a), the absolute-value function is an even function, decreasing on the interval $(-\infty, 0)$, increasing on the interval $(0, \infty)$, and is continuous.

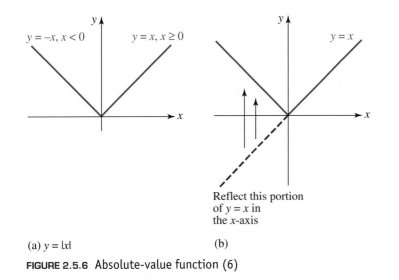

(a) $y = |x|$        (b)

**FIGURE 2.5.6** Absolute-value function (6)

In some applications we are interested in the graph of the absolute value of an arbitrary function $y = f(x)$, in other words, $y = |f(x)|$. Since $|f(x)|$ is nonnegative for all numbers $x$ in the domain of $f$, the graph of $y = |f(x)|$ does not extend below the $x$-axis. Moreover, the definition of the absolute value of $f(x)$,

$$|f(x)| = \begin{cases} -f(x), & \text{if } f(x) < 0 \\ f(x), & \text{if } f(x) \geq 0, \end{cases} \qquad (7)$$

shows that we must negate $f(x)$ whenever $f(x)$ is negative. There is no need to worry about solving the inequalities in (7); to obtain the graph of $y = |f(x)|$, we can proceed just as we did in Figure 2.5.6(b): Carefully draw the graph of $y = f(x)$ and then reflect in the $x$-axis all portions of the graph that are below the $x$-axis.

### ▮ EXAMPLE 3      Absolute Value of a Function

Graph $y = |-3x + 2|$.

    **Solution** We first draw the graph of the linear function $f(x) = -3x + 2$. Note that since the slope is negative, $f$ is decreasing and its graph crosses the $x$-axis at $\left(\frac{2}{3}, 0\right)$. We dash the graph for $x > \frac{2}{3}$ since that portion is below the $x$-axis. Finally, we reflect that

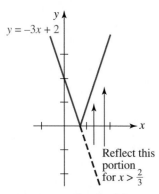

$y = -3x + 2$

Reflect this portion for $x > \frac{2}{3}$

**FIGURE 2.5.7** Graph of function in Example 3

portion upward in the $x$-axis to obtain the solid blue v-shaped graph in **FIGURE 2.5.7**. Since $f(x) = x$ is a simple linear function, it is not surprising that the graph of the absolute value of any linear function $f(x) = ax + b$, $a \neq 0$, will result in a graph similar to that of the absolute-value function shown in Figure 2.5.6(a). ∎

### ▌EXAMPLE 4       **Absolute Value of a Function**

Graph $y = |-x^2 + 2x + 3|$.

    **Solution** As in Example 3, we begin by drawing the graph of the function $f(x) = -x^2 + 2x + 3$ by finding its intercepts $(-1, 0)$, $(3, 0)$, $(0, 3)$ and, since $f$ is a quadratic function, its vertex $(1, 4)$. Observe in **FIGURE 2.5.8(a)** that $y < 0$ for $x < -1$ and for $x > 3$. These portions of the graph of $f$ are reflected in the $x$-axis to obtain the graph of $y = |-x^2 + 2x + 3|$ given in Figure 2.5.8(b).

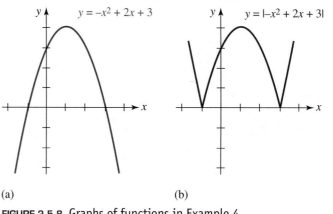

(a)                    (b)

**FIGURE 2.5.8** Graphs of functions in Example 4   ∎

---

**2.5**    Exercises    Answers to selected odd-numbered problems begin on page ANS-6.

In Problems 1–4, find the indicated values of the given piecewise-defined function $f$.

**1.** $f(x) = \begin{cases} \dfrac{x^2 - 4}{x - 2}, & x \neq 2 \\ 4, & x = 2 \end{cases}$;     $f(0), f(2), f(-7)$

**2.** $f(x) = \begin{cases} \dfrac{x^4 - 1}{x^2 - 1}, & x \neq \pm 1 \\ 3, & x = -1 \\ 5, & x = 1 \end{cases}$;     $f(-1), f(1), f(3)$

**3.** $f(x) = \begin{cases} x^2 + 2x, & x \geq 1 \\ -x^3, & x < 1 \end{cases}$;     $f(1), f(0), f(-2), f(\sqrt{2})$

**4.** $f(x) = \begin{cases} 0, & x < 0 \\ x, & 0 < x < 1 \\ x + 1, & x \geq 1 \end{cases}$;     $f(-\tfrac{1}{2}), f(\tfrac{1}{3}), f(4), f(6.2)$

**5.** If the piecewise-defined function $f$ is defined by

$$f(x) = \begin{cases} 1, & x \text{ a rational number} \\ 0, & x \text{ an irrational number,} \end{cases}$$

find each of the following values.
(a) $f(\frac{1}{3})$      (b) $f(-1)$      (c) $f(\sqrt{2})$
(d) $f(1.\overline{12})$      (e) $f(5.72)$      (f) $f(\pi)$

**6.** What is the $y$-intercept of the graph of the function $f$ in Problem 5?

**7.** Determine the values of $x$ for which the piecewise-defined function

$$f(x) = \begin{cases} x^3 + 1, & x < 0 \\ x^2 - 2, & x \geq 0, \end{cases}$$

is equal to the given number.
(a) 7      (b) 0      (c) $-1$
(d) $-2$      (e) 1      (f) $-7$

**8.** Determine the values of $x$ for which the piecewise-defined function

$$f(x) = \begin{cases} x + 1, & x < 0 \\ 2, & x = 0 \\ x^2, & x > 0, \end{cases}$$

is equal to the given number.
(a) 1      (b) 0      (c) 4
(d) $\frac{1}{2}$      (e) 2      (f) $-4$

In Problems 9–34, sketch the graph of the given piecewise-defined function. Find any $x$- and $y$-intercepts of the graph. Give any numbers at which the function is discontinuous.

**9.** $y = \begin{cases} -x, & x \leq 1 \\ -1, & x > 1 \end{cases}$      **10.** $y = \begin{cases} x - 1, & x < 0 \\ x + 1, & x \geq 0 \end{cases}$

**11.** $y = \begin{cases} -3, & x < -3 \\ x, & -3 \leq x \leq 3 \\ 3, & x > 3 \end{cases}$      **12.** $y = \begin{cases} -x^2 - 1, & x < 0 \\ 0, & x = 0 \\ x^2 + 1, & x > 0 \end{cases}$

**13.** $y = [\![x + 2]\!]$      **14.** $y = 2 + [\![x]\!]$
**15.** $y = -[\![x]\!]$      **16.** $y = [\![-x]\!]$
**17.** $y = |x + 3|$      **18.** $y = -|x - 4|$
**19.** $y = 2 - |x|$      **20.** $y = -1 - |x|$
**21.** $y = -2 + |x + 1|$      **22.** $y = 1 - \frac{1}{2}|x - 2|$
**23.** $y = -|5 - 3x|$      **24.** $y = |2x - 5|$
**25.** $y = |x^2 - 1|$      **26.** $y = |4 - x^2|$
**27.** $y = |x^2 - 2x|$      **28.** $y = |-x^2 - 4x + 5|$
**29.** $y = ||x| - 2|$      **30.** $y = |\sqrt{x} - 2|$
**31.** $y = |x^3 - 1|$      **32.** $y = |[\![x]\!]|$

**33.** $y = \begin{cases} 1, & x < 0 \\ |x - 1|, & 0 \leq x \leq 2 \\ 1, & x > 2 \end{cases}$      **34.** $y = \begin{cases} -x, & x < 0 \\ 1 - |x - 1|, & 0 \leq x \leq 2 \\ x - 2, & x > 2 \end{cases}$

**35.** Without graphing, give the range of the function $f(x) = (-1)^{[\![x]\!]}$.
**36.** Compare the graphs of $y = 2[\![x]\!]$ and $y = [\![2x]\!]$.

In Problems 37–40, find a piecewise-defined formula for the function $f$ whose graph is given. Assume that the domain of $f$ is $(-\infty, \infty)$.

**37.**

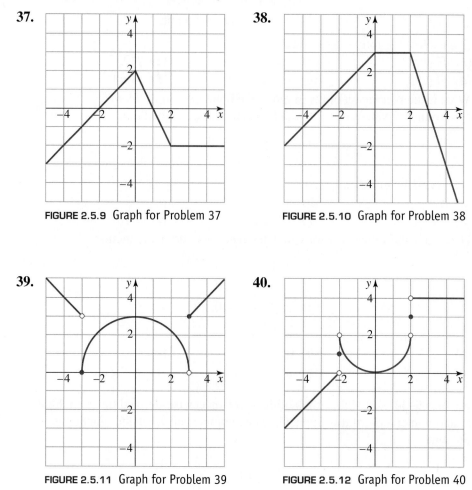

FIGURE 2.5.9 Graph for Problem 37

**38.**

FIGURE 2.5.10 Graph for Problem 38

**39.**

FIGURE 2.5.11 Graph for Problem 39

**40.**

FIGURE 2.5.12 Graph for Problem 40

In Problems 41 and 42, sketch the graph of $y = |f(x)|$.

**41.** $f$ is the function whose graph is given in Figure 2.5.9.
**42.** $f$ is the function whose graph is given in Figure 2.5.10.

In Problems 43 and 44, use the definition of absolute value and express the given function $f$ as a piecewise-defined function.

**43.** $f(x) = \dfrac{|x|}{x}$

**44.** $f(x) = \dfrac{x-3}{|x-3|}$

In Problems 45 and 46, find the value of the constant $k$ such that the given piecewise-defined function $f$ is continuous at $x = 2$. That is, the graph of $f$ has no holes, gaps, or breaks in its graph at $x = 2$.

**45.** $f(x) = \begin{cases} \frac{1}{2}x + 1, & x \le 2 \\ kx, & x > 2 \end{cases}$

**46.** $f(x) = \begin{cases} kx + 2, & x < 2 \\ x^2 + 1, & x \ge 2 \end{cases}$

**47.** The **ceiling function** $g(x) = \lceil x \rceil$ is defined to be the least integer $n$ that is greater than or equal to $x$. Fill in the blanks.

$$g(x) = \lceil x \rceil = \begin{cases} \quad\vdots \\ \underline{\hspace{1cm}}, & -3 < x \le -2 \\ \underline{\hspace{1cm}}, & -2 < x \le -1 \\ \underline{\hspace{1cm}}, & -1 < x \le 0 \\ \underline{\hspace{1cm}}, & 0 < x \le 1 \\ \underline{\hspace{1cm}}, & 1 < x \le 2 \\ \underline{\hspace{1cm}}, & 2 < x \le 3 \\ \quad\vdots \end{cases}$$

**48.** Graph the ceiling function $g(x) = \lceil x \rceil$ defined in Problem 47.

## For Discussion

In Problems 49–52, describe in words how the graphs of the given functions differ. [*Hint*: Factor and cancel.]

**49.** $f(x) = \dfrac{x^2 - 9}{x - 3}$, $\quad g(x) = \begin{cases} \dfrac{x^2 - 9}{x - 3}, & x \ne 3 \\ 4, & x = 3 \end{cases}$,

$h(x) = \begin{cases} \dfrac{x^2 - 9}{x - 3}, & x \ne 3 \\ 6, & x = 3 \end{cases}$

**50.** $f(x) = -\dfrac{x^2 - 7x + 6}{x - 1}$, $\quad g(x) = \begin{cases} -\dfrac{x^2 - 7x + 6}{x - 1}, & x \ne 1 \\ 8, & x = 1 \end{cases}$,

$h(x) = \begin{cases} -\dfrac{x^2 - 7x + 6}{x - 1}, & x \ne 1 \\ 5, & x = 1 \end{cases}$

**51.** $f(x) = \dfrac{x^4 - 1}{x^2 - 1}$, $\quad g(x) = \begin{cases} \dfrac{x^4 - 1}{x^2 - 1}, & x \ne 1 \\ 0, & x = 1 \end{cases}$,

$h(x) = \begin{cases} \dfrac{x^4 - 1}{x^2 - 1}, & x \ne 1 \\ 2, & x = 1 \end{cases}$

**52.** $f(x) = \dfrac{x^3 - 8}{x - 2}$, $\quad g(x) = \begin{cases} \dfrac{x^3 - 8}{x - 2}, & x \ne 2 \\ 5, & x = 2 \end{cases}$,

$h(x) = \begin{cases} \dfrac{x^3 - 8}{x - 2}, & x \ne 2 \\ 12, & x = 2 \end{cases}$

**53.** Using the notion of a reflection of a graph in an axis, express the ceiling function $g(x) = \lceil x \rceil$ in terms of the floor function $f(x) = \lfloor x \rfloor$ (see page 86).

**54.** Discuss how to graph the function $y = |x| + |x - 3|$. Carry out your ideas.

## 2.6 Combining Functions

☐ **Introduction**  Two functions $f$ and $g$ can be combined in several ways to create new functions. In this section we will examine two such ways in which functions can be combined: through arithmetic operations, and through the operation of function composition.

☐ **Arithmetic Combinations**  Two functions can be combined through the familiar four arithmetic operations of addition, subtraction, multiplication, and division.

---

### ARITHMETIC COMBINATIONS

If $f$ and $g$ are two functions, then the **sum** $f + g$, the **difference** $f - g$, the **product** $fg$, and the **quotient** $f/g$ are defined as follows:

$$(f + g)(x) = f(x) + g(x), \tag{1}$$
$$(f - g)(x) = f(x) - g(x), \tag{2}$$
$$(fg)(x) = f(x)g(x), \tag{3}$$
$$\left(\frac{f}{g}\right)(x) = \frac{f(x)}{g(x)}, \quad \text{provided } g(x) \neq 0. \tag{4}$$

---

### EXAMPLE 1  Sum, Difference, Product, and Quotient

Consider the functions $f(x) = x^2 + 4x$ and $g(x) = x^2 - 9$. From (1)–(4) we can produce four new functions:

$$(f + g)(x) = f(x) + g(x) = (x^2 + 4x) + (x^2 - 9) = 2x^2 + 4x - 9,$$
$$(f - g)(x) = f(x) - g(x) = (x^2 + 4x) - (x^2 - 9) = 4x + 9,$$
$$(fg)(x) = f(x)g(x) = (x^2 + 4x)(x^2 - 9) = x^4 + 4x^3 - 9x^2 - 36x,$$

and
$$\left(\frac{f}{g}\right)(x) = \frac{f(x)}{g(x)} = \frac{x^2 + 4x}{x^2 - 9}. \qquad \blacksquare$$

☐ **Domain of an Arithmetic Combination**  When combining two functions arithmetically it is necessary that both $f$ and $g$ be defined at a same number $x$. Hence the **domain** of the functions $f + g, f - g$, and $fg$ is the set of real numbers that are *common* to both domains, that is, the domain is the *intersection* of the domain of $f$ with the domain of $g$. In the case of the quotient $f/g$, the domain is also the intersection of the two domains, *but* we must also exclude any values of $x$ for which the denominator $g(x)$ is zero. In Example 1 the domain of $f$ and the domain of $g$ is the set of real numbers $(-\infty, \infty)$, and so the domain of $f + g, f - g$, and $fg$ is also $(-\infty, \infty)$. However, since $g(-3) = 0$ and $g(3) = 0$, the domain of the quotient $(f/g)(x)$ is $(-\infty, \infty)$ with $x = 3$ and $x = -3$ excluded, in other words, $(-\infty, -3) \cup (-3, 3) \cup (3, \infty)$. In summary, if the domain of $f$ is the set $X_1$ and the domain of $g$ is the set $X_2$, then:

- the domain of $f + g, f - g$, and $fg$ is $X_1 \cap X_2$, and
- the domain of $f/g$ is the set $\{x \mid x \in X_1 \cap X_2, g(x) \neq 0\}$.

### EXAMPLE 2  Domain of $f + g$

By solving the inequality $1 - x \geq 0$, it is seen that the domain of $f(x) = \sqrt{1 - x}$ is the interval $(-\infty, 1]$. Similarly, the domain of the function $g(x) = \sqrt{x + 2}$ is the interval $[-2, \infty)$. Hence, the domain of the sum

$$(f + g)(x) = f(x) + g(x) = \sqrt{1 - x} + \sqrt{x + 2}$$

is the intersection $(-\infty, 1] \cap [-2, \infty)$. You should verify by sketching these intervals on the number line that this intersection, or the set of numbers common to both domains, is the closed interval $[-2, 1]$. ∎

☐ **Composition of Functions** Another method of combining functions $f$ and $g$ is called **function composition**. To illustrate the idea, let's suppose that for a given $x$ in the domain of $g$ the function value $g(x)$ is a number in the domain of the function $f$. This means we are able to evaluate $f$ at $g(x)$, in other words, $f(g(x))$. For example, suppose $f(x) = x^2$ and $g(x) = x + 2$. Then for $x = 1$, $g(1) = 3$, and since 3 is the domain of $f$, we can write $f(g(1)) = f(3) = 3^2 = 9$. Indeed, for these two particular functions it turns out that we can evaluate $f$ at any function value $g(x)$, that is,

$$f(g(x)) = f(x + 2) = (x + 2)^2.$$

The resulting function, called the composition of $f$ and $g$, is defined next.

---

### FUNCTION COMPOSITION

If $f$ and $g$ are two functions, then the **composition** of $f$ and $g$, denoted by $f \circ g$, is the function defined by

$$(f \circ g)(x) = f(g(x)). \tag{5}$$

The **composition** of $g$ and $f$, denoted by $g \circ f$, is the function defined by

$$(g \circ f)(x) = g(f(x)). \tag{6}$$

---

When computing a composition such as $(f \circ g)(x) = f(g(x))$, be sure to substitute $g(x)$ for every $x$ that appears in $f(x)$. See part (a) of the next example.

### EXAMPLE 3        Two Compositions

If $f(x) = x^2 + 3x - 1$ and $g(x) = 2x^2 + 1$, find **(a)** $(f \circ g)(x)$ and **(b)** $(g \circ f)(x)$.

**Solution**

**(a)** For emphasis we replace $x$ by the set of parentheses ( ) and write $f$ in the form

$$f(x) = (\quad)^2 + 3(\quad) - 1.$$

Thus to evaluate $(f \circ g)(x)$ we fill each set of parentheses with $g(x)$. We find

$$
\begin{aligned}
(f \circ g)(x) = f(g(x)) &= f(2x^2 + 1) \\
&= (2x^2 + 1)^2 + 3(2x^2 + 1) - 1 \qquad \leftarrow \text{ use } (a + b)^2 = a^2 + 2ab + b^2 \\
& \qquad\qquad\qquad\qquad\qquad\qquad\qquad\quad \text{and the distributive law} \\
&= 4x^4 + 4x^2 + 1 + 3 \cdot 2x^2 + 3 \cdot 1 - 1 \\
&= 4x^4 + 10x^2 + 3.
\end{aligned}
$$

**(b)** In this case write $g$ in the form

$$g(x) = 2(\quad)^2 + 1.$$

Then

$$
\begin{aligned}
(g \circ f)(x) = g(f(x)) &= g(x^2 + 3x - 1) \\
&= 2(x^2 + 3x - 1)^2 + 1 \qquad \leftarrow \text{ use } (a + b + c)^2 = ((a + b) + c)^2 \\
&= 2(x^4 + 6x^3 + 7x^2 - 6x + 1) + 1 \qquad = (a + b)^2 + 2(a + b)c + c^2 \text{ etc.} \\
&= 2 \cdot x^4 + 2 \cdot 6x^3 + 2 \cdot 7x^2 - 2 \cdot 6x + 2 \cdot 1 + 1 \\
&= 2x^4 + 12x^3 + 14x^2 - 12x + 3.
\end{aligned}
$$
∎

Parts (a) and (b) of Example 1 illustrate that function composition is not commutative. That is, in general

$$f \circ g \neq g \circ f.$$

The next example shows that a function can be composed with itself.

### ▪ EXAMPLE 4    *f* Composed with *f*

If $f(x) = 5x - 1$, then the composition $f \circ f$ is given by

$$(f \circ f)(x) = f(f(x)) = f(5x - 1) = 5(5x - 1) - 1 = 25x - 6. \quad ▪$$

### ▪ EXAMPLE 5    Writing a Function as a Composition

Express $F(x) = \sqrt{6x^3 + 8}$ as the composition of two functions $f$ and $g$.

**Solution** If we define $f$ and $g$ as $f(x) = \sqrt{x}$ and $g(x) = 6x^3 + 8$, then

$$F(x) = (f \circ g)(x) = f(g(x)) = f(6x^3 + 8) = \sqrt{6x^3 + 8}. \quad ▪$$

There are other solutions to Example 6. For instance, if the functions $f$ and $g$ are defined by $f(x) = \sqrt{6x + 8}$ and $g(x) = x^3$, then observe $(f \circ g)(x) = f(x^3) = \sqrt{6x^3 + 8}$.

☐ **Domain of a Composition** As stated in the introductory example to this discussion, to evaluate the composition $(f \circ g)(x) = f(g(x))$ the number $g(x)$ must be in the domain of $f$. For example, the domain $f(x) = \sqrt{x}$ is $x \geq 0$ and the domain of $g(x) = x - 2$ is the set of real numbers $(-\infty, \infty)$. Observe that we cannot evaluate $f(g(1))$ because $g(1) = -1$ and $-1$ is not in the domain of $f$. In order to substitute $g(x)$ into $f(x)$, $g(x)$ must satisfy the inequality that defines the domain of $f$, namely, $g(x) \geq 0$. This last inequality is the same as $x - 2 \geq 0$ or $x \geq 2$. The domain of the composition $f(g(x)) = \sqrt{g(x)} = \sqrt{x - 2}$ is $[2, \infty)$, which is only a portion of the original domain $(-\infty, \infty)$ of $g$. In general,

Read this sentence several times. ▶

- the domain of the composition $f \circ g$ consists of the numbers $x$ in the domain of $g$ such that $g(x)$ is in the domain of $f$.

### ▪ EXAMPLE 6    Domain of a Composition

Consider the function $f(x) = \sqrt{x - 3}$. From the requirement that $x - 3 \geq 0$ we see that whatever number $x$ is substituted into $f$ must satisfy $x \geq 3$. Now suppose $g(x) = x^2 + 2$ and we want to evaluate $f(g(x))$. Although the domain of $g$ is the set of all real numbers, in order to substitute $g(x)$ into $f(x)$ we require that $x$ be a number in that domain so that $g(x) \geq 3$. From **FIGURE 2.6.1** we see that the last inequality is satisfied whenever $x \leq -1$ or $x \geq 1$. In other words, the domain of the composition

$$f(g(x)) = f(x^2 + 2) = \sqrt{(x^2 + 2) - 3} = \sqrt{x^2 - 1}$$

is $(-\infty, -1] \cup [1, \infty)$. ▪

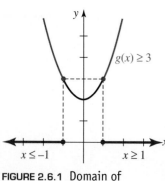

**FIGURE 2.6.1** Domain of $(f \circ g)(x)$ in Example 6

In certain applications a quantity $y$ is given as a function of a variable $x$ which in turn is a function of another variable $t$. By means of function composition we can express $y$

as a function of $t$. The next example illustrates the idea; the symbol $V$ plays the part of $y$ and $r$ plays the part of $x$.

## EXAMPLE 7     Inflating a Balloon

A weather balloon is being inflated with a gas. If the radius of the balloon is increasing at a rate of 5 cm/s, express the volume of the balloon as a function of time $t$ in seconds.

    **Solution** Let's assume that as the balloon is inflated, its shape is that of a sphere. If $r$ denotes the radius of the balloon, then $r(t) = 5t$. Since the volume of a sphere is $V = \frac{4}{3}\pi r^3$, the composition is $(V \circ r)(t) = V(r(t)) = V(5t)$ or

$$V = \frac{4}{3}\pi(5t)^3 = \frac{500}{3}\pi t^3.$$     ■

Weather balloon

☐ **Postscript** The rigid and nonrigid transformations that were studied in Section 2.2 are examples of the operations on functions just discussed. For $c > 0$ a constant, the rigid transformations defined by $y = f(x) + c$ and $y = f(x) - c$ are the *sum* and *difference* of the function $f(x)$ and the constant function $g(x) = c$. The nonrigid transformation $y = cf(x)$ is the *product* of $f(x)$ and the constant function $g(x) = c$. The rigid transformations defined by $y = f(x + c)$ and $y = f(x - c)$ are *compositions* of $f(x)$ with the linear functions $g(x) = x + c$ and $g(x) = x - c$, respectively.

---

**2.6**    **Exercises**    Answers to selected odd-numbered problems begin on page ANS-7.

In Problems 1–8, find the functions $f + g$, $f - g$, $fg$, and $f/g$, and give their domains.

1. $f(x) = x^2 + 1$,   $g(x) = 2x^2 - x$
2. $f(x) = x^2 - 4$,   $g(x) = x + 3$
3. $f(x) = x$,   $g(x) = \sqrt{x - 1}$
4. $f(x) = x - 2$,   $g(x) = \dfrac{1}{x + 8}$
5. $f(x) = 3x^3 - 4x^2 + 5x$,   $g(x) = (1 - x)^2$
6. $f(x) = \dfrac{4}{x - 6}$,   $g(x) = \dfrac{x}{x - 3}$
7. $f(x) = \sqrt{x + 2}$,   $g(x) = \sqrt{5 - 5x}$
8. $f(x) = \dfrac{1}{x^2 - 9}$,   $g(x) = \dfrac{\sqrt{x + 4}}{x}$

9. Fill in the table.

| $x$ | 0 | 1 | 2 | 3 | 4 |
|---|---|---|---|---|---|
| $f(x)$ | $-1$ | 2 | 10 | 8 | 0 |
| $g(x)$ | 2 | 3 | 0 | 1 | 4 |
| $(f \circ g)(x)$ | | | | | |

**10.** Fill in the table where $g$ is an odd function.

| $x$ | 0 | 1 | 2 | 3 | 4 |
|---|---|---|---|---|---|
| $f(x)$ | $-2$ | $-3$ | 0 | $-1$ | $-4$ |
| $g(x)$ | 9 | 7 | $-6$ | $-5$ | 13 |
| $(g \circ f)(x)$ | | | | | |

In Problems 11–14, find the functions $f \circ g$ and $g \circ f$ and give their domains.

**11.** $f(x) = x^2 + 1, \quad g(x) = \sqrt{x - 1}$
**12.** $f(x) = x^2 - x + 5, \quad g(x) = -x + 4$
**13.** $f(x) = \dfrac{1}{2x - 1}, \quad g(x) = x^2 + 1$
**14.** $f(x) = \dfrac{x + 1}{x}, \quad g(x) = \dfrac{1}{x}$

In Problems 15–20, find the functions $f \circ g$ and $g \circ f$.

**15.** $f(x) = 2x - 3, \quad g(x) = \frac{1}{2}(x + 3)$
**16.** $f(x) = x - 1, \quad g(x) = x^3$
**17.** $f(x) = x + \dfrac{1}{x^2}, \quad g(x) = \dfrac{1}{x}$
**18.** $f(x) = \sqrt{x - 4}, \quad g(x) = x^2$
**19.** $f(x) = x + 1, \quad g(x) = x + \sqrt{x - 1}$
**20.** $f(x) = x^3 - 4, \quad g(x) = \sqrt[3]{x + 3}$

In Problems 21–24, find $f \circ f$ and $f \circ (1/f)$.

**21.** $f(x) = 2x + 6$
**22.** $f(x) = x^2 + 1$
**23.** $f(x) = \dfrac{1}{x^2}$
**24.** $f(x) = \dfrac{x + 4}{x}$

In Problems 25 and 26, find $(f \circ g \circ h)(x) = f(g(h(x)))$.

**25.** $f(x) = \sqrt{x}, \quad g(x) = x^2, \quad h(x) = x - 1$
**26.** $f(x) = x^2, \quad g(x) = x^2 + 3x, \quad h(x) = 2x$
**27.** For the functions $f(x) = 2x + 7, g(x) = 3x^2$, find $(f \circ g \circ g)(x)$.
**28.** For the functions $f(x) = -x + 5, g(x) = -4x^2 + x$, find $(f \circ g \circ f)(x)$.

In Problems 29 and 30, find $(f \circ f \circ f)(x) = f(f(f(x)))$.

**29.** $f(x) = 2x - 5$
**30.** $f(x) = x^2 - 1$

In Problems 31–34, find functions $f$ and $g$ such that $F(x) = f \circ g$.

**31.** $F(x) = (x^2 - 4x)^5$
**32.** $F(x) = \sqrt{9x^2 + 16}$
**33.** $F(x) = (x - 3)^2 + 4\sqrt{x - 3}$
**34.** $F(x) = 1 + |2x + 9|$

In Problems 35 and 36, sketch the graphs of the compositions $f \circ g$ and $g \circ f$.

**35.** $f(x) = |x| - 2, \quad g(x) = |x - 2|$  **36.** $f(x) = [\![x - 1]\!], \quad g(x) = |x|$

**37.** Consider the function $y = f(x) + g(x)$, where $f(x) = x$ and $g(x) = -[\![x]\!]$. Fill in the blanks and then sketch the graph of the sum $f + g$ on the indicated intervals.

$$y = \begin{cases} \;\;\vdots \\ \underline{\hspace{2cm}}, & -3 \le x < -2 \\ \underline{\hspace{2cm}}, & -2 \le x < -1 \\ \underline{\hspace{2cm}}, & -1 \le x < 0 \\ \underline{\hspace{2cm}}, & 0 \le x < 1 \\ \underline{\hspace{2cm}}, & 1 \le x < 2 \\ \underline{\hspace{2cm}}, & 2 \le x < 3 \\ \;\;\vdots \end{cases}$$

**38.** Consider the function $y = f(x) + g(x)$, where $f(x) = |x|$ and $g(x) = [\![x]\!]$. Proceed as in Problem 37 and then sketch the graph of the sum $f + g$.

In Problems 39 and 40, sketch the graph of the sum $f + g$.

**39.** $f(x) = |x - 1|, \quad g(x) = |x|$   **40.** $f(x) = x, \quad g(x) = |x|$

In Problems 41 and 42, sketch the graph of the product $fg$.

**41.** $f(x) = x, \quad g(x) = |x|$   **42.** $f(x) = x, \quad g(x) = [\![x]\!]$

In Problems 43 and 44, sketch the graph of the reciprocal $1/f$.

**43.** $f(x) = |x|$   **44.** $f(x) = x - 3$

## Miscellaneous Calculus-Related Problems

In Problems 45 and 46,

**(a)** find the points of intersection of the graphs of the given functions,

**(b)** find the vertical distance $d$ between the graphs on the interval $I$ determined by the $x$-coordinates of their points of intersection,

**(c)** use the concept of a vertex of a parabola to find the maximum value of $d$ on the interval $I$.

**45.**

**46.**

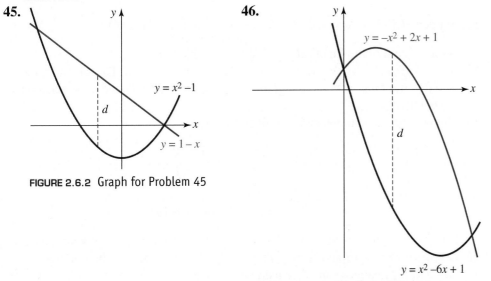

FIGURE 2.6.2 Graph for Problem 45

FIGURE 2.6.3 Graph for Problem 46

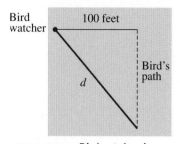

Bird watcher · 100 feet · Bird's path · d

**FIGURE 2.6.4** Birdwatcher in Problem 47

## Miscellaneous Applications

**47. For the Birds** A birdwatcher sights a bird 100 ft due east of her position. If the bird is flying due south at a rate of 500 ft/min, express the distance $d$ from the birdwatcher to the bird as a function of time $t$. Find the distance 5 minutes after the sighting. See **FIGURE 2.6.4**.

**48. Bacteria** A certain bacteria when cultured grows in a circular shape. The radius of the circle, measured in centimeters, is given by

$$r(t) = 4 - \frac{4}{t^2 + 1},$$

where time $t$ is measured in hours.
(a) Express the area covered by the bacteria as a function of time $t$.
(b) Express the circumference of the area covered as a function of time $t$.

## For Discussion

**49.** Suppose $f(x) = x^2 + 1$ and $g(x) = \sqrt{x}$. Discuss: Why is the domain of

$$(f \circ g)(x) = f(g(x)) = (\sqrt{x})^2 + 1 = x + 1$$

not $(-\infty, \infty)$?

**50.** Suppose $f(x) = \dfrac{2}{x - 1}$ and $g(x) = \dfrac{5}{x + 3}$. Discuss: Why is the domain of

$$(f \circ g)(x) = f(g(x)) = \frac{2}{g(x) - 1} = \frac{2}{\dfrac{5}{x + 3} - 1} = \frac{2x + 6}{2 - x}$$

not $\{x \mid x \neq 2\}$?

**51.** Find the error in the following reasoning: If $f(x) = 1/(x - 2)$ and $g(x) = 1/\sqrt{x + 1}$, then

$$\left(\frac{f}{g}\right)(x) = \frac{1/(x - 2)}{1/\sqrt{x + 1}} = \frac{\sqrt{x + 1}}{x - 2} \quad \text{and so} \quad \left(\frac{f}{g}\right)(-1) = \frac{\sqrt{0}}{-3} = 0.$$

**52.** Suppose $f_1(x) = \sqrt{x + 2}$, $f_2(x) = \dfrac{x}{\sqrt{x(x - 10)}}$, and $f_3(x) = \dfrac{x + 1}{x}$. What is the domain of the function $y = f_1(x) + f_2(x) + f_3(x)$?

**53.** Suppose $f(x) = x^3 + 4x$, $g(x) = x - 2$, and $h(x) = -x$. Discuss: Without actually graphing, how are the graphs of $f \circ g$, $g \circ f$, $f \circ h$, and $h \circ f$ related to the graph of $f$?

**54.** The domain of each piecewise-defined function,

$$f(x) = \begin{cases} x, & x < 0 \\ x + 1, & x \geq 0, \end{cases}$$

$$g(x) = \begin{cases} x^2, & x \leq -1 \\ x - 2, & x > -1, \end{cases}$$

is $(-\infty, \infty)$. Discuss how to find $f + g$, $f - g$, and $fg$. Carry out your ideas.

**55.** Discuss how the graph of $y = \frac{1}{2}\{f(x) + |f(x)|\}$ is related to the graph of $y = f(x)$. Illustrate your ideas using $f(x) = x^2 - 6x + 5$.

**56.** Discuss:

    **(a)** Is the sum of two even functions $f$ and $g$ even?

    **(b)** Is the sum of two odd functions $f$ and $g$ odd?

    **(c)** Is the product of an even function $f$ with an odd function $g$ even, odd, or neither?

    **(d)** Is the product of an odd function $f$ with an odd function $g$ even, odd, or neither?

**57.** The product $fg$ of two linear functions with real coefficients, $f(x) = ax + b$ and $g(x) = cx + d$, is a quadratic function. Discuss: Why must the graph of this quadratic function have at least one $x$-intercept?

**58.** Make up two different functions $f$ and $g$ so that the domain of $F(x) = f \circ g$ is $[-2, 0) \cup (0, 2]$.

---

## 2.7 Inverse Functions

☐ **Introduction** Recall that a function $f$ is a rule of correspondence that assigns to each value $x$ in its domain $X$, a single or unique value $y$ in its range. This rule does not preclude having the same number $y$ associated with several *different* values of $x$. For example, for $f(x) = x^2 + 1$, the value $y = 5$ occurs at either $x = -2$ or $x = 2$. On the other hand, for the function $g(x) = x^3$, the value $y = 64$ occurs only at $x = 4$. Indeed, for every value $y$ in the range of $g(x) = x^3$, there corresponds only one value of $x$ in the domain. Functions of this last kind are given a special name.

> ### ONE-TO-ONE FUNCTION
>
> A function $f$ is said to be **one-to-one** if each number in the range of $f$ is associated with exactly one number in its domain $X$.

☐ **Horizontal Line Test** Interpreted geometrically, this means that a horizontal line ($y = $ constant) can intersect the graph of a one-to-one function in at most one point. Furthermore, if *every* horizontal line that intersects the graph of a function does so in at most one point, then the function is necessarily one-to-one. A function is *not* one-to-one if *some* horizontal line intersects its graph more than once.

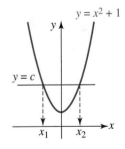

(a) Not one-to-one

### EXAMPLE 1      Horizontal Line Test

The graphs of the functions $f(x) = x^2 + 1$ and $g(x) = x^3$, and a horizontal line $y = c$ intersecting the graphs of $f$ and $g$, are shown in **FIGURE 2.7.1**. Figure 2.7.1(a) indicates that there are two numbers $x_1$ and $x_2$ in the domain of $f$ for which $f(x_1) = f(x_2) = c$. Inspection of Figure 2.7.1(b) shows that for every horizontal line $y = c$ intersecting the graph, there is only one number $x_1$ in the domain of $g$ such that $g(x_1) = c$. Hence the function $f$ is not one-to-one, whereas the function $g$ is one-to-one. ∎

A one-to-one function can be defined in several different ways. Based on the preceding discussion, the following statement should make sense.

>     *A function $f$ is* **one-to-one** *if and only if $f(x_1) = f(x_2)$ implies $x_1 = x_2$ for all $x_1$ and $x_2$ in the domain of $f$.*    (1)

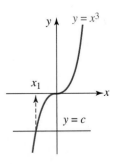

(b) One-to-one

**FIGURE 2.7.1** Two types of functions

Stated in a negative way, (1) indicates that a function $f$ is *not* one-to-one if different numbers $x_1$ and $x_2$ (that is, $x_1 \neq x_2$) can be found in the domain of $f$ such that $f(x_1) = f(x_2)$. You will see this formulation of the one-to-one concept when we solve certain kinds of equations in Chapter 5.

Consider (1) as a way of determining whether a function $f$ is one-to-one without the benefit of a graph.

■ EXAMPLE 2              **Checking for One-to-One**

(a) Consider the function $f(x) = x^4 - 8x + 6$. Now $0 \neq 2$ but observe that $f(0) = f(2) = 6$. Therefore $f$ is not one-to-one.

(b) Consider the function $f(x) = \dfrac{1}{2x - 3}$, and let $x_1$ and $x_2$ be numbers in the domain of $f$. If we assume $f(x_1) = f(x_2)$, that is, $\dfrac{1}{2x_1 - 3} = \dfrac{1}{2x_2 - 3}$, then by taking the reciprocal of both sides we see

$$2x_1 - 3 = 2x_2 - 3 \qquad \text{implies} \qquad 2x_1 = 2x_2 \qquad \text{or} \qquad x_1 = x_2.$$

We conclude from (1) that $f$ is one-to-one.  ∎

☐ **Inverse of a One-to-One Function**  Suppose $f$ is a one-to-one function with domain $X$ and range $Y$. Since every number $y$ in $Y$ corresponds to precisely one number $x$ in $X$, the function $f$ must actually determine a "reverse" function $f^{-1}$ whose domain is $Y$ and range is $X$. As shown in **FIGURE 2.7.2**, $f$ and $f^{-1}$ must satisfy

$$f(x) = y \qquad \text{and} \qquad f^{-1}(y) = x. \tag{2}$$

The equations in (2) are actually the compositions of the functions $f$ and $f^{-1}$:

$$f(f^{-1}(y)) = y \qquad \text{and} \qquad f^{-1}(f(x)) = x. \tag{3}$$

The function $f^{-1}$ is called the **inverse** of $f$ or the **inverse function** for $f$. Following convention that each domain element be denoted by the symbol $x$, the first equation in (3) is rewritten as $f(f^{-1}(x)) = x$. We summarize the results in (3).

Note of Caution: The symbol $f^{-1}$ does *not* ▶ mean the reciprocal $1/f$. The number $-1$ is **not** an exponent.

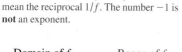
Domain of $f$      Range of $f$

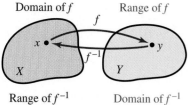
Range of $f^{-1}$      Domain of $f^{-1}$

**FIGURE 2.7.2** Functions $f$ and $f^{-1}$

---

**INVERSE FUNCTION**

Let $f$ be a one-to-one function with domain $X$ and range $Y$. The **inverse** of $f$ is the function $f^{-1}$ with domain $Y$ and range $X$ for which

$$f(f^{-1}(x)) = x \text{ for every } x \text{ in } Y, \tag{4}$$

and
$$f^{-1}(f(x)) = x \text{ for every } x \text{ in } X. \tag{5}$$

---

Of course, if a function $f$ is not one-to-one, then it has no inverse function.

☐ **Properties**  Before we actually examine methods for finding the inverse of a one-to-one function $f$, let's list some important properties about $f$ and its inverse $f^{-1}$.

---

**PROPERTIES OF INVERSE FUNCTIONS**

  (*i*) The domain of $f^{-1}$ = range of $f$.
 (*ii*) The range of $f^{-1}$ = domain of $f$.
(*iii*) $y = f(x)$ is equivalent to $x = f^{-1}(y)$.
(*iv*) An inverse function $f^{-1}$ is one-to-one.
 (*v*) The inverse of $f^{-1}$ is $f$.
(*vi*) The inverse of $f$ is unique

---

☐ **First Method for Finding $f^{-1}$** We will consider two ways of finding the inverse of a one-to-one function $f$. Both methods require that you solve an equation; the first method begins with (4).

### EXAMPLE 3      Inverse of a Function

**(a)** Find the inverse of $f(x) = \dfrac{1}{2x - 3}$. **(b)** Find the domain and range of $f^{-1}$. Find the range of $f$.

**Solution**

**(a)** We proved in part (b) of Example 2 that $f$ is one-to-one. To find the inverse of $f$ using (4), we must substitute $f^{-1}(x)$ wherever $x$ appears in $f$ and then set the expression $f(f^{-1}((x))$ equal to $x$:

$$\text{solve this equation for } f^{-1}(x)$$
$$\downarrow$$

$$f(f^{-1}(x)) = \boxed{\dfrac{1}{2f^{-1}(x) - 3} = x}$$

By taking the reciprocal of both sides of the equation in the outline box we get,

$$2f^{-1}(x) - 3 = \dfrac{1}{x}$$

$$2f^{-1}(x) = 3 + \dfrac{1}{x} = \dfrac{3x + 1}{x}. \quad \leftarrow \text{common denominator}$$

Dividing both sides of the last equation by 2 yields the inverse of $f$:

$$f^{-1}(x) = \dfrac{3x + 1}{2x}.$$

**(b)** Inspection of $f$ reveals that its domain is the set of real numbers except $\frac{3}{2}$, that is, $\{x \mid x \neq \frac{3}{2}\}$. Moreover, from the inverse just found we see that the domain of $f^{-1}$ is $\{x \mid x \neq 0\}$. Because range of $f^{-1} = $ domain of $f$ we then know that the range of $f^{-1}$ is $\{y \mid y \neq \frac{3}{2}\}$. From domain of $f^{-1} = $ range of $f$ we have also discovered that the range of $f$ is $\{y \mid y \neq 0\}$. ∎

☐ **Second Method for Finding $f^{-1}$** The inverse of a function $f$ can be found in a different manner. If $f^{-1}$ is the inverse of $f$, then $x = f^{-1}(y)$. Thus we need only do the following two things:

- Solve $y = f(x)$ for the symbol $x$ in terms of $y$ (if possible). This gives $x = f^{-1}(y)$.
- Relabel the variable $x$ as $y$ and the variable $y$ as $x$. This gives $y = f^{-1}(x)$.

### EXAMPLE 4      Inverse of a Function

Find the inverse of $f(x) = x^3$.

    **Solution** In Example 1 we saw that this function was one-to-one. To begin, we rewrite the function as $y = x^3$. Solving for $x$ then gives $x = y^{1/3}$. Next we relabel variables to obtain $y = x^{1/3}$. Thus $f^{-1}(x) = x^{1/3}$ or equivalently $f^{-1}(x) = \sqrt[3]{x}$. ∎

    Finding the inverse of a one-to-one function $y = f(x)$ is sometimes difficult and at times impossible. For example, it can be shown that the function $f(x) = x^3 + x + 3$ is one-to-one and so has an inverse $f^{-1}$, but solving the equation $y = x^3 + x + 3$ for $x$ is difficult for everyone (including your instructor). Nevertheless since $f$ only involves positive integer powers of $x$, its domain is $(-\infty, \infty)$. If you investigate $f$ graphically you are led to the fact that the range of $f$ is also $(-\infty, \infty)$. Consequently the domain and range of $f^{-1}$ are $(-\infty, \infty)$. Even though we don't know $f^{-1}$ explicitly it makes complete sense to talk about the values such as $f^{-1}(3)$ and $f^{-1}(5)$. In the case of $f^{-1}(3)$ note that $f(0) = 3$. This means that $f^{-1}(3) = 0$. Can you figure out the value of $f^{-1}(5)$?

☐ **Graphs of $f$ and $f^{-1}$** Suppose that $(a, b)$ represents any point on the graph of a one-to-one function $f$. Then $f(a) = b$ and

$$f^{-1}(b) = f^{-1}(f(a)) = a$$

implies that $(b, a)$ is a point on the graph of $f^{-1}$. As shown in **FIGURE 2.7.3(a)**, the points $(a, b)$ and $(b, a)$ are reflections of each other in line $y = x$. This means that the line $y = x$ is the perpendicular bisector of the line segment from $(a, b)$ to $(b, a)$. Because each point on one graph is the reflection of a corresponding point on the other graph, we see in Figure 2.7.3(b) that the graphs of $f^{-1}$ and $f$ are **reflections** of each other in the line $y = x$. We also say that the graphs of $f^{-1}$ and $f$ are symmetric with respect to the line $y = x$.

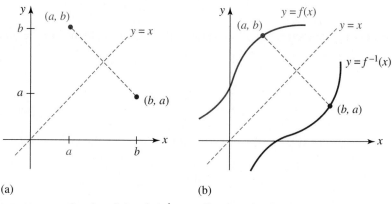

(a)                                              (b)

**FIGURE 2.7.3** Graphs of $f$ and $f^{-1}$ are reflections in the line $y = x$.

### EXAMPLE 5            Graphs of $f$ and $f^{-1}$

In Example 4 we saw that the inverse of $y = x^3$ is $y = x^{1/3}$. In **FIGURES 2.7.4(a)** and 2.7.4(b) we show the graphs of these functions; in Figure 2.7.4(c) the graphs are superimposed on the same coordinate system to illustrate that the graphs are reflections of each other in the line $y = x$.

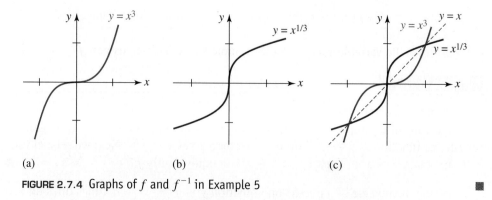

(a)                              (b)                              (c)

**FIGURE 2.7.4** Graphs of $f$ and $f^{-1}$ in Example 5

Every linear function $f(x) = ax + b$, $a \neq 0$, is one-to-one.

EXAMPLE 6                    **Inverse of a Function**

Find the inverse of the linear function $f(x) = 5x - 7$.

   **Solution** Since the graph of $y = 5x - 7$ is a nonhorizontal line, it follows from the horizontal line test that $f$ is a one-to-one function. To find $f^{-1}$ solve $y = 5x - 7$ for $x$:

$$5x = y + 7 \qquad \text{implies} \qquad x = \frac{1}{5}y + \frac{7}{5}.$$

Relabeling variables in the last equation gives $y = \frac{1}{5}x + \frac{7}{5}$. Therefore $f^{-1}(x) = \frac{1}{5}x + \frac{7}{5}$. The graphs of $f$ and $f^{-1}$ are compared in FIGURE 2.7.5. ∎

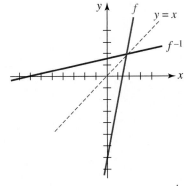

FIGURE 2.7.5 Graphs of $f$ and $f^{-1}$ in Example 6

   Every quadratic function $f(x) = ax^2 + bx + c,\ a \neq 0$, is not one-to-one.

☐ **Restricted Domains** For a function $f$ that is not one-to-one, it may be possible to restrict its domain in such a manner so that the new function consisting of $f$ defined on this restricted domain is one-to-one and so has an inverse. In most cases we want to restrict the domain so that the new function retains its original range. The next example illustrates this concept.

EXAMPLE 7                    **Restricted Domain**

In Example 1 we showed graphically that the quadratic function $f(x) = x^2 + 1$ is not one-to-one. The domain of $f$ is $(-\infty, \infty)$, and as seen in FIGURE 2.7.6(a), the range of $f$ is $[1, \infty)$. Now by defining $f(x) = x^2 + 1$ only on the interval $[0, \infty)$, we see two things in Figure 2.7.6(b): the range of $f$ is preserved and $f(x) = x^2 + 1$ confined to the domain $[0, \infty)$ passes the horizontal line test, in other words, is one-to-one. The inverse of this new one-to-one function is obtained in the usual manner. Solving $y = x^2 + 1$ implies

$$x^2 = y - 1 \qquad \text{and} \qquad x = \pm\sqrt{y - 1} \qquad \text{and so} \qquad y = \pm\sqrt{x - 1}.$$

The appropriate algebraic sign in the last equation is determined from the fact that the domain and range of $f^{-1}$ are $[1, \infty)$ and $[0, \infty)$, respectively. This forces us to choose $f^{-1}(x) = \sqrt{x - 1}$ as the inverse of $f$. See Figure 2.7.6(c). ∎

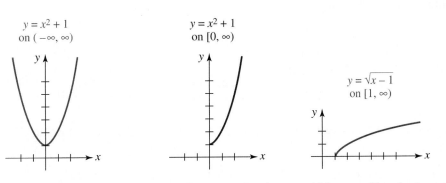

(a) Not a one-to-one function          (b) One-to-one function          (c) Inverse of function in part (b)

FIGURE 2.7.6 Inverse function in Example 7

In Problems 1–6, the graph of a function $f$ is given. Use the horizontal line test to determine whether $f$ is one-to-one.

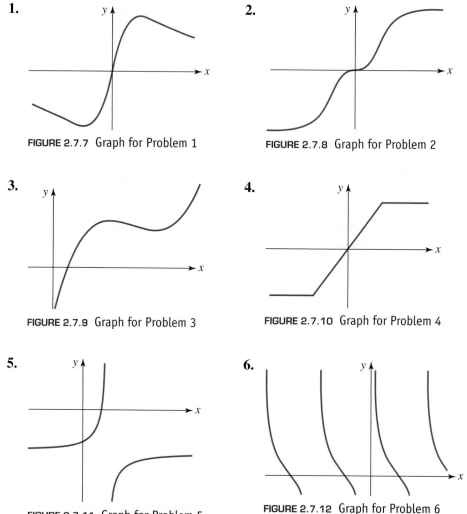

**1.**

FIGURE 2.7.7 Graph for Problem 1

**2.**

FIGURE 2.7.8 Graph for Problem 2

**3.**

FIGURE 2.7.9 Graph for Problem 3

**4.**

FIGURE 2.7.10 Graph for Problem 4

**5.**

FIGURE 2.7.11 Graph for Problem 5

**6.**

FIGURE 2.7.12 Graph for Problem 6

In Problems 7–10, sketch the graph of the given piecewise-defined function $f$ to determine whether it is one-to-one.

**7.** $f(x) = \begin{cases} x - 2, & x < 0 \\ \sqrt{x}, & x \geq 0 \end{cases}$

**8.** $f(x) = \begin{cases} -\sqrt{-x}, & x < 0 \\ \sqrt{x}, & x \geq 0 \end{cases}$

**9.** $f(x) = \begin{cases} -x - 1, & x < 0 \\ x^2, & x \geq 0 \end{cases}$

**10.** $f(x) = \begin{cases} x^2 + x, & x < 0 \\ x^2 - x, & x \geq 0 \end{cases}$

In Problems 11–14, proceed as in Example 2(a) to show that the given function $f$ is *not* one-to-one.

**11.** $f(x) = x^2 - 6x$

**12.** $f(x) = (x - 2)(x + 1)$

**13.** $f(x) = \dfrac{x^2}{4x^2 + 1}$

**14.** $f(x) = |x + 10|$

In Problems 15–18, proceed as in Example 2(b) to show that the given function $f$ is one-to-one.

**15.** $f(x) = \dfrac{2}{5x + 8}$

**16.** $f(x) = \dfrac{2x - 5}{x - 1}$

**17.** $f(x) = \sqrt{4 - x^2}$

**18.** $f(x) = \dfrac{1}{x^3 + 1}$

In Problems 19 and 20, the given function $f$ is one-to-one. Without finding $f^{-1}$ find its domain and range.

**19.** $f(x) = 4 + \sqrt{x}$

**20.** $f(x) = 5 - \sqrt{x + 8}$

In Problems 21 and 22, the given function $f$ is one-to-one. The domain and range of $f$ is given. Find $f^{-1}$ and give its domain and range.

**21.** $f(x) = \dfrac{2}{\sqrt{x}}, \quad x > 0, y > 0$

**22.** $f(x) = 2 + \dfrac{3}{\sqrt{x}}, \quad x > 0, y > 2$

In Problems 23–28, the given function $f$ is one-to-one. Find $f^{-1}$. Sketch the graph of $f$ and $f^{-1}$ on the same coordinate axes.

**23.** $f(x) = -2x + 6$

**24.** $f(x) = -2x + 1$

**25.** $f(x) = x^3 + 2$

**26.** $f(x) = 1 - x^3$

**27.** $f(x) = 2 - \sqrt{x}$

**28.** $f(x) = \sqrt{x - 7}$

In Problems 29–32, the given function $f$ is one-to-one. Find $f^{-1}$. Proceed as in Example 3(b) and find the domain and range of $f^{-1}$. Then find the range of $f$.

**29.** $f(x) = \dfrac{1}{2x - 1}$

**30.** $f(x) = \dfrac{2}{5x + 8}$

**31.** $f(x) = \dfrac{7x}{2x - 3}$

**32.** $f(x) = \dfrac{1 - x}{x - 2}$

In Problems 33–36, the given function $f$ is one-to-one. Without finding $f^{-1}$, find the point on the graph of $f^{-1}$ corresponding to the indicated value of $x$ in the domain of $f$.

**33.** $f(x) = 2x^3 + 2x; \quad x = 2$

**34.** $f(x) = 8x - 3; \quad x = 5$

**35.** $f(x) = x + \sqrt{x}; \quad x = 9$

**36.** $f(x) = \dfrac{4x}{x + 1}; \quad x = \dfrac{1}{2}$

In Problems 37 and 38, sketch the graph of $f^{-1}$ from the graph of $f$.

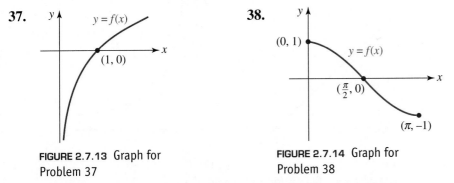

**37.**

**38.**

**FIGURE 2.7.13** Graph for Problem 37

**FIGURE 2.7.14** Graph for Problem 38

In Problems 39 and 40, sketch the graph of $f$ from the graph of $f^{-1}$.

**39.**

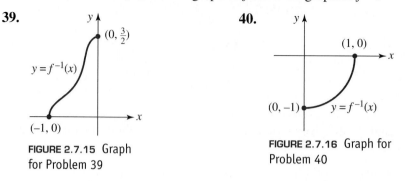

$y = f^{-1}(x)$

$(0, \frac{3}{2})$

$(-1, 0)$

**FIGURE 2.7.15** Graph for Problem 39

**40.**

$(1, 0)$

$(0, -1)$   $y = f^{-1}(x)$

**FIGURE 2.7.16** Graph for Problem 40

In Problems 41–44, the function $f$ is not one-to-one on the given domain but is one-to-one on the restricted domain (the second interval). Find the inverse of the one-to-one function and give its domain. Sketch the graph of $f$ on the restricted domain and the graph of $f^{-1}$ on the same coordinate axes.

**41.** $f(x) = 4x^2 + 2, (-\infty, \infty); \quad [0, \infty)$
**42.** $f(x) = (3 - 2x)^2, (-\infty, \infty); \quad [\frac{3}{2}, \infty)$
**43.** $f(x) = \frac{1}{2}\sqrt{4 - x^2}, [-2, 2]; \quad [0, 2]$
**44.** $f(x) = \sqrt{1 - x^2}, [-1, 1]; \quad [0, 1]$

In Problems 45 and 46, verify that $f(f^{-1}(x)) = x$ and $f^{-1}(f(x)) = x$.

**45.** $f(x) = 5x - 10, \quad f^{-1}(x) = \frac{1}{5}x + 2$
**46.** $f(x) = \dfrac{1}{x + 1}, \quad f^{-1}(x) = \dfrac{1 - x}{x}$

## Discussion Problems

**47.** Suppose $f$ is a continuous function that is increasing (or decreasing) for all $x$ in its domain. Explain why $f$ is necessarily one-to-one.
**48.** Explain why the graph of a one-to-one function $f$ can have at most one $x$-intercept.
**49.** The function $f(x) = |2x - 4|$ is not one-to-one. How should the domain of $f$ be restricted so that the new function has an inverse? Find $f^{-1}$ and give its domain and range. Sketch the graph of $f$ on the restricted domain and the graph of $f^{-1}$ on the same coordinate axes.
**50.** What property do the one-to-one functions $y = f(x)$ shown in **FIGURES 2.7.17(a)** and 2.7.17(b) have in common? Find two more explicit functions with this same property. Be very explicit about what this property has to do with $f^{-1}$.

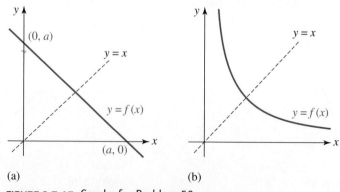

(a)

(b)

**FIGURE 2.7.17** Graphs for Problem 50

## 2.8 Translating Words into Functions

☐ **Introduction** In calculus there will be several instances when you will be expected to translate the words that describe a problem into mathematical symbols and then set up or construct either an *equation* or a *function*.

In this section we focus on problems that involve functions. We begin with a verbal description about the product of two numbers.

**EXAMPLE 1**    **Product of Two Numbers**

The sum of two nonnegative numbers is 15. Express the product of one and the square of the other as a function of one of the numbers.

**Solution** We first represent the two numbers by the symbols $x$ and $y$ and recall that "nonnegative" means that $x \geq 0$ and $y \geq 0$. The first sentence then says that $x + y = 15$; this is *not* the function we are seeking. The second sentence describes the function we want; it is called "the product." Let's denote "the product" by the symbol $P$. Now $P$ is the product of one of the numbers, say, $x$ and the square of the other, that is, $y^2$:

$$P = xy^2. \tag{1}$$

No, we are not finished because $P$ is supposed to be a "function of *one* of the numbers." We now use the fact that the numbers $x$ and $y$ are related by $x + y = 15$. From this last equation we substitute $y = 15 - x$ into (1) to obtain the desired result:

$$P(x) = x(15 - x)^2. \tag{2} \blacksquare$$

Here is a symbolic summary of the analysis of the problem given in Example 1:

$$x + y = 15$$

let the numbers be $x \geq 0$ and $y \geq 0$                    $P$

The sum of two nonnegative numbers is 15. Express the product of    (3)

$x$            $y^2$                    use $x$

one and the square of the other as a function of one of the numbers.

Notice that the second sentence is vague about which number is squared. This means that it really doesn't matter; (1) could also be written as $P = yx^2$. Also, we could have used $x = 15 - y$ in (1) to arrive at $P(y) = (15 - y)y^2$. In a calculus setting it would not have mattered whether we worked with $P(x)$ or with $P(y)$ because by finding *one* of the numbers we automatically find the other from the equation $x + y = 15$. This last equation is commonly called a **constraint**. A constraint not only defines the relationship between the variables $x$ and $y$ but often puts a limitation on how $x$ and $y$ can vary. As we see in the next example, the constraint helps in determining the domain of the function that you have just constructed.

**EXAMPLE 2**    **Example 1 Continued**

What is the domain of the function $P(x)$ in (2)?

**Solution** Taken out of the context of the statement of the problem in Example 1, one would have to conclude from the discussion on page 50 of Section 2.1 that the domain of

$$P(x) = x(15 - x)^2 = 225x - 30x^2 + x^3$$

is the set of real numbers $(-\infty, \infty)$. *But* in the context of the original problem, the numbers were to be nonnegative. From the requirement that $x \geq 0$ *and* $y = 15 - x \geq 0$ we get $x \geq 0$ and $x \leq 15$, which means that $x$ must satisfy the simultaneous inequality $0 \leq x \leq 15$. Using interval notation, the domain of the product function $P$ in (2) is the closed interval $[0, 15]$. ∎

Another way of looking at the conclusion of Example 2 is this: The constraint $x + y = 15$ dictates that $y = 15 - x$. Thus *if* $x$ were allowed to be larger than 15 (say, $x = 17.5$), then $y = 15 - x$ would be a negative number, which contradicts the initial assumption that $y \geq 0$.

☐ **Optimization Problems** For the remainder of this section we are going to examine "word problems" taken directly from a calculus text. These problems, variously called "optimization problems" or "applied maximum and minimum problems," consist of two parts, the "precalculus part" where you set up the function to be optimized and the "calculus part" where you perform calculus-specific operations on the function that you have just found to find its maximum or minimum value. The calculus part is usually identifiable by words such as "maximum (or minimum)," "least," "greatest," "large as possible," "find the dimensions," and so on. For example, the actual statement of Example 1 as it appears in a calculus text is:

> *Find two nonnegative numbers whose sum is* 15 *such that the product of one and the square of the other is a maximum.*

The big hurdle for many students is separating out the words that define the function to be optimized from the all words contained in the statement of the problem.

Before proceeding with the examples, you are encouraged to read the *Notes from the Classroom* at the end of this section.

The next example describes a geometric problem that asks for a "largest rectangle." Remember, you are not expected to work the entire problem by trying to actually find the "largest rectangle," which you would do in a calculus course. Right now your only job is to pick out the words, as we illustrated in (3), that tell you what the function is and then construct it using the variables introduced. In calculus, the function to be optimized is called the **objective function**.

◄ Please note.

**EXAMPLE 3**          **Largest Rectangle**

Find the objective function in the following calculus problem:

> *A rectangle has two vertices on the x-axis and two vertices on the semicircle whose equation is* $y = \sqrt{25 - x^2}$. *See* **FIGURE 2.8.1(a)**. *Find the dimensions of the largest rectangle.*

**Solution** In calculus the words "largest rectangle" mean that we are seeking *the* rectangle, of the many that can be drawn in the semicircle, that has the greatest or maximum *area*. Hence, the function we must construct is the area $A$ of the rectangle. If $(x, y)$, $x > 0$, $y > 0$, denotes the vertex of the rectangle on the circle in the first quadrant, then as shown in Figure 2.8.1(b) the area $A$ is length × width, or

$$A = (2x) \times y = 2xy. \tag{4}$$

The constraint in this problem is the equation $y = \sqrt{25 - x^2}$ of the semicircle. We use the constraint equation to eliminate $y$ in (4) and obtain the area of the rectangle or the objective function,

$$A(x) = 2x\sqrt{25 - x^2}. \tag{5}$$

This ends the "precalculus part" of the problem.

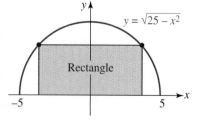

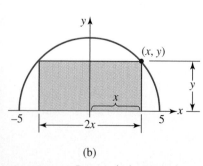

FIGURE 2.8.1 Rectangle in Example 3

The next step would be calculus procedures to determine the value of $x$ for which the objective function $A(x)$ takes on its largest value. ∎

Were we again to consider the function $A(x)$ out of the context of the problem in Example 3, its domain would be $[-5, 5]$. Because we assumed that $x > 0$ the domain of $A(x)$ in (4) is actually the open interval $(0, 5)$. But in calculus we would use the closed interval $[0, 5]$ even though when $x = 0$ and $x = 5$ the area would be $A(0) = 0$ and $A(5) = 0$, respectively. Do not worry about this last technicality.

## ■ EXAMPLE 4     Least Amount of Fencing

Find the objective function in the following calculus problem:

> *A rancher intends to mark off a rectangular plot of land that will have an area of 1000 m². The plot will be fenced and divided into two equal portions by an additional fence parallel to two sides. Find the dimensions of the land that require the least amount of fence.*

**Solution** Your drawing should be a rectangle with a line drawn down its middle, similar to that given in **FIGURE 2.8.2**. As shown in the figure, let $x > 0$ be the length of the rectangular plot of land and let $y > 0$ denote its width. The function we seek is the "amount of fence." If the symbol $F$ represents this amount, then the sum of the lengths of the *five* portions—two horizontal and three vertical—of the fence is

$$F = 2x + 3y. \qquad (6)$$

But the fenced-in land is to have an area of 1000 m², and so $x$ and $y$ must be related by the constraint $xy = 1000$. From the last equation we get $y = 1000/x$ which can be used to eliminate $y$ in (6). Thus, the amount of fence $F$ as a function of $x$ is $F(x) = 2x + 3(1000/x)$ or

$$F(x) = 2x + \frac{3000}{x}. \qquad (7)$$

Since $x$ represents a physical dimension that satisfies $xy = 1000$, we conclude that it is positive. But other than that, there is no restriction on $x$. Thus, unlike the previous example, the objective function (7) is not defined on a closed interval. The domain of $f(x)$ is $(0, \infty)$. ∎

As can be seen from the graph of (7) given in **FIGURE 2.8.3**, $F$ has a minimum at some value of $x$, say $x = c$. With a graphing calculator or computer we can approximate $c$ and $F(c)$, but with calculus we can find their exact values.

If a problem involves triangles, you should study the problem carefully and determine whether the Pythagorean Theorem, similar triangles, or trigonometry is applicable.

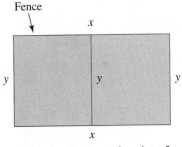

**FIGURE 2.8.2** Rectangular plot of land in Example 4

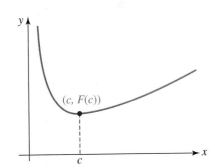

**FIGURE 2.8.3** In Example 4, $F(c)$ is the smallest value of $F$ for $x > 0$

## ■ EXAMPLE 5     Shortest Ladder

Find the objective function in the following calculus problem:

> *A 10-ft wall stands 5 ft from a building. Find the length of the shortest ladder, supported by the wall, that reaches from the ground to the building.*

**Solution** The words "shortest ladder" indicate that we want a function that describes the length of the ladder. Let $L$ denote this length. With $x$ and $y$ defined in **FIGURE 2.8.4**, we see that there are two right triangles, the larger triangle has three sides with lengths

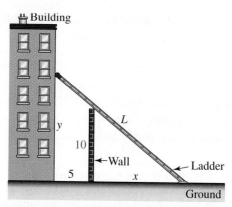

Building

$L$

$y$

$10$

←Wall

Ladder

$5$

$x$

Ground

**FIGURE 2.8.4** Ladder in Example 5

$L$, $y$, and $x + 5$, and the smaller triangle has two sides of lengths $x$ and 10. Now the ladder is the hypotenuse of the larger right triangle, so by the Pythagorean Theorem,

$$L^2 = (x + 5)^2 + y^2. \tag{8}$$

The right triangles in Figure 2.8.4 are similar because they both contain a right angle and share the common acute angle the ladder makes with the ground. We then use the fact that the ratios of corresponding sides of similar triangles are equal. This enables us to write

$$\frac{y}{x + 5} = \frac{10}{x} \qquad \text{so that} \qquad y = \frac{10(x + 5)}{x}.$$

Using the last result, (8) becomes

$$L^2 = (x + 5)^2 + \left(\frac{10(x + 5)}{x}\right)^2$$

$$= (x + 5)^2 \left(1 + \frac{100}{x^2}\right) \qquad \leftarrow \text{factoring } (x + 5)^2$$

$$= (x + 5)^2 \left(\frac{x^2 + 100}{x^2}\right) \qquad \leftarrow \text{common denominator}$$

Taking the square root gives us $L$ as a function of $x$,

$$L(x) = \frac{x + 5}{x}\sqrt{x^2 + 100}. \qquad \leftarrow \begin{array}{l} \text{square root of a product} \\ \text{is the product of the square roots} \end{array}$$

The domain of the objective function $L(x)$ is $(0, \infty)$. ∎

## EXAMPLE 6      Closest Point

Find the objective function in the following calculus problem:

> *Find the point in the first quadrant on the circle $x^2 + y^2 = 1$ that is closest to the point $(2, 4)$.*

**Solution** Let $(x, y)$ denote the point in the first quadrant on the circle closest to $(2, 4)$ and let $d$ represent the distance from $(x, y)$ to $(2, 4)$. See **FIGURE 2.8.5**. Then from the distance formula, (2) of Section 1.3,

$$d = \sqrt{(x - 2)^2 + (y - 4)^2} = \sqrt{x^2 + y^2 - 4x - 8y + 20}. \tag{9}$$

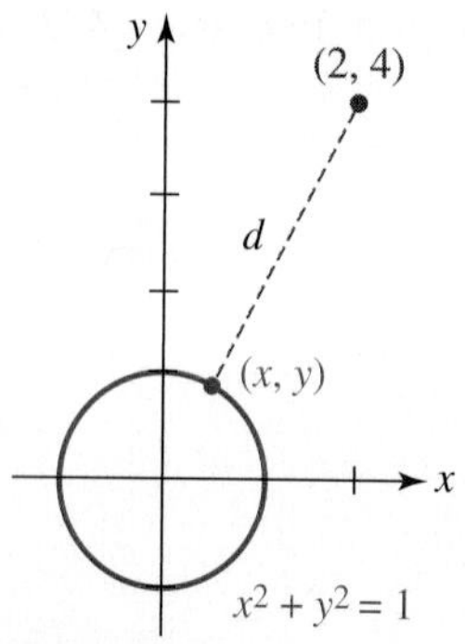

The constraint in this problem is the equation of the circle $x^2 + y^2 = 1$. From this we can immediately replace $x^2 + y^2$ in (9) by the number 1. Moreover, using the con-

$y$

$(2, 4)$

$d$

$(x, y)$

$x$

$x^2 + y^2 = 1$

**FIGURE 2.8.5** Distance $d$ in Example 6

CHAPTER 2 FUNCTIONS

straint to write $y = \sqrt{1 - x^2}$ allows us to eliminate $y$ in (9). Thus the distance $d$ as a function of $x$ is:

$$d(x) = \sqrt{21 - 4x - 8\sqrt{1 - x^2}}. \tag{10}$$

Since $(x, y)$ is a point on the circle in the first quadrant the variable $x$ can range from 0 to 1, that is, the domain of the objective function in (10) is the closed interval $[0, 1]$. ∎

## NOTES FROM THE CLASSROOM

When we get to the sections in a calculus text devoted to word problems students often react with groans, ambivalence, and dismay. While not guaranteeing anything, the following suggestions might help you to get through the problems in Exercises 2.8.

- At least try to develop a positive attitude. Try to be neat and organized.
- Read the problem slowly. Then read the problem several more times.
- Pay attention to words such as "maximum," "least," "greatest," and "closest" because they may provide a clue about the nature of the function you are seeking. For example, if a problem asks for "closest," then the function you are seeking most probably involves *distance*; if a problem asks for "least material," then the function you want may be *surface area*. See Problems 35 and 42 in Exercises 2.8.
- Whenever possible, sketch a curve or a picture and identify given quantities in your sketch. Keep your sketch simple.
- Introduce variables and note any constraint or relationship between the variables (such as $x + y = 15$ in Example 1).
- Identify the domain of the function just constructed. Keep in mind that if the problem mentions "dimensions" then the variables representing those quantities must be nonnegative.

**2.8** Exercises    Answers to selected odd-numbered problems begin on page ANS–8.

In Problems 1–26, proceed as in Example 1 and translate the words into an appropriate function. Give the domain of the function.

1. The product of two positive numbers is 50. Express their sum as a function of one of the numbers.
2. Express the sum of a nonzero number and its reciprocal as a function of the number.
3. The sum of two nonnegative numbers is 1. Express the sum of the square of one and twice the square of the other as a function of one of the numbers.
4. Let $m$ and $n$ be positive integers. The sum of two nonnegative numbers is $S$. Express the product of the $m$th power of one and the $n$th power of the other as a function of one of the numbers.
5. A rectangle has a perimeter of 200 in. Express the area of the rectangle as a function of the length of one of its sides.
6. A rectangle has an area of 400 in². Express the perimeter of the rectangle as a function of the length of one of its sides.
7. Express the area of the rectangle shaded in **FIGURE 2.8.6** as a function of $x$.

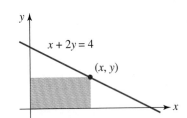

**FIGURE 2.8.6** Rectangle in Problem 7

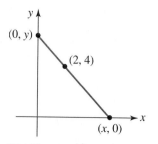

**FIGURE 2.8.7** Line segment in Problem 8

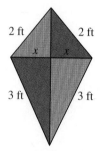

**FIGURE 2.8.8** Kite in Problem 20

**8.** Express the length of the line segment containing the point $(2, 4)$ shown in **FIGURE 2.8.7** as a function of $x$.

**9.** Express the distance from a point $(x, y)$ on the graph of $x + y = 1$ to the point $(2, 3)$ as a function of $x$.

**10.** Express the distance from a point $(x, y)$ on the graph of $y = 4 - x^2$ to the point $(0, 1)$ as a function of $x$.

**11.** Express the perimeter of a square as a function of its area $A$.

**12.** Express the area of a circle as a function of its diameter $d$.

**13.** Express the diameter of a circle as a function of its circumference $C$.

**14.** Express the volume of a cube as a function of the area $A$ of its base.

**15.** Express the area of an equilateral triangle as a function of its height $h$.

**16.** Express the area of an equilateral triangle as a function of the length $s$ of one of its sides.

**17.** A wire of length $x$ is bent into the shape of a circle. Express the area of the circle as a function of $x$.

**18.** A wire of length $L$ is cut $x$ units from one end. One piece of the wire is bent into a square and the other piece is bent into a circle. Express the sum of the areas as a function of $x$.

**19.** A tree is planted 30 ft from the base of a street lamp that is 25 ft tall. Express the length of the tree's shadow as a function of its height.

**20.** The frame of a kite consists of six pieces of lightweight plastic. The outer frame of the kite consists of four precut pieces, two pieces of length 2 ft, and two pieces of length 3 ft. Express the area of the kite as a function of $x$, where $2x$ is the length of the horizontal cross bar piece shown in **FIGURE 2.8.8**.

**21.** A company wants to construct an open rectangular box with a volume of 450 in³ so that the length of its base is 3 times its width. Express the surface area of the box as a function of the width.

**22.** A conical tank, with vertex down, has a radius of 5 ft and a height of 15 ft. Water is pumped into the tank. Express the volume of the water as a function of its depth. [*Hint*: The volume of a cone is $V = \frac{1}{3}\pi r^2 h$. Although the tank is a three-dimensional object, examine it in cross section as a two-dimensional triangle.]

**23.** Car $A$ passes point $O$ heading east at a constant rate of 40 mi/h; car $B$ passes the same point 1 hour later heading north at a constant rate of 60 mi/h. Express the distance between the cars as a function of time $t$, where $t$ is measured starting when car $B$ passes point $O$.

**24.** At time $t = 0$ (measured in hours), two airliners with a vertical separation of 1 mile pass each other going in opposite directions. If the planes are flying horizontally at rates of 500 mi/h and 550 mi/h, express the horizontal distance between them as a function of $t$. [*Hint*: distance = rate × time.]

**25.** The swimming pool shown in **FIGURE 2.8.9** is 3 ft deep at the shallow end, 8 ft deep at the deepest end, 40 ft long, 30 ft wide, and the bottom is an inclined plane. Water is pumped into the pool. Express the volume of the water in the pool as a function of height $h$ of the water above the deep end. [*Hint*: The volume will be a piecewise-defined function with domain $0 \le h \le 8$.]

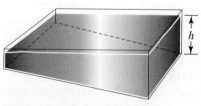

**FIGURE 2.8.9** Swimming pool in Problem 25

**26.** USPS regulations for parcel post stipulate that the length plus girth (the perimeter of one end) of a package must not exceed 108 inches. Express the volume of the package as a function of the width $x$ shown in FIGURE 2.8.10.

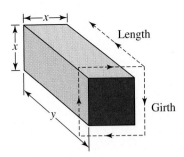

FIGURE 2.8.10 Package in Problem 26

In Problems 27–46, proceed as in Examples 3–5 and find the objective function for the given calculus problem. Give the domain of the objective function, but **do not** attempt to solve the problem.

**27.** Find a number that exceeds its square by the greatest amount.

**28.** Of all rectangles with perimeter 20 inches, find the one with the shortest diagonal.

**29.** A rectangular plot of land will be fenced into three equal portions by two dividing fences parallel to two sides. If the area to be enclosed is 4000 m$^2$, find the dimensions of the land that require the least amount of fence.

**30.** A rectangular plot of land will be fenced into three equal portions by two dividing fences parallel to two sides. If the total fence to be used is 8000 m, find the dimensions of the land that has the greatest area.

**31.** A rancher wishes to build a rectangular corral with an area of 128,000 ft$^2$ with one side along a straight river. The fencing along the river costs $1.50 per foot, whereas along the other three sides the fencing costs $2.50 per foot. Find the dimensions of the corral so that the cost of construction is a minimum. [*Hint*: Along the river the cost of $x$ ft of fence is $1.50x$.]

**32.** A rectangular yard is to be enclosed with a fence by attaching it to a house whose length is 40 feet. See FIGURE 2.8.11. The amount of fencing to be used is 160 feet. Find the dimensions of the yard so that the greatest area is enclosed.

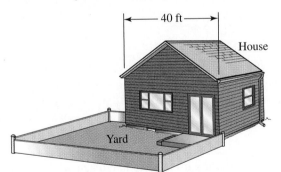

FIGURE 2.8.11 House and yard in Problem 32

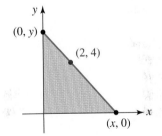

FIGURE 2.8.12 Line segment in Problem 34

**33.** Consider all rectangles that have the same perimeter $p$. (Here $p$ represents a constant.) Of these rectangles, show that the one with the largest area is a square.

**34.** Find the vertices $(x, 0)$ and $(0, y)$ of the shaded triangular region in FIGURE 2.8.12 so that its area is a minimum.

**35. (a)** An open rectangular box is to be constructed with a square base and a volume of 32,000 cm$^3$. Find the dimensions of the box that require the least amount of material. See FIGURE 2.8.13.

**(b)** If the rectangular box in part (a) is closed, find the dimensions that require the least amount of material.

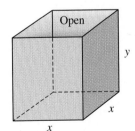

FIGURE 2.8.13 Box in Problem 35

**36.** A closed rectangular box is to be constructed with a square base. The material for the top costs $2 per square foot whereas the material for the remaining sides costs $1 per square foot. If the total cost to construct each box is $36, find the dimensions of the box of greatest volume that can be made.

**37.** A rain gutter with a rectangular cross section is made from a 1 ft $\times$ 20 ft piece of metal by bending up equal amounts from the 1-ft side. See FIGURE 2.8.14. How should the metal be bent up on each side in order to make the capacity of the gutter a maximum? [*Hint*: Capacity = volume.]

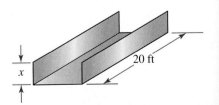

FIGURE 2.8.14 Rain gutter in Problem 37

**38.** A Norman window consists of a rectangle surmounted by a semicircle as shown in FIGURE 2.8.15. If the total perimeter of the window is 10 m, find the dimensions of the window with the largest area.

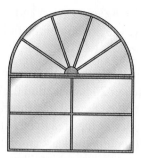

FIGURE 2.8.15 Norman window in Problem 38

**39.** A printed page will have 2-in. margins of white space on the sides and 1-in. margins of white space on the top and bottom. The area of the printed portion is 32 in². Determine the dimensions of the page so that the least amount of paper is used.

**40.** Find the dimensions of the right circular cylinder with greatest volume that can be inscribed in a right circular cone of radius 8 in. and height 12 in. See FIGURE 2.8.16.

**41.** Find the maximum length $L$ of a thin board that can be carried horizontally around the right-angle corner shown in FIGURE 2.8.17. [*Hint:* Use similar triangles.]

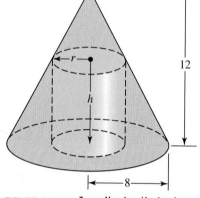

FIGURE 2.8.16 Inscribed cylinder in Problem 40

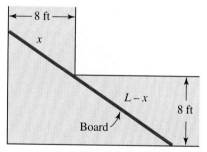

FIGURE 2.8.17 Board in Problem 41

**42.** A juice can is to be made in the form of a right circular cylinder and have a volume of 32 in³. See FIGURE 2.8.18. Find the dimensions of the can so that the least amount of material is used in its construction. [*Hint:* Material = total surface area of can = area of top + area of bottom + area of lateral side. If the circular top and bottom covers are removed and the cylinder is cut straight up its side and flattened out, the result is the rectangle shown in Figure 2.8.18(c).]

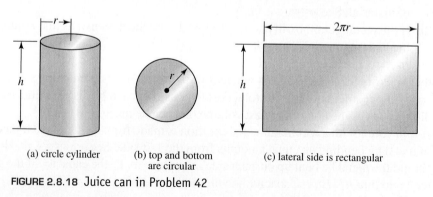

(a) circle cylinder        (b) top and bottom         (c) lateral side is rectangular
                               are circular

FIGURE 2.8.18 Juice can in Problem 42

**43.** The lateral side of a cylinder is to be made from a rectangle of flimsy sheet plastic. Because the plastic material cannot support itself, a thin, stiff wire is embedded in the material as shown in FIGURE 2.8.19(a). Find the dimensions of the cylinder of largest volume that can be constructed if the wire has a fixed length $L$. [*Hint*: There are two constraints in this problem. In Figure 2.8.19(b), the circumference of a circular end of the cylinder is $y$.]

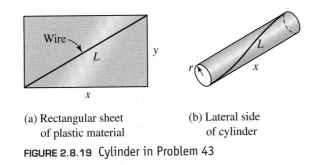

(a) Rectangular sheet     (b) Lateral side
of plastic material        of cylinder

FIGURE 2.8.19 Cylinder in Problem 43

**44.** Many medications are packaged in capsules as shown in the accompanying photo. Assume that a capsule is formed by adjoining two hemispheres to the ends of a right circular cylinder as shown in FIGURE 2.8.20. If the total volume of the capsule is to be 0.007 in.$^3$, find the dimensions of the capsule so that the least amount material is used in its construction. [*Hint*: The volume of a sphere is $\frac{4}{3}\pi r^3$ and its surface area is $4\pi r^2$.]

Capsule

Hemisphere

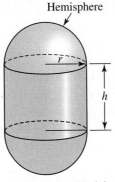

FIGURE 2.8.20 Model of
a capsule in Problem 44

**45.** A 20-ft long water trough has ends in the form of isosceles triangles with sides that are 4 ft long. See FIGURE 2.8.21. Determine the dimension across the top of the triangular end so that the volume of the trough is a maximum. [*Hint*: A *right cylinder* is not necessarily a *circular cylinder* where the top and bottom are circles. The top and bottom of a right cylinder are the same but could be a triangle, a pentagon, a trapezoid, and so on. The volume of a right cylinder is the area of the base $\times$ the height.]

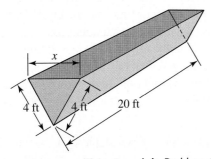

FIGURE 2.8.21 Water trough in Problem 45

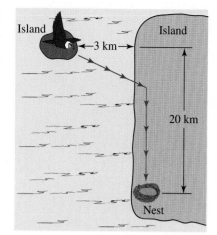

**FIGURE 2.8.22** The bird in Problem 46

**46.** Some birds fly more slowly over water than over land. A bird flies at constant rates 6 km/h over water and 10 km/h over land. Use the information in **FIGURE 2.8.22** to find the path the bird should take to minimize the total flying time between the shore of one island and its nest on the shore of another island. [*Hint*: distance = rate × time.]

### For Discussion

**47.** In Problem 19, what happens to the length of the tree's shadow as its height approaches 25 ft?

**48.** In an engineering text, the area of the octagon shown in **FIGURE 2.8.23** is given as $A = 3.31r^2$. Show that this formula is actually an approximation to the area; that is, find the exact area $A$ of the octagon as a function of $r$.

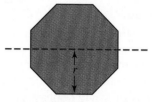

**FIGURE 2.8.23** Octagon in Problem 48

## 2.9 The Tangent Line Problem

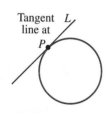

∫**Calculus PREVIEW**

☐ **Introduction** In a calculus course you will study many different things, but roughly, the subject "calculus" is divided into two broad but related areas known as **differential calculus** and **integral calculus**. The discussion of each of these topics invariably begins with a motivating problem involving the graph of a function. Differential calculus is motivated by the problem

*Find a tangent line to the graph of a function f,*

whereas integral calculus is motivated by the problem

*Find the area under the graph of a function f.*

The first problem will be addressed in this section; the second problem will be discussed in Section 3.7.

☐ **Tangent Line to a Graph** The word *tangent* stems from the Latin verb *tangere*, meaning "to touch." You might remember from the study of plane geometry that a tangent to a circle is a line $L$ that intersects, or touches, the circle in exactly one point $P$. See **FIGURE 2.9.1**. It is not quite as easy to define a tangent line to the graph of a function $f$. The idea of *touching* carries over to the notion of a tangent line to the graph of a function, but the idea of *intersecting the graph in one point* does not carry over.*

Suppose $y = f(x)$ is a continuous function. If, as shown in **FIGURE 2.9.2**, $f$ possesses a line $L$ tangent to its graph at a point $P$, then what is the equation of this line? To answer

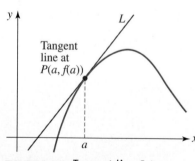

**FIGURE 2.9.1**
Tangent line $L$ touches a circle at point $P$

**FIGURE 2.9.2** Tangent line $L$ to a graph at point $P$

---

* We leave the discussion of the many subtleties and questions surrounding the tangent line problem to a course in calculus.

this question, we need the coordinates of $P$ and the slope $m_{\text{tan}}$ of $L$. The coordinates of $P$ pose no difficulty, since a point on the graph of a function $f$ is obtained by specifying a value of $x$ in the domain of $f$. The coordinates of the point of tangency at $x = a$ are then $(a, f(a))$. As a means of approximating the slope $m_{\text{tan}}$, we can readily find the slopes $m_{\text{sec}}$ of *secant lines* that pass through the point $P$ and any other point $Q$ on the graph. See FIGURE 2.9.3.

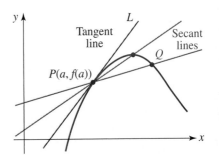

FIGURE 2.9.3 Slopes of secant lines approximate the slope $m_{\text{tan}}$ of $L$

☐ **Slope of Secant Lines** If $P$ has coordinates $(a, f(a))$ and if $Q$ has coordinates $(a + h, f(a + h))$, then as shown in FIGURE 2.9.4, the slope of the secant line through $P$ and $Q$ is

$$m_{\text{sec}} = \frac{\text{rise}}{\text{run}} = \frac{f(a + h) - f(a)}{(a + h) - a}$$

or

$$m_{\text{sec}} = \frac{f(a + h) - f(a)}{h}. \tag{1}$$

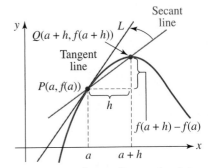

FIGURE 2.9.4 Secant lines swing into the tangent line $L$ as $h \to 0$

The expression on the right-hand side of the equality in (1) is called a **difference quotient**. When we let $h$ take on values that are closer and closer to zero, that is, as $h \to 0$, the sequence of points $Q(a + h, f(a + h))$ move along the curve closer and closer to the point $P(a, f(a))$. Intuitively, we expect the secant lines to approach the tangent line $L$, and that $m_{\text{sec}} \to m_{\text{tan}}$ as $h \to 0$. Using the idea of a limit introduced in Section 1.5 we write $m_{\text{tan}} = \lim\limits_{h \to 0} m_{\text{sec}}$, or equivalently from (1),

$$m_{\text{tan}} = \lim_{h \to 0} \frac{f(a + h) - f(a)}{h}, \tag{2}$$

provided the limit exists. Just like the problems discussed in Section 1.5, observe that the limit in (2) has the indeterminate form $0/0$ as $h \to 0$.

We are not going to delve into any theoretical details about when the limit (2) exists or does not exist—that discussion properly belongs in a calculus course. So to simplify the discussion, we will drop the phase "provided the limit exists." For this course it suffices simply to be aware of the fact that the limit (2) may not exist for certain values of $a$.

It is very likely that early on in your calculus course you will be asked to compute the limit of a difference quotient such as (2). The computation of (2) is essentially a *four-step process*, and three of these steps involve only precalculus mathematics: algebra and trigonometry. Getting over the hurdles of algebraic or trigonometric manipulations in these first three steps is your primary goal. If done accurately, the fourth step, or the calculus step, may be the easiest part of the problem. In preparation for calculus we recommend that you be able to carry out the calculation of (2) for functions involving

- positive integer powers of $x$ such as $x^n$ for $n = 1, 2,$ and 3,

◀ review $(a + b)^n$ for $n = 2$ and 3

- division of functions such as $\dfrac{1}{x}$ and $\dfrac{x}{x + 1}$, and

◀ review adding symbolic fractions

- radicals such as $\sqrt{x}$.

◀ review rationalization of numerators and denominators

See Problems 1–10 in Exercises 2.9.

### EXAMPLE 1      The Four-Step Process

Find the slope of the tangent line to the graph of $y = x^2 + 2$ at $x = 1$.

    **Solution** We first compute the difference quotient in (2) with the identification that $a = 1$.

(*i*) The initial step is the computation of $f(a + h)$. Because functions can be complicated, it might help in this step to think of $x$ wherever it appears in the function $f(x)$ as a set of parentheses ( ). For the given function we write $f(\ ) = (\ )^2 + 2$. The idea is to substitute $1 + h$ into those parentheses and carry out the required algebra:

$$\begin{aligned} f(1 + h) &= (1 + h)^2 + 2 \\ &= (1 + 2h + h^2) + 2 \\ &= 3 + 2h + h^2. \end{aligned}$$

(*ii*) The computation of the difference $f(a + h) - f(a)$ is the most important step. It is imperative that you simplify this step as much as possible. Here is a tip: In many of the problems that you will be required to do in calculus you will be able to factor $h$ from the difference $f(a + h) - f(a)$. To begin, compute $f(a)$, which in this case is $f(1) = 1^2 + 2 = 3$. Next, you can use the result from the preceding step:

$$\begin{aligned} f(1 + h) - f(1) &= 3 + 2h + h^2 - 3 \\ &= 2h + h^2 \\ &= h(2 + h). \qquad \leftarrow \text{ notice the factor of } h \end{aligned}$$

(*iii*) The computation of the difference quotient $\dfrac{f(a + h) - f(a)}{h}$ is now straightforward. Again, we use the results from the preceding step:

$$\frac{f(1 + h) - f(1)}{h} = \frac{h(2 + h)}{h} = 2 + h. \quad \leftarrow \text{ cancel the } h\text{'s}$$

(*iv*) The calculus step is now easy. From (2) we have

from the preceding line
$$\downarrow \qquad\qquad\qquad \downarrow$$
$$m_{\tan} = \lim_{h \to 0} \frac{f(1 + h) - f(1)}{h} = \lim_{h \to 0} (2 + h) = 2.$$

The slope of the tangent line to the graph of $y = x^2 + 2$ at $(1, 3)$ is 2. ∎

### EXAMPLE 2  Equation of Tangent Line

Find an equation of the tangent line whose slope was found in Example 1.

**Solution** We know a point $(1, 3)$ and a slope $m_{\tan} = 2$, and so from the point-slope equation of a line we find

$$y - 3 = 2(x - 1) \qquad \text{or} \qquad y = 2x + 1.$$

Observe that the last equation is consistent with the $x$- and $y$-intercepts of the red line in **FIGURE 2.9.5**. ∎

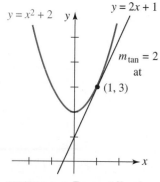

$y = x^2 + 2$  $y = 2x + 1$

$m_{\tan} = 2$ at $(1, 3)$

**FIGURE 2.9.5** Tangent line in Example 2

□ **The Derivative** As you inspect Figure 2.9.5, imagine tangent lines at various points on the graph of $f(x) = x^2 + 2$. This particular function is known to have a tangent line at every point on its graph. The tangent lines to the left of the origin have negative slope, the tangent line at $(0, 2)$ has zero slope, and the tangent lines to the right of the origin have positive slope (as seen in Example 1). In other words, for a function $f$ the value of $m_{\tan}$ at a point $(a, f(a))$ depends on the choice of the number $a$. Roughly speaking, there is at most *one* value of $m_{\tan}$ for each number $a$ in the domain of a function $f$. More specifically, $m_{\tan}$ is itself a *function* with a domain that is a subset of the domain of the function $f$. Furthermore, it is usually possible to obtain a formula for this *slope function*. This is accomplished by computing the limit of the difference quotient $\dfrac{f(x + h) - f(x)}{h}$ as $h \to 0$. We then substitute a value of $x$ *after* the limit has been found. The slope function derived in this manner from $f$ is said to be the derivative of $f$ and (instead of $m_{\text{tan}}$) is denoted by the symbol $f'$.

---

**THE DERIVATIVE**

The **derivative** of a function $y = f(x)$ is the function $f'$ defined by

$$f'(x) = \lim_{h \to 0} \frac{f(x + h) - f(x)}{h}. \qquad (3)$$

---

**EXAMPLE 3**  **Example 1 Revisited**

Compute the derivative of $f(x) = x^2 + 2$.

    **Solution**  We proceed exactly as in Example 1 except that we find $f(x + h)$ instead of $f(1 + h)$. In the first three steps we calculate the difference quotient; in steps (*ii*) and (*iii*) we use the results in the preceding step. In step (*iv*) we compute the limit of the difference quotient.

    (*i*) $f(x + h) = (x + h)^2 + 2 = x^2 + 2xh + h^2 + 2$

    (*ii*) $\begin{aligned}f(x + h) - f(x) &= x^2 + 2xh + h^2 + 2 - (x^2 + 2) \\ &= x^2 + 2xh + h^2 + 2 - x^2 - 2 \\ &= 2xh + h^2 \\ &= h(2x + h)\end{aligned}$

    (*iii*) $\dfrac{f(x + h) - f(x)}{h} = \dfrac{h(2x + h)}{h} = 2x + h$    ← cancel $h$'s

    (*iv*) From (3) the derivative of $f$ is the limit as $h \to 0$ of the result in (*iii*). During the process of shrinking $h$ smaller and smaller, $x$ is held fixed. Hence

$$f'(x) = \lim_{h \to 0}(2x + h) = 2x.$$

So now we have two functions; from $f(x) = x^2 + 2$ we have obtained the derivative $f'(x) = 2x$. When evaluated at a number $x$, the function $f$ gives the $y$-coordinate of a point on the graph and the derived function $f'$ gives the slope of the tangent line at that point. We have already seen in Example 1 that $f(1) = 3$ and $f'(1) = 2$. ∎

    With the aid of the derivative $f'(x) = 2x$ we can find slopes at other points on the graph of $f(x) = x^2 + 2$. For example,

at $x = 0$,   $\begin{cases} f(0) = 2 \\ f'(0) = 0 \end{cases}$    ← point of tangency is $(0, 2)$
                                 ← slope of tangent line at $(0, 2)$ is $m = 0$

at $x = -3$,   $\begin{cases} f(-3) = 11 \\ f'(-3) = -6 \end{cases}$    ← point of tangency is $(-3, 11)$
                                 ← slope of tangent line at $(-3, 11)$ is $m = -6$

The fact that $f'(0) = 0$ means that the tangent line is horizontal at $(0, 2)$.

**EXAMPLE 4**  **Derivative of a Function**

Compute the derivative of $f(x) = 2x^3 - 4x + 5$.

    **Solution**

    (*i*) The function is $f(\quad) = 2(\quad)^3 - 4(\quad) + 5$ and so

$$f(x + h) = 2(x + h)^3 - 4(x + h) + 5.$$

See (7) on page 36. ▶

The algebra here is a bit more complicated than in the previous example. We will use the binomial expansion for $(a + b)^3$ and the distributive law. Continuing,

$$f(x + h) = 2(x^3 + 3x^2h + 3xh^2 + h^3) - 4(x + h) + 5$$
$$= 2x^3 + 6x^2h + 6xh^2 + 2h^3 - 4x - 4h + 5 \qquad \leftarrow \begin{array}{l}\text{two applications}\\ \text{of the distributive law}\end{array}$$

(*ii*) As mentioned previously, in this step we are looking for a factor of $h$:

$$f(x + h) - f(x) = 2x^3 + 6x^2h + 6xh^2 + 2h^3 - 4x$$
$$- 4h + 5 - (2x^3 - 4x + 5)$$
$$= 2x^3 + 6x^2h + 6xh^2 + 2h^3 - 4x$$
$$- 4h + 5 - 2x^3 + 4x - 5 \qquad \leftarrow \text{terms in color add to 0}$$
$$= 6x^2h + 6xh^2 + 2h^3 - 4h$$
$$= h(6x^2 + 6xh + 2h^2 - 4) \qquad \leftarrow \text{factor out } h$$

(*iii*) We use the last result:

$$\frac{f(x + h) - f(x)}{h} = \frac{h(6x^2 + 6xh + 2h^2 - 4)}{h} \qquad \leftarrow \text{cancel } h\text{'s}$$
$$= 6x^2 + 6xh + 2h^2 - 4$$

(*iv*) From (3) and the preceding step the derivative of $f$ is

$$f'(x) = \lim_{h \to 0}(6x^2 + 6xh + 2h^2 - 4) = 6x^2 - 4. \qquad \blacksquare$$

### EXAMPLE 5        Equation of Tangent Line

Find an equation of the tangent line to the graph of $f(x) = 2/x$ at $x = 2$.

**Solution** We start by finding the derivative of $f$. In the second of the four steps we will have to combine two symbolic fractions by means of a common denominator.

(*i*) $f(x + h) = \dfrac{2}{x + h}$

(*ii*) $f(x + h) - f(x) = \dfrac{2}{x + h} - \dfrac{2}{x}$

$$= \frac{2}{x + h}\frac{x}{x} - \frac{2}{x}\frac{x + h}{x + h} \qquad \leftarrow \text{a common denominator is } x(x + h)$$
$$= \frac{2x - 2x - 2h}{x(x + h)} \qquad \leftarrow 2x - 2x = 0$$
$$= \frac{-2h}{x(x + h)} \qquad \leftarrow \text{there is the factor of } h$$

(*iii*) The last result is to be divided by $h$, or more precisely $\dfrac{h}{1}$. We invert and multiply by $\dfrac{1}{h}$:

$$\frac{f(x + h) - f(x)}{h} = \frac{\dfrac{-2h}{x(x + h)}}{\dfrac{h}{1}} = \frac{-2h}{x(x + h)}\frac{1}{h} = \frac{-2}{x(x + h)} \qquad \leftarrow \text{cancel } h\text{'s}$$

(*iv*) From (3) the derivative of $f$ is

$$f'(x) = \lim_{h \to 0}\frac{-2}{x(x + h)} = \frac{-2}{x^2}.$$

We are now in a position to find an equation of the tangent line at the point corresponding to $x = 2$. From $f(2) = 2/2 = 1$, we get the point of tangency $(2, 1)$. Then from the derivative $f'(x) = -2/x^2$ we see that $f'(2) = -2/4$, and so the slope of the tangent line at $(2, 1)$ is $-\frac{1}{2}$. From the point-slope equation of a line, the tangent line is

$$y - 1 = -\tfrac{1}{2}(x - 2) \qquad \text{or} \qquad y = -\tfrac{1}{2}x + 2.$$

The graph of $y = 2/x$ is the graph of $y = 1/x$ stretched vertically. (See Figure 2.2.1(e).) The tangent line at $(2, 1)$ is shown in red in **FIGURE 2.9.6**. ∎

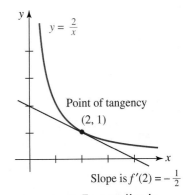

**FIGURE 2.9.6** Tangent line in Example 5

☐ **Postscript** There is an alternative definition of the derivative. If we let $x = a + h$ in (2), then $h = x - a$. Consequently the slope of secant line through $P(a, f(a))$ and $Q(x, f(x))$, as shown in **FIGURE 2.9.7**, is $\dfrac{f(x) - f(a)}{x - a}$. As $h \to 0$ we must have $x \to a$, and so the derivative (3) takes on the form

$$f'(a) = \lim_{x \to a} \frac{f(x) - f(a)}{x - a}. \tag{4}$$

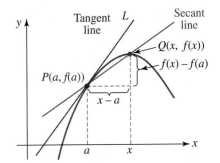

**FIGURE 2.9.7** Secant line and tangent line at $(a, f(a))$

◀ Review (1) and (2) of Section 1.5.

### EXAMPLE 6    Using (4)

Use (4) to compute the derivative of $f(x) = 4x^2 - 5x + 9$.

**Solution** We use the four-step process exactly as in Examples 3 and 4. The algebra is a slightly different; the analogue of the tip in (*ii*) of Example 1 is that we look for the factor $x - a$ in the difference $f(x) - f(a)$. Thus step (*ii*) will often require factoring the difference of two squares, the difference of two cubes, and so on.

(*i*) $f(a) = 4a^2 - 5a + 9$

(*ii*) $f(x) - f(a) = 4x^2 - 5x + 9 - (4a^2 - 5a + 9)$

$\qquad\qquad\qquad = 4x^2 - 5x + 9 - 4a^2 + 5a - 9$

$\qquad\qquad\qquad = 4x^2 - 5x - 4a^2 + 5a \qquad \leftarrow$ regroup terms in preparation for factoring

$\qquad\qquad\qquad = 4x^2 - 4a^2 - 5x + 5a$

$\qquad\qquad\qquad = 4(x^2 - a^2) - 5(x - a) \qquad \leftarrow$ first term is the difference of two squares

$\qquad\qquad\qquad = 4(x - a)(x + a) - 5(x - a) \qquad \leftarrow$ notice the factor of $x - a$

$\qquad\qquad\qquad = (x - a)[4(x + a) - 5]$

$\qquad\qquad\qquad = (x - a)(4x + 4a - 5)$

(*iii*) $\dfrac{f(x) - f(a)}{x - a} = \dfrac{(x - a)(4x + 4a - 5)}{x - a} \qquad \leftarrow$ cancel $x - a$

$\qquad\qquad\qquad = 4x + 4a - 5$

(*iv*) In the limit process indicated in (4), $a$ is held fixed. Hence

$$f'(a) = \lim_{x \to a}(4x + 4a - 5) = 8a - 5. \qquad \leftarrow \text{the limit of } 4x \text{ as } x \to a \text{ is } 4a \quad ∎$$

As you can see in (4) and the final line in Example 6, the derivative comes out a function of the symbol $a$ rather than $x$, that is, $f'(a) = 8a - 5$. As a consequence, (4) is not used as often as (3) to compute a derivative. See Problems 33–40 in Exercises 2.9. Nevertheless, (4) is important because it is convenient to use in some theoretical aspects of differential calculus.

In Problems 1–10, proceed as in Example 1.

(a) Compute the difference quotient $\dfrac{f(a + h) - f(a)}{h}$ at the given value of $a$.

(b) Then, if instructed, compute $m_{\tan} = \lim\limits_{h \to 0} \dfrac{f(a + h) - f(a)}{h}$.

(c) Use the result of part (b) to find an equation of the tangent line at the point of tangency.

1. $f(x) = x^2 - 6, a = 3$

2. $f(x) = -3x^2 + 10, a = -1$

3. $f(x) = x^2 - 3x, a = 1$

4. $f(x) = -x^2 + 5x - 3, a = -2$

5. $f(x) = -2x^3 + x, a = 2$

6. $f(x) = 8x^3 - 4, a = \frac{1}{2}$

7. $f(x) = \dfrac{1}{2x}, a = -1$

8. $f(x) = \dfrac{4}{x - 1}, a = 2$

9. $f(x) = \sqrt{x}, a = 4$

10. $f(x) = \dfrac{1}{\sqrt{x}}, a = 1$

In Problems 11–26, proceed as in Examples 3 and 4.

(a) Compute the difference quotient $\dfrac{f(x + h) - f(x)}{h}$ for the given function.

(b) Then, if instructed, compute the derivative $f'(x) = \lim\limits_{h \to 0} \dfrac{f(x + h) - f(x)}{h}$.

11. $f(x) = 10$

12. $f(x) = -3x + 8$

13. $f(x) = -4x^2$

14. $f(x) = x^2 - x$

15. $f(x) = 3x^2 - x + 7$

16. $f(x) = 2x^2 + x - 1$

17. $f(x) = x^3 + 5x - 4$

18. $f(x) = 2x^3 + x^2$

19. $f(x) = \dfrac{1}{4 - x}$

20. $f(x) = \dfrac{3}{2x - 4}$

21. $f(x) = \dfrac{x}{x - 1}$

22. $f(x) = \dfrac{2x + 3}{x + 5}$

23. $f(x) = x + \dfrac{1}{x}$

24. $f(x) = \dfrac{1}{x^2}$

25. $f(x) = 2\sqrt{x}$

26. $f(x) = \sqrt{2x + 1}$

In Problems 27–32, use the appropriate derivatives obtained in Problems 11–26. For the given function, find the point of tangency and slope of the tangent line at the indicated value of $x$. Find an equation of the tangent line at that point.

27. $f(x) = 3x^2 - x + 7, \quad x = 2$

28. $f(x) = x^2 - x, \quad x = 3$

29. $f(x) = x^3 + 5x - 4, \quad x = 1$

30. $f(x) = 2x^3 + x^2, \quad x = -\frac{1}{2}$

31. $f(x) = x + \dfrac{1}{x}, \quad x = \frac{1}{2}$

32. $f(x) = \dfrac{3}{2x - 4}; \quad x = -1$

In Problems 33–40, proceed as in Example 6.

(a) Compute the difference quotient $\dfrac{f(x) - f(a)}{x - a}$ for the given function.

(b) Then, if instructed, compute the derivative $f'(a) = \lim\limits_{x \to a} \dfrac{f(x) - f(a)}{x - a}$.

**33.** $f(x) = 3x^2 + 1$       **34.** $f(x) = x^2 - 8x - 3$

**35.** $f(x) = 10x^3$       **36.** $f(x) = x^4$

**37.** $f(x) = \dfrac{1}{x}$       **38.** $f(x) = \dfrac{3x - 1}{x}$

**39.** $f(x) = \sqrt{7x}$       **40.** $f(x) = -\sqrt{x + 9}$

## For Discussion

**41.** Use either (3) or (4) to compute the derivative of $f(x) = x^3 - 3x^2 - 9x$. Find the points on the graph of $f$ at which $f'(x) = 0$. Interpret your answers geometrically.

**42.** Use either (3) or (4) to compute the derivative of $f(x) = x^{1/3}$. [*Hint*: Recall from Section 1.5, $a^3 - b^3 = (a - b)(a^2 + ab + b^2)$.]

**43.** What is the tangent line to the graph of a linear function $f(x) = ax + b$?

**44.** If $f'(x) > 0$ for every $x$ in an interval, then what can be said about $f$ on the interval? If $f'(x) < 0$ for every $x$ in an interval, then what can be said about $f$ on the interval? [*Hint*: Draw a graph.]

**45.** If $f$ is an even function and if $(x, y)$ is on the graph of $f$, then $(-x, y)$ is also on the graph of $f$. How are the slopes of the tangent lines at $(x, y)$ and $(-x, y)$ related?

**46.** If $f$ is an odd function and if $(x, y)$ is on the graph of $f$, then $(-x, -y)$ is also on the graph of $f$. How are the slopes of the tangent lines at $(x, y)$ and $(-x, -y)$ related?

**47.** Consider the semicircle whose equation is $f(x) = \sqrt{1 - x^2}$. Discuss: How can the derivative $f'(x)$ be found using only the geometric fact that the radius of a circle is perpendicular to the tangent line at a point $(x, y)$ on the circle?

**48.** Consider the semicircle whose equation is $f(x) = \sqrt{1 - x^2}$. Use (3) to find the derivative $f'(x)$ and compare your result with that in Problem 47.

---

| **CHAPTER 2** | **Review Exercises** | Answers to selected odd-numbered problems begin on page ANS–8. |

In Problems 1–20, fill in the blanks.

**1.** If $f(x) = \dfrac{2x^3 - 1}{x^2 + 2}$, then $\left(\tfrac{1}{2}, \underline{\hspace{1cm}}\right)$ is a point on the graph of $f$.

**2.** If $f(x) = \dfrac{Ax}{10x - 2}$ and $f(2) = 3$, then $A = \underline{\hspace{1cm}}$.

**3.** The domain of the function $f(x) = \dfrac{1}{\sqrt{5 - x}}$ is $\underline{\hspace{1cm}}$.

**4.** The range of the function $f(x) = |x| - 10$ is _____.

**5.** The zeros of the function $f(x) = \sqrt{x^2 - 2x}$ are _____.
**6.** If the graph of $f$ is symmetric with respect to the $y$-axis, $f(-x) =$ _____.
**7.** The lines $2x - 5y = 1$ and $kx + 3y + 3 = 0$ are parallel if $k =$ _____.
**8.** The $x$- and $y$-intercepts of the line $-4x + 3y - 48 = 0$ are _____.
**9.** The graph of a linear function for which $f(-2) = 0$ and $f(0) = -3$ has slope $m =$ _____.
**10.** An equation of a line through $(1, 2)$ that is perpendicular to $y = 3x - 5$ is

_____.

**11.** The $x$- and $y$-intercepts of the parabola $f(x) = x^2 - 2x - 1$ are _____.
**12.** The range of the function $f(x) = -x^2 + 6x - 21$ is _____.
**13.** The quadratic function $f(x) = ax^2 + bx + c$ for which $f(0) = 7$ and whose only $x$-intercept is $(-2, 0)$ is $f(x) =$ _____.
**14.** If $f(x) = x + 2$ and $g(x) = x^2 - 2x$, then $(f \circ g)(-1) =$ _____.
**15.** The vertex of the graph of $f(x) = x^2$ is $(0, 0)$. Therefore, the vertex of the graph of $y = -5(x - 10)^2 + 2$ is _____.
**16.** Given that $f^{-1}(x) = \sqrt{x - 4}$ is the inverse of a one-to-one function $f$, and without finding $f$, the domain of $f$ is _____ and range of $f$ is _____.
**17.** The $x$-intercept of a one-to-one function $f$ is $(5, 0)$, and so the $y$-intercept of $f^{-1}$ is _____.

**18.** The inverse of $f(x) = \dfrac{x - 5}{2x + 1}$ is $f^{-1} =$ _____.

**19.** The point $(a, 16a)$ lies on the graph of

$$f(x) = \begin{cases} 4x - 3, & x < 0 \\ x^3, & 0 \le x \le 1 \\ x^2 + 64, & x > 1 \end{cases}$$

for $a =$ _____.
**20.** For $f(x) = [\![x + 2]\!] - 4$, $f(-5.3) =$ _____.

In Problems 21–40, answer true or false.

**21.** The points $(0, 3)$, $(2, 2)$, and $(6, 0)$ are collinear. _____
**22.** The graph of a function can have only one $y$-intercept. _____
**23.** If $f$ is a function such that $f(a) = f(b)$, then $a = b$. _____
**24.** No nonzero function $f$ can be symmetric with respect to the $x$-axis. _____
**25.** The domain of $f(x) = (x - 1)^{1/3}$ is $(-\infty, \infty)$. _____
**26.** If $f(x) = x$ and $g(x) = \sqrt{x + 2}$, then the domain of $g/f$ is $[-2, \infty)$. _____
**27.** A function $f$ is one-to-one if it never takes on the same value twice. _____
**28.** Two lines with positive slopes cannot be perpendicular. _____
**29.** The equation of a vertical line through $(2, -5)$ is $x = 2$. _____
**30.** A point of intersection of the graphs of $f$ and $f^{-1}$ must lie on the line $y = x$.

_____

**31.** The one-to-one function $f(x) = 1/x$ has the property that $f = f^{-1}$. _____
**32.** The function $f(x) = 2x^2 + 16x - 2$ decreases on the interval $[-7, -5]$.

_____

**33.** No even function can be one-to-one. _____
**34.** All odd functions are one-to-one. _____

**35.** If a function $f$ is one-to-one, then $f^{-1}(x) = \dfrac{1}{f(x)}$. _____

**36.** If $f$ is an increasing function on an interval containing $x_1 < x_2$, then
$f(x_1) < f(x_2)$._____

**37.** The function $f(x) = |x| - 1$ is decreasing on the interval $[0, \infty)$. _____

**38.** For function composition, $f \circ (g + h) = f \circ g + f \circ h$. _____

**39.** If the $y$-intercept for the graph of a function $f$ is $(0, 1)$, then the $y$-intercept for
the graph of $y = 4 - 3f(x)$ is $(0, 1)$. _____

**40.** For any function $f$, $f(x_1 + x_2) = f(x_1) + f(x_2)$. _____

In Problems 41 and 42 , identify two functions $f$ and $g$ so that $h = f \circ g$.

**41.** $h(x) = \dfrac{(3x - 5)^2}{x^2}$

**42.** $h(x) = 4(x + 1) - \sqrt{x + 1}$

**43.** Write the equation of each new function if the graph of $f(x) = x^3 - 2$ is
  **(a)** shifted to the left 3 units.
  **(b)** shifted down 5 units.
  **(c)** shifted to the right 1 unit and up 2 units.
  **(d)** reflected in the $x$-axis.
  **(e)** reflected in the $y$-axis.
  **(f)** vertically stretched by a factor of 3.

**44.** FIGURE 2.R.1 shows the graph of a function $f$ whose domain is $(-\infty, \infty)$. Sketch
the graph of the following functions.
  **(a)** $y = f(x) - \pi$
  **(b)** $y = f(x - 2)$
  **(c)** $y = f(x + 3) + \dfrac{\pi}{2}$
  **(d)** $y = -f(x)$
  **(e)** $y = f(-x)$
  **(f)** $y = 2f(x)$

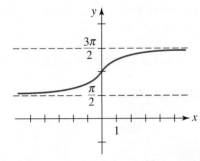

FIGURE 2.R.1 Graph for Problem 44

In Problems 45 and 46, use the graph of the one-to-one function $f$ in Figure 2.R.1.

**45.** Give the domain and range of $f^{-1}$.

**46.** Sketch the graph of $f^{-1}$.

**47.** Express $y = x - |x| + |x - 1|$ as a piecewise-defined function. Sketch the
graph of the function.

**48.** Sketch the graph of the function $y = [\![x]\!] + [\![-x]\!]$. Give the numbers at which
the function is discontinuous.

In Problems 49 and 50, by examining the graph of the function $f$ give the domain of
the function $g$.

**49.** $f(x) = x^2 - 6x + 10$, $\quad g(x) = \sqrt{x^2 - 6x + 10}$

**50.** $f(x) = -x^2 + 7x - 6$, $\quad g(x) = \dfrac{1}{\sqrt{-x^2 + 7x - 6}}$

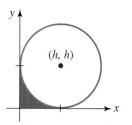

FIGURE 2.R.2 Circle in
Problem 53

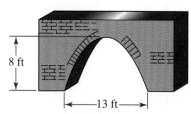

FIGURE 2.R.3 Arch in Problem 54

In Problems 51 and 52, the given function $f$ is one-to-one. Find $f^{-1}$.

**51.** $f(x) = (x + 1)^3$  **52.** $f(x) = x + \sqrt{x}$

**53.** Express the area of the shaded region in FIGURE 2.R.2 as a function of $h$.

**54.** Determine a quadratic function that describes the parabolic arch shown in FIGURE 2.R.3.

**55.** The diameter $d$ of a cube is the distance between opposite vertices as shown in FIGURE 2.R.4. Express the diameter $d$ as a function of the length $s$ of a side of the cube by first expressing the length $y$ of the diagonal in Figure 2.R.4 as a function of $s$.

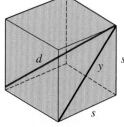

FIGURE 2.R.4 Cube in
Problem 55

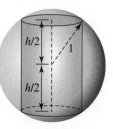

FIGURE 2.R.5
Inscribed cylinder in
Problem 56

**56.** A circular cylinder of height $h$ is inscribed in a sphere of radius 1 as shown in FIGURE 2.R.5. Express the volume of the cylinder as a function of $h$.

**57.** A baseball diamond is a square that is 90 ft on a side. See FIGURE 2.R.6. After a player hits a home run, he jogs around the bases at a rate of 6 ft/s.

    **(a)** As the player jogs between home base and first base, express his distance from home base as a function of time $t$, where $t = 0$ corresponds to the time he left home base—that is, $0 \le t \le 15$.

    **(b)** As the player jogs between home base and first base, express his distance from second base as a function of time $t$, where $0 \le t \le 15$.

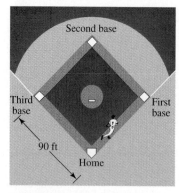

FIGURE 2.R.6 Baseball player in
Problem 57

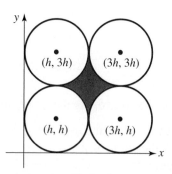

FIGURE 2.R.7 Circles in
Problem 58

**58.** Consider the four circles shown in FIGURE 2.R.7. Express the area of the shaded region between them as a function of $h$.

In Problems 59–62, find the objective function for the given calculus problem. Do *not* actually attempt to solve the problem.

**59.** Find the minimum value of the sum of 20 times a positive number and 5 times the reciprocal of that number.

**60.** A rancher wants to use 100 m of fence to construct a diagonal fence connecting two existing walls that meet at a right angle. How should this be done so that the area enclosed by the walls and fence is a maximum?

**61.** The running track shown as the black outline curve in FIGURE 2.R.8 is to consist of two parallel straight parts and two semicircular parts. The length of the track is to be 2 km. Find the design of the track so that the rectangular plot of land enclosed by the track is a maximum.

**62.** A pipeline is to be constructed from a refinery across a swamp to storage tanks. See FIGURE 2.R.9. The cost of construction is $25,000 per mile over the swamp and $20,000 per mile over land. How should the pipeline be made so that the cost of construction is a minimum?

FIGURE 2.R.8 Running track in Problem 61

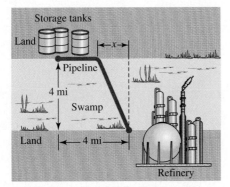

FIGURE 2.R.9 Pipeline in Problem 62

In Problems 63–66, compute $f'(x) = \lim\limits_{h \to 0} \dfrac{f(x + h) - f(x)}{h}$ for the given function. Find an equation of the tangent line at the indicated value of $x$.

**63.** $f(x) = -3x^2 + 16x + 12, \quad x = 2$

**64.** $f(x) = x^3 - x^2, \quad x = -1$

**65.** $f(x) = \dfrac{-1}{2x^2}, \quad x = \frac{1}{2}$

**66.** $f(x) = x + 4\sqrt{x}, \quad x = 4$

## Chapter Outline

# Polynomial and Rational Functions

**3**

## 3.1 Polynomial Functions

☐ **Introduction** In Chapter 2 we graphed functions such as $y = 3$, $y = 2x - 1$, $y = 5x^2 - 2x + 4$, and $y = x^3$. These functions, in which the variable $x$ is raised to a *nonnegative integer power*, are examples of a more general type of function called a **polynomial function**. Our goal in this section is to present some general guidelines for graphing such functions. First we state the formal definition of a polynomial function.

> ### POLYNOMIAL FUNCTION
>
> A **polynomial function** $y = f(x)$ is a function of the form
>
> $$f(x) = a_n x^n + a_{n-1} x^{n-1} + \cdots + a_2 x^2 + a_1 x + a_0, \qquad (1)$$
>
> where the coefficients $a_n, a_{n-1}, \ldots, a_2, a_1, a_0$ are real constants and $n$ is a nonnegative integer.

The **domain** of any polynomial function $f$ is the set of all real numbers $(-\infty, \infty)$.

The following functions are *not* polynomials:

not a nonnegative integer ↓        ↓ not a nonnegative integer

$$y = 5x^2 - 3x^{-1} \qquad \text{and} \qquad y = 2x^{1/2} - 4.$$

The function

nonnegative integer powers

$$y = 8x^5 - \tfrac{1}{2}x^4 - 10x^3 + 7x^2 + 6x + 4$$

is a polynomial, where we interpret the number 4 as the coefficient of $x^0$. Since 0 is a nonnegative integer, a constant function such as $y = 3$ is a polynomial function because it is the same as $y = 3x^0$.

☐ **Degree** Polynomial functions are classified by their degree. The highest power of $x$ in a polynomial is said to be its **degree**. So if $a_n \neq 0$, then we say that $f(x)$ in (1) has **degree** $n$. The number $a_n$ in (1) is called the **leading coefficient** and $a_0$ is called the **constant term** of the polynomial. For example,

degree ↓

$$f(x) = 3x^5 - 4x^3 - 3x + 8,$$

↑                  ↑<br>leading coefficient         constant term is a polynomial function of degree 5. We have already studied special polynomial functions in Sections 2.3 and 2.4. Polynomials of degrees $n = 0$, $n = 1$, and $n = 2$ are, respectively,

$$f(x) = a_0, \qquad \textbf{constant function}$$
$$f(x) = a_1 x + a_0, \qquad \textbf{linear function} \qquad \text{Section 2.3}$$
$$f(x) = a_2 x^2 + a_1 x + a_0, \qquad \textbf{quadratic function} \qquad \text{Section 2.4}$$

Polynomials of degrees $n = 3$, $n = 4$, and $n = 5$ are, in turn, commonly referred to as **cubic**, **quartic**, and **quintic functions**. The constant function $f(x) = 0$ is called the **zero polynomial**.

☐ **Graphs** Recall that the graph of a constant function $f(x) = a_0$ is a **horizontal line**, the graph of a linear function $f(x) = a_1 x + a_0$ is a **line with slope $m = a_1$**, and the graph of a quadratic function $f(x) = a_2 x^2 + a_1 x + a_0$ is a **parabola**. See Sections 2.3 and 2.4. Such descriptive statements cannot be made about the graph of a higher-degree polynomial function. What is the shape of the graph of a fifth-degree polynomial function? It turns out that the graph of a polynomial function of degree $n \geq 3$ can have several possible shapes. In general, graphing a polynomial function $f$ of degree $n \geq 3$ often demands the use of either calculus or a graphing utility. However, we will see in the discussion that follows that by determining

- shifting,
- end behavior,
- symmetry,
- intercepts, and
- local behavior

of the function we can, in some instances, quickly sketch a reasonable graph of a higher-degree polynomial function while keeping point-plotting to a minimum. Before elaborating on each of these concepts we return to the notion of a power function first introduced in Section 2.2.

☐ **Power Function** A special case of the power function (see Section 2.2) is the **single-term polynomial function** or **monomial**,

$$f(x) = x^n, \qquad n \text{ a positive integer.} \qquad (2)$$

The graphs of $f$ for degrees $n = 1, 2, 3, 4, 5,$ and $6$ are given in Figure 3.1.1. The interesting fact about (2) is that all the graphs for $n$ odd are basically the same. The notable characteristics are that the graphs are symmetric about the origin and become increasingly flatter near the origin as the degree $n$ increases. See FIGURES 3.1.1(a)–3.1.1(c). A similar observation is true for the graphs of (2) for $n$ even, except, of course, the graphs are symmetric with respect to the $y$-axis. See Figures 3.1.1(d)–3.1.1(f).

☐ **Shifted Graphs** Recall from Section 2.2 that for $c > 0$, the graphs of polynomial functions of the form

$$y = ax^n + c, \qquad y = ax^n - c$$
$$\text{and} \qquad y = a(x + c)^n, \qquad y = a(x - c)^n$$

can be obtained by vertical and horizontal shifts of the graph of $y = ax^n$. Also, if the leading coefficient $a$ is positive, the graph of $y = ax^n$ is either a vertical stretch or a vertical compression of the graph of the basic single-term polynomial function $f(x) = x^n$. When $a$ is negative we also carry out a reflection in the $x$-axis.

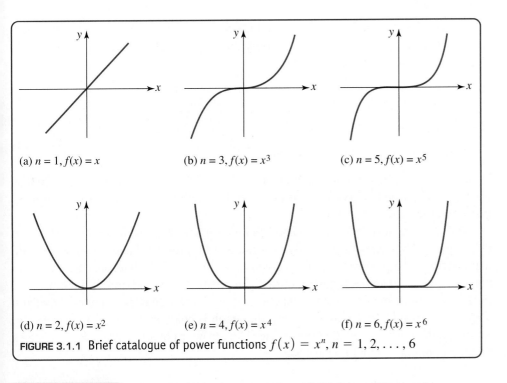

(a) $n = 1, f(x) = x$

(b) $n = 3, f(x) = x^3$

(c) $n = 5, f(x) = x^5$

(d) $n = 2, f(x) = x^2$

(e) $n = 4, f(x) = x^4$

(f) $n = 6, f(x) = x^6$

**FIGURE 3.1.1** Brief catalogue of power functions $f(x) = x^n$, $n = 1, 2, \ldots, 6$

## EXAMPLE 1　Graphing a Shifted Polynomial Function

The graph of $y = -(x + 2)^3 - 1$ is the graph of $f(x) = x^3$ reflected in the $x$-axis, shifted 2 units to the left, and then shifted vertically downward 1 unit. First review Figure 3.1.1(b) and then see **FIGURE 3.1.2**. ■

□ **End Behavior** The knowledge of the shape of a single-term polynomial function $f(x) = x^n$ is important for another reason. First, examine the computer-generated graphs given in **FIGURES 3.1.3** and 3.1.4. Although the graph in Figure 3.1.3 certainly resembles the graphs in Figures 3.1.1(b) and 3.1.1(c), and the graph in **FIGURE 3.1.4** resembles the graphs in Figure 3.1.1(d)–(f), the functions graphed in these two figures are *not* power functions $f(x) = x^n$, odd, or $f(x) = x^n$, even. We will not tell you at this point what the specific functions are except to say that they were both graphed on the interval $[-1000, 1000]$. The point is this: the function whose graph is given in Figure 3.1.3 could be almost *any* polynomial function

$$f(x) = a_n x^n \boxed{+ a_{n-1}x^{n-1} + \cdots + a_1 x + a_0} \qquad (3)$$

$a_n > 0$, of *odd* degree $n$, $n = 3, 5, \ldots$ when graphed on $[-1000, 1000]$. Similarly, the graph in Figure 3.1.4 could be that of any polynomial function given in (1), with $a_n > 0$, of *even* degree $n$, $n = 2, 4, \ldots$ when graphed on a large interval around the origin. As the next theorem indicates, the terms enclosed in the colored rectangle in (3) are irrelevant when we look at a graph of a polynomial globally—that is, for $|x|$ large. How a polynomial function $f$ behaves when $|x|$ is very large is said to be its **end behavior**.

## END BEHAVIOR

For $|x|$ very large, that is, for $x \to -\infty$ and $x \to \infty$, the graph of the polynomial function $f(x) = a_n x^n + a_{n-1}x^{n-1} + \cdots + a_2 x^2 + a_1 x + a_0$, resembles the graph of $y = a_n x^n$.

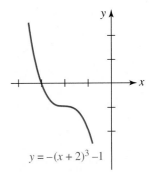

$y = -(x + 2)^3 - 1$

**FIGURE 3.1.2** Reflected and shifted graph in Example 1

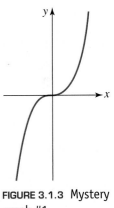

**FIGURE 3.1.3** Mystery graph #1

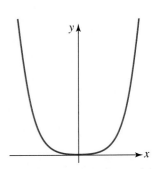

**FIGURE 3.1.4** Mystery graph #2

To see why the graph of a polynomial function such as $f(x) = -2x^3 + 4x^2 + 5$ resembles the graph of the single-term polynomial $y = -2x^3$ when $|x|$ is large, let's factor out the highest power of $x$, that is, $x^3$:

both these terms become
negligible when $|x|$ is large
↓    ↓

$$f(x) = x^3\left(-2 + \frac{4}{x} + \frac{5}{x^3}\right). \qquad (4)$$

By letting $|x|$ increase without bound, both $4/x$ and $5/x^3$ can be made as close to 0 as we want. Thus when $|x|$ is large, the values of the function $f$ in (4) are closely approximated by the values of $y = -2x^3$.

There can be only four types of end behavior for a polynomial function $f$. Although two of the end behaviors are already illustrated in Figures 3.1.3 and 3.1.4, we include them again in the pictorial summary given in Figure 3.1.5. To interpret the arrows in **FIGURE 3.1.5** let's examine Figure 3.1.5(a). The position and direction of the left arrow (left arrow points down) indicates that as $x \to -\infty$, the values $f(x)$ are negative and large in magnitude. Stated another way, the graph is heading downward as $x \to -\infty$. Similarly, the position and direction of the right arrow (right arrow points up) indicates that the graph is heading upward as $x \to \infty$.

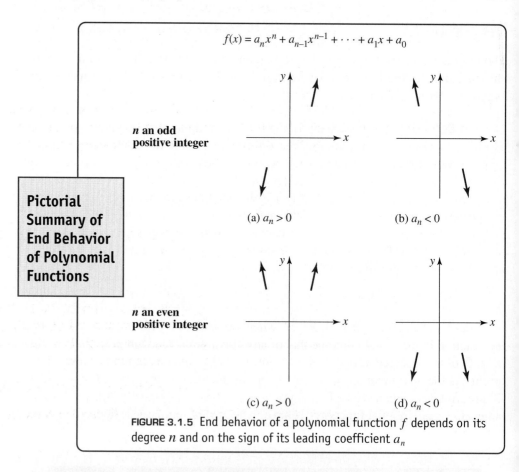

**Pictorial Summary of End Behavior of Polynomial Functions**

$$f(x) = a_n x^n + a_{n-1}x^{n-1} + \cdots + a_1 x + a_0$$

*n* an odd positive integer

(a) $a_n > 0$    (b) $a_n < 0$

*n* an even positive integer

(c) $a_n > 0$    (d) $a_n < 0$

**FIGURE 3.1.5** End behavior of a polynomial function $f$ depends on its degree $n$ and on the sign of its leading coefficient $a_n$

The gaps between the arrows in Figure 3.1.5 correspond to some interval around the origin. In these gaps the graph of $f$ exhibits **local behavior**, in other words, the graph of $f$ shows the characteristics of a polynomial function of a particular degree. This local behavior includes the $x$- and $y$-intercepts of the graph, the behavior of the graph at an $x$-intercept, the turning points of the graph, and observable symmetry of the graph (if any). A **turning point** is a point $(c, f(c))$ at which the graph of a polynomial function $f$ changes direction, that is, the function $f$ changes from increasing to decreasing or vice versa. The graph of a polynomial function of degree $n$ can have up to $n - 1$ turning points. In calculus a turning point is called a **relative**, or **local**, **extremum**. A relative extremum is classified as either a **maximum** or a **minimum**. If $(c, f(c))$ is a turning point or a relative extremum, then in some neighborhood of $x = c$, $f(c)$ is either the *largest* (relative maximum) or *smallest* (relative minimum) function value. At a relative maximum $(c, f(c))$ the graph of a polynomial function $f$ must change from increasing immediately to the left of $x = c$ to decreasing immediately to the right of $x = c$, whereas at a relative minimum $(c, f(c))$ the function $f$ changes from decreasing to increasing. These concepts will be illustrated in Example 2.

☐ **Symmetry** It is easy to tell by inspection those polynomial functions whose graphs possess symmetry with respect to either the $y$-axis or the origin. The words "even" and "odd" functions have special meaning for polynomial functions. Recall that an even function is one for which $f(-x) = f(x)$ and an odd function is one for which $f(-x) = -f(x)$. These two conditions hold for polynomial functions in which all the powers of $x$ are even integers and odd integers, respectively. For example,

even powers ⟶ ↓    ↓     odd powers ⟶ ↓   ↓   ↓    mixed powers ⟶ ↓   ↓   ↓   ↓

$$f(x) = 5x^4 - 7x^2 \qquad f(x) = 10x^5 + 7x^3 + 4x \qquad f(x) = -3x^7 + 2x^4 + x^3 + 2$$

     even function            odd function             neither even nor odd

A function such as $f(x) = 3x^6 - x^4 + 6$ is an even function because the obvious powers are even integers; the constant term 6 is actually $6x^0$, and 0 is an even nonnegative integer.

☐ **Intercepts** The graph of every polynomial function $f$ passes through the $y$-axis since $x = 0$ is the domain of the function. The $y$-intercept is the point $(0, f(0))$. Recall that a number $c$ is a **zero** of a function $f$ if $f(c) = 0$. In this discussion we assume $c$ is a real zero. If $x - c$ is a factor of a polynomial function $f$, that is, $f(x) = (x - c)q(x)$ where $q(x)$ is another polynomial, then clearly $f(c) = 0$ and the corresponding point on the graph is $(c, 0)$. Thus the real zeros of a polynomial function are the $x$-coordinates of the $x$-intercepts of its graph. If $(x - c)^m$ is a factor of $f$, where $m > 1$ is a positive integer, and $(x - c)^{m+1}$ is *not* a factor of $f$, then $c$ is said to be a **repeated zero**, or more precisely, a **zero of multiplicity $m$**. For example, $f(x) = x^2 - 10x + 25$ is equivalent to $f(x) = (x - 5)^2$. Hence 5 is a repeated zero or a zero of multiplicity 2. When $m = 1$, $c$ is called a **simple zero**. For example, $-\frac{1}{3}$ and $\frac{1}{2}$ are simple zeros of $f(x) = 6x^2 - x - 1$ since $f$ can be written as $f(x) = 6(x + \frac{1}{3})(x - \frac{1}{2})$. The behavior of the graph of $f$ at an $x$-intercept $(c, 0)$ depends on whether $c$ is a simple zero or a zero of multiplicity $m > 1$, where $m$ is either an even or an odd integer.

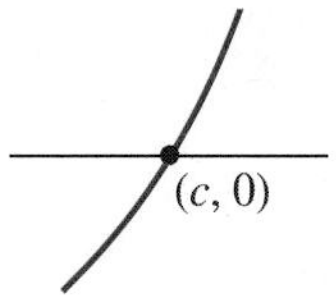

- If $c$ is a simple zero, then the graph of $f$ passes directly through the $x$-axis at $(c, 0)$. See **FIGURE 3.1.6(a)**.

(a) Simple zero

- If $c$ is a zero of odd multiplicity $m = 3, 5, \ldots$, then the graph of $f$ passes through the $x$-axis but is flattened at $(c, 0)$. See Figure 3.1.6(b).

(b) Zero of odd multiplicity $m = 3, 5, \ldots$

- If $c$ is a zero of even multiplicity $m = 2, 4, \ldots$, then the graph of $f$ is tangent to, or touches, the $x$-axis at $(c, 0)$. See Figure 3.1.6(c).

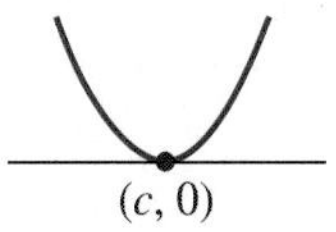

(c) Zero of even multiplicity $m = 2, 4, \ldots$

In the case when $c$ is either a simple zero or a zero of odd multiplicity $n = 3, 5, \ldots$, $f(x)$ changes sign *at* $(c, 0)$, whereas if $c$ is a zero of even multiplicity $n = 2, 4, \ldots$, $f(x)$ does not change sign at $(c, 0)$. We note that depending on the sign of the leading coefficient of the polynomial, the graphs in Figure 3.1.6 could be reflected

**FIGURE 3.1.6** $x$-intercepts of a polynomial function $f(x)$

in the *x*-axis. For example, at a zero of even multiplicity the graph of *f* could be tangent to the *x*-axis from below that axis.

### ■ EXAMPLE 2　　Graphing a Polynomial Function

Graph $f(x) = x^3 - 9x$.

**Solution**　Here are some of the things we look at to sketch the graph of *f*:

*End Behavior*:　By ignoring all terms but the first, we see that the graph of *f* resembles the graph of $y = x^3$ for large $|x|$. That is, the graph goes down to the left as $x \to -\infty$ and up to the right as $x \to \infty$, as illustrated in Figure 3.1.5(a).

*Symmetry*:　Since all the powers are odd integers, *f* is an odd function. The graph of *f* is symmetric with respect to the origin.

*Intercepts*:　$f(0) = 0$, and so the *y*-intercept is $(0, 0)$. Setting $f(x) = 0$, we see that we must solve $x^3 - 9x = 0$. Factoring

$$\overset{\text{difference of two squares}}{\overset{\downarrow}{x(x^2 - 9) = 0}} \quad \text{or} \quad x(x-3)(x+3) = 0$$

shows that the zeros of *f* are $x = 0$ and $x = \pm 3$. The *x*-intercepts are $(0, 0)$, $(-3, 0)$, and $(3, 0)$.

*The Graph*:　From left to right, the graph rises ( *f* is increasing) from the third quadrant and passes straight through $(-3, 0)$ since $-3$ is a simple zero. Although the graph is rising as it passes through this intercept it must turn back downward ( *f* decreasing) at some point in the second quadrant to get through the intercept $(0, 0)$. Since the graph is symmetric with respect to the origin, its behavior is just the opposite in the first and fourth quadrants. See **FIGURE 3.1.7**.　■

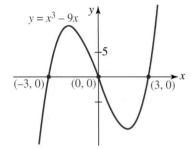

**FIGURE 3.1.7** Graph of function in Example 2

In Example 2, the graph of *f* has two turning points. On the interval $[-3, 0]$ there is a relative maximum and on the interval $[0, 3]$ there is a relative minimum. We made no attempt to locate these points precisely; this is something that would, in general, require techniques from calculus. The best we can do using precalculus mathematics to refine the graph is to resort to plotting additional points on the intervals of interest. By the way, $f(x) = x^3 - 9x$ is the function whose graph on the interval $[-1000, 1000]$ is given in Figure 3.1.3.

### ■ EXAMPLE 3　　Graphing a Polynomial Function

Graph $f(x) = (1 - x)(x + 1)^2$.

**Solution**　Multiplying out, *f* is the same as $f(x) = -x^3 - x^2 + x + 1$.

*End Behavior*:　From the preceding line we see that the graph of *f* resembles the graph of $y = -x^3$ for large $|x|$, just the opposite of the end behavior of the function in Example 2. See Figure 3.1.5(b).

*Symmetry*:　As we see from $f(x) = -x^3 - x^2 + x + 1$, there are both even and odd powers of *x* present. Hence *f* is neither even nor odd; its graph possesses no *y*-axis or origin symmetry.

*Intercepts*:　$f(0) = 1$ so the *y*-intercept is $(0, 1)$. From the given factored form of $f(x)$, we see that $(-1, 0)$ and $(1, 0)$ are the *x*-intercepts.

*The Graph*:　From left to right, the graph falls ( *f* decreasing) from the second quadrant and then, because $-1$ is a zero of multiplicity 2, the graph is tangent to the *x*-axis at $(-1, 0)$. The graph then rises ( *f* increasing) as it passes through the *y*-intercept $(0, 1)$. At some point within the interval $[0, 1]$ the graph turns downward ( *f* decreasing) and, since 1 is a simple zero, passes through the *x*-axis at $(1, 0)$, heading downward into the fourth quadrant. See **FIGURE 3.1.8**.　■

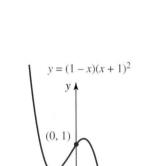
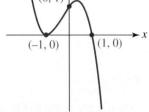

**FIGURE 3.1.8** Graph of function in Example 3

In Example 3, there are again two turning points. It should be clear that the point $(-1, 0)$ is a turning point ($f$ changes from decreasing to increasing at this point) and is a relative minimum of $f$. There is a turning point ($f$ changes from increasing to decreasing at this point) on the interval $[0, 1]$ and is a relative maximum of $f$.

## ■ EXAMPLE 4  Zeros of Multiplicity Two

Graph $f(x) = x^4 - 4x^2 + 4$.

**Solution** Before proceeding, note that the right-hand side of $f$ is a perfect square. That is, $f(x) = (x^2 - 2)^2$. Since $x^2 - 2 = (x - \sqrt{2})(x + \sqrt{2})$, by the laws of exponents we can write

$$f(x) = (x - \sqrt{2})^2(x + \sqrt{2})^2. \tag{5}$$

*End Behavior*: Inspection of $f(x)$ shows that its graph resembles the graph of $y = x^4$ for large $|x|$. That is, the graph goes up to the left as $x \to -\infty$ and up to the right as $x \to \infty$, as shown in Figure 3.1.5(c).

*Symmetry*: Because $f(x)$ contains only even powers of $x$, it is an even function and so its graph is symmetric with respect to the $y$-axis.

*Intercepts*: $f(0) = 4$, so the $y$-intercept is $(0, 4)$. Inspection of (5) shows the $x$-intercepts are $(-\sqrt{2}, 0)$ and $(\sqrt{2}, 0)$.

*The Graph*: From left to right, the graph falls from the second quadrant and then, because $-\sqrt{2}$ is a zero of multiplicity 2, the graph touches the $x$-axis at $(-\sqrt{2}, 0)$. The graph then rises from this point to the $y$-intercept $(0, 4)$. We then use the $y$-axis symmetry to finish the graph in the first quadrant. See **FIGURE 3.1.9**. ■

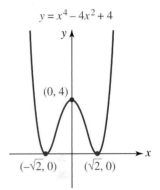

**FIGURE 3.1.9** Graph of function in Example 4

In Example 4, the graph of $f$ has three turning points. From the even multiplicity of the zeros, along with the $y$-axis symmetry, it can be deduced that the $x$-intercepts $(-\sqrt{2}, 0)$ and $(\sqrt{2}, 0)$ are turning points and are relative minima, and that the $y$-intercept $(0, 4)$ is a turning point and is a relative maximum.

## ■ EXAMPLE 5  Zero of Multiplicity Three

Graph $f(x) = -(x + 4)(x - 2)^3$.

**Solution**

*End Behavior*: Inspection of $f$ shows that its graph resembles the graph of $y = -x^4$ for large $|x|$. This end behavior of $f$ is shown in Figure 3.1.5(d).

*Symmetry*: The function $f$ is neither even nor odd. It is straightforward to show that $f(-x) \neq f(x)$ and $f(-x) \neq -f(x)$.

*Intercepts*: $f(0) = (-4)(-2)^3 = 32$, so the $y$-intercept is $(0, 32)$. From the factored form of $f(x)$, we see that $(-4, 0)$ and $(2, 0)$ are the $x$-intercepts.

*The Graph*: From left to right, the graph rises from the third quadrant and then, because $-4$ is a simple zero, the graph of $f$ passes directly through the $x$-axis at $(-4, 0)$. Somewhere within the interval $[-4, 0]$ the function $f$ must change from increasing to decreasing to enable its graph to pass through the $y$-intercept $(0, 32)$. After its graph passes through the $y$-intercept, the function $f$ continues to decrease but, since 2 is a zero of order three, its graph flattens as it passes through $(2, 0)$, heading downward into the fourth quadrant. See **FIGURE 3.1.10**. ■

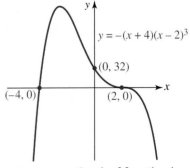

**FIGURE 3.1.10** Graph of function in Example 5

Note in Example 5 that since $f$ is of degree 4, its graph could have up to three turning points. But as can be seen from Figure 3.1.10, the graph of $f$ possesses only one turning point and this point is a relative maximum.

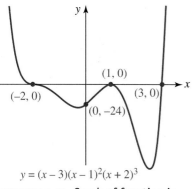

$y = (x - 3)(x - 1)^2(x + 2)^3$

**FIGURE 3.1.11** Graph of function in Example 6

EXAMPLE 6     **Zeros of Multiplicity Two and Three**

Graph $f(x) = (x - 3)(x - 1)^2(x + 2)^3$.

**Solution** The function $f$ is of degree 6 and so its end behavior resembles the graph of $y = x^6$ for large $|x|$. See Figure 3.1.5(c). Also, the function $f$ is neither even nor odd; its graph possesses no $y$-axis or origin symmetry. The $y$-intercept is $(0, f(0)) = (0, -24)$. From the factors of $f$ we see that $x$-intercepts of the graph are $(-2, 0)$, $(1, 0)$, and $(3, 0)$. Since $-2$ is a zero of multiplicity 3, the graph of $f$ is flattened as it passes through $(-2, 0)$. Since 1 is a zero of multiplicity 2, the graph of $f$ is tangent to the $x$-axis at $(1, 0)$. Since 3 is a simple zero, the graph of $f$ passes directly through the $x$-axis at $(3, 0)$. Putting all these facts together we obtain the graph in **FIGURE 3.1.11**. ∎

In Example 6, since the function $f$ is of degree 6 its graph could have up to five turning points. But as the graph in Figure 3.1.11 shows, there are only three turning points. Two of these points are relative minima and the remaining point, which is $(1, 0)$, is a relative maximum.

☐ **Intermediate Value Theorem** A polynomial function $f$ is a continuous function. Recall that this means that the graph of $y = f(x)$ has no breaks, gaps, or holes in it. The following result is a direct consequence of continuity.

### INTERMEDIATE VALUE THEOREM

Suppose $y = f(x)$ is a continuous function on the closed interval $[a, b]$. If $f(a) \neq f(b)$ for $a < b$, and if $N$ is any number between $f(a)$ and $f(b)$, then there exists a number $c$ in the open interval $(a, b)$ for which $f(c) = N$.

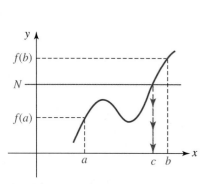

**FIGURE 3.1.12** $f(x)$ takes on all values between $f(a)$ and $f(b)$

As we see in **FIGURE 3.1.12**, the Intermediate Value Theorem simply states that $f(x)$ takes on all values between the numbers $f(a)$ and $f(b)$. In particular, if the function values $f(a)$ and $f(b)$ have opposite signs, then by identifying $N = 0$, we can say that there is at least one number in $(a, b)$ for which $f(c) = 0$. In other words, if either $f(a) > 0, f(b) < 0$ or $f(a) < 0, f(b) > 0$, then $f(x)$ has at least one zero $c$ in the interval $(a, b)$. The plausibility of this conclusion is illustrated in **FIGURE 3.1.13**.

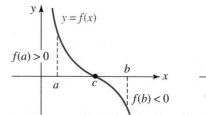

(a) One zero $c$ in $(a, b)$

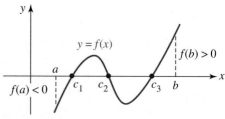

(b) Three zeros $c_1, c_2, c_3$ in $(a, b)$

**FIGURE 3.1.13** Locating zeros using the Intermediate Value Theorem

EXAMPLE 7     **Using the Intermediate Value Theorem**

Consider the function $f(x) = x^3 + x - 1$. Since $f$ is continuous on the interval $[-1, 1]$,

$$f(-1) = -3 < 0 \quad \text{and} \quad f(1) = 1 > 0,$$

and 0 satisfies $-3 < 0 < 1$, we know from the Intermediate Value Theorem that the graph of $f$ must cross the line $y = 0$ (the $x$-axis). More precisely, there is a real number $c$ in the open interval $(-1, 1)$ such that $f(c) = 0$. ∎

In Problems 1–8, proceed as in Example 1 and use transformations to sketch the graph of the given polynomial function.

**1.** $y = x^3 - 3$
**2.** $y = -(x + 2)^3$
**3.** $y = (x - 2)^3 + 2$
**4.** $y = 3 - (x + 2)^3$
**5.** $y = (x - 5)^4$
**6.** $y = x^4 - 1$
**7.** $y = 1 - (x - 1)^4$
**8.** $y = 4 + (x + 1)^4$

In Problems 9–12, determine whether the given polynomial function $f$ is even, odd, or neither even nor odd. Do not graph.

**9.** $f(x) = -2x^3 + 4x$
**10.** $f(x) = x^6 - 5x^2 + 7$
**11.** $f(x) = x^5 + 4x^3 + 9x + 1$
**12.** $f(x) = x^3(x + 2)(x - 2)$

In Problems 13–18, match the given graph with one of the polynomial functions in (a)–(f).

**(a)** $f(x) = x^2(x - 1)^2$
**(b)** $f(x) = -x^3(x - 1)$
**(c)** $f(x) = x^3(x - 1)^3$
**(d)** $f(x) = -x(x - 1)^3$
**(e)** $f(x) = -x^2(x - 1)$
**(f)** $f(x) = x^3(x - 1)^2$

**13.**

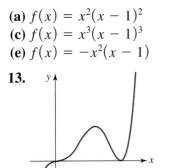

FIGURE 3.1.14 Graph for Problem 13

**14.**

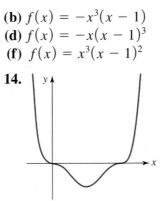

FIGURE 3.1.15 Graph for Problem 14

**15.**

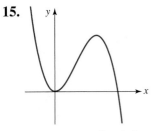

FIGURE 3.1.16 Graph for Problem 15

**16.**

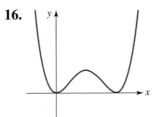

FIGURE 3.1.17 Graph for Problem 16

**17.**

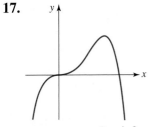

FIGURE 3.1.18 Graph for Problem 17

**18.**

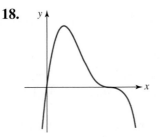

FIGURE 3.1.19 Graph for Problem 18

In Problems 19–40, proceed as in Example 2 and sketch the graph of the given polynomial function $f$.

**19.** $f(x) = x^3 - 4x$　　　　　　　**20.** $f(x) = 9x - x^3$
**21.** $f(x) = -x^3 + x^2 + 6x$　　　**22.** $f(x) = x^3 + 7x^2 + 12x$
**23.** $f(x) = (x + 1)(x - 2)(x - 4)$　**24.** $f(x) = (2 - x)(x + 2)(x + 1)$
**25.** $f(x) = x^4 - 4x^3 + 3x^2$　　　**26.** $f(x) = x^2(x - 2)^2$
**27.** $f(x) = (x^2 - x)(x^2 - 5x + 6)$　**28.** $f(x) = x^2(x^2 + 3x + 2)$
**29.** $f(x) = (x^2 - 1)(x^2 + 9)$　　**30.** $f(x) = x^4 + 5x^2 - 6$
**31.** $f(x) = -x^4 + 2x^2 - 1$　　　**32.** $f(x) = x^4 - 6x^2 + 9$
**33.** $f(x) = x^4 + 3x^3$　　　　　　**34.** $f(x) = x(x - 2)^3$
**35.** $f(x) = x^5 - 4x^3$　　　　　　**36.** $f(x) = (x - 2)^5 - (x - 2)^3$
**37.** $f(x) = 3x(x + 1)^2(x - 1)^2$　**38.** $f(x) = (x + 1)^2(x - 1)^3$
**39.** $f(x) = -\frac{1}{2}x^2(x + 2)^3(x - 2)^2$　**40.** $f(x) = x(x + 1)^2(x - 2)(x - 3)$

**41.** The graph of $f(x) = x^3 - 3x$ is given in FIGURE 3.1.20.
   **(a)** Use the figure to obtain the graph of $g(x) = f(x) + 2$.
   **(b)** Using only the graph obtained in part (a) write an equation, in *factored* form, for $g(x)$. Then verify by multiplying out the factors that your equation for $g(x)$ is the same as $f(x) + 2 = x^3 - 3x + 2$.

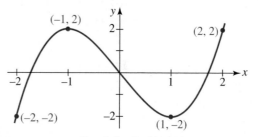

FIGURE 3.1.20　Graph for Problem 41

**42.** Find a polynomial function $f$ of lowest possible degree whose graph is consistent with the graph given in FIGURE 3.1.21.

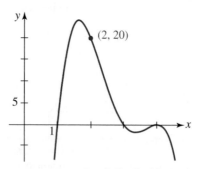

FIGURE 3.1.21　Graph for Problem 42

**43.** Find the value of $k$ such that $(2, 0)$ is an $x$-intercept for the graph of $f(x) = kx^5 - x^2 + 5x + 8$.
**44.** Find the values of $k_1$ and $k_2$ such that $(-1, 0)$ and $(1, 0)$ are $x$-intercepts for the graph of $f(x) = k_1 x^4 - k_2 x^3 + x - 4$.
**45.** Find the value of $k$ such that $(0, 10)$ is the $y$-intercept for the graph of $f(x) = x^3 - 2x^2 + 14x - 3k$.
**46.** Consider the polynomial function $f(x) = (x - 2)^{n+1}(x + 5)$, where $n$ is a positive integer. For what values of $n$ does the graph of $f$ touch, but not cross, the $x$-axis at $(2, 0)$?

47. Consider the polynomial function $f(x) = (x - 1)^{n+2}(x + 1)$, where $n$ is a positive integer. For what values of $n$ does the graph of $f$ cross the $x$-axis at $(1, 0)$?

48. Consider the polynomial function $f(x) = (x - 5)^{2m}(x + 1)^{2n-1}$, where $m$ and $n$ are positive integers.
    (a) For what values of $m$ does the graph of $f$ cross the $x$-axis at $(5, 0)$?
    (b) For what values of $n$ does the graph of $f$ cross the $x$-axis at $(-1, 0)$?

In Problems 49 and 50, use the Intermediate Value Theorem to determine whether there is a zero of the given function $f$ in the indicated intervals. Do not graph.

49. $f(x) = 60x^3 - 13x^2 - 145x - 28$; $[-2, -1], [-1, 0], [0, 1], [1, 2]$
50. $f(x) = 8x^4 - 23x^3 + 23x^2 - 26x + 15$; $[-1, 0], [0, 1], [1, 2], [2, 3]$

## Miscellaneous Calculus-Related Problems

51. **Constructing a Box** An open box can be made from a rectangular piece of cardboard by cutting a square of length $x$ from each corner and bending up the sides. See FIGURE 3.1.22. If the cardboard measures 30 cm by 40 cm, show that the volume of the resulting box is given by

$$V(x) = x(30 - 2x)(40 - 2x).$$

Sketch the graph of $V(x)$ for $x > 0$. What is the domain of the function $V$?

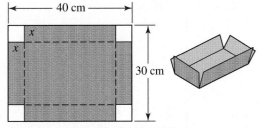

FIGURE 3.1.22  Box in Problem 51

52. **Another Box** In order to hold its shape, the box in Problem 51 will require tape or some other fastener at the corners. An open box that holds itself together can be made by cutting out a square of length $x$ from each corner of a rectangular piece of cardboard, cutting on the solid line, and folding on the dashed lines, as shown in FIGURE 3.1.23. Find a polynomial function $V(x)$ that gives the volume of the resulting box if the original cardboard measures 30 cm by 40 cm. Sketch the graph of $V(x)$ for $x > 0$.

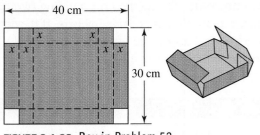

FIGURE 3.1.23  Box in Problem 52

## For Discussion

53. Examine Figure 3.1.5. Then discuss whether there can exist cubic polynomial functions that have no real zeros.
54. Suppose a polynomial function $f$ has three zeros, $-3$, $2$, and $4$, and has the end behavior that its graph goes down to the left as $x \to -\infty$ and down to the right as $x \to \infty$. Discuss possible equations for $f$.

## Calculator/Computer Problems

In Problems 55 and 56, use a graphing utility to examine the graph of the given polynomial function on the indicated intervals.

**55.** $f(x) = -(x - 8)(x + 10)^2$;   $[-15, 15], [-100, 100], [-1000, 1000]$
**56.** $f(x) = (x - 5)^2(x + 5)^2$;   $[-10, 10], [-100, 100], [-1000, 1000]$

# 3.2 Division of Polynomial Functions

☐ **Introduction** If $p > 0$ and $s > 0$ are integers such that $p \geq s$, then $p/s$ is called an **improper fraction**. By dividing $p$ by $s$, we obtain unique numbers $q$ and $r$ that satisfy

$$\frac{p}{s} = q + \frac{r}{s} \quad \text{or} \quad p = sq + r, \tag{1}$$

where $0 \leq r < s$. The number $p$ is called the **dividend**, $s$ is the **divisor**, $q$ is the **quotient**, and $r$ is the **remainder**. For example, consider the improper fraction $\frac{1052}{23}$. Performing long division gives

$$
\begin{array}{r}
45 \quad \leftarrow \text{quotient} \\
\text{divisor} \rightarrow \quad 23\overline{)1052} \quad \leftarrow \text{dividend} \\
\underline{92} \quad\quad \leftarrow \text{subtract} \\
132 \quad\quad\quad \\
\underline{115} \quad\quad\quad \\
17. \quad \leftarrow \text{remainder}
\end{array}
\tag{2}
$$

The result in (2) can be written as $\frac{1052}{23} = 45 + \frac{17}{23}$, where $\frac{17}{23}$ is a **proper fraction** since the numerator is less than the denominator; in other words, the fraction is less than 1. If we multiply this result by the divisor 23 we obtain the special way of writing the dividend $p$ illustrated in the second equation in (1):

$$1052 = \overset{\text{quotient}}{\underset{\downarrow}{23}} \cdot \overset{}{\underset{\downarrow}{45}} + \overset{\text{remainder}}{\underset{\downarrow}{17}}. \tag{3}$$

$$\underset{\underset{\text{divisor}}{\uparrow}}{}$$

☐ **Division of Polynomials** The method for dividing two polynomial functions $f(x)$ and $d(x)$ is similar to division of positive integers. If the degree of a polynomial $f(x)$ is greater than or equal to the degree of the polynomial $d(x)$, then $f(x)/d(x)$ is also called an **improper fraction**. A result analogous to (1) is called the **Division Algorithm** for polynomials.

### DIVISION ALGORITHM

Let $f(x)$ and $d(x) \neq 0$ be polynomials where the degree of $f(x)$ is greater than or equal to the degree of $d(x)$. Then there exist unique polynomials $q(x)$ and $r(x)$ such that

$$\frac{f(x)}{d(x)} = q(x) + \frac{r(x)}{d(x)} \quad \text{or} \quad f(x) = d(x)q(x) + r(x), \tag{4}$$

where $r(x)$ has degree less than the degree of $d(x)$.

The polynomial $f(x)$ is called the **dividend**, $d(x)$ the **divisor**, $q(x)$ the **quotient**, and $r(x)$ the **remainder**. Because $r(x)$ has degree less than the degree of $d(x)$, the rational expression $r(x)/d(x)$ is called a **proper fraction**.

Observe in (4) when $r(x) = 0$, then $f(x) = d(x)q(x)$, and so the divisor $d(x)$ is a factor of $f(x)$. In this case, we say that $f(x)$ is **divisible** by $d(x)$ or, in older terminology, $d(x)$ **divides evenly** into $f(x)$.

## ◼ EXAMPLE 1    Division of Two Polynomials

Use long division to find the quotient $q(x)$ and remainder $r(x)$ when the polynomial $f(x) = 3x^3 - x^2 - 2x + 6$ is divided by the polynomial $d(x) = x^2 + 1$.

**Solution** By long division,

$$
\begin{array}{r}
\phantom{x^2+1)}3x \;-\; 1 \qquad \leftarrow \text{quotient} \\
\text{divisor} \rightarrow \quad x^2 + 1 \overline{)3x^3 - \phantom{x}x^2 - 2x + 6} \qquad \leftarrow \text{dividend} \\
\underline{3x^3 + 0x^2 + 3x} \qquad \leftarrow \text{subtract} \\
-x^2 - 5x + 6 \\
\underline{-x^2 + 0x - 1} \\
-5x + 7 \qquad \leftarrow \text{remainder}
\end{array}
\tag{5}
$$

The result of the division (5) can be written

$$
\frac{3x^3 - x^2 - 2x + 6}{x^2 + 1} = 3x - 1 + \frac{-5x + 7}{x^2 + 1}.
$$

If we multiply both sides of the last equation by the divisor $x^2 + 1$, we get the second form given in (4):

$$
3x^3 - x^2 - 2x + 6 = (x^2 + 1)(3x - 1) + (-5x + 7). \tag{6} \quad \blacksquare
$$

If the divisor $d(x)$ is a linear polynomial $x - c$, it follows from the Division Algorithm that the degree of the remainder $r$ is 0, that is to say, $r$ is a constant. Thus (4) becomes

$$
f(x) = (x - c)q(x) + r. \tag{7}
$$

When the number $x = c$ is substituted into (7), we discover an alternative way of evaluating a polynomial function:

$$
f(c) = (c - c)\overset{\underset{\displaystyle 0}{\downarrow}}{q(c)} + r = r.
$$

The foregoing result is called the **Remainder Theorem**.

---

### REMAINDER THEOREM

If a polynomial $f(x)$ is divided by a linear polynomial $x - c$, then the remainder $r$ is the value of $f(x)$ at $x = c$, that is, $f(c) = r$.

---

## ◼ EXAMPLE 2    Finding the Remainder

Use the Remainder Theorem to find $r$ when $f(x) = 4x^3 - x^2 + 4$ is divided by $x - 2$.

**Solution** From the Remainder Theorem, the remainder $r$ is the value of the function $f$ evaluated at $x = 2$:

$$
r = f(2) = 4(2)^3 - (2)^2 + 4 = 32. \tag{8} \quad \blacksquare
$$

Example 2, where a remainder $r$ is determined by calculating a function value $f(c)$, is more interesting than important. What *is* important is the reverse problem: determine the function value $f(c)$ by finding the remainder $r$ by division of $f$ by $x - c$. The next two examples illustrate this concept.

### ■ EXAMPLE 3        Evaluation by Division

Use the Remainder Theorem to find $f(c)$ for $f(x) = x^5 - 4x^3 + 2x - 10$ when $c = -3$.

**Solution** The value $f(-3)$ is the remainder when $f(x) = x^5 - 4x^3 + 2x - 10$ is divided by $x - (-3) = x + 3$. For the purposes of long division we must account for the missing $x^4$ and $x^2$ terms by rewriting the dividend as

$$f(x) = x^5 + 0x^4 - 4x^3 + 0x^2 + 2x - 10.$$

Then,

$$
\begin{array}{r}
x^4 - 3x^3 + 5x^2 - 15x + 47 \\
x + 3 \overline{)x^5 + 0x^4 - 4x^3 + 0x^2 + 2x - 10} \\
\underline{x^5 + 3x^4} \\
-3x^4 - 4x^3 + 0x^2 + 2x - 10 \\
\underline{-3x^4 - 9x^3} \\
5x^3 + 0x^2 + 2x - 10 \\
\underline{5x^3 + 15x^2} \\
-15x^2 + 2x - 10 \\
\underline{-15x^2 - 45x} \\
47x - 10 \\
\underline{47x + 141} \\
-151
\end{array}
\tag{9}
$$

The remainder $r$ in the division is the value of the function $f$ at $x = -3$, that is, $f(-3) = -151$.   ■

☐ **Synthetic Division** After working through Example 3 one could justifiably ask the question: Why would anyone want to calculate the value of a polynomial function $f$ by division? The answer is: We would not bother do this were it not for **synthetic division**. Synthetic division is a shorthand method of dividing a polynomial $f(x)$ by a *linear* polynomial $x - c$; it does not require writing down the various powers of the variable $x$ but only the coefficients of these powers in the dividend $f(x)$ (which must include all 0 coefficients). It is also a very efficient and quick way of evaluating $f(c)$, since the process utilizes only the arithmetic operations of multiplication and addition. No exponentiations such as $(2)^3$ and $(2)^2$ in (8) are involved. Here is the same division in (9) done synthetically:

For a review of synthetic division please see the *Student Resource Manual* that accompanies this text. ▶

$$
\begin{array}{r|rrrrrr}
-3 & 1 & 0 & -4 & 0 & 2 & -10 \\
   &   & -3 & 9 & -15 & 45 & -141 \\
\hline
   & 1 & -3 & 5 & -15 & 47 & \boxed{-151} = r = f(-3)
\end{array}
\tag{10}
$$

Recall that the bottom line of numbers in (10) are the coefficients of the various powers of $x$ in the quotient $q(x)$ when $f(x) = x^5 - 4x^3 + 2x - 10$ is divided by $x + 3$. You should compare this with the quotient obtained by the long division in (9).

### ■ EXAMPLE 4        Using Synthetic Division to Evaluate a Function

Use the remainder theorem to find $f(c)$ for

$$f(x) = -3x^6 + 4x^5 + x^4 - 8x^3 - 6x^2 + 9$$

when $c = 2$.

**Solution** We will use synthetic division to find the remainder $r$ in the division of $f$ by $x - 2$. We begin by writing down all the coefficients in $f(x)$, including 0 as the coefficient of $x$. From

$$
\begin{array}{r|rrrrrrr}
2 & -3 & 4 & 1 & -8 & -6 & 0 & 9 \\
  &    & -6 & -4 & -6 & -28 & -68 & -136 \\
\hline
  & -3 & -2 & -3 & -14 & -34 & -68 & \boxed{-127} = r
\end{array}
$$

we see that $f(2) = -127$. ∎

### ▌EXAMPLE 5     Using Synthetic Division to Evaluate a Function

Use synthetic division to evaluate $f(x) = x^3 - 7x^2 + 13x - 15$ at $x = 5$.

**Solution** From the synthetic division

$$
\begin{array}{r|rrrr}
5 & 1 & -7 & 13 & -15 \\
  &   & 5 & -10 & 15 \\
\hline
  & 1 & -2 & 3 & \boxed{0} = r
\end{array}
$$

we see that $f(5) = 0$. ∎

The result in Example 5 that $f(5) = 0$ shows that 5 is a zero of the given function $f$. Moreover, we have found additionally that $f$ is divisible by the linear polynomial $x - 5$, or put another way, $x - 5$ is a factor of $f$. The synthetic division shows that $f(x) = x^3 - 7x^2 + 13x - 15$ is equivalent to

$$f(x) = (x - 5)(x^2 - 2x + 3).$$

In the next section we will further explore the use of the Division Algorithm and the Remainder Theorem as a help in finding zeros and factors of a polynomial function.

---

### ▌3.2  Exercises     Answers to selected odd-numbered problems begin on page ANS–10.

In Problems 1–10, use long division to find the quotient $q(x)$ and remainder $r(x)$ when the polynomial $f(x)$ is divided by the given polynomial $d(x)$. In each case write your answer in the form $f(x) = d(x)q(x) + r(x)$.

**1.** $f(x) = 8x^2 + 4x - 7;$   $d(x) = x^2$
**2.** $f(x) = x^2 + 2x - 3;$   $d(x) = x^2 + 1$
**3.** $f(x) = 5x^3 - 7x^2 + 4x + 1;$   $d(x) = x^2 + x - 1$
**4.** $f(x) = 14x^3 - 12x^2 + 6;$   $d(x) = x^2 - 1$
**5.** $f(x) = 2x^3 + 4x^2 - 3x + 5;$   $d(x) = (x + 2)^2$
**6.** $f(x) = x^3 + x^2 + x + 1;$   $d(x) = (2x + 1)^2$
**7.** $f(x) = 27x^3 + x - 2;$   $d(x) = 3x^2 - x$
**8.** $f(x) = x^4 + 8;$   $d(x) = x^3 + 2x - 1$
**9.** $f(x) = 6x^5 + 4x^4 + x^3;$   $d(x) = x^3 - 2$
**10.** $f(x) = 5x^6 - x^5 + 10x^4 + 3x^2 - 2x + 4;$   $d(x) = x^2 + x - 1$

In Problems 11–16, proceed as in Example 2 and use the Remainder Theorem to find $r$ when $f(x)$ is divided by the given linear polynomial.

**11.** $f(x) = 2x^2 - 4x + 6;$   $x - 2$
**12.** $f(x) = 3x^2 + 7x - 1;$   $x + 3$
**13.** $f(x) = x^3 - 4x^2 + 5x + 2;$   $x - \frac{1}{2}$

**14.** $f(x) = 5x^3 + x^2 - 4x - 6;\quad x + 1$

**15.** $f(x) = x^4 - x^3 + 2x^2 + 3x - 5;\quad x - 3$

**16.** $f(x) = 2x^4 - 7x^2 + x - 1;\quad x + \frac{3}{2}$

In Problems 17–22, proceed as in Example 3 and use the Remainder Theorem to find $f(c)$ for the given value of $c$.

**17.** $f(x) = 4x^2 - 10x + 6;\quad c = 2$

**18.** $f(x) = 6x^2 + 4x - 2;\quad c = \frac{1}{4}$

**19.** $f(x) = x^3 + 3x^2 + 6x + 6;\quad c = -5$

**20.** $f(x) = 15x^3 + 17x^2 - 30;\quad c = \frac{1}{5}$

**21.** $f(x) = 3x^4 - 5x^2 + 20;\quad c = \frac{1}{2}$

**22.** $f(x) = 14x^4 - 60x^3 + 49x^2 - 21x + 19;\quad c = 1$

In Problems 23–32, use synthetic division to find the quotient $q(x)$ and remainder $r(x)$ when $f(x)$ is divided by the given linear polynomial.

**23.** $f(x) = 2x^2 - x + 5;\quad x - 2$

**24.** $f(x) = 4x^2 - 8x + 6;\quad x - \frac{1}{2}$

**25.** $f(x) = x^3 - x^2 + 2;\quad x + 3$

**26.** $f(x) = 4x^3 - 3x^2 + 2x + 4;\quad x - 7$

**27.** $f(x) = x^4 + 16;\quad x - 2$

**28.** $f(x) = 4x^4 + 3x^3 - x^2 - 5x - 6;\quad x + 3$

**29.** $f(x) = x^5 + 56x^2 - 4;\quad x + 4$

**30.** $f(x) = 2x^6 + 3x^3 - 4x^2 - 1;\quad x + 1$

**31.** $f(x) = x^3 - (2 + \sqrt{3})x^2 + 3\sqrt{3}x - 3;\quad x - \sqrt{3}$

**32.** $f(x) = x^8 - 3^8;\quad x - 3$

In Problems 33–38, use synthetic division and the Remainder Theorem to find $f(c)$ for the given value of $c$.

**33.** $f(x) = 4x^2 - 2x + 9;\quad c = -3$

**34.** $f(x) = 3x^4 - 5x^2 + 27;\quad c = \frac{1}{2}$

**35.** $f(x) = 14x^4 - 60x^3 + 49x^2 - 21x + 19;\quad c = 1$

**36.** $f(x) = 3x^5 + x^2 - 16;\quad c = -2$

**37.** $f(x) = 2x^6 - 3x^5 + x^4 - 2x + 1;\quad c = 4$

**38.** $f(x) = x^7 - 3x^5 + 2x^3 - x + 10;\quad c = 5$

In Problems 39 and 40, use long division to find a value of $k$ such that $f(x)$ is divisible by $d(x)$.

**39.** $f(x) = x^4 + x^3 + 3x^2 + kx - 4;\quad d(x) = x^2 - 1$

**40.** $f(x) = x^5 - 3x^4 + 7x^3 + kx^2 + 9x - 5;\quad d(x) = x^2 - x + 1$

In Problems 41 and 42, use synthetic division to find a value of $k$ such that $f(x)$ is divisible by $d(x)$.

**41.** $f(x) = kx^4 + 2x^2 + 9k;\quad d(x) = x - 1$

**42.** $f(x) = x^3 + kx^2 - 2kx + 4;\quad d(x) = x + 2$

**43.** Find a value of $k$ such that the remainder in the division of
$f(x) = 3x^2 - 4kx + 1$ by $d(x) = x + 3$ is $r = -20$.

**44.** When $f(x) = x^2 - 3x - 1$ is divided by $x - c$, the remainder is $r = 3$. Determine $c$.

# Zeros and Factors of Polynomial Functions

☐ **Introduction** In Section 2.1 we saw that a zero of a function $f$ is a number $c$ for which $f(c) = 0$. A zero $c$ of a function $f$ can be a *real* or a *complex number*. Recall that a **complex number** is a number of the form

$$z = a + bi, \quad \text{where} \quad i^2 = -1,$$

and $a$ and $b$ are real numbers. The number $a$ is called the **real part** of $z$ and $b$ is called the **imaginary part** of $z$. The symbol $i$ is called the **imaginary unit** and it is common practice to write it as $i = \sqrt{-1}$. If $z = a + bi$ is a complex number, then $\bar{z} = a - bi$ is called its **conjugate**. Thus the simple polynomial function $f(x) = x^2 + 1$ has two complex zeros since the solutions of $x^2 + 1 = 0$ are $\pm\sqrt{-1}$, that is, $i$ and $-i$.

◄ The arithmetic of complex numbers is reviewed in the *Student Resource Manual*.

In this section we explore the connection between the zeros of a polynomial function $f$, the operation of division, and the factors of $f$.

---

**EXAMPLE 1**    **A Real Zero**

Consider the polynomial function $f(x) = 2x^3 - 9x^2 + 6x - 1$. The real number $\frac{1}{2}$ is a zero of the function since

$$f(\tfrac{1}{2}) = 2(\tfrac{1}{2})^3 - 9(\tfrac{1}{2})^2 + 6(\tfrac{1}{2}) - 1$$
$$= 2(\tfrac{1}{8}) - \tfrac{9}{4} + 3 - 1$$
$$= \tfrac{1}{4} - \tfrac{9}{4} + \tfrac{8}{4} = 0. \qquad \blacksquare$$

---

**EXAMPLE 2**    **A Complex Zero**

Consider the polynomial function $f(x) = x^3 - 5x^2 + 8x - 6$. The complex number $1 + i$ is a zero of the function. To verify this we use the binomial expansion of $(a + b)^3$ and the fact that $i^2 = -1$ and $i^3 = -i$:

◄ See (7) in Section 1.5.

$$f(1 + i) = (1 + i)^3 - 5(1 + i)^2 + 8(1 + i) - 6$$
$$= (1^3 + 3 \cdot 1^2 \cdot i + 3 \cdot 1 \cdot i^2 + i^3) - 5(1^2 + 2i + i^2) + 8(1 + i) - 6$$
$$= (-2 + 2i) - 5(2i) + (2 + 8i)$$
$$= (-2 + 2) + (10 - 10)i = 0 + 0i = 0. \qquad \blacksquare$$

☐ **Factor Theorem** We can now relate the notion of a zero of a polynomial function $f$ with division of polynomials. From the Remainder Theorem we know that when $f(x)$ is divided by the linear polynomial $x - c$ the remainder is $r = f(c)$. If $c$ is a zero of $f$, then $f(c) = 0$ implies $r = 0$. From the form of the Division Algorithm given in (4) of Section 3.2 we can then write $f$ as

$$f(x) = (x - c)q(x). \qquad (1)$$

Thus, if $c$ is a zero of a polynomial function $f$, then $x - c$ is a factor of $f(x)$. Conversely, if $x - c$ is a factor of $f(x)$, then $f$ has the form given in (1). In this case, we see immediately that $f(c) = (c - c)q(c) = 0$. These results are summarized in the **Factor Theorem** given next.

## FACTOR THEOREM

A number $c$ is a zero of a polynomial function $f$ if and only if $x - c$ is a factor of $f(x)$.

If a polynomial function $f$ is of degree $n$ and $(x - c)^m$, $m \leq n$, is a factor of $f(x)$, then $c$ is said to be a **zero of multiplicity $m$**. When $m = 1$, $c$ is a **simple zero**. Equivalently, we say that the number $c$ is a **root of multiplicity $m$** of the equation $f(x) = 0$. We have already examined the graphical significance of repeated real zeros of a polynomial function $f$ in Section 3.1. See Figure 3.1.6.

### EXAMPLE 3    Factors of a Polynomial

Determine whether

**(a)** $x + 1$ is a factor of $f(x) = x^4 - 5x^2 + 6x - 1$,
**(b)** $x - 2$ is a factor of $f(x) = x^3 - 3x^2 + 4$.

**Solution**  We use synthetic division to divide $f(x)$ by the given linear term.

**(a)** From the division

$$\begin{array}{r|rrrrr} -1 & 1 & 0 & -5 & 6 & -1 \\ & & -1 & 1 & 4 & -10 \\ \hline & 1 & -1 & -4 & 10 & \boxed{-11} = r = f(-1) \end{array}$$

we see that $f(-1) = -11$ and so $-1$ is not a zero of $f$. We conclude that $x - (-1) = x + 1$ is not a factor of $f(x)$.

**(b)** From the division

$$\begin{array}{r|rrrr} 2 & 1 & -3 & 0 & 4 \\ & & 2 & -2 & -4 \\ \hline & 1 & -1 & -2 & \boxed{0} = r = f(2) \end{array}$$

we see that $f(2) = 0$. This means that 2 is a zero and that $x - 2$ is a factor of $f(x)$. From the division we see that the quotient is $q(x) = x^2 - x - 2$ and so $f(x) = (x - 2)(x^2 - x - 2)$. ■

☐ **Number of Zeros**  In Example 6 of Section 3.1 we graphed the polynomial function

$$f(x) = (x - 3)(x - 1)^2(x + 2)^3. \qquad (2)$$

The number 3 is a zero of multiplicity one, or a simple zero of $f$; the number 1 is a zero of multiplicity two; and $-2$ is a zero of multiplicity three. Although the function $f$ has three *distinct* zeros (different from one another), it is, nevertheless, standard practice to say that $f$ has *six zeros* because we count the multiplicities of each zero. Hence for the function $f$ in (2), the number of zeros is $1 + 2 + 3 = 6$. The question *How many zeros does a polynomial function $f$ have?* is answered next.

## FUNDAMENTAL THEOREM OF ALGEBRA

A polynomial function $f$ of degree $n > 0$ has at least one zero.

The foregoing theorem, first proved by the German mathematician Carl Friedrich Gauss (1777–1855) in 1799, is considered one of the major milestones in the history of mathematics. At first reading this theorem does not appear to say much, but when combined with the Factor Theorem, the Fundamental Theorem of Algebra shows:

> *Every polynomial function $f$ of degree $n > 0$ has exactly $n$ zeros.*     (3)

Of course if a zero is repeated—say, it has multiplicity $k$—we count that zero $k$ times. To prove (3), we know from the Fundamental Theorem of Algebra that $f$ has a zero (call it $c_1$). By the Factor Theorem we can write

$$f(x) = (x - c_1)q_1(x), \qquad (4)$$

where $q_1$ is a polynomial function of degree $n - 1$. If $n - 1 \neq 0$, then in like manner we know that $q_1$ must have a zero (call it $c_2$) and so (4) becomes

$$f(x) = (x - c_1)(x - c_2)q_2(x),$$

where $q_2$ is a polynomial function of degree $n - 2$. If $n - 2 \neq 0$, we continue and arrive at

$$f(x) = (x - c_1)(x - c_2)(x - c_3)q_3(x), \qquad (5)$$

and so on. Eventually we arrive at a factorization of $f(x)$ with $n$ linear factors and the last factor $q_n(x)$ of degree 0. In other words, $q_n(x) = a_n$, where $a_n$ is a constant. We have arrived at the *complete* factorization of $f(x)$. Bear in mind that some or all the zeros $c_1, \ldots, c_n$ in (6) may be complex numbers $a + ib$, where $b \neq 0$.

---

## COMPLETE FACTORIZATION THEOREM

Let $c_1, c_2, \ldots, c_n$ be the $n$ (not necessarily distinct) zeros of the polynomial function of degree $n > 0$:

$$f(x) = a_n x^n + a_{n-1} x^{n-1} + \cdots + a_2 x^2 + a_1 x + a_0.$$

Then $f(x)$ can be written as a product of $n$ linear factors

$$f(x) = a_n(x - c_1)(x - c_2)\cdots(x - c_n). \qquad (6)$$

---

In the case of a second-degree, or quadratic, polynomial function $f(x) = ax^2 + bx + c$, where the coefficients $a$, $b$, and $c$ are real numbers, the zeros $c_1$ and $c_2$ of $f$ can be found using the quadratic formula:

$$c_1 = \frac{-b - \sqrt{b^2 - 4ac}}{2a} \qquad \text{and} \qquad c_2 = \frac{-b + \sqrt{b^2 - 4ac}}{2a}. \qquad (7)$$

The results in (7) tell the whole story about the zeros of the quadratic function: the zeros are real and distinct when $b^2 - 4ac > 0$, real with multiplicity two when $b^2 - 4ac = 0$, and complex and distinct when $b^2 - 4ac < 0$. It follows from (6) that the complete factorization of a quadratic polynomial function is

$$f(x) = a(x - c_1)(x - c_2). \qquad (8)$$

### EXAMPLE 4     Example 1 Revisited

In Example 1 we demonstrated that $\frac{1}{2}$ is a zero of $f(x) = 2x^3 - 9x^2 + 6x - 1$. We now know that $x - \frac{1}{2}$ is a factor of $f(x)$ and that $f(x)$ has three zeros. The synthetic division

$$
\begin{array}{r|rrrr}
\frac{1}{2} & 2 & -9 & 6 & -1 \\
           &   & 1  & -4 & 1 \\
\hline
           & 2 & -8 & 2 & \boxed{0} = r
\end{array}
$$

again demonstrates that $\frac{1}{2}$ is a zero of $f(x)$ (the 0 remainder is the value of $f(\frac{1}{2})$) and, additionally, gives us the quotient $q(x)$ obtained in the division of $f(x)$ by $x - \frac{1}{2}$, that is, $f(x) = (x - \frac{1}{2})(2x^2 - 8x + 2)$. As shown in (8), we can now factor the quadratic quotient $q(x) = 2x^2 - 8x + 2$ by finding the roots of $2x^2 - 8x + 2 = 0$ by the quadratic formula:

$$\overset{\downarrow \sqrt{48} = \sqrt{16 \cdot 3} = \sqrt{16}\sqrt{3} = 4\sqrt{3}}{}$$

$$x = \frac{-(-8) \pm \sqrt{(-8)^2 - 4(2)(2)}}{4} = \frac{8 \pm \sqrt{48}}{4} = \frac{8 \pm 4\sqrt{3}}{4}$$

$$= \frac{4(2 \pm \sqrt{3})}{4} = 2 \pm \sqrt{3}.$$

Thus the remaining zeros of $f(x)$ are the irrational numbers $2 + \sqrt{3}$ and $2 - \sqrt{3}$. With the identification of the leading coefficient as $a_3 = 2$, it follows from (8) that the complete factorization of $f(x)$ is then

$$f(x) = 2(x - \tfrac{1}{2})(x - (2 + \sqrt{3}))(x - (2 - \sqrt{3}))$$
$$= 2(x - \tfrac{1}{2})(x - 2 - \sqrt{3})(x - 2 + \sqrt{3}). \qquad \blacksquare$$

### ▮EXAMPLE 5  Using Synthetic Division

Find the complete factorization of

$$f(x) = x^4 - 12x^3 + 47x^2 - 62x + 26$$

given that 1 is a zero of $f$ of multiplicity two.

**Solution** We know that $x - 1$ is a factor of $f(x)$, so by the division

$$\begin{array}{r|rrrrr}
1\! & 1 & -12 & 47 & -62 & 26 \\
 &  & 1 & -11 & 36 & -26 \\
\hline
 & 1 & -11 & 36 & -26 & \boxed{0} = r
\end{array}$$

we find $\qquad\qquad f(x) = (x - 1)(x^3 - 11x^2 + 36x - 26).$

Since 1 is a zero of multiplicity two, $x - 1$ must also be a factor of the quotient $q(x) = x^3 - 11x^2 + 36x - 26$. By the division,

$$\begin{array}{r|rrrr}
1\! & 1 & -11 & 36 & -26 \\
 &  & 1 & -10 & 26 \\
\hline
 & 1 & -10 & 26 & \boxed{0} = r
\end{array}$$

we conclude that $q(x)$ can be written $q(x) = (x - 1)(x^2 - 10x + 26)$. Therefore,

$$f(x) = (x - 1)^2(x^2 - 10x + 26).$$

The remaining two zeros, found by solving $x^2 - 10x + 26 = 0$ by the quadratic formula, are the complex numbers $5 + i$ and $5 - i$. Since the leading coefficient is $a_4 = 1$ the complete factorization of $f(x)$ is

$$f(x) = (x - 1)^2(x - (5 + i))(x - (5 - i))$$
$$= (x - 1)^2(x - 5 - i)(x - 5 + i). \qquad \blacksquare$$

### ▮EXAMPLE 6  Complete Linear Factorization

Find a polynomial function $f$ of degree three, with zeros $1$, $-4$, and $5$ such that its graph possesses the $y$-intercept $(0, 5)$.

**Solution** Because we have three zeros 1, −4, and 5 we know $x - 1$, $x + 4$, and $x - 5$ are factors of $f$. However, the function we seek is *not*

$$f(x) = (x - 1)(x + 4)(x - 5). \qquad (9)$$

The reason for this is that any nonzero constant multiple of $f$ is a different polynomial with the same zeros. Notice, too, that the function in (9) gives $f(0) = 20$ but we want $f(0) = 5$. Hence we must assume that $f$ has the form

$$f(x) = a(x - 1)(x + 4)(x - 5), \qquad (10)$$

where $a$ is some real constant. Using (10), $f(0) = 5$ gives

$$f(0) = a(0 - 1)(0 + 4)(0 - 5) = 20a = 5$$

and so $a = \frac{5}{20} = \frac{1}{4}$. The desired function is then

$$f(x) = \tfrac{1}{4}(x - 1)(x + 4)(x - 5). \qquad \blacksquare$$

We have seen in the introduction that the complex zeros of $f(x) = x^2 + 1$ are $i$ and $-i$. In Example 5 the complex zeros are $5 + i$ and $5 - i$. In each case the complex zeros of the polynomial function are conjugate pairs. In other words, one complex zero is the conjugate of the other. This is no coincidence; complex zeros of polynomials with *real* coefficients *always* appear in conjugate pairs.

## CONJUGATE ZEROS THEOREM

Let $f(x)$ be a polynomial function of degree $n > 1$ with real coefficients. If $z$ is a complex zero of $f(x)$, then the conjugate $\bar{z}$ is also a zero of $f(x)$.

**EXAMPLE 7**     **Example 2 Revisited**

In Example 2 we demonstrated that $1 + i$ is a complex zero of $f(x) = x^3 - 5x^2 + 8x - 6$. Since the coefficients of $f$ are real numbers we conclude that another zero is the conjugate of $1 + i$, namely, $1 - i$. Thus we know two factors of $f(x)$, $x - (1 + i)$ and $x - (1 - i)$. Carrying out the multiplication, we find

$$(x - 1 - i)(x - 1 + i) = x^2 - 2x + 2.$$

Thus we can write

$$f(x) = (x - 1 - i)(x - 1 + i)q(x) = (x^2 - 2x + 2)q(x).$$

We determine $q(x)$ by performing the *long division* of $f(x)$ by $x^2 - 2x + 2$. (We can't do synthetic division because we are not dividing by a linear factor.) From

$$
\begin{array}{r}
x - 3 \\
x^2 - 2x + 2 \overline{\smash{\big)}\, x^3 - 5x^2 + 8x - 6} \\
\underline{x^3 - 2x^2 + 2x} \\
-3x^2 + 6x - 6 \\
\underline{-3x^2 + 6x - 6} \\
0
\end{array}
$$

we see that the complete factorization of $f(x)$ is

$$f(x) = (x - 1 - i)(x - 1 + i)(x - 3).$$

The three zeros of $f(x)$ are $1 + i$, $1 - i$, and 3. $\qquad \blacksquare$

In Problems 1–6, determine whether the indicated real number is a zero of the given polynomial function $f$. If yes, find all other zeros and then give the complete factorization of $f(x)$.

**1.** 1;  $f(x) = 4x^3 - 9x^2 + 6x - 1$     **2.** $\frac{1}{2}$;  $f(x) = 2x^3 - x^2 + 32x - 16$
**3.** 5;  $f(x) = x^3 - 6x^2 + 6x + 5$     **4.** 3;  $f(x) = x^3 - 3x^2 + 4x - 12$
**5.** $-\frac{2}{3}$;  $f(x) = 3x^3 - 10x^2 - 2x + 4$   **6.** $-2$;  $f(x) = x^3 - 4x^2 - 2x + 20$

In Problems 7–10, verify that each of the indicated numbers are zeros of the given polynomial function $f$. Find all other zeros and then give the complete factorization of $f(x)$.

**7.** $-3, 5$;  $f(x) = 4x^4 - 8x^3 - 61x^2 + 2x + 15$
**8.** $\frac{1}{4}, \frac{3}{2}$;  $f(x) = 8x^4 - 30x^3 + 23x^2 + 8x - 3$
**9.** $1, -\frac{1}{3}$ (multiplicity 2);   $f(x) = 9x^4 + 69x^3 - 29x^2 - 41x - 8$
**10.** $-\sqrt{5}$,  $\sqrt{5}$;  $f(x) = 3x^4 + x^3 - 17x^2 - 5x + 10$

In Problems 11–16, use synthetic division to determine whether the indicated linear polynomial is a factor of the given polynomial function $f$. If yes, find all other zeros and then give the complete factorization of $f(x)$.

**11.** $x - 5$;  $f(x) = 2x^2 + 6x - 25$
**12.** $x + \frac{1}{2}$;  $f(x) = 10x^2 - 27x + 11$
**13.** $x - 1$;  $f(x) = x^3 + x - 2$
**14.** $x + \frac{1}{2}$;  $f(x) = 2x^3 - x^2 + x + 1$
**15.** $x - \frac{1}{3}$;  $f(x) = 3x^3 - 3x^2 + 8x - 2$
**16.** $x - 2$;  $f(x) = x^3 - 6x^2 - 16x + 48$

In Problems 17–20, use division to show that the indicated polynomial is a factor of the given polynomial function $f$. Find all other zeros and then give the complete factorization of $f(x)$.

**17.** $(x - 1)(x - 2)$;  $f(x) = x^4 - 3x^3 + 6x^2 - 12x + 8$
**18.** $x(3x - 1)$;  $f(x) = 3x^4 - 7x^3 + 5x^2 - x$
**19.** $(x - 1)^2$;  $f(x) = 2x^4 + x^3 - 5x^2 - x + 3$
**20.** $(x + 3)^2$;  $f(x) = x^4 - 4x^3 - 22x^2 + 84x + 261$

In Problems 21–26, verify that the indicated complex number is a zero of the given polynomial function $f$. Proceed as in Example 7 to find all other zeros and then give the complete factorization of $f(x)$.

**21.** $2i$;  $f(x) = 3x^3 - 5x^2 + 12x - 20$
**22.** $\frac{1}{2}i$;  $f(x) = 12x^3 + 8x^2 + 3x + 2$
**23.** $-1 + i$;  $f(x) = 5x^3 + 12x^2 + 14x + 4$
**24.** $-i$;  $f(x) = 4x^4 - 8x^3 + 9x^2 - 8x + 5$
**25.** $1 + 2i$;  $f(x) = x^4 - 2x^3 - 4x^2 + 18x - 45$
**26.** $1 + i$;  $f(x) = 6x^4 - 11x^3 + 9x^2 + 4x - 2$

In Problems 27–32, find a polynomial function $f$ with real coefficients of the indicated degree that possesses the given zeros.

**27.** degree 4; $\quad 2, 1, -3$ (multiplicity 2)
**28.** degree 5; $\quad -4i, -\frac{1}{3}, \frac{1}{2}$ (multiplicity 2)
**29.** degree 5; $\quad 3 + i, 0$ (multiplicity 3)
**30.** degree 4; $\quad 5i, 2 - 3i$
**31.** degree 2; $\quad 1 - 6i$
**32.** degree 2; $\quad 4 + 3i$

In Problems 33–36, find the zeros of the given polynomial function $f$. State the multiplicity of each zero.

**33.** $f(x) = x(4x - 5)^2(2x - 1)^3$ **34.** $f(x) = x^4 + 6x^3 + 9x^2$
**35.** $f(x) = (9x^2 - 4)^2$ **36.** $f(x) = (x^2 + 25)(x^2 - 5x + 4)^2$

In Problems 37 and 38, find the value(s) of $k$ such that the indicated number is a zero of $f(x)$. Then give the complete factorization of $f(x)$.

**37.** $3; f(x) = 2x^3 - 2x^2 + k$ **38.** $1; f(x) = x^3 + 5x^2 - k^2x + k$

In Problems 39 and 40, find a polynomial function $f$ of the indicated degree whose graph is given in the figure.

**39.** degree 3                            **40.** degree 5

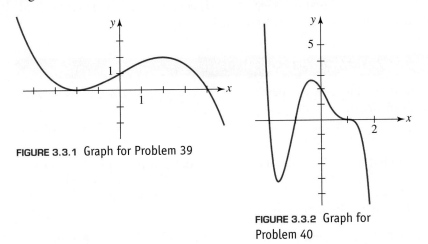

FIGURE 3.3.1 Graph for Problem 39

FIGURE 3.3.2 Graph for Problem 40

## For Discussion

**41.** Discuss:
    **(a)** For what positive-integer values of $n$ is $x - 1$ a factor of $f(x) = x^n - 1$?
    **(b)** For what positive-integer values of $n$ is $x + 1$ a factor of $f(x) = x^n + 1$?
**42.** Suppose $f(x)$ is a polynomial function of degree three with real coefficients. Why can't $f(x)$ have three complex zeros? Put another way, why must at least one zero of a cubic polynomial function be a real number? Can you generalize this result?
**43.** What is the smallest degree that a polynomial function $f(x)$ with real coefficients can have such that $1 + i$ is a complex zero of multiplicity two? Of multiplicity three?
**44.** Let $z = a + bi$. Show that $z + \bar{z}$ and $z\bar{z}$ are real numbers.

**45.** Let $z = a + bi$. Use the results of Problem 44 to show that

$$f(x) = (x - z)(x - \bar{z})$$

is a polynomial function with real coefficients.

<br>

## ▨ 3.4 ▨  Real Zeros of Polynomial Functions

☐ **Introduction**  In the preceding section we saw that as a consequence of the Fundamental Theorem of Algebra, a polynomial function $f$ of degree $n$ has $n$ zeros when the multiplicities of the zeros are counted. We also saw that a zero of a polynomial function could be either a real or a complex number. In this section we confine our attention to *real zeros* of polynomial functions with real coefficients.

☐ **Real Zeros**  If a polynomial function $f$ of degree $n > 0$ has $m$ (not necessarily distinct) real zeros $c_1, c_2, \ldots, c_m$, then by the Factor Theorem each of the linear polynomials $x - c_1, x - c_2, \ldots, x - c_m$ are factors of $f(x)$. That is,

$$f(x) = (x - c_1)(x - c_2)\cdots(x - c_m)q(x),$$

where $q(x)$ is a polynomial. Thus $n$, the degree of $f$, must be greater than or possibly equal to $m$, the number of real zeros when each is counted according to its multiplicity. Using slightly different words, we restate the last sentence.

> ### NUMBER OF REAL ZEROS
>
> A polynomial function $f$ of degree $n > 0$ has at most $n$ real zeros (not necessarily distinct).

Let's summarize some facts about real zeros of a polynomial function $f$ of degree $n$:

- $f$ may not have any real zeros.

For example, the fourth degree polynomial function $f(x) = x^4 + 9$ has no real zeros, since there exists no real number $x$ satisfying $x^4 + 9 = 0$ or $x^4 = -9$.

- $f$ may have $m$ real zeros where $m < n$.

For example, the third degree polynomial function $f(x) = (x - 1)(x^2 + 1)$ has one real zero.

- $f$ may have $n$ real zeros.

For example, by factoring the third degree polynomial function $f(x) = x^3 - x$ as $f(x) = x(x^2 - 1) = x(x + 1)(x - 1)$, we see that it has three real zeros.

- $f$ has at least one real zero when $n$ is odd.

This is a consequence of the fact that complex zeros of a polynomial function $f$ with real coefficients must appear in conjugate pairs. Thus if we write down an arbitrary cubic polynomial function such as $f(x) = x^3 + x + 1$, we know that $f$ cannot have just one complex zero nor can it have three complex zeros. Put another way, $f(x) = x^3 + x + 1$ either has exactly one real zero or it has exactly three real zeros.

- If the coefficients of $f(x)$ are positive and the constant term $a_0 \neq 0$, then any real zeros of $f$ must be negative.

☐ **Finding Real Zeros** It is one thing to talk about the existence of real and complex zeros of a polynomial function; it is an entirely different problem to actually find these zeros. The problem of finding a *formula* that expresses the zeros of a general $n$th degree polynomial function $f$ in terms of its coefficients perplexed mathematicians for centuries. We have seen in Sections 2.4 and 3.3 that in the case of a second-degree, or quadratic, polynomial function $f(x) = ax^2 + bx + c$, where the coefficients $a$, $b$, and $c$ are real numbers, the zeros $c_1$ and $c_2$ of $f$ can be found using the quadratic formula.

The problem of finding zeros of third-degree, or cubic, polynomial functions was solved in the sixteenth century through the pioneering work of the Italian mathematician Nicolo Fontana (1499–1557), also known as Tartaglia—"the stammerer." Around 1540 another Italian mathematician, Lodovico Farrari (1522–1565) discovered an algebraic formula for determining the zeros for fourth degree, or quartic, polynomial functions. Since these formulas are complicated and difficult to use, they are seldom discussed in elementary courses.

For the next 284 years no one discovered any formulas for zeros for general polynomial functions of degrees five, six, . . . . For good reason! In 1824, at age 22, the Norwegian mathematician Niels Henrik Abel (1802–1829) proved it was impossible to find such formulas for the zeros of all general polynomials of degrees $n \geq 5$ in terms of their coefficients.

Niels Henrik Abel

☐ **Rational Zeros** Real zeros of a polynomial function are either rational or irrational numbers. A rational number is a number of the form $p/s$, where $p$ and $s$ are integers and $s \neq 0$. An irrational number is one that is not rational. For example, $\frac{1}{4}$ and $-9$ are rational numbers, but $\sqrt{2}$ and $\pi$ are irrational, that is, neither $\sqrt{2}$ nor $\pi$ can be written as a fraction $p/s$ where $p$ and $s$ are integers. So how do we find real zeros for polynomial functions of degree $n > 2$? The bad news: For irrational real zeros, we *may* have to be content to use an accurate graph to "eyeball" their location on the $x$-axis and then use one of the many sophisticated methods for *approximating* the zero that have been developed over the years. The good news: We can always find the rational real zeros of *any* polynomial function with rational coefficients. We have already seen that synthetic division is a useful method for determining whether a given number $c$ is a zero of a polynomial function $f(x)$. When the remainder in the division of $f(x)$ by $x - c$ is $r = 0$, we have found a zero of the polynomial function $f$, since $r = f(c) = 0$. For example, $\frac{2}{3}$ is a zero of $f(x) = 18x^3 - 15x^2 + 14x - 8$, since

$$\begin{array}{r|rrrr} \frac{2}{3} & 18 & -15 & 14 & -8 \\ & & 12 & -2 & 8 \\ \hline & 18 & -3 & 12 & \boxed{0} = r. \end{array}$$

Hence by the Factor Theorem, both $x - \frac{2}{3}$ and the quotient $18x^2 - 3x + 12$ are factors of $f$ and so we can write the polynomial function as the product

$$\begin{aligned} f(x) &= (x - \tfrac{2}{3})(18x^2 - 3x + 12) \quad \leftarrow \text{factor 3 from the quadratic polynomial} \\ &= (x - \tfrac{2}{3})(3)(6x^2 - x + 4) \\ &= (3x - 2)(6x^2 - x + 4). \end{aligned} \tag{1}$$

As discussed in the preceding section, if we can factor the polynomial to the point where the remaining factor is a quadratic polynomial, we can then find the remaining two zeros by the quadratic formula. For this example, the factorization in (1) is as far as we can go using real numbers since the zeros of the quadratic factor $6x^2 - x + 4$ are complex (verify). But the indicated multiplication in (1) illustrates something important about rational zeros. The leading coefficient 18 and the constant term $-8$ of $f(x)$ are obtained from the products

$$\overbrace{(3x - 2)}^{}\overbrace{(6x^2 - x + 4)}^{-8}.$$
$$\underbrace{\phantom{(3x - 2)(6x^2}}_{18}$$

Thus we see that the denominator 3 of the rational zero $\frac{2}{3}$ is a *factor* of the leading coefficient 18 of $f(x) = 18x^3 - 15x^2 + 14x - 8$, and the numerator 2 of the rational zero is a factor of the constant term $-8$.

This example illustrates the following general principle for determining the rational zeros of a polynomial function. Read the following theorem carefully; the coefficients of $f$ are not only real numbers—they must be *integers*.

---

### RATIONAL ZEROS THEOREM

Let $p/s$ be a rational number in lowest terms and a zero of the polynomial function

$$f(x) = a_nx^n + a_{n-1}x^{n-1} + \cdots + a_2x^2 + a_1x + a_0,$$

where the coefficients $a_n, a_{n-1}, \ldots, a_2, a_1, a_0$ are integers with $a_n \neq 0$. Then $p$ is an integer factor of the constant term $a_0$ and $s$ is an integer factor of the leading coefficient $a_n$.

---

The Rational Zeros Theorem deserves to be read several times. Note that the theorem *does not* assert that a polynomial function $f$ with integer coefficients *must* have a rational zero; rather, it states that *if* a polynomial function $f$ with integer coefficients has a rational zero $p/s$, then necessarily:

$$\frac{p}{s} \quad \begin{array}{l} \leftarrow \text{is an integer factor of } a_0 \\ \leftarrow \text{is an integer factor of } a_n \end{array}$$

By forming all possible quotients of each integer factor of $a_0$ to each integer factor of $a_n$, we can construct a list of *potential* rational zeros of $f$.

### EXAMPLE 1    Rational Zeros

Find all rational zeros of $f(x) = 3x^4 - 10x^3 - 3x^2 + 8x - 2$.

**Solution** We identify the constant term $a_0 = -2$ and leading coefficient $a_4 = 3$, and then list all the integer factors of $a_0$ and $a_4$, respectively:

$$p: \pm 1, \pm 2,$$
$$s: \pm 1, \pm 3.$$

Now we form a list of all possible rational zeros $p/s$ by dividing all the factors of $p$ by $\pm 1$ and then by $\pm 3$:

$$\frac{p}{s}: \pm 1, \pm 2, \pm \tfrac{1}{3}, \pm \tfrac{2}{3}. \tag{2}$$

We know that the given fourth-degree polynomial function $f$ has four zeros; if any of these zeros is a real number and is rational, then it must appear in the list (2).

To determine which, if any, of the numbers in (2) are zeros, we could use direct substitution into $f(x)$. Synthetic division, however, is usually a more efficient means of evaluating $f(x)$. We begin by testing $-1$:

$$
\begin{array}{r|rrrrr}
-1 & 3 & -10 & -3 & 8 & -2 \\
   &   & -3 & 13 & -10 & 2 \\
\hline
   & 3 & -13 & 10 & -2 & \boxed{0} = r.
\end{array}
\tag{3}
$$

The zero remainder shows $r = f(-1) = 0$, and so $-1$ is a zero of $f$. Hence $x - (-1) = x + 1$ is a factor of $f$. Using the quotient found in (3) we can write

$$
f(x) = (x + 1)(3x^3 - 13x^2 + 10x - 2).
\tag{4}
$$

From (4) we see that any other rational zero of $f$ must be a zero of the quotient $3x^3 - 13x^2 + 10x - 2$. Since the latter polynomial is of lower degree, it will be easier to use synthetic division on it rather than on $f(x)$ to check the next rational zero. At this point in the process you should check to see whether the zero just found is a repeated zero. This is done by determining whether the found zero is also a zero of the quotient. A quick check, using synthetic division, shows that $-1$ is *not* a repeated zero of $f$ since it is not a zero of $3x^3 - 13x^2 + 10x - 2$. So we move on and determine whether the number 1 is a rational zero of $f$. Indeed, it is *not* because the division

$$
\begin{array}{r|rrrrl}
1 & 3 & -13 & 10 & -2 & \quad\leftarrow \text{coefficients of the quotient} \\
  &   & 3 & -10 & 0 & \\
\hline
  & 3 & -10 & 0 & \boxed{-2} = r &
\end{array}
\tag{5}
$$

shows that the remainder is $r = -2 \neq 0$. Checking $\frac{1}{3}$, we have

$$
\begin{array}{r|rrrr}
\frac{1}{3} & 3 & -13 & 10 & -2 \\
  &   & 1 & -4 & 2 \\
\hline
  & 3 & -12 & 6 & \boxed{0} = r.
\end{array}
\tag{6}
$$

Thus $\frac{1}{3}$ is a zero. At this point we can stop using synthetic division since (6) indicates that the remaining factor of $f$ is the quadratic polynomial $3x^2 - 12x + 6$. From the quadratic formula we find that the remaining real zeros are $2 + \sqrt{2}$ and $2 - \sqrt{2}$. Therefore the given polynomial function $f$ has two rational zeros, $-1$ and $\frac{1}{3}$, and two irrational zeros, $2 + \sqrt{2}$ and $2 - \sqrt{2}$. ∎

If you have access to technology, your selection of rational numbers to test in Example 1 can be motivated by a graph of the function $f(x) = 3x^4 - 10x^3 - 3x^2 + 8x - 2$. With the aid of a graphing utility we obtain the graphs in **FIGURE 3.4.1**. In Figure 3.4.1(a) it would appear that $f$ has at least three real zeros. But by "zooming-in" on

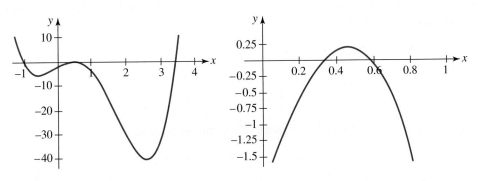

(a) Graph of $f$ on the interval $[-1, 4]$      (b) Zoom-in of graph on the interval $[0, 1]$

**FIGURE 3.4.1** Graph of function $f$ in Example 1

the graph on the interval $[0, 1]$, Figure 3.4.1(b) reveals that $f$ actually has four real zeros: one negative and three positive. Thus, once you have determined one negative rational zero of $f$ you may disregard all other negative numbers as potential zeros.

### ◼ EXAMPLE 2                    Complete Factorization

Since the function $f$ in Example 1 is of degree 4 and we have found four real zeros, we can give its complete factorization. Using the leading coefficient $a_4 = 3$, it follows from (6) of Section 3.3 that

$$f(x) = 3(x + 1)(x - \tfrac{1}{3})(x - (2 - \sqrt{2}))(x - (2 + \sqrt{2}))$$
$$= 3(x + 1)(x - \tfrac{1}{3})(x - 2 + \sqrt{2})(x - 2 - \sqrt{2}).$$                    ◼

### ◼ EXAMPLE 3                    Rational Zeros

Find all rational zeros of $f(x) = x^4 + 4x^3 + 5x^2 + 4x + 4$.

   **Solution** In this case the constant term is $a_0 = 4$ and the leading coefficient is $a_4 = 1$. The integer factors of $a_0$ and $a_4$ are, respectively:

$$p\colon \pm 1, \pm 2, \pm 4,$$
$$s\colon \pm 1.$$

The list of all possible rational zeros $p/s$ is:

$$\frac{p}{s}\colon \pm 1, \pm 2, \pm 4.$$

Since all the coefficients of $f$ are positive, substituting a positive number from the foregoing list into $f(x)$ can never result in $f(x) = 0$. Thus the only numbers that are potential rational zeros are $-1$, $-2$, and $-4$. From the synthetic division

$$
\begin{array}{r|rrrrr}
-1 & 1 & 4 & 5 & 4 & 4 \\
   &   & -1 & -3 & -2 & -2 \\
\hline
   & 1 & 3 & 2 & 2 & 2 = r
\end{array}
$$

we see that $-1$ is not a zero. However, from

$$
\begin{array}{r|rrrrr}
-2 & 1 & 4 & 5 & 4 & 4 \\
   &   & -2 & -4 & -2 & -4 \\
\hline
   & 1 & 2 & 1 & 2 & 0 = r
\end{array}
$$

we see $-2$ is a zero. We now test to see whether $-2$ is a repeated zero. Using the coefficients in the quotient,

$$
\begin{array}{r|rrrr}
-2 & 1 & 2 & 1 & 2 \\
   &   & -2 & 0 & -2 \\
\hline
   & 1 & 0 & 1 & 0 = r
\end{array}
$$

it follows that $-2$ is a zero of multiplicity 2. So far we have shown that

$$f(x) = (x + 2)^2(x^2 + 1).$$

Since the zeros of $x^2 + 1$ are the complex conjugates $i$ and $-i$, we can conclude that $-2$ is the only real zero of $f(x)$.                    ◼

## EXAMPLE 4      No Rational Zeros

Consider the polynomial function $f(x) = x^5 - 4x - 1$. The only possible rational zeros are $-1$ and $1$, and it is easy to see that neither $f(-1)$ nor $f(1)$ are 0. Thus $f$ has no rational zeros. Since $f$ is of odd degree we know that it has at least one real zero, and so that zero must be an irrational number. With the aid of a graphing utility we obtain the graph in FIGURE 3.4.2. Note in the figure that the graph to the right of $x = 2$ cannot turn back down *and* the graph to the left of $x = -2$ cannot turn back up, so that the graph crosses the $x$-axis five times because that shape of the graph would be inconsistent with the end behavior of $f$. Thus we can conclude that the function $f$ possesses three irrational real zeros and two complex conjugate zeros. The best we can do here is to approximate these zeros. Using a computer algebra system such as *Mathematica* we can approximate both the real and the complex zeros. We find these approximations to be $-1.34$, $-0.25, 1.47, 0.061 + 1.42i$, and $0.061 - 1.42i$. ■

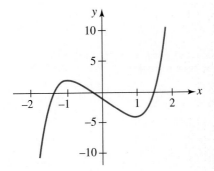

FIGURE 3.4.2 Graph of $f$ in Example 4

Although the Rational Zeros Theorem requires that the coefficients of a polynomial function $f$ be integers, in some circumstances we can apply the theorem to a polynomial function with some real *noninteger* coefficients. The next example illustrates the concept.

## EXAMPLE 5      Noninteger Coefficients

Find the rational zeros of $f(x) = \frac{5}{6}x^4 - \frac{23}{12}x^3 + \frac{10}{3}x^2 - 3x - \frac{3}{4}$.

**Solution** By multiplying $f$ by the least common denominator 12 of all the rational coefficients, we obtain a new function $g$ with integer coefficients:

$$g(x) = 10x^4 - 23x^3 + 40x^2 - 36x - 9.$$

In other words, $g(x) = 12f(x)$. If $c$ is a zero of the function $g$, then $c$ is also zero of $f$ because $g(c) = 0 = 12f(c)$ implies $f(c) = 0$. After working through the numbers in the list of potential rational zeros

$$\frac{p}{s}: \pm 1, \ \pm 3, \ \pm 9, \ \pm \tfrac{1}{2}, \ \pm \tfrac{3}{2}, \ \pm \tfrac{9}{2}, \ \pm \tfrac{1}{5}, \ \pm \tfrac{3}{5}, \ \pm \tfrac{9}{5}, \ \pm \tfrac{1}{10}, \ \pm \tfrac{3}{10}, \ \pm \tfrac{9}{10},$$

we find that $-\frac{1}{5}$ and $\frac{3}{2}$ are zeros of $g$, and hence are zeros of $f$. ■

---

**3.4**    **Exercises**    Answers to selected odd-numbered problems begin on page ANS–10.

In Problems 1–20, find all rational zeros of the given polynomial function $f$.

**1.** $f(x) = 5x^3 - 3x^2 + 8x + 4$
**2.** $f(x) = 2x^3 + 3x^2 - x + 2$
**3.** $f(x) = x^3 - 8x - 3$
**4.** $f(x) = 2x^3 - 7x^2 - 17x + 10$
**5.** $f(x) = 4x^4 - 7x^2 + 5x - 1$
**6.** $f(x) = 8x^4 - 2x^3 + 15x^2 - 4x - 2$
**7.** $f(x) = x^4 + 2x^3 + 10x^2 + 14x + 21$
**8.** $f(x) = 3x^4 + 5x^2 + 1$
**9.** $f(x) = 6x^4 - 5x^3 - 2x^2 - 8x + 3$
**10.** $f(x) = x^4 + 2x^3 - 2x^2 - 6x - 3$
**11.** $f(x) = x^4 + 6x^3 - 7x$
**12.** $f(x) = x^5 - 2x^2 - 12x$

3.4 Real Zeros of Polynomial Functions      **157**

**13.** $f(x) = x^5 + x^4 - 5x^3 + x^2 - 6x$

**14.** $f(x) = 128x^6 - 2$

**15.** $f(x) = \frac{1}{2}x^3 - \frac{9}{4}x^2 + \frac{17}{4}x - 3$

**16.** $f(x) = 0.2x^3 - x + 0.8$

**17.** $f(x) = 2.5x^3 + x^2 + 0.6x + 0.1$

**18.** $f(x) = \frac{3}{4}x^3 + \frac{9}{4}x^2 + \frac{5}{3}x + \frac{1}{3}$

**19.** $f(x) = 6x^4 + 2x^3 - \frac{11}{6}x^2 - \frac{1}{3}x + \frac{1}{6}$

**20.** $f(x) = x^4 + \frac{5}{2}x^3 + \frac{3}{2}x^2 - \frac{1}{2}x - \frac{1}{2}$

In Problems 21–30, find all real zeros of the given polynomial function $f$. Then factor $f(x)$ using only real numbers.

**21.** $f(x) = 8x^3 + 5x^2 - 11x + 3$

**22.** $f(x) = 6x^3 + 23x^2 + 3x - 14$

**23.** $f(x) = 10x^4 - 33x^3 + 66x - 40$

**24.** $f(x) = x^4 - 2x^3 - 23x^2 + 24x + 144$

**25.** $f(x) = x^5 + 4x^4 - 6x^3 - 24x^2 + 5x + 20$

**26.** $f(x) = 18x^5 + 75x^4 + 47x^3 - 52x^2 - 11x + 3$

**27.** $f(x) = 4x^5 - 8x^4 - 24x^3 + 40x^2 - 12x$

**28.** $f(x) = 6x^5 + 11x^4 - 3x^3 - 2x^2$

**29.** $f(x) = 16x^5 - 24x^4 + 25x^3 + 39x^2 - 23x + 3$

**30.** $f(x) = x^6 - 12x^4 + 48x^2 - 64$

In Problems 31–36, find all real solutions of the given equation.

**31.** $2x^3 + 3x^2 + 5x + 2 = 0$

**32.** $x^3 - 3x^2 = -4$

**33.** $2x^4 + 7x^3 - 8x^2 - 25x - 6 = 0$

**34.** $9x^4 + 21x^3 + 22x^2 + 2x - 4 = 0$

**35.** $x^5 - 2x^4 + 2x^3 - 4x^2 + 5x - 2 = 0$

**36.** $8x^4 - 6x^3 - 7x^2 + 6x - 1 = 0$

In Problems 37 and 38, find a polynomial function $f$ of the indicated degree with integer coefficients that possesses the given rational zeros.

**37.** degree 4; $-4, \frac{1}{3}, 1, 3$

**38.** degree 5; $-2, -\frac{2}{3}, \frac{1}{2}, 1$ (multiplicity two)

**39.** Use the Intermediate Value Theorem (see Section 3.1) to show that the polynomial function $f(x) = 4x^3 - 11x^2 + 14x - 6$ has a zero in the interval $[0, 1]$. Find the zero.

**40.** List, but do not test, all possible rational zeros of

$$f(x) = 24x^3 - 14x^2 + 36x + 105.$$

In Problems 41 and 42, find a cubic polynomial function $f$ that satisfies the given conditions.

**41.** rational zeros 1 and 2, $f(0) = 1$ and $f(-1) = 4$

**42.** rational zero $\frac{1}{2}$, irrational zeros $1 + \sqrt{3}$ and $1 - \sqrt{3}$, coefficient of $x$ is 2

## Miscellaneous Calculus-Related Problems

**43. Construction of a Box** A box with no top is made from a square piece of cardboard by cutting square pieces from each corner and then folding up the sides. See **FIGURE 3.4.3**. The length of one side of the cardboard is 10 inches. Find the length of one side of the squares that were cut from the corners if the volume of the box is 48 in³.

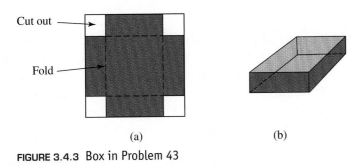

(a)                                          (b)

**FIGURE 3.4.3** Box in Problem 43

**44. Deflection of a Beam** A cantilever beam 20 ft long with a load of 600 lb at its right end is deflected by an amount $d(x) = \frac{1}{16,000}(60x^2 - x^3)$, where $d$ is measured in inches and $x$ in feet. See **FIGURE 3.4.4**. Find $x$ when the deflection is 0.1215 in. When the deflection is 1 in.

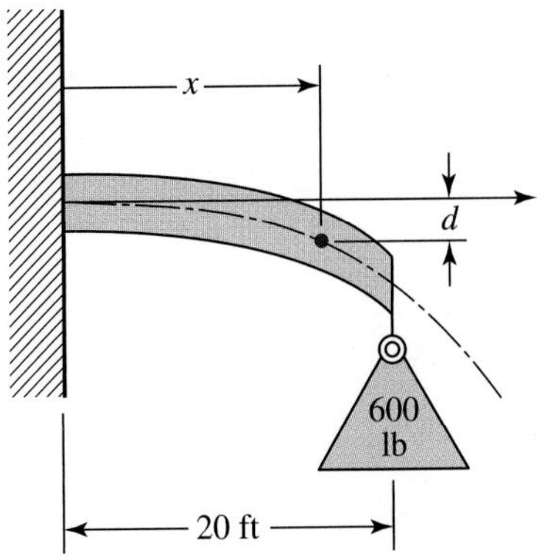

**FIGURE 3.4.4** Cantilever beam in Problem 44

## For Discussion

**45.** Discuss: What is the maximum number of times the graphs of the polynomial functions

$$f(x) = a_3x^3 + a_2x^2 + a_1x + a_0 \quad \text{and} \quad g(x) = b_2x^2 + b_1x + b_0$$

can intersect?

**46.** Consider the polynomial function $f(x) = x^n + a_{n-1}x^{n-1} + \cdots + a_1x + a_0$, where the coefficients $a_{n-1}, \ldots, a_1, a_0$ are nonzero even integers. Discuss why $-1$ and $1$ cannot be zeros of $f(x)$.

# 3.5 Rational Functions

☐ **Introduction** Many functions are built up out of polynomial functions by means of arithmetic operations and function composition (see Section 2.6). In this section we construct a class of functions by forming the quotient of two polynomial functions.

> **RATIONAL FUNCTION**
>
> A **rational function** $y = f(x)$ is a function of the form
>
> $$f(x) = \frac{P(x)}{Q(x)}, \tag{1}$$
>
> where $P$ and $Q$ are polynomial functions.

For example, the following functions are rational functions:

$$y = \frac{x}{x^2 + 5}, \qquad y = \frac{\overset{\text{polynomial}}{\overset{\downarrow}{x^3 - x + 7}}}{\underset{\uparrow}{\underset{\text{polynomial}}{x + 3}}}, \qquad y = \frac{1}{x}.$$

The function

$$y = \frac{\sqrt{x}}{x^2 - 1} \qquad \leftarrow \text{not a polynomial}$$

is not a rational function. In (1) we cannot allow the denominator to be zero. So the **domain** of a rational function $f(x) = P(x)/Q(x)$ is the set of all real numbers *except* those numbers for which the denominator $Q(x)$ is zero. For example, the domain of the rational function $f(x) = (2x^3 - 1)/(x^2 - 9)$ is $\{x \mid x \neq -3, x \neq 3\}$ or $(-\infty, -3) \cup (-3, 3) \cup (3, \infty)$. It goes without saying that we also disallow the zero polynomial $Q(x) = 0$ as a denominator.

□ **Graphs** Graphing a rational function $f$ is a little more complicated than graphing a polynomial function because in addition to paying attention to

- intercepts,
- symmetry, and
- shifting/reflecting/stretching of known graphs,

you should also keep an eye on

- the domain of $f$, and
- the degrees of $P(x)$ and $Q(x)$.

The latter two topics are important in determining whether a graph of a rational function possesses *asymptotes*.

The $y$-intercept is the point $(0, f(0))$, provided the number 0 is in the domain of $f$. For example, the graph of the rational function $f(x) = (1 - x)/x$ does not cross the $y$-axis since $f(0)$ is not defined. If the polynomials $P(x)$ and $Q(x)$ have no common factors, then the $x$-intercepts of the graph of the rational function $f(x) = P(x)/Q(x)$ are the points whose $x$-coordinates are the real zeros of the numerator $P(x)$. In other words, the only way we can have $f(x) = P(x)/Q(x) = 0$ is to have $P(x) = 0$. The graph of a rational function $f$ is symmetric with respect to the $y$-axis if $f(-x) = f(x)$, and symmetric with respect to the origin if $f(-x) = -f(x)$. Since it is easy to spot an even or an odd polynomial function (see page 135), here is an easy way to determine symmetry of the graph of a rational function.

- The quotient of two even functions is even. (2)
- The quotient of two odd functions is even. (3)
- The quotient of an even and an odd function is odd. (4)

See Problem 48 in Exercises 3.5.

We have already seen the graphs of two simple rational functions, $y = 1/x$ and $y = 1/x^2$, in Figures 2.2.1(e) and 2.2.1(f). You are encouraged to review those graphs at this time. Note that $P(x) = 1$ is an even function and $Q(x) = x$ is an odd function, so $y = 1/x$ is an odd function by (4). On the other hand, $P(x) = 1$ is an even function and $Q(x) = x^2$ is an even function, so $y = 1/x^2$ is an even function by (2).

**EXAMPLE 1**  **Shifted Reciprocal Function**

Graph the function $f(x) = \dfrac{2}{x - 1}$.

**Solution**  The graph possesses no symmetry since $Q(x) = x - 1$ is neither even nor odd. Since $f(0) = -2$, the $y$-intercept is $(0, -2)$. Because $P(x) = 2$ is never 0, there are no $x$-intercepts. You might also recognize that the graph of this rational function is the graph of the reciprocal function $y = 1/x$ stretched vertically by a factor of 2 and shifted 1 unit to the right. The point $(1, 1)$ is on the graph of $y = 1/x$; in **FIGURE 3.5.1**, after

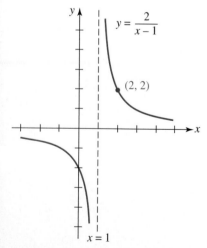

**FIGURE 3.5.1** Graph of function in Example 1

the vertical stretch and horizontal shift, the corresponding point on the graph of $y = 2/(x - 1)$ is $(2, 2)$.

The vertical line $x = 1$ and the horizontal line $y = 0$ (the equation of the $x$-axis) are of special importance for this graph.

The vertical dashed line $x = 1$ in Figure 3.5.1 is the $y$-axis in Figure 2.2.1(e) shifted 1 unit to the right. Although the number 1 is not in the domain of the given function, we can evaluate $f$ at values of $x$ that are *near* 1. For example, you should verify that

| $x$ | 0.999 | 1.001 |
|---|---|---|
| $f(x)$ | $-2000$ | 2000 |

(5)

The table in (5) shows that for values of $x$ close to 1, the corresponding function values $f(x)$ are large in absolute value. On the other hand, for values of $x$ for which $|x|$ is large, the corresponding function values $f(x)$ are near 0. For example, you should verify that

| $x$ | $-999$ | 1001 |
|---|---|---|
| $f(x)$ | $-0.002$ | 0.002 |

(6)

Geometrically, as $x$ approaches 1, the graph of the function approaches the vertical line $x = 1$, and as $|x|$ increases without bound the graph of the function approaches the horizontal line $y = 0$. ∎

☐ **Asymptotes** Recall from Section 1.5 that to indicate that $x$ is approaching a number $a$, we use the notation

- $x \to a^-$ to mean that $x$ is approaching $a$ from the *left*, that is, through numbers that are less than $a$;
- $x \to a^+$ to mean that $x$ is approaching $a$ from the *right*, that is, through numbers that are greater than $a$; and
- $x \to a$ to mean that $x$ is approaching $a$ from both the *left* and the *right*.

We also use the infinity symbols and the notation

- $x \to -\infty$ to mean that $x$ becomes *unbounded in the negative direction*, and
- $x \to \infty$ to mean that $x$ becomes *unbounded in the positive direction*.

Similar interpretations are given to the symbols $f(x) \to -\infty$ and $f(x) \to \infty$. These notational devices are a convenient way of describing the behavior of a function either near a number $x = a$ or as $x$ increases to the right or decreases to the left. Thus, in Example 1 it is apparent from (5) and Figure 3.5.1 that

$$f(x) \to -\infty \text{ as } x \to 1^- \qquad \text{and} \qquad f(x) \to \infty \text{ as } x \to 1^+.$$

In words, the notation in the preceding line signifies that the function values are decreasing without bound as $x$ approaches 1 from the left, and the function values are increasing without bound as $x$ approaches 1 from the right. From (6) and Figure 3.5.1 it should also be apparent that

$$f(x) \to 0 \text{ as } x \to -\infty \qquad \text{and} \qquad f(x) \to 0 \text{ as } x \to \infty.$$

In Figure 3.5.1, the vertical line whose equation is $x = 1$ is called a **vertical asymptote** for the graph of $f$, and the horizontal line whose equation is $y = 0$ is called a **horizontal asymptote** for the graph of $f$.

In this section we will examine three types of asymptotes, which correspond to the three types of lines studied in Section 2.3: *vertical lines*, *horizontal lines*, and *slant* (or oblique) *lines*. The characteristic of any asymptote is that the graph of a function $f$ must get close to, or approach, the line.

## VERTICAL ASYMPTOTE

A line $x = a$ is said to be a **vertical asymptote** for the graph of a function $f$ if at least one of the following six statements is true:

$$
\begin{array}{llll}
f(x) \to -\infty & \text{as} & x \to a^-, & f(x) \to \infty & \text{as} & x \to a^-, \\
f(x) \to -\infty & \text{as} & x \to a^+, & f(x) \to \infty & \text{as} & x \to a^+, \quad (7) \\
f(x) \to -\infty & \text{as} & x \to a, & f(x) \to \infty & \text{as} & x \to a.
\end{array}
$$

**FIGURE 3.5.2** illustrates four of the possibilities listed in (7) for the unbounded behavior of a function $f$ near a vertical asymptote $x = a$. If the function exhibits the *same kind of unbounded behavior from both sides of $x = a$*, then we write either

$$
f(x) \to \infty \qquad \text{as} \qquad x \to a, \qquad (8)
$$

or

$$
f(x) \to -\infty \qquad \text{as} \qquad x \to a. \qquad (9)
$$

In Figure 3.5.2(d) we see that $f(x) \to \infty$ as $x \to a^-$ *and* $f(x) \to \infty$ as $x \to a^+$, and so we write $f(x) \to \infty$ as $x \to a$.

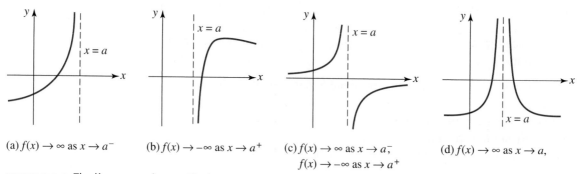

(a) $f(x) \to \infty$ as $x \to a^-$

(b) $f(x) \to -\infty$ as $x \to a^+$

(c) $f(x) \to \infty$ as $x \to a^-$,
$f(x) \to -\infty$ as $x \to a^+$

(d) $f(x) \to \infty$ as $x \to a$,

**FIGURE 3.5.2** The line $x = a$ is a vertical asymptote

If $x = a$ is a vertical asymptote for the graph of a *rational function* $f(x) = P(x)/Q(x)$, then the function values $f(x)$ become unbounded as $x$ approaches $a$ from *both sides*, that is, from the right $(x \to a^+)$ *and* from the left $(x \to a^-)$. The graphs in Figures 3.5.2(c) and 3.5.2(d) (or the reflection of these graphs in the $x$-axis) are typical graphs of a rational function with a single vertical asymptote. As can be seen from these figures, a rational function with a vertical asymptote is a **discontinuous function**. There is an infinite break in each graph at $x = a$. As seen in Figures 3.5.2(c) and 3.5.2(d), a single vertical asymptote divides the $xy$-plane into two regions, and within each region there is a single piece or **branch** of the graph of the rational function $f$.

## HORIZONTAL ASYMPTOTE

A line $y = c$ is said to be a **horizontal asymptote** for the graph of a function $f$ if

$$
f(x) \to c \text{ as } x \to -\infty \qquad \text{or} \qquad f(x) \to c \text{ as } x \to \infty. \qquad (10)
$$

CHAPTER 3 POLYNOMIAL AND RATIONAL FUNCTIONS

In **FIGURE 3.5.3** we have illustrated some typical horizontal asymptotes. We note, in conjunction with Figure 3.5.3(d) that, in general, the graph of a function can have at most *two* horizontal asymptotes, but the graph of a *rational function* $f(x) = P(x)/Q(x)$ can have at most *one*. If the graph of a rational function $f$ possesses a horizontal asymptote $y = c$, then as shown in Figure 3.5.3(c),

◀ Remember this.

$$f(x) \to c \text{ as } x \to -\infty \qquad \text{and} \qquad f(x) \to c \text{ as } x \to \infty.$$

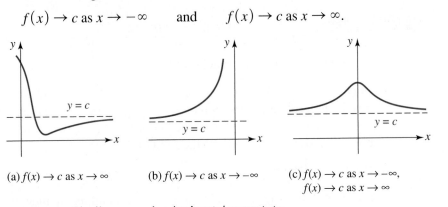

| (a) $f(x) \to c$ as $x \to \infty$ | (b) $f(x) \to c$ as $x \to -\infty$ | (c) $f(x) \to c$ as $x \to -\infty$, $f(x) \to c$ as $x \to \infty$ | (d) $f(x) \to c_1$ as $x \to -\infty$, $f(x) \to c_2$ as $x \to \infty$ |

**FIGURE 3.5.3** The line $y = c$ is a horizontal asymptote

The last line is a mathematical description of the end behavior of the graph of a rational function with a horizontal asymptote. Also, the graph of a function can *never* cross a vertical asymptote but, as suggested in Figure 3.5.3(a), a graph can cross a horizontal asymptote.

## SLANT ASYMPTOTE

A line $y = mx + b$, $m \neq 0$, is said to be a **slant asymptote** for the graph of a function $f$ if

◀ A slant asymptote is also called an oblique asymptote.

$$f(x) \to mx + b \text{ as } x \to -\infty$$

or

$$f(x) \to mx + b \text{ as } x \to \infty. \tag{11}$$

The notation in (11) means that the graph of $f$ possesses a slant asymptote whenever the function values $f(x)$ become closer and closer to the values of $y$ on the line $y = mx + b$ as $x$ becomes large in absolute value. Another way of stating (11) is: A line $y = mx + b$ is a slant asymptote for the graph of $f$ if the vertical distance $d(x)$ between points with the same $x$-coordinate on the two graphs satisfies

$$d(x) = f(x) - (mx + b) \to 0 \text{ as } x \to -\infty \text{ or as } x \to \infty.$$

See **FIGURE 3.5.4**. We note that if a graph of a rational function $f(x) = P(x)/Q(x)$ possesses a slant asymptote it can have vertical asymptotes, but the graph *cannot* have a horizontal asymptote.

On a practical level, vertical and horizontal asymptotes of the graph of a rational function $f$ can be determined by inspection. So for the sake of discussion let us suppose that

$$f(x) = \frac{P(x)}{Q(x)} = \frac{a_n x^n + a_{n-1} x^{n-1} + \cdots + a_1 x + a_0}{b_m x^m + b_{m-1} x^{m-1} + \cdots + b_1 x + b_0}, \quad a_n \neq 0, b_m \neq 0, \quad (12)$$

represents a general rational function.

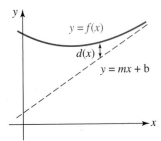

**FIGURE 3.5.4** Slant asymptote is $y = mx + b$

☐ **Finding Vertical Asymptotes** Let us assume that the polynomial functions $P(x)$ and $Q(x)$ in (12) have no common factors. In that case:

• If $a$ is a real number such that $Q(a) = 0$, then the line $x = a$ is a vertical asymptote for the graph of $f$.

Since $Q(x)$ is a polynomial function of degree $m$, it can have up to $m$ real zeros, and so the graph of a rational function $f$ can have up to $m$ vertical asymptotes. If the graph of a rational function $f$ has, say, $k$ $(k \le m)$ vertical asymptotes, then the $k$ vertical lines divide the $xy$-plane into $k + 1$ regions. Thus the graph of this rational function would have $k + 1$ branches.

### ◼ EXAMPLE 2          Vertical Asymptotes

**(a)** Inspection of the rational function $f(x) = \dfrac{2x + 1}{x^2 - 4}$ shows that the denominator $Q(x) = x^2 - 4 = (x + 2)(x - 2) = 0$ at $x = -2$ and $x = 2$. These are equations of vertical asymptotes for the graph of $f$. The graph of $f$ has three branches: one to the left of the line $x = -2$, one between the lines $x = -2$ and $x = 2$, and one to the right of the line $x = 2$.

**(b)** The graph of the rational function $f(x) = \dfrac{1}{x^2 + x + 4}$ has no vertical asymptotes, since $Q(x) = x^2 + x + 4 \ne 0$ for all real numbers.    ◼

☐ **Finding Horizontal Asymptotes** When we discussed end behavior of a polynomial function $P(x)$ of degree $n$, we pointed out that $P(x)$ behaves like $y = a_n x^n$, that is, $P(x) \approx a_n x^n$, for values of $x$ large in absolute value. As a consequence, we see from

$$
\begin{array}{c}
\text{lower powers of } x \text{ are} \\
\text{irrelevant as } x \to \pm\infty \\
\downarrow
\end{array}
$$

$$f(x) = \dfrac{a_n x^n \boxed{+\, a_{n-1}x^{n-1} + \cdots + a_1 x + a_0}}{b_m x^m \boxed{+\, b_{m-1}x^{m-1} + \cdots + b_1 x + b_0}}$$

that $f(x)$ behaves like $y = \dfrac{a_n}{b_m} x^{n-m}$ because $f(x) \approx \dfrac{a_n x^n}{b_m x^m} = \dfrac{a_n}{b_m} x^{n-m}$ for $x \to \pm\infty$.

Therefore:

$$
\begin{array}{c}
0 \\
\downarrow
\end{array}
$$

If $n = m$,    $f(x) \approx \dfrac{a_n}{b_m} x^{n-n} \to \dfrac{a_n}{b_m}$    as $x \to \pm\infty$.    (13)

$$
\begin{array}{c}
\text{negative} \\
\downarrow
\end{array}
$$

If $n < m$,    $f(x) \approx \dfrac{a_n}{b_m} x^{n-m} = \dfrac{a_n}{b_m}\dfrac{1}{x^{m-n}} \to 0$    as $x \to \pm\infty$.    (14)

$$
\begin{array}{c}
\text{positive} \\
\downarrow
\end{array}
$$

If $n > m$,    $f(x) \approx \dfrac{a_n}{b_m} x^{n-m} \to \infty$    as $x \to \pm\infty$.    (15)

From (13), (14), and (15) we glean the following three facts about horizontal asymptotes for the graph of $f(x) = P(x)/Q(x)$:

- If degree of $P(x)$ = degree of $Q(x)$, then $y = a_n/b_m$ (the quotient of the leading coefficients) is a horizontal asymptote.    (16)
- If degree of $P(x)$ < degree of $Q(x)$, then $y = 0$ is a horizontal asymptote.    (17)
- If degree of $P(x)$ > degree of $Q(x)$, then the graph of $f$ has *no* horizontal asymptote.    (18)

EXAMPLE 3 **Horizontal Asymptotes**

Determine whether the graph of each of the following rational functions possesses a horizontal asymptote.

**(a)** $f(x) = \dfrac{3x^2 + 4x - 1}{8x^2 + x}$    **(b)** $f(x) = \dfrac{4x^3 + 7x + 8}{2x^4 + 3x^2 - x + 6}$    **(c)** $f(x) = \dfrac{5x^3 + x^2 + 1}{2x + 3}$

**Solution**

**(a)** Since the degree of the numerator $3x^2 + 4x - 1$ is the same as the degree of the denominator $8x^2 + x$ (both degrees are 2), we see from (13) that

$$f(x) \approx \frac{3}{8}x^{2-2} = \frac{3}{8} \quad \text{as} \quad x \to \pm\infty.$$

As summarized in (16), $y = \frac{3}{8}$ is a horizontal asymptote for the graph of $f$.

**(b)** Since the degree of the numerator $4x^3 + 7x + 8$ is 3 and the degree of the denominator $2x^4 + 3x^2 - x + 6$ is 4 (and $3 < 4$), we see from (14) that

$$f(x) \approx \frac{4}{2}x^{3-4} = \frac{2}{x} \to 0 \quad \text{as} \quad x \to \pm\infty.$$

As summarized in (17), $y = 0$ (the $x$-axis) is a horizontal asymptote for the graph of $f$.

**(c)** Since the degree of the numerator $5x^3 + x^2 - 1$ is 3 and the degree of the denominator $2x + 3$ is 1 (and $3 > 1$), we see from (15) that

$$f(x) \approx \frac{5}{2}x^{3-1} = \frac{5}{2}x^2 \to \infty \quad \text{as} \quad x \to \pm\infty.$$

As summarized in (18), the graph of $f$ has no horizontal asymptote. ∎

In the graphing examples that follow we will assume that $P(x)$ and $Q(x)$ in (12) have no common factors.

EXAMPLE 4 **Graph of a Rational Function**

Graph the function $f(x) = \dfrac{3 - x}{x + 2}$.

**Solution** Here are some things we look at to sketch the graph of $f$.

*Symmetry*: No symmetry. $P(x) = 3 - x$ and $Q(x) = x + 2$ are neither even nor odd.

*Intercepts*: $f(0) = \frac{3}{2}$ and so the $y$-intercept is $(0, \frac{3}{2})$. Setting $P(x) = 0$ or $3 - x = 0$ implies 3 is a zero of $P$. The single $x$-intercept is $(3, 0)$.

*Vertical Asymptotes*: Setting $Q(x) = 0$ or $x + 2 = 0$ gives $x = -2$. The line $x = -2$ is a vertical asymptote.

*Branches*: Because there is only a single vertical asymptote, the graph of $f$ consists of two distinct branches, one to the left of $x = -2$ and one to the right of $x = -2$.

*Horizontal Asymptote*: The degree of $P$ and the degree of $Q$ are the same (namely, 1), and so the graph of $f$ has a horizontal asymptote. By rewriting $f$ as

$f(x) = \dfrac{-x + 3}{x + 2}$ we see that the ratio of leading coefficients is $-1/1 = -1$.

From (16) we see that the line $y = -1$ is a horizontal asymptote.

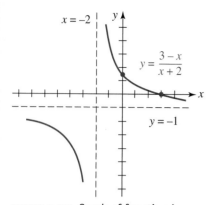

FIGURE 3.5.5 Graph of function in Example 4

*The Graph*: We draw the vertical and horizontal asymptotes using dashed lines. The right branch of the graph of $f$ is drawn through the intercepts $\left(0, \frac{3}{2}\right)$ and $(3, 0)$ in such a manner that it approaches both asymptotes. The left branch is drawn *below* the horizontal asymptote $y = -1$. Were we to draw this branch above the horizontal asymptote it would have to be near the horizontal asymptote from above and near the vertical asymptote from the left. In order to do this the branch of the graph would have to cross the $x$-axis, but since there are no more $x$-intercepts this is impossible. See FIGURE 3.5.5. ∎

**EXAMPLE 5**     **Example 4 Using Transformations**

Long division and rigid transformations can sometimes be an aid in graphing rational functions. Note that if we carry out the long division for the function $f$ in Example 4, then we see that

$$f(x) = \frac{3 - x}{x + 2} \quad \text{is the same as} \quad f(x) = -1 + \frac{5}{x + 2}.$$

Thus, starting with the graph of $y = 1/x$, we stretch it vertically by a factor of 5. Next, shift the graph of $y = 5/x$ two units to the left. Finally, shift $y = 5/(x + 2)$ one unit vertically downward. You should verify that the net result is the graph in Figure 3.5.5. ∎

**EXAMPLE 6**     **Graph of a Rational Function**

Graph the function $f(x) = \dfrac{x}{1 - x^2}$.

**Solution**

*Symmetry*: Since $P(x) = x$ is odd and $Q(x) = 1 - x^2$ is even, the quotient $P(x)/Q(x)$ is odd. The graph of $f$ is symmetric with respect to the origin.

*Intercepts*: $f(0) = 0$, and so the $y$-intercept is $(0, 0)$. Setting $P(x) = x = 0$ gives $x = 0$. Thus the only intercept is $(0, 0)$.

*Vertical Asymptotes*: Setting $Q(x) = 0$ or $1 - x^2 = 0 = 0$ gives $x = -1$ and $x = 1$. The lines $x = -1$ and $x = 1$ are vertical asymptotes.

*Branches*: Because there are two vertical asymptotes, the graph of $f$ consists of three distinct branches, one to the left of the line $x = -1$, one between the lines $x = -1$ and $x = 1$, and one to the right of the line $x = 1$.

*Horizontal Asymptote*: Since the degree of the numerator $x$ is 1 and the degree of the denominator $1 - x^2$ is 2 (and $1 < 2$), it follows from (14) and (17) that $y = 0$ is a horizontal asymptote for the graph of $f$.

*The Graph*: We can plot the graph for $x \geq 0$ and then use symmetry to obtain the remaining part of the graph of $x < 0$. We begin by drawing the vertical asymptotes using dashed lines. The half-branch of the graph of $f$ on the interval $[0, 1)$ is drawn starting at $(0, 0)$. The function $f$ must then increase because $P(x) = x > 0$, and $Q(x) = 1 - x^2 > 0$ indicates that $f(x) > 0$ for $0 < x < 1$. This implies that near the vertical asymptote $x = 1$, $f(x) \to \infty$ as $x \to 1^-$. The branch of the graph for $x > 1$ is drawn below the horizontal asymptote $y = 0$, since $P(x) = x > 0$ and $Q(x) = 1 - x^2 < 0$ imply $f(x) < 0$. Thus $f(x) \to -\infty$ as $x \to 1^+$ and $f(x) \to 0$ as $x \to \infty$. The remainder of the graph for $x < 0$ is obtained by reflecting the graph for $x > 0$ through the origin. See FIGURE 3.5.6. ∎

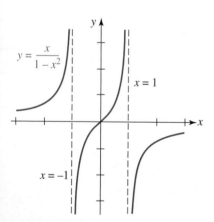

FIGURE 3.5.6 Graph of function in Example 6

EXAMPLE 7          **Graph of a Rational Function**

Graph the function $f(x) = \dfrac{x}{1 + x^2}$.

**Solution** The given function $f$ is similar to the function in Example 6 in that $f$ is an odd function, $(0, 0)$ is the only intercept of its graph, and its graph has the horizontal asymptote $y = 0$. However, note that since $1 + x^2 > 0$ for all real numbers, there are no vertical asymptotes. Thus there are no branches; the graph is one continuous curve. For $x \geq 0$, the graph passes through $(0, 0)$ and then must increase since $f(x) > 0$ for $x > 0$. Also, $f$ must attain a relative maximum and then decrease in order to satisfy the condition $f(x) \to 0$ as $x \to \infty$. As mentioned in Section 3.1, the exact location of this relative maximum can be obtained through calculus techniques. Finally, we reflect the portion of the graph for $x > 0$ through the origin. The graph must look something like that shown in **FIGURE 3.5.7**. ∎

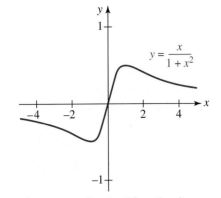

**FIGURE 3.5.7** Graph of function in Example 7

☐ **Finding Slant Asymptotes** Let us again assume that the polynomials $P(x)$ and $Q(x)$ in (12) have no common factors. In that case we can recognize the existence of a slant asymptote in the following manner:

- If the degree of $P(x)$ is precisely one greater than the degree of $Q(x)$, that is, if the degree of $Q(x)$ is $m$ and the degree of $P(x)$ is $n = m + 1$, then the graph of $f$ possesses a slant asymptote.

We find the slant asymptote by division. Using long division to divide $P(x)$ by $Q(x)$ yields a quotient that is a linear polynomial $mx + b$ and a polynomial remainder $R(x)$:

$$f(x) = \frac{P(x)}{Q(x)} = mx + b + \frac{R(x)}{Q(x)}. \tag{19}$$

Because the degree of $R(x)$ must be less than the degree of the divisor $Q(x)$, we have $R(x)/Q(x) \to 0$ as $x \to -\infty$ and as $x \to \infty$, and consequently

$$f(x) \to mx + b \text{ as } x \to -\infty \qquad \text{and} \qquad f(x) \to mx + b \text{ as } x \to \infty.$$

In other words, an equation of the slant asymptote is $y = mx + b$, where $mx + b$ is the quotient in (19).

If the denominator $Q(x)$ is a *linear* polynomial, we can then use synthetic division to carry out the long division.

EXAMPLE 8          **Graph with a Slant Asymptote**

Graph the function $f(x) = \dfrac{x^2 - x - 6}{x - 5}$.

**Solution**

*Symmetry*: No symmetry. $P(x) = x^2 - x - 6$ and $Q(x) = x - 5$ are neither even nor odd.

*Intercepts*: $f(0) = \frac{6}{5}$, and so the $y$-intercept is $\left(0, \frac{6}{5}\right)$. Setting $P(x) = 0$ or $x^2 - x - 6 = 0$ or $(x + 2)(x - 3) = 0$ shows that $-2$ and $3$ are zeros of $P(x)$. The $x$-intercepts are $(-2, 0)$ and $(3, 0)$.

*Vertical Asymptotes*: Setting $Q(x) = 0$ or $x - 5 = 0$ gives $x = 5$. The line $x = 5$ is a vertical asymptote.

*Branches*: The graph of $f$ consists of two branches, one to the left of $x = 5$ and one to the right of $x = 5$.

*Horizontal Asymptote*: None.

*Slant Asymptote*: Since the degree of $P(x) = x^2 - x - 6$ (which is 2) is exactly one greater than the degree of $Q(x) = x - 5$ (which is 1), the graph of $f(x)$ has a slant asymptote. To find it, we divide $P(x)$ by $Q(x)$. Because $Q(x)$ is a linear polynomial we can use synthetic division:

$$
\begin{array}{r|rrr}
5 & 1 & -1 & -6 \\
  &   & 5 & 20 \\
\hline
  & 1 & 4 & \underline{14.}
\end{array}
$$

Recall that the latter notation means that

$$y = x + 4 \text{ is the slant asymptote}$$
$$\downarrow$$
$$\frac{x^2 - x - 6}{x - 5} = x + 4 + \frac{14}{x - 5}.$$

Note again that $14/(x - 5) \to 0$ as $x \to \pm\infty$. Hence the line $y = x + 4$ is a slant asymptote.

*The Graph*: Using the foregoing information we obtain the graph in **FIGURE 3.5.8**. The asymptotes are the dashed lines in the figure. ∎

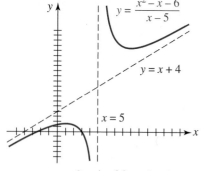

**FIGURE 3.5.8** Graph of function in Example 8

---

**EXAMPLE 9**     **Graph with a Slant Asymptote**

By inspection it should be apparent that the graph of the rational function $f(x) = \dfrac{x^3 - 8x + 12}{x^2 + 1}$ possesses a slant asymptote but no vertical asymptotes. Since the denominator is a quadratic polynomial we resort to long division to obtain

$$\frac{x^3 - 8x + 12}{x^2 + 1} = x + \frac{-9x + 12}{x^2 + 1}.$$

The slant asymptote is the line $y = x$. The graph has no symmetry. The $y$-intercept is $(0, 12)$. The lack of vertical asymptotes indicates that the function $f$ is continuous; its graph consists of an unbroken curve. Because the numerator is a polynomial of odd degree, we know that it has at least one real zero. Since $x^3 - 8x + 12 = 0$ has no rational roots, we use approximation or graphical techniques to show that the equation possesses only one real irrational root. Thus the $x$-intercept is approximately $(-3.4, 0)$. The graph of $f$ is given in **FIGURE 3.5.9**. Notice in the figure that the graph of $f$ crosses the slant asymptote. ∎

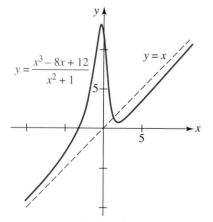

**FIGURE 3.5.9** Graph of function in Example 9

---

☐ **Postscript—Graph with a Hole**  We assumed throughout the preceding discussion of asymptotes of rational functions that the polynomial functions $P(x)$ and $Q(x)$ in (1) have no common factors. We now know that if $a$ is a real number such that $Q(a) = 0$, and $P(x)$ and $Q(x)$ have no common factors, then the line $x = a$ is a vertical asymptote for the graph of $f$. Because $Q$ is a polynomial function, it follows from the Factor Theorem that $Q(x) = (x - a)q(x)$. The assumption that the numerator $P$ and denominator $Q$ have no common factors tells us that $x - a$ is not a factor of $P$ and so $P(a) \neq 0$. When $P(a) = 0$ *and* $Q(a) = 0$, then $x = a$ *may not* be a vertical asymptote. For example, when $a$ is a *simple zero* of both $P$ and $Q$, then $x = a$ is *not* a vertical asymptote for the graph of $f(x) = P(x)/Q(x)$. To see this, we know from the Factor Theorem that if $P(a) = 0$ and $Q(a) = 0$, then $x - a$ is a common factor of $P$ and $Q$:

$$P(x) = (x - a)p(x) \qquad \text{and} \qquad Q(x) = (x - a)q(x),$$

where $p$ and $q$ are polynomials such that $p(a) \neq 0$ and $q(a) \neq 0$. After canceling

$$f(x) = \frac{P(x)}{Q(x)} = \frac{(x - a)p(x)}{(x - a)q(x)} = \frac{p(x)}{q(x)}, \quad x \neq a,$$

we see that $f(x)$ is undefined at $a$, but the function values $f(x)$ do not become unbounded as $x \to a^-$ or as $x \to a^+$ because $q(x)$ is not approaching 0. As an example, we saw in Section 2.5 that the graph of the rational function

$$f(x) = \frac{x^2 - 4}{x - 2} = \frac{(x-2)(x+2)}{x-2} = x+2, \quad x \neq 2,$$

is basically a straight line. But since $f(2)$ is undefined there is no point $(2, f(2))$ on the line. Instead, there is a **hole** in the graph at the point $(2, 4)$. See Figure 2.5.5(a).

## ■ EXAMPLE 10    Graph with a Hole

Graph the function $f(x) = \dfrac{x^2 - 2x - 3}{x^2 - 1}$.

**Solution** Although $x^2 - 1 = 0$ for $x = -1$ and $x = 1$, only $x = 1$ is a vertical asymptote. Note that the numerator $P(x)$ and denominator $Q(x)$ have the common factor $x + 1$, which we cancel provided $x \neq -1$:

equality is true for $x \neq -1$
↓

$$f(x) = \frac{(x+1)(x-3)}{(x+1)(x-1)} = \frac{x-3}{x-1}. \qquad (20)$$

Thus we see from (20) that there is no infinite break in the graph at $x = -1$. We graph $y = \dfrac{x-3}{x-1}$, $x \neq -1$, by observing that the $y$-intercept is $(0, 3)$, an $x$-intercept is $(3, 0)$, a vertical asymptote is $x = 1$, and a horizontal asymptote is $y = 1$. The graph of this function has two branches, but the branch to the left of the vertical asymptote $x = 1$ has a hole in it corresponding to the point $(-1, 2)$. See **FIGURE 3.5.10**. ■

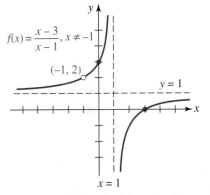

$f(x) = \dfrac{x-3}{x-1}, \; x \neq -1$

$(-1, 2)$

$y = 1$

$x = 1$

**FIGURE 3.5.10** Graph of function in Example 10

## NOTES FROM THE CLASSROOM

When asked whether they have ever heard the statement "*An asymptote is a line that the graph approaches but does not cross,*" a surprising number of students will raise their hands. First, let's make it clear that the statement is false; a graph *can* cross a horizontal asymptote and *can* cross a slant asymptote. A graph can never cross a *vertical* asymptote $x = a$, since the function is inherently undefined at $x = a$. We can even find the points where a graph crosses a horizontal or slant asymptote. For example, the rational function $f(x) = \dfrac{x^2 + 2x}{x^2 - 1}$ has the horizontal asymptote $y = 1$. Determining whether the graph of $f$ crosses the horizontal line $y = 1$ is equivalent to asking whether $y = 1$ is in the range of the function $f$. Setting $f(x)$ equal to 1, that is,

$$\frac{x^2 + 2x}{x^2 - 1} = 1$$

implies $\qquad x^2 + 2x = x^2 - 1 \quad$ and $\quad x = -\frac{1}{2}.$

Since $x = -\frac{1}{2}$ is in the domain of $f$, the graph of $f$ crosses the horizontal asymptote at $\left(-\frac{1}{2}, f\left(-\frac{1}{2}\right)\right) = \left(-\frac{1}{2}, 1\right)$. Observe in Example 9 we can find the point where the slant asymptote crosses the graph of $y = x$ by solving $f(x) = x$. You should verify that the point of intersection is $\left(\frac{4}{3}, \frac{4}{3}\right)$. See Problems 31–36 in Exercises 3.5.

In Problems 1 and 2, use a calculator to fill out the given table for the rational function $f(x) = \dfrac{2x}{x-3}$.

**1.** $x = 3$ is a vertical asymptote for the graph of $f$

| $x$ | 3.1 | 3.01 | 3.001 | 3.0001 | 3.00001 |
|---|---|---|---|---|---|
| $f(x)$ | | | | | |
| $x$ | 2.9 | 2.99 | 2.999 | 2.9999 | 2.99999 |
| $f(x)$ | | | | | |

**2.** $y = 2$ is a horizontal asymptote for the graph of $f$

| $x$ | 10 | 100 | 1000 | 10,000 | 100,000 |
|---|---|---|---|---|---|
| $f(x)$ | | | | | |
| $x$ | $-10$ | $-100$ | $-1000$ | $-10,000$ | $-100,000$ |
| $f(x)$ | | | | | |

In Problems 3–22, find the vertical and horizontal asymptotes for the graph of the given rational function. Find $x$- and $y$-intercepts of the graph. Sketch the graph of $f$.

**3.** $f(x) = \dfrac{1}{x-2}$

**4.** $f(x) = \dfrac{4}{x+3}$

**5.** $f(x) = \dfrac{x}{x+1}$

**6.** $f(x) = \dfrac{x}{2x-5}$

**7.** $f(x) = \dfrac{4x-9}{2x+3}$

**8.** $f(x) = \dfrac{2x+4}{x-2}$

**9.** $f(x) = \dfrac{1-x}{x+1}$

**10.** $f(x) = \dfrac{2x-3}{x}$

**11.** $f(x) = \dfrac{1}{(x-1)^2}$

**12.** $f(x) = \dfrac{4}{(x+2)^3}$

**13.** $f(x) = \dfrac{1}{x^3}$

**14.** $f(x) = \dfrac{8}{x^4}$

**15.** $f(x) = \dfrac{x}{x^2-1}$

**16.** $f(x) = \dfrac{x^2}{x^2-4}$

**17.** $f(x) = \dfrac{1}{x(x-2)}$

**18.** $f(x) = \dfrac{1}{x^2-2x-8}$

**19.** $f(x) = \dfrac{1-x^2}{x^2}$

**20.** $f(x) = \dfrac{16}{x^2+4}$

**21.** $f(x) = \dfrac{-2x^2+8}{(x-1)^2}$

**22.** $f(x) = \dfrac{x(x-5)}{x^2-9}$

In Problems 23–30, find the vertical and slant asymptotes for the graph of the given rational function. Find $x$- and $y$-intercepts of the graph. Sketch the graph $f$.

**23.** $f(x) = \dfrac{x^2 - 9}{x}$

**24.** $f(x) = \dfrac{x^2 - 3x - 10}{x}$

**25.** $f(x) = \dfrac{x^2}{x + 2}$

**26.** $f(x) = \dfrac{x^2 - 2x}{x + 2}$

**27.** $f(x) = \dfrac{x^2 - 2x - 3}{x - 1}$

**28.** $f(x) = \dfrac{-(x - 1)^2}{x + 2}$

**29.** $f(x) = \dfrac{x^3 - 8}{x^2 - x}$

**30.** $f(x) = \dfrac{5x(x + 1)(x - 4)}{x^2 + 1}$

In Problems 31–34, find the point where the graph of $f$ crosses its horizontal asymptote. Sketch the graph of $f$.

**31.** $f(x) = \dfrac{x - 3}{x^2 + 3}$

**32.** $f(x) = \dfrac{(x - 3)^2}{x^2 - 5x}$

**33.** $f(x) = \dfrac{4x(x - 2)}{(x - 3)(x + 4)}$

**34.** $f(x) = \dfrac{2x^2}{x^2 + x + 1}$

In Problems 35 and 36, find the point where the graph of $f$ crosses its slant asymptote. Use a graphing utility to obtain the graph of $f$ and the slant asymptote in the same coordinate plane.

**35.** $f(x) = \dfrac{x^3 - 3x^2 + 2x}{x^2 + 1}$

**36.** $f(x) = \dfrac{x^3 + 2x - 4}{x^2}$

In Problems 37–40, find a rational function that satisfies the given conditions. There is no unique answer.

**37.** vertical asymptote: $x = 2$
horizontal asymptote: $y = 1$
$x$-intercept: $(5, 0)$

**38.** vertical asymptote: $x = 1$
horizontal asymptote: $y = -2$
$y$-intercept: $(0, -1)$

**39.** vertical asymptotes: $x = -1, x = 2$
horizontal asymptote: $y = 3$
$x$-intercept: $(3, 0)$

**40.** vertical asymptote: $x = 4$
slant asymptote: $y = x + 2$

In Problems 41–44, find the asymptotes and any holes in the graph of the given rational function. Find $x$- and $y$-intercepts of the graph. Sketch the graph $f$.

**41.** $f(x) = \dfrac{x^2 - 1}{x - 1}$

**42.** $f(x) = \dfrac{x - 1}{x^2 - 1}$

**43.** $f(x) = \dfrac{x + 1}{x(x^2 + 4x + 3)}$

**44.** $f(x) = \dfrac{x^3 + 8}{x + 2}$

## Miscellaneous Applications

**45. Parallel Resistors** A 5-ohm resistor and a variable resistor are placed in parallel as shown in **FIGURE 3.5.11**. The resulting resistance $R$ (in ohms) is related to the resistance $r$ (in ohms) of the variable resistor by the equation

$$R = \frac{5r}{5 + r}.$$

Sketch the graph of $R$ as a function of $r$ for $r > 0$. What is the resulting resistance $R$ as $r$ becomes very large?

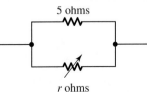

**FIGURE 3.5.11** Parallel resistors in Problem 45

**46. Power** The electrical power $P$ produced by a certain source is given by

$$P = \frac{E^2 r}{R^2 + 2Rr + r^2},$$

where $E$ is the voltage of the source, $R$ is the resistance of the source, and $r$ is the resistance in the circuit. Sketch the graph of $P$ as a function of $r$ using the values $E = 5$ volts and $R = 1$ ohm.

**47. Illumination Intensity** The intensity of illumination from a light source at any point is directly proportional to the strength of the source and inversely proportional to the square of the distance from the source. Given two sources of strengths 16 units and 2 units that are 100 cm apart, as shown in FIGURE 3.5.12, the intensity $I$ at any point $P$ between them is given by

$$I(x) = \frac{16}{x^2} + \frac{2}{(100 - x)^2},$$

where $x$ is the distance from the 16-unit source. Sketch the graph of $I(x)$ on the interval $0 < x < 100$. Describe the behavior of $I(x)$ as $x \to 0^+$. As $x \to 100^-$.

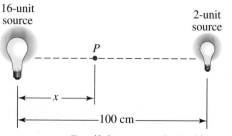

**FIGURE 3.5.12** Two light sources in Problem 47

## For Discussion

**48.** Suppose $f(x) = P(x)/Q(x)$. Prove the symmetry rules (2), (3), and (4) for rational functions.

**49.** Construct a rational function $f(x) = P(x)/Q(x)$ whose graph crosses its slant asymptote twice.

**50.** If you have studied Section 1.5, then discuss how topics in this section and Section 3.2 can be used to help find the limit: $\displaystyle\lim_{x \to 1} \frac{x^5 - 1}{x - 1}$.

## 3.6 Partial Fractions

☐ **Introduction** When two rational functions, say, $f(x) = \dfrac{2}{x + 5}$ and $g(x) = \dfrac{1}{x + 1}$, are added, the terms are combined by means of a common denominator:

$$\frac{2}{x + 5} + \frac{1}{x + 1} = \frac{2}{x + 5}\left(\frac{x + 1}{x + 1}\right) + \frac{1}{x + 1}\left(\frac{x + 5}{x + 5}\right). \qquad (1)$$

Adding numerators on the right-hand side of (1) yields the single rational expression

$$\frac{3x + 7}{(x + 5)(x + 1)}. \tag{2}$$

An important procedure in the study of integral calculus requires that we be able to reverse the process. In other words, we would start with a rational expression such as (2) and then break it down, or *decompose* it, into simpler component fractions $2/(x + 5)$ and $1/(x + 1)$, called **partial fractions**.

☐ **Partial Fractions** The algebraic process for breaking down a rational expression such as (2) into partial fractions is known as **partial fraction decomposition**. For convenience we will assume that the rational function $P(x)/Q(x)$, $Q(x) \neq 0$, is a **proper fraction** or **proper rational expression**, that is, the degree of $P(x)$ is less than the degree of $Q(x)$. We will also assume once again that the polynomials $P(x)$ and $Q(x)$ have no common factors.

In the discussion that follows we consider four cases of partial fraction decomposition of $P(x)/Q(x)$. The cases depend on the factors in the denominator $Q(x)$. When the polynomial $Q(x)$ is factored as a product of $(ax + b)^n$ and $(ax^2 + bx + c)^m$, $n = 1, 2, \ldots$, $m = 1, 2, \ldots$, where the coefficients $a$, $b$, and $c$ are real numbers and the quadratic polynomial $ax^2 + bx + c$ is **irreducible** over the real numbers (that is, does not factor using real numbers), the rational expression $P(x)/Q(x)$ can be decomposed into a sum of partial fractions of the form

$$\frac{C_k}{(ax + b)^k} \quad \text{and} \quad \frac{A_k x + B_k}{(ax^2 + bx + c)^k}.$$

## CASE 1: $Q(x)$ Contains Only Nonrepeated Linear Factors

We state the following fact from algebra without proof. If the denominator can be factored completely into linear factors,

$$Q(x) = (a_1 x + b_1)(a_2 x + b_2) \cdots (a_n x + b_n),$$

where all the $a_i x + b_i$, $i = 1, 2, \ldots, n$ are distinct (that is, no two factors are the same), then unique real constants $C_1, C_2, \ldots, C_n$ can be found such that

$$\frac{P(x)}{Q(x)} = \frac{C_1}{a_1 x + b_1} + \frac{C_2}{a_2 x + b_2} + \cdots + \frac{C_n}{a_n x + b_n}. \tag{3}$$

In practice we will use the letters $A, B, C, \ldots$, in place of the subscripted coefficients $C_1, C_2, C_3, \ldots$. The next example illustrates this first case.

### EXAMPLE 1    Distinct Linear Factors

To decompose $\dfrac{2x + 1}{(x - 1)(x + 3)}$ into individual partial fractions we make the assumption, based on the form given in (3), that the rational function can be written as

$$\frac{2x + 1}{(x - 1)(x + 3)} = \frac{A}{x - 1} + \frac{B}{x + 3}. \tag{4}$$

We now clear (4) of fractions; this can be done by either combining the terms on the right-hand side of the equality over a least common denominator and equating numerators or by simply multiplying both sides of the equality by the denominator $(x - 1)(x + 3)$ on the left-hand side. Either way, we arrive at

$$2x + 1 = A(x + 3) + B(x - 1). \tag{5}$$

Multiplying out the right-hand side of (5) and grouping by powers of $x$ gives

$$2x + 1 = A(x + 3) + B(x - 1) = (A + B)x + (3A - B). \tag{6}$$

Each of the equations (5) and (6) is an identity, which means that the equality is true for *all* real values of $x$. As a consequence, the coefficients of $x$ on the left-hand side of (6) must be the same as the coefficients of the corresponding powers of $x$ on the right-hand side, that is,

$$2x + 1x^0 = (A + B)x + (3A - B)x^0.$$

The result is a system of two equations in two unknowns $A$ and $B$:

$$\begin{aligned} 2 &= A + B \\ 1 &= 3A - B. \end{aligned} \tag{7}$$

By adding the two equations we get $3 = 4A$, and so we find that $A = \frac{3}{4}$. Substituting this value into either equation in (7) then yields $B = \frac{5}{4}$. Hence the desired decomposition is

$$\frac{2x + 1}{(x - 1)(x + 3)} = \frac{\frac{3}{4}}{x - 1} + \frac{\frac{5}{4}}{x + 3}.$$

You are encouraged to verify the foregoing result by combining the terms on the right-hand side of the last equation by means of a common denominator. ∎

☐ **A Shortcut Worth Knowing** If the denominator contains, say, three linear factors such as in $\dfrac{4x^2 - x + 1}{(x - 1)(x + 3)(x - 6)}$, then the partial fraction decomposition looks like this

$$\frac{4x^2 - x + 1}{(x - 1)(x + 3)(x - 6)} = \frac{A}{x - 1} + \frac{B}{x + 3} + \frac{C}{x - 6}.$$

By following the same steps as in Example 1, we would find that the analogue of (7) is now three equations in the three unknowns $A$, $B$, and $C$. The point is this: the more linear factors in the denominator, the larger the system of equations we must solve. There is an algebraic procedure worth learning that can cut down on some of the algebra. To illustrate, let's return to the identity (5). Since the equality is true for every value of $x$, it holds for $x = 1$ and $x = -3$, *the zeros of the denominator*. Setting $x = 1$ in (5) gives $3 = 4A$, from which it follows immediately that $A = \frac{3}{4}$. Similarly, by setting $x = -3$ in (5), we obtain $-5 = (-4)B$ or $B = \frac{5}{4}$.

### CASE 2: $Q(x)$ Contains Repeated Linear Factors

If the denominator $Q(x)$ contains a repeated linear factor $(ax + b)^n, n > 1$, then unique real constants $C_1, C_2, \ldots, C_n$ can be found such that the partial fraction decomposition of $P(x)/Q(x)$ contains the terms

$$\frac{C_1}{ax + b} + \frac{C_2}{(ax + b)^2} + \cdots + \frac{C_n}{(ax + b)^n}. \tag{8}$$

**EXAMPLE 2**      **Repeated Linear Factors**

To decompose $\dfrac{6x - 1}{x^3(2x - 1)}$ into partial fractions, we first observe that the denominator consists of the repeated linear factor $x$ and the nonrepeated linear factor $2x - 1$. Based on the forms in (3) and (8), we assume that

$$\frac{6x - 1}{x^3(2x - 1)} = \overbrace{\frac{A}{x} + \frac{B}{x^2} + \frac{C}{x^3}}^{\text{according to Case 2}} + \overbrace{\frac{D}{2x - 1}}^{\text{according to Case 1}}. \tag{9}$$

Multiplying (9) by $x^3(2x - 1)$ clears it of fractions and yields

$$6x - 1 = Ax^2(2x - 1) + Bx(2x - 1) + C(2x - 1) + Dx^3 \tag{10}$$

or $\quad 6x - 1 = (2A + D)x^3 + (-A + 2B)x^2 + (-B + 2C)x - C.$ (11)

Now the zeros of the denominator in the original expression are $x = 0$ and $x = \frac{1}{2}$. If we then set $x = 0$ and $x = \frac{1}{2}$ in (10), we find, in turn, that $C = 1$ and $D = 16$. Because the denominator of the original expression has only two distinct zeros, we can find $A$ and $B$ by equating the corresponding coefficients of $x^3$ and $x^2$ in (11):

$$0 = 2A + D, \qquad 0 = -A + 2B.$$

◀ The coefficients of $x^3$ and $x^2$ on the left-hand side of (11) are both 0.

Using the known value of $D$, the first equation yields $A = -D/2 = -8$. The second then gives $B = A/2 = -4$. The partial fraction decomposition is

$$\frac{6x - 1}{x^3(2x - 1)} = -\frac{8}{x} - \frac{4}{x^2} + \frac{1}{x^3} + \frac{16}{2x - 1}. \quad ■$$

### CASE 3: $Q(x)$ Contains Nonrepeated Irreducible Quadratic Factors

If the denominator $Q(x)$ contains nonrepeated irreducible quadratic factors $a_i x^2 + b_i x + c_i$, then unique real constants $A_1, A_2, \ldots, A_n, B_1, B_2, \ldots, B_n$ can be found such that the partial fraction decomposition of $P(x)/Q(x)$ contains the terms

$$\frac{A_1 x + B_1}{a_1 x^2 + b_1 x + c} + \frac{A_2 x + B_2}{a_2 x^2 + b_2 x + c_2} + \cdots + \frac{A_n x + B_n}{a_n x^2 + b_n x + c_n}. \tag{12}$$

**EXAMPLE 3**      **Irreducible Quadratic Factors**

To decompose $\dfrac{4x}{(x^2 + 1)(x^2 + 2x + 3)}$ into partial fractions, we first observe that the quadratic polynomials $x^2 + 1$ and $x^2 + 2x + 3$ are irreducible over the real numbers. Hence by (12) we assume that

◀ Use the quadratic formula. For either factor you will find that $b^2 - 4ac < 0$.

$$\frac{4x}{(x^2 + 1)(x^2 + 2x + 3)} = \frac{Ax + B}{x^2 + 1} + \frac{Cx + D}{x^2 + 2x + 3}.$$

After clearing fractions in the preceding line, we find

$$\begin{aligned} 4x &= (Ax + B)(x^2 + 2x + 3) + (Cx + D)(x^2 + 1) \\ &= (A + B)x^3 + (2A + B + D)x^2 + (3A + 2B + C)x + (3B + D). \end{aligned}$$

Because the denominator of the original fraction has no real zeros, we have no recourse except to form a system of equations by comparing coefficients of all powers of $x$:

$$\begin{aligned} 0 &= A + C \\ 0 &= 2A + B + D \\ 4 &= 3A + 2B + C \\ 0 &= 3B + D. \end{aligned}$$

Using $C = -A$ and $D = -3B$ from the first and fourth equations, we can eliminate $C$ and $D$ in the second and third equations:

$$0 = A - B$$
$$2 = A + B.$$

Solving this simpler system of equations yields $A = 1$ and $B = 1$. Hence, $C = -1$ and $D = -3$. The partial fraction decomposition is

$$\frac{4x}{(x^2 + 1)(x^2 + 2x + 3)} = \frac{x + 1}{x^2 + 1} - \frac{x + 3}{x^2 + 2x + 3}.$$  ■

### CASE 4: $Q(x)$ Contains Repeated Irreducible Quadratic Factors

If the denominator $Q(x)$ contains a repeated irreducible quadratic factor $(ax^2 + bx + c)^n$, $n > 1$, then unique real constants $A_1, A_2, \ldots, A_n, B_1, B_2, \ldots, B_n$ can be found such that the partial fraction decomposition of $P(x)/Q(x)$ contains the terms

$$\frac{A_1 x + B}{ax^2 + bx + c} + \frac{A_2 x + B_2}{(ax^2 + bx + c)^2} + \cdots + \frac{A_n x + B_n}{(ax^2 + bx + c)^n}. \qquad (13)$$

### ▌EXAMPLE 4          Repeated Quadratic Factor

Decompose $\dfrac{x^2}{(x^2 + 4)^2}$ into partial fractions.

   **Solution**  The denominator contains only the repeated irreducible quadratic factor $x^2 + 4$. As indicated in (13), we assume a decomposition of the form

$$\frac{x^2}{(x^2 + 4)^2} = \frac{Ax + B}{x^2 + 4} + \frac{Cx + D}{(x^2 + 4)^2}.$$

Clearing fractions by multiplying both sides of the preceding equality by $(x^2 + 4)^2$ gives

$$x^2 = (Ax + B)(x^2 + 4) + Cx + D. \qquad (14)$$

As in Example 3, the denominator of the original has no real zeros, and so we must solve a system of four equations for $A$, $B$, $C$, and $D$. To that end we rewrite (14) as

$$0x^3 + 1x^2 + 0x + 0x^0 = Ax^3 + Bx^2 + (4A + C)x + (4B + D)x^0$$

and compare coefficients of like powers (match the colors) to obtain

$$0 = A$$
$$1 = B$$
$$0 = 4A + C$$
$$0 = 4B + D.$$

From this system we find that $A = 0$, $B = 1$, $C = 0$, and $D = -4$. The required partial fraction decomposition is then

$$\frac{x^2}{(x^2 + 4)^2} = \frac{1}{x^2 + 4} - \frac{4}{(x^2 + 4)^2}.$$  ■

EXAMPLE 5    **Combination of Cases**

Determine the form of the decomposition of $\dfrac{x^2 + 3x + 5}{(x - 5)(x + 2)^2(x^2 + 1)^2}$.

**Solution** The denominator contains a single linear factor $x - 5$, a repeated linear factor $x + 2$, and a repeated irreducible quadratic factor $x^2 + 1$. By Cases 1, 2, and 4 the assumed form of the partial fraction decomposition is

$$\frac{x + 3}{(x - 5)(x + 2)^2(x^2 + 1)^2} = \overbrace{\frac{A}{x - 5}}^{\text{Case 1}} + \overbrace{\frac{B}{x + 2} + \frac{C}{(x + 2)^2}}^{\text{Case 2}} + \overbrace{\frac{Dx + E}{x^2 + 1} + \frac{Fx + G}{(x^2 + 1)^2}}^{\text{Case 4}}$$

∎

## NOTES FROM THE CLASSROOM

We assumed throughout the foregoing discussion that the degree of the numerator $P(x)$ was less than the degree of the denominator $Q(x)$. If, however, the degree of $P(x)$ is greater than or equal to the degree of $Q(x)$, then $P(x)/Q(x)$ is an **improper fraction**. We can still do partial fraction decomposition but the process starts with long division until a polynomial quotient and a proper fraction is attained. For example, long division gives

$$\underset{\text{improper fraction}}{\underbrace{\frac{x^3 + x - 1}{x^2 - 3x}}} = x + 3 + \underset{\text{proper fraction}}{\underbrace{\frac{10x - 1}{x(x - 3)}}}.$$

Then by using Case 1 we finish the problem with the decomposition of the proper-fraction term in the last equality:

$$\frac{x^3 + x - 1}{x^2 - 3x} = x + 3 + \frac{10x - 1}{x(x - 3)} = x + 3 + \frac{\frac{1}{3}}{x} + \frac{\frac{29}{3}}{x - 3}.$$

See Problems 25–30 in Exercises 3.6.

**3.6** **Exercises**   Answers to selected odd-numbered problems begin on page ANS–11.

In Problems 1–24, find the partial fraction decomposition of the given rational expression.

**1.** $\dfrac{1}{x(x + 2)}$

**2.** $\dfrac{2}{x(4x - 1)}$

**3.** $\dfrac{-9x + 27}{x^2 - 4x - 5}$

**4.** $\dfrac{-5x + 18}{x^2 + 2x - 63}$

**5.** $\dfrac{2x^2 - x}{(x + 1)(x + 2)(x + 3)}$

**6.** $\dfrac{1}{x(x - 2)(2x - 1)}$

**7.** $\dfrac{3x}{x^2 - 16}$

**8.** $\dfrac{10x - 5}{25x^2 - 1}$

**9.** $\dfrac{5x - 6}{(x - 3)^2}$

**10.** $\dfrac{5x^2 - 25x + 28}{x^2(x - 7)}$

**11.** $\dfrac{1}{x^2(x + 2)^2}$

**12.** $\dfrac{-4x + 6}{(x - 2)^2(x - 1)^2}$

**13.** $\dfrac{3x - 1}{x^3(x - 1)(x + 3)}$

**14.** $\dfrac{x^2 - x}{x(x + 4)^3}$

**15.** $\dfrac{6x^2 - 7x + 11}{(x - 1)(x^2 + 9)}$

**16.** $\dfrac{2x + 10}{2x^3 + x}$

**17.** $\dfrac{4x^2 + 4x - 6}{(2x - 3)(x^2 - x + 1)}$

**18.** $\dfrac{2x^2 - x + 7}{(x - 6)(x^2 + x + 5)}$

**19.** $\dfrac{t + 8}{t^4 - 1}$

**20.** $\dfrac{y^2 + 1}{y^3 - 1}$

**21.** $\dfrac{x^3}{(x^2 + 2)(x^2 + 1)}$

**22.** $\dfrac{x - 15}{(x^2 + 2x + 5)(x^2 + 6x + 10)}$

**23.** $\dfrac{(x + 1)^2}{(x^2 + 1)^2}$

**24.** $\dfrac{2x^2}{(x - 2)(x^2 + 4)^2}$

In Problems 25–30, first use long division followed by partial fraction decomposition.

**25.** $\dfrac{x^5}{x^2 - 1}$

**26.** $\dfrac{(x + 2)^2}{x(x + 3)}$

**27.** $\dfrac{x^2 - 4x + 1}{2x^2 + 5x + 2}$

**28.** $\dfrac{x^4 + 3x}{x^2 + 2x + 1}$

**29.** $\dfrac{x^6}{x^3 - 2x^2 + x - 2}$

**30.** $\dfrac{x^3 + x^2 - x + 1}{x^3 + 3x^2 + 3x + 1}$

---

## 3.7 The Area Problem

$\int$ **Calculus PREVIEW** □ **Introduction** As we saw in Section 2.9, the fundamental motivating problem of differential calculus, *Find a tangent line to the graph of a function f*, is answered by the notion of the **derivative** of the function. Differential calculus is the study of the properties and applications of the derivative of a function $y = f(x)$. Integral calculus, on the other hand, is the study of the properties and the applications of the **definite integral** of a function $y = f(x)$. As mentioned in Section 2.9, the historical problem that leads to the concept of the definite integral is, *Find the area under the graph of a function f*. We examine the area problem in this section.

□ **Area Under a Graph** Throughout the discussion that follows we will assume that $y = f(x)$ is a function that is continuous and nonnegative on an interval $[a, b]$. Recall that the concept of continuity has been mentioned several times in previous sections; in this case, the graph of $f$ has no breaks, gaps, or holes in it anywhere on the interval $[a, b]$. The requirement that $f$ be nonnegative, that is, $f(x) \geq 0$ for all $x$ in $[a, b]$, means that no portion of its graph on the interval is below the $x$-axis. Specifically, then,

> By the **area under a graph** we mean the area $A$ of the region in the plane bounded by the graph of $f$, the lines $x = a$ and $x = b$, and the $x$-axis.

See **FIGURE 3.7.1**.

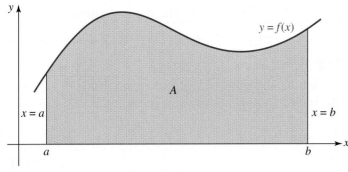

FIGURE 3.7.1 Area $A$ under a graph

To get to the answer of the question, What is the *exact* value of $A$? we begin with a method for systematically *approximating* $A$. The basic idea is simply this: build rectangles across the interval $[a, b]$ and use the sum of the areas of the rectangles as an approximation for $A$.

☐ **Approximating the Area** One possible systematic procedure for approximating the value of the area $A$ under a graph is summarized next.

(*i*) Subdivide the interval $[a, b]$ into $n$ subintervals $[x_{k-1}, x_k]$, where

$$a = x_0 < x_1 < x_2 < \cdots < x_{n-1} < x_n = b,$$

so that each subinterval has the same width $\Delta x = \dfrac{b-a}{n}$. This is called a **regular partition** of the interval $[a, b]$.

(*ii*) Choose a number $x_k^*$ in each of the $n$ subintervals $[x_{k-1}, x_k]$ and form the $n$ products $f(x_k^*) \Delta x$. Since the area of a rectangle is *length* $\times$ *width*, $f(x_k^*) \Delta x$ is the area of the rectangle of length $f(x_k^*)$ and width $\Delta x$ built up on the $k$th subinterval $[x_{k-1}, x_k]$. The $n$ numbers $x_1^*, x_2^*, x_3^*, \ldots, x_n^*$ are called **sample points**. See FIGURE 3.7.2.

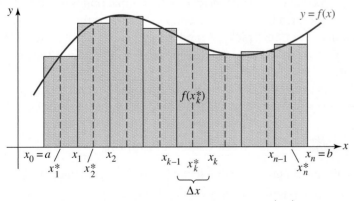

FIGURE 3.7.2 $n$ rectangles of width $\Delta x$ and length $f(x_k^*)$

(*iii*) The sum of the areas of the $n$ rectangles represents an approximation to the value of the area,

$$A \approx f(x_1^*)\Delta x + f(x_2^*)\Delta x + f(x_3^*)\Delta x + \cdots + f(x_n^*)\Delta x. \qquad (1)$$

To simplify the hand calculations, the sample points $x_k^*, k = 1, 2, \ldots, n$, are generally chosen to be either the left-hand endpoint or the right-hand endpoint of each subinterval $[x_{k-1}, x_k]$.

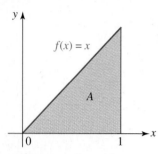

$f(x) = x$

$A$

0    1

**FIGURE 3.7.3** Area $A$ in Example 1

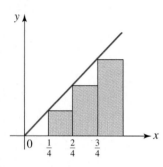

0   $\frac{1}{4}$   $\frac{2}{4}$   $\frac{3}{4}$

(a) Using left-hand endpoints

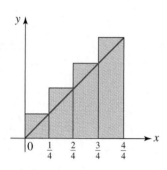

0   $\frac{1}{4}$   $\frac{2}{4}$   $\frac{3}{4}$   $\frac{4}{4}$

(b) Using right-hand endpoints

**FIGURE 3.7.4** Approximating the area $A$ in Example 1

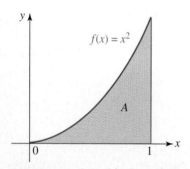

$f(x) = x^2$

$A$

0    1

**FIGURE 3.7.5** Area $A$ in Example 2

**■ EXAMPLE 1**   **Area of a Triangular Region**

Approximate the area $A$ under the graph of $f(x) = x$ on the interval $[0, 1]$ using four subintervals of equal width and choosing

**(a)** $x_k^*$ as the left-hand endpoint of each subinterval, and
**(b)** $x_k^*$ as the right-hand endpoint of each subinterval.

See **FIGURE 3.7.3**.

**Solution**  By dividing $[0, 1]$ into four subintervals, the width of each subinterval is $\Delta x = \frac{1 - 0}{4} = \frac{1}{4}$.

**(a)** If $x_k^*$ is the left-hand endpoint of each of the four subintervals, then $x_1^* = 0$, $x_2^* = \frac{1}{4}$, $x_3^* = \frac{2}{4} = \frac{1}{2}$, and $x_4^* = \frac{3}{4}$. See **FIGURE 3.7.4(a)**. We have from (1),

$$A \approx f(0)\tfrac{1}{4} + f(\tfrac{1}{4})\tfrac{1}{4} + f(\tfrac{1}{2})\tfrac{1}{4} + f(\tfrac{3}{4})\tfrac{1}{4}$$
$$= 0 \cdot \tfrac{1}{4} + \tfrac{1}{4} \cdot \tfrac{1}{4} + \tfrac{1}{2} \cdot \tfrac{1}{4} + \tfrac{3}{4} \cdot \tfrac{1}{4}$$
$$= \tfrac{3}{8} = 0.375.$$

**(b)** If $x_k^*$ is the right-hand endpoint of each of the four subintervals, then $x_1^* = \frac{1}{4}$, $x_2^* = \frac{2}{4} = \frac{1}{2}$, $x_3^* = \frac{3}{4}$, and $x_4^* = \frac{4}{4} = 1$. See Figure 3.7.4(b). We have from (1),

$$A \approx f(\tfrac{1}{4})\tfrac{1}{4} + f(\tfrac{1}{2})\tfrac{1}{4} + f(\tfrac{3}{4})\tfrac{1}{4} + f(1)\tfrac{1}{4}$$
$$= \tfrac{1}{4} \cdot \tfrac{1}{4} + \tfrac{1}{2} \cdot \tfrac{1}{4} + \tfrac{3}{4} \cdot \tfrac{1}{4} + 1 \cdot \tfrac{1}{4}$$
$$= \tfrac{5}{8} = 0.625.$$   ■

As can be seen in Figures 3.7.4(a) and 3.7.4(b), the value obtained in part (a) of Example 1 underestimates the area $A$, whereas the value in part (b) overestimates $A$, that is, $0.375 < A < 0.625$. We can compare these approximations with the actual area. Since the area under the graph of $f(x) = x$ on the interval $[0, 1]$ is the area of a right triangle of base $= 1$ and height $= 1$, the exact area is $A = \frac{1}{2} \cdot$ base $\cdot$ height $= \frac{1}{2} \cdot 1 \cdot 1 = \frac{1}{2} = 0.5$.

There is no special reason that we chose the sample points $x_k^*$, $k = 1, 2, \ldots, n$, to be the left-hand and then the right-hand endpoints of the subintervals $[x_{k-1}, x_k]$, other than *convenience*. We could pick $x_k^*$ randomly in each subinterval. In Problem 3 of Exercises 3.7 you are asked to approximate the area in Example 1 using the midpoint of each subinterval.

Intuitively, the more rectangles we use the better (1) approximates the area $A$ under a graph. The trade-off, of course, is that we must do more calculations.

**■ EXAMPLE 2**   **Area Under a Parabola**

Approximate the area $A$ under the graph of $f(x) = x^2$ on the interval $[0, 1]$ using eight subintervals of equal width and choosing

**(a)** $x_k^*$ as the left-hand endpoint of each subinterval, and
**(b)** $x_k^*$ as the right-hand endpoint of each subinterval.

See **FIGURE 3.7.5**.

**Solution**  By dividing $[0, 1]$ into eight subintervals, the width of each subinterval is $\Delta x = \frac{1 - 0}{8} = \frac{1}{8}$.

**(a)** If $x_k^*$ is the left-hand endpoint of each of the four subintervals, then

$$x_1^* = 0, \ x_2^* = \tfrac{1}{8}, \ x_3^* = \tfrac{2}{8} = \tfrac{1}{4}, \ x_4^* = \tfrac{3}{8}, \ x_5^* = \tfrac{4}{8} = \tfrac{1}{2}, \ x_6^* = \tfrac{5}{8}, \ x_7^* = \tfrac{6}{8} = \tfrac{3}{4}, \ x_8^* = \tfrac{7}{8}.$$

See **FIGURE 3.7.6(a)**. We have from (1),

$$A \approx f(0) \cdot \tfrac{1}{8} + f\left(\tfrac{1}{8}\right) \cdot \tfrac{1}{8} + f\left(\tfrac{1}{4}\right) \cdot \tfrac{1}{8} + f\left(\tfrac{3}{8}\right) \cdot \tfrac{1}{8} + f\left(\tfrac{1}{2}\right) \cdot \tfrac{1}{8} + f\left(\tfrac{5}{8}\right) \cdot \tfrac{1}{8} + f\left(\tfrac{3}{4}\right) \cdot \tfrac{1}{8} + f\left(\tfrac{7}{8}\right) \cdot \tfrac{1}{8}$$
$$= 0 \cdot \tfrac{1}{8} + \tfrac{1}{64} \cdot \tfrac{1}{8} + \tfrac{1}{16} \cdot \tfrac{1}{8} + \tfrac{9}{64} \cdot \tfrac{1}{8} + \tfrac{1}{4} \cdot \tfrac{1}{8} + \tfrac{25}{64} \cdot \tfrac{1}{8} + \tfrac{9}{16} \cdot \tfrac{1}{8} + \tfrac{49}{64} \cdot \tfrac{1}{8}$$
$$= \tfrac{35}{128} = 0.2734375.$$

**(b)** If $x_k^*$ is the right-hand endpoint of each of the four subintervals, then

$$x_1^* = \tfrac{1}{8}, \; x_2^* = \tfrac{2}{8} = \tfrac{1}{4}, \; x_3^* = \tfrac{3}{8}, \; x_4^* = \tfrac{4}{8} = \tfrac{1}{2}, \; x_5^* = \tfrac{5}{8}, \; x_6^* = \tfrac{6}{8} = \tfrac{3}{4}, \; x_7^* = \tfrac{7}{8}, \; x_8^* = \tfrac{8}{8} = 1.$$

See Figure 3.7.6(b). We have from (1),

$$A \approx f\left(\tfrac{1}{8}\right) \cdot \tfrac{1}{8} + f\left(\tfrac{1}{4}\right) \cdot \tfrac{1}{8} + f\left(\tfrac{3}{8}\right) \cdot \tfrac{1}{8} + f\left(\tfrac{1}{2}\right) \cdot \tfrac{1}{8} + f\left(\tfrac{5}{8}\right) \cdot \tfrac{1}{8} + f\left(\tfrac{3}{4}\right) \cdot \tfrac{1}{8} + f\left(\tfrac{7}{8}\right) \cdot \tfrac{1}{8} + f(1) \cdot \tfrac{1}{8}$$
$$= \tfrac{1}{64} \cdot \tfrac{1}{8} + \tfrac{1}{16} \cdot \tfrac{1}{8} + \tfrac{9}{64} \cdot \tfrac{1}{8} + \tfrac{1}{4} \cdot \tfrac{1}{8} + \tfrac{25}{64} \cdot \tfrac{1}{8} + \tfrac{9}{16} \cdot \tfrac{1}{8} + \tfrac{49}{64} \cdot \tfrac{1}{8} + 1 \cdot \tfrac{1}{8}$$
$$= \tfrac{51}{128} = 0.3984375. \quad\blacksquare$$

From Figure 3.7.6(a) we see that the area of the seven rectangles underestimates $A$ in Example 2, whereas the eight rectangles in Figure 3.7.6(b) overestimates $A$. From the calculations in Example 2 we can write $0.2734375 < A < 0.3984375$. But an observation is in order at this point. Don't assume that by using left-hand endpoints followed by the right-hand endpoints of the subintervals for $x_k^*$ that we *always* get, in turn, a lower estimate followed by an upper estimate of the area $A$ under the graph of $f$ on $[a, b]$. This occurred in Examples 1 and 2 simply because, in both cases, the function $f$ was increasing on the interval $[0, 1]$.

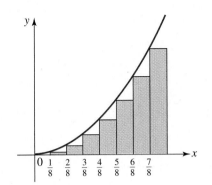

(a) Using left-hand endpoints

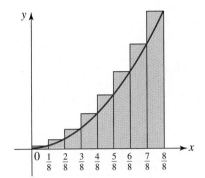

(b) Using right-hand endpoints

**FIGURE 3.7.6** Approximating the area $A$ in Example 2

□ **Summation Notation** Writing out sums such as (1) can become very tedious. To facilitate the discussion of the area problem, a special notation for summation is used in calculus. Suppose $a_k$ denotes a real number that depends on an integer $k$. The sum of $n$ such real numbers $a_k$, $a_1 + a_2 + a_3 + \cdots + a_n$, is denoted by the symbol $\sum_{k=1}^{n} a_k$, that is,

$$\sum_{k=1}^{n} a_k = a_1 + a_2 + a_3 + \cdots + a_n. \tag{2}$$

Since $\Sigma$ is the capital Greek letter sigma, (2) is called **sigma notation** or **summation notation**. The integer $k$ is called the **index of summation** and takes on consecutive integer values starting with $k = 1$ and ending with $k = n$. For example, the sum of the first 100 squared positive integers,

$$1^2 + 2^2 + 3^2 + 4^2 + \cdots + 98^2 + 99^2 + 100^2,$$

can be written compactly as

sum ends with this number
↓
$$\sum_{k=1}^{100} k^2.$$
↑
sum starts with this number

Using sigma notation, the sum of the areas in (1) can be written as

$$\sum_{k=1}^{n} f(x_k^*)\Delta x.$$

☐ **Area** It should seem believable that we can reduce the error inherent in the method of approximating an area $A$ under a graph by summing areas of rectangles by using more and more rectangles $(n \to \infty)$ of decreasing width $\left( \Delta x = \dfrac{b - a}{n} \to 0 \right)$. Thus the 32 rectangles in **FIGURE 3.7.7** should give us a better approximation to area $A$ in Figure 3.7.1 than the eight rectangles shown in Figure 3.7.2. Indeed that is the case. It can be proved that when $f$ is continuous on $[a, b]$ and $f(x) \geq 0$ for all $x$ in the interval, the area $A$ under the graph of the function $y = f(x)$ on the interval is given by the limit

$$A = \lim_{n \to \infty} \sum_{k=1}^{n} f(x_k^*) \Delta x. \tag{3}$$

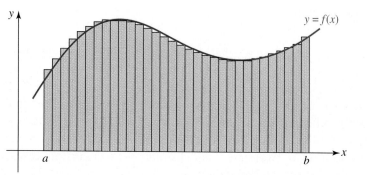

**FIGURE 3.7.7** Using more rectangles improves the approximation to area A.

The limit (3) exists regardless of how the sample points $x_1^*, x_2^*, x_3^*, \ldots, x_n^*$ are chosen in the subintervals $[x_0, x_1], [x_1, x_2], [x_2, x_3], \ldots, [x_{n-1}, x_n]$. Thus in (3), each sample point $x_k^*$ could always be chosen, say, to be the right-hand endpoint of each subinterval. Since we are in no position to deal, in general terms, with limits of the kind given in (3), we leave that aspect of the area problem to a course in calculus. But if you are willing to put in the time to work Problems 15–22, then Problems 23 and 24 will give you a small taste of what is involved in computing area $A$ by the limiting process given in (3).

We said at the start that the area problem is the motivating problem for the definite integral. You ask: So what is a definite integral? It is now just a small jump from (3) to the concept of the definite integral.

### DEFINITE INTEGRAL

Let the function $f$ be continuous on $[a, b]$. The **definite integral** of $f$ from $x = a$ to $x = b$, denoted by $\int_a^b f(x)dx$, is

$$\int_a^b f(x)dx = \lim_{n \to \infty} \sum_{k=1}^{n} f(x_k^*) \Delta x. \tag{4}$$

The integral symbol $\int$ in (4), as used by Wilhelm Gottfried Leibniz (who is considered the co-inventor of calculus, along with Isaac Newton), is simply an elongated S for the word "sum."

# NOTES FROM THE CLASSROOM

(*i*) If read quickly, you might conclude that formula (4) is the same as (3). In a way this is correct; however, (4) is a more general concept (notice that we are not requiring $f$ to be nonnegative on the interval $[a, b]$). Thus, *a definite integral need not be area.* Also, in its most general setting, even the conditions of continuity of $f$ and the use of a regular partition are dropped in the definition of the definite integral. What, then, is a definite integral? For now, accept the fact that a definite integral is simply a real number that can be negative, zero, or positive. When the conditions of continuity and nonnegativity are imposed on $y = f(x)$ on the interval $[a, b]$, then the area under the graph is $A = \int_a^b f(x)\,dx$. Also, you should be aware that the interpretations of derivative and the definite integral are much broader that just slopes of tangent lines and areas under graphs. As you progress through courses in mathematics, sciences, and engineering you will see many diverse applications of the derivative and the definite integral.

(*ii*) In this chapter we worked principally with polynomial functions. Polynomial functions are the fundamental building blocks of a class known as **algebraic functions**. In Section 3.5 we saw that a rational function is the quotient of two polynomial functions. In general, an algebraic function $f$ involves a finite number of additions, subtractions, multiplications, divisions, and roots of polynomial functions. Thus

$$y = 2x^2 - 5x,\ y = \sqrt[3]{x^2},\ y = x^4 + \sqrt{x^2 + 5},\ \text{and}\ y = \frac{\sqrt{x}}{x^3 - 2x^2 + 7}$$

are algebraic functions. Indeed, all the functions in Chapters 2 and 3 are algebraic functions. Starting with the next chapter we consider functions that belong to a different class known as **transcendental functions**. A transcendental function $f$ is defined to be one that is not algebraic. The six trigonometric functions (Chapter 4) and the exponential and logarithmic functions (Chapter 5) are examples of transcendental functions.

---

## 3.7 | Exercises

Answers to selected odd-numbered problems begin on page ANS–11.

In Problems 1–4, the function $f$ and the interval are given in Example 1.

1. Approximate the area $A$, this time using eight subintervals of equal width and choosing $x_k^*$ as the left-hand endpoint of each subinterval. Draw the eight rectangles.

2. Approximate the area $A$, this time using eight subintervals of equal width and choosing $x_k^*$ as the right-hand endpoint of each subinterval. Draw the eight rectangles.

3. Approximate the area $A$, this time using four subintervals of equal width but choosing $x_k^*$ as the midpoint of each subinterval. Draw the four rectangles.

4. Compare the approximation obtained in Problem 3 of the exact area $A = 0.5$. Explain why your answer in Problem 3 is not surprising.

5. Approximate the area under the graph of $f(x) = x + 2$ on the interval $[-1, 2]$ using six subintervals of equal width and choosing:
   (a) $x_k^*$ as the left-hand endpoint of each subinterval, and
   (b) $x_k^*$ as the right-hand endpoint of each subinterval.
6. Repeat Problem 5 using twelve subintervals of equal width.
7. Approximate the area under the graph of $f(x) = -x^2 + 5x$ on the interval $[0, 5]$ using five subintervals of equal width and choosing:
   (a) $x_k^*$ as the left-hand endpoint of each subinterval, and
   (b) $x_k^*$ as the right-hand endpoint of each subinterval.
8. Repeat Problem 7 using ten subintervals of equal width.
9. Approximate the area under the graph of $f(x) = -x^2 + 5x$ on the interval $[0, 5]$ using five subintervals of equal width and choosing $x_k^*$ as the midpoints of each subinterval.
10. Approximate the area under the graph of $f(x) = -x^3 + 2x^2$ on the interval $[0, 2]$ using ten subintervals of equal width and choosing $x_k^*$ as the right-hand endpoint of each subinterval.
11. Find two different approximations for the area $A$ under the graph $y = f(x)$ on the interval $[1, 4]$ shown in **FIGURE 3.7.8**.

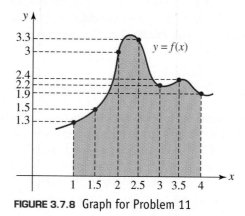

**FIGURE 3.7.8** Graph for Problem 11

12. Find two different approximations for the area $A$ under the graph $y = f(x)$ on the interval shown in **FIGURE 3.7.9**.

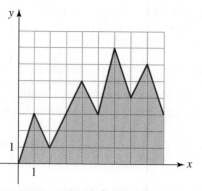

**FIGURE 3.7.9** Graph for Problem 12

## Miscellaneous Applications

13. **Lakefront Property** Suppose a realtor wants to find the area of an irregularly shaped piece of land that is bounded between a 1 mile-long segment of a straight road and the shore of a lake. Measurements (in feet) of the perpendicular dis-

tances from the road to the lake are taken at equally spaced intervals along the road as shown in FIGURE 3.7.10. Find two different approximations of the area of the land. Express your answer in acres using the fact that 1 acre = 43,560 ft$^2$.

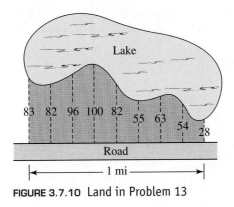

FIGURE 3.7.10 Land in Problem 13

14. **For the Fish** The large irregularly shaped fish pond shown in FIGURE 3.7.11 is filled with water to a uniform depth of 4 ft. Find an approximation to the number of gallons of water in the pond. Measurements are in feet and the vertical spacing between the horizontal measurements is 1.86 ft. There are 7.48 gallons in 1 cubic foot of water. [*Hint*: The volume of water is the area of the surface × depth.]

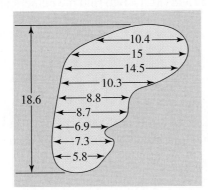

FIGURE 3.7.11 Fish pond in Problem 14

## For Discussion

If $c$ denotes a constant—that is, independent of the summation index $k$—then $\sum_{k=1}^{n} c$ means $c + c + c + \cdots + c$. Since there are $n$ $c$'s in this sum, we have

$$\sum_{k=1}^{n} c = nc. \tag{5}$$

In Problems 15 and 16, use (5) to find the numerical value of the given sum.

**15.** $\displaystyle\sum_{k=1}^{75} 6$          **16.** $\displaystyle\sum_{k=1}^{25} 10$

The sum of the first $n$ positive integers can be written $\sum_{k=1}^{n} k$. If this sum is denoted by $S$, then

$$S = 1 + 2 + 3 + \cdots + (n - 1) + n \tag{6}$$

can also be written as

$$S = n + (n - 1) + \cdots + 3 + 2 + 1. \tag{7}$$

If we add (6) and (7), then

$$2S = \underbrace{(n + 1) + (n + 1) + (n + 1) + \cdots + (n + 1)}_{n \text{ terms of } n + 1} = n(n + 1).$$

Solving for $S$ gives $S = \frac{1}{2}n(n + 1)$, or

$$\sum_{k=1}^{n} k = \tfrac{1}{2}n(n + 1). \tag{8}$$

In Problems 17 and 18, use (8) to find the numerical value of the given sum.

**17.** $\displaystyle\sum_{k=1}^{50} k$          **18.** $\displaystyle\sum_{k=1}^{1000} k$

Here are two properties of summation notation:

$$\sum_{k=1}^{n} ca_k = c \sum_{k=1}^{n} a_k, \ c \text{ a constant}, \tag{9}$$

$$\sum_{k=1}^{n} (a_k \pm b_k) = \sum_{k=1}^{n} a_k \pm \sum_{k=1}^{n} b_k. \tag{10}$$

In Problems 19–22, use (5) and (8)–(10) to find the numerical value of the given sum.

**19.** $\displaystyle\sum_{k=1}^{20} 2k$          **20.** $\displaystyle\sum_{k=1}^{15} (-6k)$

**21.** $\displaystyle\sum_{k=1}^{10} (4k + 5)$          **22.** $\displaystyle\sum_{k=1}^{20} (4k - 3)$

In Problems 23 and 24, use the results in (5) and (8)–(10) and the limit definition of area given in (3) to find the exact value of the area $A$. In each case, partition the given interval into $n$ subintervals of width $\Delta x = \dfrac{b - a}{n}$ and use $x_k^*$ as the right-hand endpoint of each subinterval.

**23.** $A$ is the area under the graph of $f(x) = 2x + 1$ on the interval $[0, 4]$
**24.** $A$ is the area under the graph of $f(x) = -3x + 12$ on the interval $[1, 3]$
**25.** Consider the trapezoid given in **FIGURE 3.7.12**.
  **(a)** Discuss how the area $A$ can be approximated using (1) of this section.
  **(b)** Using well-known area formulas, find a formula that expresses $A$ in terms of $h_1$, $h_2$, and $b$.

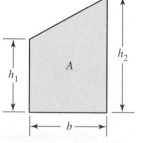

**FIGURE 3.7.12** Trapezoid in Problem 25

---

| **CHAPTER 3** | Review Exercises | Answers to selected odd-numbered problems begin on page ANS–12. |

In Problems 1–12, fill in the blanks.

**1.** The graph of the polynomial function $f(x) = x^3(x - 1)^2(x - 5)$ is tangent to the $x$-axis at _____ and passes through the $x$-axis at _____.
**2.** A third-degree polynomial function with zeros 1 and $3i$ is _____.

3. The end behavior of the graph of $f(x) = x^2(x+3)(x-5)$ resembles the graph of the power function $f(x) =$ _____.

4. The polynomial function $f(x) = x^4 - 3x^3 + 17x^2 - 2x + 2$ has _____ (how many) possible rational zeros.

5. For $f(x) = kx^2(x-2)(x-3)$, $f(-1) = 8$ if $k =$ _____.

6. The $y$-intercept of the graph of the rational function $f(x) = \dfrac{2x+8}{x^2 - 5x + 4}$ is _____.

7. The vertical asymptotes for the graph of the rational function
$f(x) = \dfrac{2x+8}{x^2 - 5x + 4}$ are _____.

8. The $x$-intercepts of the graph of the rational function $f(x) = \dfrac{x^3 - x}{4 - 2x^3}$ are

   is _____.

9. The horizontal asymptote for the graph of the rational function $f(x) = \dfrac{x^3 - x}{4 - 2x^3}$

   _____.

10. A rational function whose graph has the horizontal asymptote $y = 1$ and $x$-intercept $(3, 0)$ is _____.

11. The graph of the rational function $f(x) = \dfrac{x^n}{x^3 + 1}$, where $n$ is nonnegative integer, has the horizontal asymptote $y = 0$ when $n =$ _____.

12. The graph of the polynomial function $f(x) = 3x^5 - 4x^2 + 5x - 2$ has at most _____ turning points.

In Problems 13–28, answer true or false.

13. $f(x) = 2x^3 - 8x^{-2} + 5$ is not a polynomial. _____

14. $f(x) = x + \dfrac{1}{x}$ is a rational function. _____

15. The graph of a polynomial function can have no holes in it. _____

16. A polynomial function of degree 4 has exactly four real zeros. _____

17. When a polynomial of degree greater than one is divided by $x - 1$, the remainder is always a constant. _____

18. If the coefficients $a$, $b$, $c$, and $d$ of the polynomial function $f(x) = ax^3 + bx^2 + cx + d$ are positive integers, then $f$ has no positive real zeros. _____

19. The polynomial equation $2x^7 = 1 - x$ has a solution in the interval $[0, 1]$.

   _____

20. The graph of the rational function $f(x) = (x^2 + 1)/x$ has a slant asymptote.

   _____

21. The graph of the polynomial function $f(x) = 4x^6 + 3x^2$ is symmetric with respect to the $y$-axis. _____

22. The graph of a polynomial function that is an odd function must pass through the origin. _____

23. An asymptote is a line that the graph of a function approaches but never crosses.

   _____

24. The point $\left(\frac{1}{3}, \frac{7}{4}\right)$ is on the graph of $f(x) = \dfrac{2x+4}{3-x}$. _____

25. The graph of a rational function $f(x) = P(x)/Q(x)$ has a slant asymptote when the degree of $P$ is greater than the degree of $Q$. _____

**26.** If $3 - 4i$ is a zero of a polynomial function $f(x)$ with real coefficients, then $3 + 4i$ is also a zero of $f(x)$. _____

**27.** A polynomial function must have at least one rational zero. _____

**28.** The graph of $f(x) = x^4 + 5x^2 + 2$ does not cross the $x$-axis. _____

In Problems 29 and 30, use long division to divide $f(x)$ by $d(x)$.

**29.** $f(x) = 6x^5 - 4x^3 + 2x^2 + 4, \quad d(x) = 2x^2 - 1$

**30.** $f(x) = 15x^4 - 2x^3 + 8x + 6, \quad d(x) = 5x^3 + x + 2$

In Problems 31 and 32, use synthetic division to divide $f(x)$ by $d(x)$.

**31.** $f(x) = 7x^4 - 6x^2 + 9x + 3, \quad d(x) = x - 2$

**32.** $f(x) = 4x^3 + 7x^2 - 8x, \quad d(x) = x + 1$

**33.** Without actually performing the division, determine the remainder when $f(x) = 5x^3 - 4x^2 + 6x - 9$ is divided by $d(x) = x + 3$.

**34.** Use synthetic division and the Remainder Theorem to find $f(c)$ for

$$f(x) = x^6 - 3x^5 + 2x^4 + 3x^3 - x^2 + 5x - 1$$

when $c = 2$.

**35.** Determine the values of the positive integer $n$ such that $f(x) = x^n + c^n$ is divisible by $d(x) = x + c$.

**36.** Suppose that

$$f(x) = 36x^{98} - 40x^{25} + 18x^{14} - 3x^7 + 40x^4 + 5x^2 - x + 2$$

is divided by $d(x) = x - 1$. What is the remainder?

**37.** List, but do not test, all possible rational zeros of

$$f(x) = 8x^4 + 19x^3 + 31x^2 + 38x - 15.$$

**38.** Find the complete factorization of $f(x) = 12x^3 + 16x^2 + 7x + 1$.

In Problems 39 and 40, verify that each of the indicated numbers is a zero of the given polynomial function $f(x)$. Find all other zeros and then give the complete factorization of $f(x)$.

**39.** $2; \quad f(x) = (x - 3)^3 + 1$         **40.** $-1; \quad f(x) = (x + 2)^4 - 1$

In Problems 41–44, find the real value of $k$ so that the given condition is satisfied.

**41.** the remainder in the division of $f(x) = x^4 - 3x^3 - x^2 + kx - 1$ by $g(x) = x - 4$ is $r = 5$

**42.** $x + \frac{1}{2}$ is a factor of $f(x) = 8x^2 - 4kx + 9$

**43.** $x - k$ is a factor of $f(x) = 2x^3 + x^2 + 2x - 12$

**44.** the graph of $f(x) = \dfrac{x - k}{x^2 + 5x + 6}$ has a hole at $x = k$

In Problems 45–48, find the partial fraction decomposition of the given rational expression.

**45.** $\dfrac{2x - 1}{x(x^2 + 2x - 3)}$         **46.** $\dfrac{1}{x^4(x^2 + 5)}$

**47.** $\dfrac{x^2}{(x^2 + 4)^2}$         **48.** $\dfrac{x^5 - x^4 + 2x^3 + 5x - 1}{(x - 1)^2}$

In Problems 49 and 50, find a polynomial function $f$ of indicated degree whose graph is given in the figure.

**49.** fifth degree

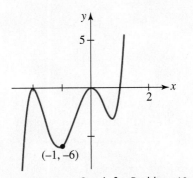

FIGURE 3.R.1 Graph for Problem 49

**50.** sixth degree

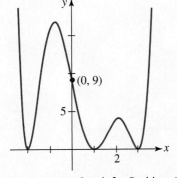

FIGURE 3.R.2 Graph for Problem 50

In Problems 51–60, match the given rational function with one of the graphs (a)–(j).

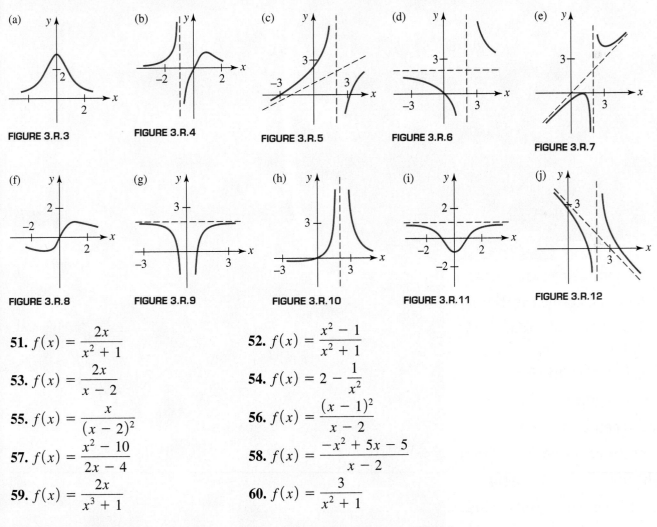

(a) FIGURE 3.R.3

(b) FIGURE 3.R.4

(c) FIGURE 3.R.5

(d) FIGURE 3.R.6

(e) FIGURE 3.R.7

(f) FIGURE 3.R.8

(g) FIGURE 3.R.9

(h) FIGURE 3.R.10

(i) FIGURE 3.R.11

(j) FIGURE 3.R.12

**51.** $f(x) = \dfrac{2x}{x^2 + 1}$

**52.** $f(x) = \dfrac{x^2 - 1}{x^2 + 1}$

**53.** $f(x) = \dfrac{2x}{x - 2}$

**54.** $f(x) = 2 - \dfrac{1}{x^2}$

**55.** $f(x) = \dfrac{x}{(x - 2)^2}$

**56.** $f(x) = \dfrac{(x - 1)^2}{x - 2}$

**57.** $f(x) = \dfrac{x^2 - 10}{2x - 4}$

**58.** $f(x) = \dfrac{-x^2 + 5x - 5}{x - 2}$

**59.** $f(x) = \dfrac{2x}{x^3 + 1}$

**60.** $f(x) = \dfrac{3}{x^2 + 1}$

In Problems 61 and 62, find the asymptotes for the graph of the given rational function. Find $x$- and $y$-intercepts of the graph. Sketch the graph of $f$.

**61.** $f(x) = \dfrac{x + 2}{x^2 + 2x - 8}$

**62.** $f(x) = \dfrac{-x^3 + 2x^2 + 9}{x^2}$

## Chapter Outline

# Trigonometric Functions

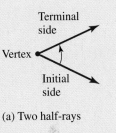

**4**

---

**Angles and Their Measurement**

☐ **Introduction** We begin our study of trigonometry by discussing angles and two methods of measuring them: degrees and radians. As we will see in Section 4.2, it is the radian measure of an angle that enables us to define trigonometric functions on sets of real numbers.

☐ **Angles** An **angle** is formed by two rays or half-lines, which have a common end-point called the **vertex**. We designate one ray the **initial side** of the angle and the other the **terminal side**. It is useful to consider the angle as having been formed by a rotation from the initial side to the terminal side, as shown in FIGURE 4.1.1(a). We can place the angle in a Cartesian coordinate plane with its vertex at the origin and its initial side coinciding with the positive $x$-axis, as shown in Figure 4.1.1(b). We then say the angle is in **standard position**.

☐ **Degree Measure** The **degree** measure of an angle is based on the assignment of 360 degrees (written 360°) to the angle formed by one complete counterclockwise rotation, as shown in FIGURE 4.1.2. Other angles are then measured in terms of a 360° angle, with a 1° angle being formed by $\frac{1}{360}$ of a complete rotation. If the rotation is counter-clockwise, the measure will be *positive*; if clockwise, the measure is *negative*. For example, the angle in FIGURE 4.1.3(a) obtained by one-fourth of a complete counterclockwise rotation will be

$$\tfrac{1}{4}(360°) = 90°.$$

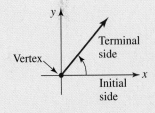

(a) Two half-rays

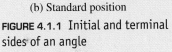

(b) Standard position

**FIGURE 4.1.1** Initial and terminal sides of an angle

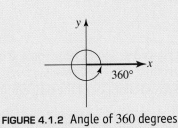

**FIGURE 4.1.2** Angle of 360 degrees

(a) 90° angle  (b) –270° angle

**FIGURE 4.1.3** Positive measure in (a); negative measure in (b)

Also shown in Figure 4.1.3(b) is the angle formed by three-fourths of a complete clock-wise rotation. This angle has measure

$$\tfrac{3}{4}(-360°) = -270°.$$

191

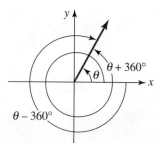

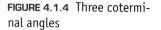

FIGURE 4.1.4 Three coterminal angles

□ **Coterminal Angles** Comparison of Figure 4.1.3(a) with Figure 4.1.3(b) shows that the terminal side of a 90° angle coincides with the terminal side of a −270° angle. When two angles in standard position have the same terminal sides we say they are **coterminal**. For example, the angles $\theta$, $\theta + 360°$, and $\theta - 360°$ shown in FIGURE 4.1.4 are coterminal. In fact, the addition of any integer multiple of 360° to a given angle results in a coterminal angle. Conversely, any two coterminal angles have degree measures that differ by an integer multiple of 360°.

### EXAMPLE 1    Angles and Coterminal Angles

For a 960° angle,
  (a) locate the terminal side and sketch the angle.
  (b) Find a coterminal angle between 0° and 360°.
  (c) Find a coterminal angle between −360° and 0°.

**Solution**

  (a) We first determine how many full rotations are made in forming this angle. Dividing 960 by 360, we obtain a quotient of 2 and a remainder of 240; that is,

$$960 = 2(360) + 240.$$

Thus, this angle is formed by making two counterclockwise rotations before completing $\frac{240}{360} = \frac{2}{3}$ of another rotation. As illustrated in FIGURE 4.1.5(a), the terminal side of 960° lies in the third quadrant.
  (b) Figure 4.1.5(b) shows that the angle 240° is coterminal with a 960° angle.
  (c) Figure 4.1.5(c) shows that the angle −120° is coterminal with a 960° angle.

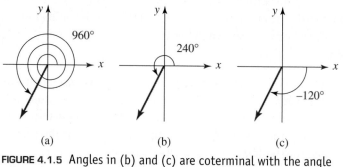

(a)              (b)              (c)

FIGURE 4.1.5 Angles in (b) and (c) are coterminal with the angle in (a)

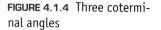

$x^2 + y^2 = 1$

FIGURE 4.1.6 Angle of $t$ radians

□ **Radian Measure** For calculus, the most convenient unit of measure for an angle is the **radian**. The radian measure of an angle is based on the length of an arc of the unit circle

$$x^2 + y^2 = 1.$$

As we know, an angle $\theta$ in standard position can be viewed as having been formed by the initial side rotating from the positive $x$-axis to the terminal side. As shown in FIGURE 4.1.6, the initial side of $\theta$ traverses a distance $t$ along the circumference of the unit circle. We say that the measure of $\theta$ is $t$ **radians**.

In radian measure we have the same convention as with degree measure: an angle formed by a counterclockwise rotation is considered positive, whereas an angle formed by a clockwise rotation is negative. Since the circumference of the unit circle is $2\pi$, an angle formed by one counterclockwise rotation is $2\pi$ radians. In FIGURE 4.1.7 we have illustrated angles of $\pi/2$, $-\pi/2$, $\pi$, and $3\pi$ radians, respectively. From Figures 4.1.7(c) and 4.1.7(d) we see that an angle of $\pi$ radians is coterminal with an angle

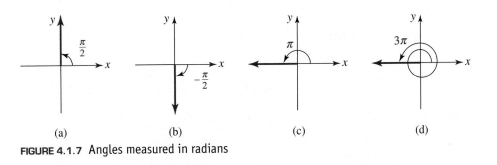

**FIGURE 4.1.7** Angles measured in radians

of $3\pi$ radians. In general, the addition of any integer multiple of $2\pi$ radians to an angle measured in radians results in a coterminal angle. Conversely, any two coterminal angles measured in radians will differ by an integer multiple of $2\pi$.

### EXAMPLE 2      A Coterminal Angle

Find an angle between $0$ and $2\pi$ radians that is coterminal with $\theta = 11\pi/4$ radians. Sketch the angle.

**Solution** Since $2\pi < 11\pi/4 < 3\pi$, we subtract the equivalent of one rotation, or $2\pi$ radians, to obtain

$$\frac{11\pi}{4} - 2\pi = \frac{11\pi}{4} - \frac{8\pi}{4} = \frac{3\pi}{4}.$$

Alternatively, we can proceed as in part (a) of Example 1 and divide: $11\pi/4 = 2\pi + 3\pi/4$. Thus, an angle of $3\pi/4$ radians is coterminal with $\theta$, as illustrated in **FIGURE 4.1.8**. ∎

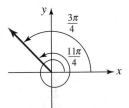

**FIGURE 4.1.8** Coterminal angles in Example 2

☐ **Conversion Formulas** While many scientific calculators have keys that convert between degree and radian measure, there is an easy way to remember the relationship between the two measures. Since the circumference of a unit circle is $2\pi$, one complete rotation has measure $2\pi$ radians as well as $360°$. It follows that $360° = 2\pi$ radians or

$$180° = \pi \text{ radians.} \tag{1}$$

If we interpret (1) as $180(1°) = \pi(1 \text{ radian})$, then we obtain the following two formulas for converting between degree and radian measure.

---

**CONVERSION BETWEEN DEGREES AND RADIANS**

$$1° = \frac{\pi}{180} \text{ radian} \tag{2}$$

$$1 \text{ radian} = \left(\frac{180}{\pi}\right)° \tag{3}$$

---

Using a calculator to carry out the divisions in (2) and (3), we find that

$$1° \approx 0.0174533 \text{ radian} \quad \text{and} \quad 1 \text{ radian} \approx 57.29578°.$$

### EXAMPLE 3      Conversion between Degrees and Radians

Convert

**(a)** $20°$ to radians,     **(b)** $7\pi/6$ radians to degrees,     **(c)** 2 radians to degrees.

**Solution**

**(a)** To convert from degrees to radians we use (2):

$$20° = 20(1°) = 20 \cdot \left( \frac{\pi}{180} \text{ radian} \right) = \frac{\pi}{9} \text{ radian.}$$

**(b)** To convert from radians to degrees we use (3):

$$\frac{7\pi}{6} \text{ radians} = \frac{7\pi}{6} \cdot (1 \text{ radian}) = \frac{7\pi}{6} \left( \frac{180}{\pi} \right)° = 210°.$$

**(c)** We again use (3):

$$2 \text{ radians} = 2 \cdot (1 \text{ radian}) = 2 \cdot \left( \frac{180}{\pi} \right)° = \left( \frac{360}{\pi} \right)° \approx 114.59°. \quad ■$$

The following table provides the radian and degree measure of the most commonly used angles.

**TABLE 1**

| Degrees | 0 | 30 | 45 | 60 | 90 | 180 |
|---------|---|----|----|----|----|-----|
| Radians | 0 | $\frac{\pi}{6}$ | $\frac{\pi}{4}$ | $\frac{\pi}{3}$ | $\frac{\pi}{2}$ | $\pi$ |

You may recall from geometry that a 90° angle is called a **right angle** and a 180° angle is called a **straight angle**. In radian measure, $\pi/2$ is a right angle and $\pi$ is a straight angle. An **acute angle** has measure between 0° and 90° (or between 0 and $\pi/2$ radians), and an **obtuse angle** has measure between 90° and 180° (or between $\pi/2$ and $\pi$ radians). Two acute angles are said to be **complementary** if their sum is 90° (or $\pi/2$ radians). Two positive angles are **supplementary** if their sum is 180° (or $\pi$ radians). A triangle that contains a right angle is called a **right triangle**. The lengths $a$, $b$, and $c$ of the sides of a right triangle satisfy the Pythagorean relationship $a^2 + b^2 = c^2$, where $d$ is the length of the side opposite the right angle (the hypotenuse).

**■ EXAMPLE 4**　　　　**Complementary and Supplementary Angles**

**(a)** Find the angle that is complementary to $\theta = 74.23°$.
**(b)** Find the angle that is supplementary to $\phi = \pi/3$ radians.

**Solution**

**(a)** Since two angles are complementary if their sum is 90°, we find the angle that is complementary to $\theta = 74.23°$ is

$$90° - \theta = 90° - 74.23° = 15.77°.$$

**(b)** Since two angles are supplementary if their sum is $\pi$ radians, we find the angle that is supplementary to $\phi = \pi/3$ radians is

$$\pi - \phi = \pi - \frac{\pi}{3} = \frac{3\pi}{3} - \frac{\pi}{3} = \frac{2\pi}{3} \text{ radians.} \quad ■$$

☐ **Arc Length** An angle $\theta$ with its vertex placed at the center of a circle of radius $r$ is called a **central angle**. The region inside the circle contained within the central angle $\theta$ is called a **sector**. As shown in FIGURE 4.1.9, the length of the arc of the circle subtended (or cut off) by the angle $\theta$ is denoted by $s$. When measured in radians, the central angle $\theta$ corresponds to $\theta/2\pi$ of one complete rotation. Hence the arc subtended by $\theta$ is $\theta/2\pi$ of the circumference of the circle. Therefore the length $s$ of the arc is

$$s = \frac{\theta}{2\pi}(2\pi r) = r\theta,$$

provided that $\theta$ is measured in radians. We summarize the preceding result.

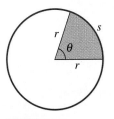

FIGURE 4.1.9
Length of arc $s$ determined by a central angle $\theta$

---

**ARC LENGTH FORMULA**

A central angle of $\theta$ radians in a circle of radius $r$ subtends an arc of length

$$s = r\theta. \tag{4}$$

---

From equation (4) we can express the radian measure $\theta$ of a central angle of a circle in terms of the length of subtended arc $s$ and radius $r$ of the circle:

$$\theta \text{ (in radians)} = \frac{s}{r}. \tag{5}$$

In equation (5) any convenient unit of length may be used for $s$ and $r$, but the same unit must be used for *both* $s$ and $r$. Thus,

$$\theta \text{ (in radians)} = \frac{s \text{ (units of length)}}{r \text{ (units of length)}}$$

appears to be a "dimensionless" quantity. This is the reason why sometimes the word *radians* is omitted when an angle is measured in radians.

**EXAMPLE 5**     **Finding Arc Length**

Find the arc length subtended by a central angle of: **(a)** 2 radians in a circle of radius 6 inches, **(b)** 30° in a circle of radius 12 feet.

**Solution**

**(a)** From the arc length formula (4) with $\theta = 2$ radians and $r = 6$ inches, we have $s = r\theta = 2 \cdot 6 = 12$. So the arc length is 12 inches.

**(b)** We must first express 30° in radians. Recall that $30° = \pi/6$ radians. Then from the arc length formula (4) we have $s = r\theta = (12)(\pi/6) = 2\pi$. So the arc length is $2\pi$ feet. ■

◀ Students often apply the arc length formula incorrectly by using degree measure. Remember $s = r\theta$ is valid only if $\theta$ is measured in radians.

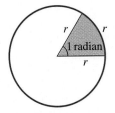

FIGURE 4.1.10
1 radius subtends an angle of 1 radian

From equation (5) it follows that an arc of length $r$ in a circle of radius $r$ will subtend an angle of $r/r = 1$ radian. FIGURE 4.1.10 illustrates an angle of 1 radian in a circle of radius $r$. We have already seen from conversion formula (3) that an angle of 1 radian is approximately 57°.

In Problems 1–16, draw the given angle in standard position. Bear in mind that the lack of a degree symbol ($°$) in an angular measurement indicates that the angle is measured in radians.

**1.** $60°$      **2.** $-120°$      **3.** $135°$      **4.** $150°$

**5.** $1140°$      **6.** $-315°$      **7.** $-240°$      **8.** $-210°$

**9.** $\dfrac{\pi}{3}$      **10.** $\dfrac{5\pi}{4}$      **11.** $\dfrac{7\pi}{6}$      **12.** $-\dfrac{2\pi}{3}$

**13.** $-\dfrac{\pi}{6}$      **14.** $-3\pi$      **15.** $3$      **16.** $4$

In Problems 17–24, convert from degrees to radians.

**17.** $10°$      **18.** $15°$      **19.** $45°$      **20.** $215°$

**21.** $270°$      **22.** $-120°$      **23.** $-230°$      **24.** $540°$

In Problems 25–32, convert from radians to degrees.

**25.** $\dfrac{2\pi}{9}$      **26.** $\dfrac{11\pi}{6}$      **27.** $\dfrac{2\pi}{3}$      **28.** $\dfrac{5\pi}{12}$

**29.** $\dfrac{5\pi}{4}$      **30.** $7\pi$      **31.** $3.1$      **32.** $12$

In Problems 33–36, for each given angle find a coterminal angle **(a)** between $0°$ and $360°$, and **(b)** between $-360°$ and $0°$.

**33.** $875°$      **34.** $400°$      **35.** $-610°$      **36.** $-150°$

In Problems 37–42, for each given angle find a coterminal angle **(a)** between $0$ and $2\pi$ radians, and **(b)** between $-2\pi$ and $0$ radians.

**37.** $-\dfrac{9\pi}{4}$      **38.** $\dfrac{17\pi}{2}$

**39.** $5.3\pi$      **40.** $-\dfrac{9\pi}{5}$

**41.** $-4$      **42.** $7.5$

In Problems 43–50, find an angle that is **(a)** complementary and **(b)** supplementary to the given angle, or state why no such angle can be found.

**43.** $48.25°$      **44.** $93°$      **45.** $98.4°$      **46.** $63.08°$

**47.** $\dfrac{\pi}{4}$      **48.** $\dfrac{\pi}{6}$      **49.** $\dfrac{2\pi}{3}$      **50.** $\dfrac{5\pi}{6}$

**51.** Find both the degree and the radian measures of the angle formed by **(a)** three-fifths of a counterclockwise rotation and **(b)** five and one-eighth clockwise rotations.

**52.** Find both the degree and the radian measures of the obtuse angle formed by the hands of a clock **(a)** at 8:00, **(b)** at 1:00, and **(c)** at 7:30.

**53.** Find both the degree and the radian measures of the angle through which the hour hand on a clock rotates in 2 hours.

**54.** Answer the question in Problem 53 for the minute hand.

**55.** The Earth rotates on its axis once every 24 hours. How long does it take the Earth to rotate through an angle of (a) 240° and (b) $\pi/6$ radians?

**56.** The planet Mercury completes one rotation on its axis every 59 days. Through what angle (measured in degrees) does it rotate in (a) 1 day, (b) 1 hour, and (c) 1 minute?

**57.** Find the arc length subtended by a central angle of 3 radians in a circle of (a) radius 3 and (b) radius 5.

**58.** Find the arc length subtended by a central angle of 30° in a circle of (a) radius 2 and (b) radius 4.

**59.** Find the measure of a central angle $\theta$ in a circle of radius 5 if $\theta$ subtends an arc of length 7.5. Give $\theta$ in (a) radians and (b) degrees.

**60.** Find the measure of a central angle $\theta$ in a circle of radius 1 if $\theta$ subtends an arc of length $\pi/3$. Give $\theta$ in (a) radians and (b) degrees.

**61.** Show that the area $A$ of a sector formed by a central angle of $\theta$ radians in a circle of radius $r$ is given by $A = \frac{1}{2}r^2\theta$. [*Hint*: Use the proportionality property from geometry that the ratio of the area $A$ of a circular sector to the total area $\pi r^2$ of the circle equals the ratio of the central angle $\theta$ to one complete revolution $2\pi$.]

**62.** What is the area of the red circular band shown in **FIGURE 4.1.11** if $\theta$ is measured (a) in radians and (b) in degrees? [*Hint*: Use the result of Problem 61.]

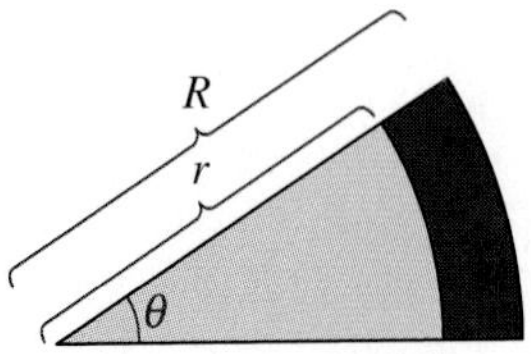

**FIGURE 4.1.11** Circular band in Problem 62

In Problems 63 and 64, make the indicated conversions using DMS notation (**Degrees, Minutes, Seconds**) and the fact that **one degree** can be divided into 60 equal parts called **minutes** (written $1° = 60'$), and one minute can be divided into 60 equal parts called **seconds** (written $1' = 60''$). It follows that $1' = \left(\frac{1}{60}\right)°$ and $1'' = \left(\frac{1}{60}\right)' = \left(\frac{1}{3600}\right)°$. For example, $37°15'55'' = 37° + 15' + 55'' = 37° + \left(\frac{15}{60}\right)° + \left(\frac{55}{3600}\right)° = 37.2653°$ in decimal degrees.

**63.** Convert to decimal degrees: (a) $5°10'$ (b) $10°25'$ (c) $10°39'17''$ (d) $143°7'2''$

**64.** Convert to degrees, minutes, and seconds: (a) $210.78°$ (b) $15.45°$ (c) $30.81°$ (d) $110.5°$

## Miscellaneous Applications

**65. Navigation at Sea** A nautical mile is defined as the arc length subtended on the surface of the Earth by an angle of measure 1 minute. If the diameter of the Earth is 7927 miles, find how many statute (land) miles there are in a nautical mile.

**66. Circumference of the Earth** Around 230 B.C., Eratosthenes calculated the circumference of the Earth from the following observations. At noon on the longest day of the year, the sun was directly overhead in Syene, while it was inclined 7.2° from the vertical in Alexandria. He believed the two cities to be on the same longitudinal line and assumed that the rays of the sun are parallel. Thus he concluded that the arc from Syene to Alexandria was subtended by a central angle of 7.2° at the center of the Earth. See **FIGURE 4.1.12**. At that time the distance from

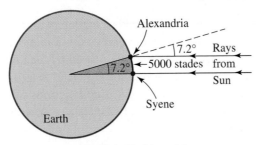

**FIGURE 4.1.12** Earth in Problem 66

Syene to Alexandria was measured as 5000 stades. If one stade = 559 feet, find the circumference of the Earth in (a) stades and (b) miles. Show that Eratosthenes' data gives a result that is within 7% of the correct value if the polar diameter of the Earth is 7900 miles (to the nearest mile).

67. **Pendulum clock** A clock pendulum is 1.3 m long and swings back and forth along a 15-cm arc. Find (a) the central angle and (b) the area of the sector through which the pendulum sweeps in one swing. [*Hint*: To answer part (b), use the result of Problem 61.]

68. **Circular Motion of a Yo-Yo** A yo-yo is whirled around in a circle at the end of its 100-cm string. (a) If it makes 6 revolutions in 4 seconds, find its rate of turning (**angular speed**) in radians per second. (b) Find the speed at which the yo-yo travels in centimeters per second. (This is called its **linear speed**.)

69. **More Yo-Yos** If there is a knot in the yo-yo string described in Problem 68 at a point 40 cm from the yo-yo, find (a) the angular speed of the knot and (b) the linear speed.

70. **Circular Motion of a Tire** If an automobile with 26-inch diameter tires is traveling at a rate of 55 mph, find (a) the number of revolutions per minute that its tires are making, and (b) the angular speed of its tires in radians per minute.

## 4.2 The Sine and Cosine Functions

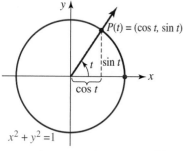

**FIGURE 4.2.1** Coordinates of $P(t)$ are $(\cos t, \sin t)$

☐ **Introduction** Originally, the trigonometric functions were defined using angles in right triangles. A more modern approach—one that is used in calculus—is to define the trigonometric functions on sets of real numbers. As we will see, the radian measure for angles is key in making these definitions.

☐ **Sine and Cosine** For each real number $t$ there corresponds an angle of $t$ radians in standard position. As shown in **FIGURE 4.2.1**, we denote the point of intersection of the terminal side of the angle $t$ with the unit circle by $P(t)$. The $x$ and $y$ coordinates of this point give us the values of the two basic trigonometric functions. The $y$-coordinate of $P(t)$ is called the **sine of $t$**, while the $x$-coordinate of $P(t)$ is called the **cosine of $t$**.

> **SINE AND COSINE FUNCTIONS**
>
> Let $t$ be any real number and $P(t) = (x, y)$ be the point of intersection of the unit circle with the terminal side of the angle of $t$ radians in standard position. Then the **sine of $t$**, denoted $\sin t$, and the **cosine of $t$**, denoted $\cos t$, are
>
> $$\sin t = y \qquad (1)$$
>
> and $$\cos t = x. \qquad (2)$$

Since to each real number $t$ there corresponds a unique point $P(t) = (\cos t, \sin t)$, we have just defined two functions—the sine and cosine functions—each with domain the set $R$ of real numbers. Because of the role played by the unit circle in this definition, the trigonometric functions are sometimes referred to as the **circular functions**.

A number of properties of the sine and cosine functions follow from the fact that $P(t) = (\cos t, \sin t)$ lies on the unit circle. For instance, the coordinates of $P(t)$ must satisfy the equation of the circle:

$$x^2 + y^2 = 1. \qquad (3)$$

Substituting $x = \cos t$ and $y = \sin t$ gives an important relationship between the sine and the cosine called the **Pythagorean identity**:

$$(\cos t)^2 + (\sin t)^2 = 1.$$

From now on we will follow two standard practices in writing this identity: $(\cos t)^2$ and $(\sin t)^2$ will be written as $\cos^2 t$ and $\sin^2 t$, respectively, and the $\sin^2 t$ term will be written first.

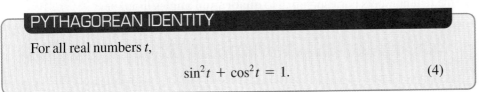

### PYTHAGOREAN IDENTITY

For all real numbers $t$,

$$\sin^2 t + \cos^2 t = 1. \tag{4}$$

Again, if $P(x, y)$ denotes a point on the unit circle (3), it follows that the coordinates of $P$ must satisfy the inequalities $-1 \le x \le 1$ and $-1 \le y \le 1$. Because $x = \cos t$ and $y = \sin t$ we have the following bounds on the values of the sine and cosine functions.

### BOUNDS ON THE VALUES OF SINE AND COSINE

For all real numbers $t$,

$$-1 \le \sin t \le 1 \qquad \text{and} \qquad -1 \le \cos t \le 1.$$

The foregoing inequalities can also be expressed as $|\sin t| \le 1$ and $|\cos t| \le 1$. Thus, for example, there is no number $t$ such that $\sin t = \frac{3}{2}$.

☐ **Domain and Range** From the preceding observations we have the sine and cosine functions $f(t) = \sin t$ and $g(t) = \cos t$ each with **domain** $R$ the set of real numbers and **range** the interval $[-1, 1]$.

☐ **Signs of the Circular Functions** The signs of the function values $\sin t$ and $\cos t$ are determined by the quadrant in which the point $P(t)$ lies, and conversely. For example, if $\sin t$ and $\cos t$ are both negative, then the point $P(t)$ and terminal side of the corresponding angle of $t$ radians must lie in quadrant III. **FIGURE 4.2.2** displays the signs of the cosine and sine functions in each of the four quadrants.

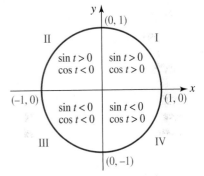

**FIGURE 4.2.2** Algebraic signs of $\sin t$ and $\cos t$ in the four quadrants

### ◼ EXAMPLE 1     Using the Pythagorean Identity

Given that $\cos t = \frac{1}{3}$ and that $P(t)$ is a point in the fourth quadrant, find $\sin t$.

    **Solution** Substitution of $\cos t = \frac{1}{3}$ into the Pythagorean identity (4) gives $\sin^2 t + \left(\frac{1}{3}\right)^2 = 1$ or $\sin^2 t = \frac{8}{9}$. Since $\sin t$ is the $y$-coordinate of $P(t)$, a point in the fourth quadrant, we must take the negative square root for $\sin t$:

$$\sin t = -\sqrt{\frac{8}{9}} = -\frac{2\sqrt{2}}{3}. \qquad \blacksquare$$

### ◼ EXAMPLE 2     Sine and Cosine of a Real Number

Use a calculator to approximate $\sin 3$ and $\cos 3$ and give a geometric interpretation of these values.

    **Solution** From a calculator set in *radian mode*, we obtain $\cos 3 \approx 0.9899925$ and $\sin 3 \approx 0.1411200$. These values represent the $x$- and $y$-coordinates, respectively, of the

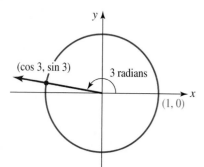

FIGURE 4.2.3 The point $P(3)$

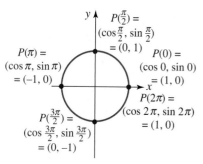

FIGURE 4.2.4 Sine and cosine for quadrantal angles

point of intersection of the terminal side of the angle of 3 radians in standard position with the unit circle. As shown in **FIGURE 4.2.3**, this point lies in the second quadrant because $\pi/2 < 3 < \pi$. This would also be expected in view of Figure 4.2.2 since $\cos 3$, the $x$-coordinate, is *negative* and $\sin 3$, the $y$-coordinate, is *positive*. ∎

☐ **Values Corresponding to Unit Circle Intercepts** As shown in **FIGURE 4.2.4**, the $x$ and $y$ intercepts of the unit circle give us the values of the sine and cosine functions for the real numbers corresponding to **quadrantal angles** listed next.

### VALUES OF THE SINE AND COSINE

For $t = 0$:     $\sin 0 = 0$    and    $\cos 0 = 1$

For $t = \dfrac{\pi}{2}$:     $\sin \dfrac{\pi}{2} = 1$    and    $\cos \dfrac{\pi}{2} = 0$

For $t = \pi$:     $\sin \pi = 0$    and    $\cos \pi = -1$

For $t = \dfrac{3\pi}{2}$:     $\sin \dfrac{3\pi}{2} = -1$    and    $\cos \dfrac{3\pi}{2} = 0$

☐ **Periodicity** In Section 4.1 we saw that for any real number $t$, the angles of $t$ radians and $t \pm 2\pi$ radians are coterminal. Thus they determine the same point $(x, y)$ on the unit circle. Therefore

$$\sin t = \sin(t \pm 2\pi) \quad \text{and} \quad \cos t = \cos(t \pm 2\pi). \tag{5}$$

In other words, the sine and cosine functions repeat their values every $2\pi$ units.

### PERIODIC FUNCTIONS

A nonconstant function $f$ is said to be **periodic** if there is a positive number $p$ such that

$$f(t) = f(t + p) \tag{6}$$

for every $t$ in the domain of $f$. If $p$ is the smallest positive number for which (6) is true, then $p$ is called the **period** of the function $f$.

The equations in (5) imply that the sine and the cosine functions are periodic with period $p \leq 2\pi$. To see that the period of $\sin t$, is actually $2\pi$, we observe that there is only one point on the unit circle with $y$ coordinate 1, namely, $P(\pi/2) = (\cos(\pi/2), \sin(\pi/2)) = (0, 1)$. Therefore,

$$\sin t = 1 \quad \text{only for} \quad t = \frac{\pi}{2}, \frac{\pi}{2} \pm 2\pi, \frac{\pi}{2} \pm 4\pi,$$

and so on. Thus the smallest possible positive value of $p$ is $2\pi$.

### PERIOD OF THE SINE AND COSINE

The sine and cosine functions are periodic with **period** $2\pi$. Therefore,

$$\sin(t + 2\pi) = \sin t \quad \text{and} \quad \cos(t + 2\pi) = \cos t \tag{7}$$

for every real number $t$.

□ **Even-Odd Properties** The symmetry of the unit circle endows the circular functions with several additional properties. For any real number $t$, the points $P(t)$ and $P(-t)$ on the unit circle are located on the terminal side of an angle of $t$ and $-t$ radians, respectively. These two points will always be symmetric with respect to the $x$ axis. **FIGURE 4.2.5** illustrates the situation for a point $P(t)$ lying in the first quadrant: the $x$-coordinates of the two points are identical; however, the $y$ coordinates have equal magnitudes but opposite signs. The same symmetries will hold regardless of which quadrant contains $P(t)$. Thus, for any real number $t$, $\cos(-t) = \cos t$ and $\sin(-t) = -\sin t$. Applying the definitions of **even** and **odd functions** from Section 2.2 we have the following result.

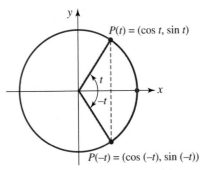

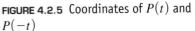

**FIGURE 4.2.5** Coordinates of $P(t)$ and $P(-t)$

---

**EVEN AND ODD FUNCTIONS**

The cosine function is **even** and the sine function is **odd**. That is, for every real number $t$,

$$\cos(-t) = \cos t \quad \text{and} \quad \sin(-t) = -\sin t. \tag{8}$$

---

The following additional properties of the sine and cosine functions can be verified by considering the symmetries of appropriately chosen points on the unit circle.

---

**ADDITIONAL PROPERTIES**

$$\cos\left(\frac{\pi}{2} - t\right) = \sin t \quad \text{and} \quad \sin\left(\frac{\pi}{2} - t\right) = \cos t \tag{9}$$

$$\cos(t + \pi) = -\cos t \quad \text{and} \quad \sin(t + \pi) = -\sin t \tag{10}$$

$$\cos(\pi - t) = -\cos t \quad \text{and} \quad \sin(\pi - t) = \sin t \tag{11}$$

---

These special properties of the sine and cosine functions become quite useful as soon as we determine additional values for $\sin t$ and $\cos t$ for $t$ in the interval $[0, 2\pi)$. Using results from plane geometry we will now find the values of the sine and cosine for $t = \pi/6, \pi/4,$ and $\pi/3$.

□ **Finding $\sin(\pi/4)$ and $\cos(\pi/4)$** We draw an angle of $\pi/4$ radians $(45°)$ in standard position and locate and label $P(\pi/4) = (\cos \pi/4, \sin \pi/4)$ on the unit circle. As shown in **FIGURE 4.2.6**, we form a right triangle by dropping a perpendicular from $P(\pi/4)$ to the $x$-axis. Since the sum of the angles in any triangle is $\pi$ radians $(180°)$, the third angle of this triangle is also $\pi/4$ radians, hence the triangle is isosceles. Therefore the coordinates of $P(\pi/4)$ are equal; that is, $\cos(\pi/4) = \sin(\pi/4)$. It follows from the Pythagorean identity (4)

$$\sin^2\frac{\pi}{4} + \cos^2\frac{\pi}{4} = 1 \quad \text{that} \quad 2\cos^2\frac{\pi}{4} = 1.$$

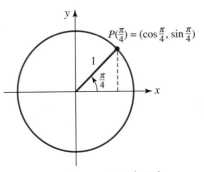

**FIGURE 4.2.6** The point $P(\pi/4)$

Dividing by 2 and taking the square root, we obtain $\cos(\pi/4) = \pm\sqrt{2}/2$. Since $P(\pi/4)$ lies in the first quadrant, both coordinates must be positive. So we have found the (equal) coordinates of $P(\pi/4)$:

$$\cos\frac{\pi}{4} = \frac{\sqrt{2}}{2} \quad \text{and} \quad \sin\frac{\pi}{4} = \frac{\sqrt{2}}{2}.$$

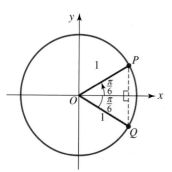

**FIGURE 4.2.7** The point $P(\pi/6)$

□ **Finding $\sin(\pi/6)$ and $\cos(\pi/6)$** We construct two angles of $\pi/6$ radians $(30°)$ in the first and fourth quadrants, as shown in **FIGURE 4.2.7**, and label the points of intersection with the unit circle $P(\pi/6)$ and $Q$, respectively. By drawing perpendicular line segments from $P$ and $Q$ to the $x$-axis, we obtain two *congruent* right triangles because each triangle has a hypotenuse of length 1 and angles of $30°$, $60°$, and $90°$. Since the $90°$ angles form a straight angle, these two right triangles form an *equilateral* triangle $\triangle POQ$ with sides of length 1. Since $\sin(\pi/6)$ is equal to half of the vertical side of $\triangle POQ$, we have

$$\sin\frac{\pi}{6} = \frac{1}{2}.$$

From this result and the Pythagorean identity (4) we find the value of $\cos\dfrac{\pi}{6}$:

$$\left(\frac{1}{2}\right)^2 + \cos^2\frac{\pi}{6} = 1 \quad \text{implies} \quad \cos^2\frac{\pi}{6} = \frac{3}{4}$$

▶ We take the positive square root here because $P(\pi/6)$ lies in the first quadrant.

or

$$\cos\frac{\pi}{6} = \frac{\sqrt{3}}{2}.$$

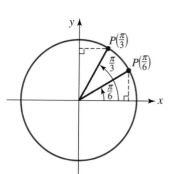

**FIGURE 4.2.8** The point $P(\pi/3)$

□ **Finding $\sin(\pi/3)$ and $\cos(\pi/3)$** We draw angles of $\pi/6$ and $\pi/3$ in standard position and locate and label the points $P(\pi/6)$ and $P(\pi/3)$, as shown in **FIGURE 4.2.8**. We then construct two congruent $30°$-$60°$-$90°$ triangles by dropping perpendiculars to the $x$- and $y$-axes, respectively. It follows from the congruence of these triangles that

$$\cos\frac{\pi}{3} = \sin\frac{\pi}{6} = \frac{1}{2} \quad \text{and} \quad \sin\frac{\pi}{3} = \cos\frac{\pi}{6} = \frac{\sqrt{3}}{2}.$$

We summarize the values of the sine and cosine functions corresponding to the basic fractions of $\pi$ that we have determined so far.

---

**VALUES OF THE SINE AND COSINE (CONTINUED)**

For $t = \dfrac{\pi}{6}$: $\quad \sin\dfrac{\pi}{6} = \dfrac{1}{2}$ $\quad$ and $\quad \cos\dfrac{\pi}{6} = \dfrac{\sqrt{3}}{2}$

For $t = \dfrac{\pi}{4}$: $\quad \sin\dfrac{\pi}{4} = \dfrac{\sqrt{2}}{2}$ $\quad$ and $\quad \cos\dfrac{\pi}{4} = \dfrac{\sqrt{2}}{2}$

For $t = \dfrac{\pi}{3}$: $\quad \sin\dfrac{\pi}{3} = \dfrac{\sqrt{3}}{2}$ $\quad$ and $\quad \cos\dfrac{\pi}{3} = \dfrac{1}{2}$

---

□ **Reference Angle** As we noted at the beginning of this section, for each real number $t$ there is a unique angle of $t$ radians in standard position that determines the point $P(t)$, with coordinates $(\cos t, \sin t)$, on the unit circle. As shown in **FIGURE 4.2.9**, the terminal side of any angle of $t$ radians (with $P(t)$ not on an axis) will form an acute angle with the $x$-axis. We can then locate an angle of $t'$ radians in the first quadrant that is congruent to this acute angle. The angle of $t'$ radians is called the **reference angle** for $t$. Because of the symmetry of the unit circle, the coordinates of $P(t')$ will be equal *in absolute value* to the respective coordinates of $P(t)$. Hence

$$\sin t = \pm\sin t' \quad \text{and} \quad \cos t = \pm\cos t'.$$

As the following examples will show, reference angles can be used to find the trigonometric function values of any integer multiple of $\pi/6$, $\pi/4$, and $\pi/3$.

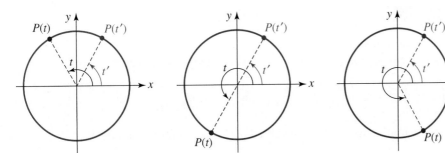

FIGURE 4.2.9 Reference angle $t'$ is an acute angle

## EXAMPLE 3    Using Reference Angles

Find exact values of $\sin t$ and $\cos t$ for the given real number $t$:

(a) $t = 5\pi/3$,    and    (b) $t = -3\pi/4$.

**Solution** In each part we begin by finding the reference angle corresponding to the given value of $t$.

(a) From FIGURE 4.2.10 we find that an angle of $t = 5\pi/3$ radians determines a point $P(5\pi/3)$ in the fourth quadrant and has the reference angle $t' = \pi/3$ radians. After adjusting the signs of the coordinates of $P(\pi/3) = (1/2, \sqrt{3}/2)$ to obtain the fourth quadrant point $P(5\pi/3) = (1/2, -\sqrt{3}/2)$, we find that

$$\sin\frac{5\pi}{3} = -\sin\frac{\pi}{3} = -\frac{\sqrt{3}}{2} \quad \text{and} \quad \cos\frac{5\pi}{3} = \cos\frac{\pi}{3} = \frac{1}{2}.$$

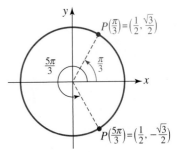

FIGURE 4.2.10 Reference angle in part (a) of Example 3

(b) The point $P(-3\pi/4)$ lies in the third quadrant and has reference angle $\pi/4$, as shown in FIGURE 4.2.11. Therefore,

$$\sin\left(-\frac{3\pi}{4}\right) = -\sin\frac{\pi}{4} = -\frac{\sqrt{2}}{2} \quad \text{and} \quad \cos\left(-\frac{3\pi}{4}\right) = -\cos\frac{\pi}{4} = -\frac{\sqrt{2}}{2}. \quad ■$$

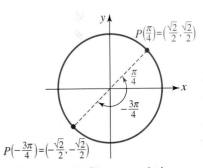

FIGURE 4.2.11 Reference angle in part (b) of Example 3

Sometimes, in order to find the trigonometric values of multiples of our basic fractions of $\pi$ we must use periodicity or the even-odd function properties in addition to reference numbers.

## EXAMPLE 4    Using Periodicity and a Reference Angle

Find exact values of $\sin t$ and $\cos t$ for $t = 29\pi/6$.

**Solution** Since $17\pi/6$ is greater than $2\pi$, we rewrite $29\pi/6$ as an integer multiple of $2\pi$ plus a number less than $2\pi$:

$$\frac{29\pi}{6} = 4\pi + \frac{5\pi}{6} = 2(2\pi) + \frac{5\pi}{6}.$$

From the periodicity equations (7) we know that $\sin(29\pi/6) = \sin(5\pi/6)$ and $\cos(29\pi/6) = \cos(5\pi/6)$. Next we see from FIGURE 4.2.12 that the reference angle for $5\pi/6$ is $\pi/6$. Since $P(5\pi/6)$ is a second quadrant point, we have

$$\sin\frac{29\pi}{6} = \sin\frac{5\pi}{6} = \sin\frac{\pi}{6} = \frac{1}{2}$$

and

$$\cos\frac{29\pi}{6} = \cos\frac{5\pi}{6} = -\cos\frac{\pi}{6} = -\frac{\sqrt{3}}{2}. \quad ■$$

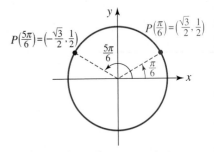

FIGURE 4.2.12 Reference angle in Example 4

 EXAMPLE 5          **Using the Even–Odd Properties**

Find exact values of $\sin t$ and $\cos t$ for $t = -\pi/6$.

**Solution** Since sine is an odd function and cosine is an even function,

$$\sin\left(-\frac{\pi}{6}\right) = -\sin\left(\frac{\pi}{6}\right) = -\frac{1}{2} \quad \text{and} \quad \cos\left(-\frac{\pi}{6}\right) = \cos\left(\frac{\pi}{6}\right) = \frac{\sqrt{3}}{2}.$$

This problem could also have been solved by using a reference angle. ∎

☐ **Trigonometric Functions of Angles** In this section we have defined sine and cosine functions of the real number $t$ by using the coordinates of a point $P(t)$ on the unit circle. It is now possible to define the **trigonometric functions of any angle $\theta$**. For any angle $\theta$, we simply let

$$\sin\theta = \sin t \quad \text{and} \quad \cos\theta = \cos t,$$

where the real number $t$ is the radian measure of $\theta$. As mentioned in Section 4.1, it is common to omit the word radians when measuring an angle. So we write $\sin(\pi/6)$ for both the sine of the real number $\pi/6$ and for the sine of the angle of $\pi/6$ radians. Furthermore, since the values of the trigonometric functions are determined by the coordinates of the point $P(t)$ on the unit circle, it really does not matter whether $\theta$ is measured in radians or in degrees. For example, regardless of whether we are given $\theta = \pi/6$ radians or $\theta = 30°$, the point on the unit circle corresponding to this angle in standard position is $\left(\sqrt{3}/2, 1/2\right)$. Thus,

$$\sin\frac{\pi}{6} = \sin 30° = \frac{1}{2} \quad \text{and} \quad \cos\frac{\pi}{6} = \cos 30° = \frac{\sqrt{3}}{2}.$$

---

**4.2**  **Exercises** Answers to selected odd-numbered problems begin on page ANS–12.

---

**1.** Given that $\cos t = -\frac{2}{5}$ and that $P(t)$ is a point in the second quadrant, find $\sin t$.
**2.** Given that $\sin t = \frac{1}{4}$ and that $P(t)$ is a point in the second quadrant, find $\cos t$.
**3.** Given that $\sin t = -\frac{2}{3}$ and that $P(t)$ is a point in the third quadrant, find $\cos t$.
**4.** Given that $\cos t = \frac{3}{4}$ and that $P(t)$ is a point in the fourth quadrant, find $\sin t$.
**5.** If $\sin t = -\frac{2}{7}$, find all possible values of $\cos t$.
**6.** If $\cos t = \frac{3}{10}$, find all possible values of $\sin t$.
**7.** If $\cos t = -0.2$, find all possible values of $\sin t$.
**8.** If $\sin t = 0.4$, find all possible values of $\cos t$.
**9.** If $2\sin t - \cos t = 0$, find all possible values of $\sin t$ and $\cos t$.
**10.** If $3\sin t - 2\cos t = 0$, find all possible values of $\sin t$ and $\cos t$.

In Problems 11–14, find the exact value of **(a)** $\sin t$ and **(b)** $\cos t$ for the given value of $t$. Do not use a calculator.

**11.** $t = -\pi/2$                      **12.** $t = 3\pi$
**13.** $t = 8\pi$                      **14.** $t = -3\pi/2$

In Problems 15–26, for the given value of $t$ determine the reference angle $t'$ and the exact values of $\sin t$ and $\cos t$. Do not use a calculator.

**15.** $t = 2\pi/3$                 **16.** $t = 4\pi/3$
**17.** $t = 5\pi/4$                 **18.** $t = 3\pi/4$

**19.** $t = 11\pi/6$

**20.** $t = 7\pi/6$

**21.** $t = -\pi/4$

**22.** $t = -7\pi/4$

**23.** $t = -5\pi/6$

**24.** $t = -11\pi/6$

**25.** $t = -5\pi/3$

**26.** $t = -2\pi/3$

In Problems 27–32, find the given trigonometric function value. Do not use a calculator.

**27.** $\sin(-11\pi/3)$

**28.** $\cos(17\pi/6)$

**29.** $\cos(-7\pi/4)$

**30.** $\sin(-19\pi/2)$

**31.** $\cos(5\pi)$

**32.** $\sin(23\pi/3)$

In Problems 33–38, justify the given statement with one of the properties of the trigonometric functions.

**33.** $\sin\pi = \sin 3\pi$

**34.** $\cos(\pi/4) = \sin(\pi/4)$

**35.** $\sin(-3 - \pi) = -\sin(3 + \pi)$

**36.** $\cos 16.8\pi = \cos 14.8\pi$

**37.** $\cos 0.43 = \cos(-0.43)$

**38.** $\sin(2\pi/3) = \sin(\pi/3)$

In Problems 39–46, find the given trigonometric function value. Do not use a calculator.

**39.** $\sin 135°$

**40.** $\cos 150°$

**41.** $\cos 210°$

**42.** $\sin 270°$

**43.** $\cos 330°$

**44.** $\sin(-180°)$

**45.** $\sin(-60°)$

**46.** $\cos(-300°)$

In Problems 47–50, find all angles $t$, where $0 \leq t < 2\pi$, that satisfy the given condition.

**47.** $\sin t = 0$

**48.** $\cos t = -1$

**49.** $\cos t = \sqrt{2}/2$

**50.** $\sin t = \frac{1}{2}$

In Problems 51–54, find all angles $\theta$, where $0° \leq \theta < 360°$, that satisfy the given condition.

**51.** $\cos\theta = \sqrt{3}/2$

**52.** $\sin\theta = -\frac{1}{2}$

**53.** $\sin\theta = -\sqrt{2}/2$

**54.** $\cos\theta = 1$

## Miscellaneous Applications

**55. Free Throw** Under certain conditions the maximum height $y$ attained by a basketball released from a height $h$ at an angle $\alpha$ measured from the horizontal with an initial velocity $v_0$ is given by $y = h + (v_0^2\sin^2\alpha)/2g$, where $g$ is the acceleration due to gravity. Compute the maximum height reached by a free throw if $h = 2.15$ m, $v_0 = 8$ m/s, $\alpha = 64.47°$, and $g = 9.81$ m/s$^2$.

Free throw

**56. Putting the Shot** The range of a shot put released from a height $h$ above the ground with an initial velocity $v_0$ at an angle $\alpha$ to the horizontal can be approximated by

$$R = \frac{v_0\cos\alpha}{g}\left[v_0\sin\alpha + \sqrt{v_0^2\sin^2\alpha + 2gh}\right],$$

where $g$ is the acceleration due to gravity. If $v_0 = 13.7$ m/s, $\alpha = 40°$, and $g = 9.81$ m/s$^2$, compare the ranges achieved for the release heights **(a)** $h = 2.0$ m and

**(b)** $h = 2.4$ m. **(c)** Explain why an increase in $h$ yields an increase in $R$ if the other parameters are held fixed. **(d)** What does this imply about the advantage that height gives a shot-putter?

**57. Acceleration Due to Gravity** Because of its rotation, the Earth bulges at the equator and is flattened at the poles. As a result, the acceleration due to gravity is not a constant 980 cm/s², but varies with latitude $\theta$. Satellite studies have shown that the acceleration due to gravity $g_{\text{sat}}$ is approximated by the function

$$g_{\text{sat}} = 978.0309 + 5.18552\sin^2\theta - 0.00570\sin^2 2\theta.$$

**(a)** Find $g_{\text{sat}}$ at the equator $(\theta = 0°)$, **(b)** at the north pole, and **(c)** at 45° north latitude.

## For Discussion

**58.** Discuss how it is possible to determine without a calculator that the point $P(6) = (\cos 6, \sin 6)$ lies in the fourth quadrant.

**59.** Discuss how it is possible to determine without the aid of a calculator that both $\sin 4$ and $\cos 4$ are negative.

**60.** Is there a real number $t$ satisfying $3\sin t = 5$? Explain why or why not.

**61.** Is there an angle $\theta$ satisfying $\cos\theta = -2$? Explain why or why not.

# 4.3 Graphs of Sine and Cosine Functions

☐ **Introduction** One way to further your understanding of the trigonometric functions is to examine their graphs. In this section we consider the graphs of the sine and cosine functions.

☐ **Graphs of Sine and Cosine** In Section 4.2 we saw that the domain of the sine function $f(t) = \sin t$ is the set of real numbers $(-\infty, \infty)$ and the interval $[-1, 1]$ is its range. Since the sine function has period $2\pi$, we begin by sketching its graph on the interval $[0, 2\pi]$. We obtain a rough sketch of the graph given in **FIGURE 4.3.1(b)** by considering various positions of the point $P(t)$ on the unit circle, as shown in Figure 4.3.1(a). As $t$ varies from 0 to $\pi/2$, the value $\sin t$ increases from 0 to its maximum value 1. But as $t$ varies from $\pi/2$ to $3\pi/2$, the value $\sin t$ decreases from 1 to its minimum value $-1$. We note that $\sin t$ changes from positive to negative at $t = \pi$. For $t$ between $3\pi/2$ and $2\pi$, we see that the corresponding values of $\sin t$ increase from $-1$ to 0. The graph of *any* periodic function over an interval of length equal to its period

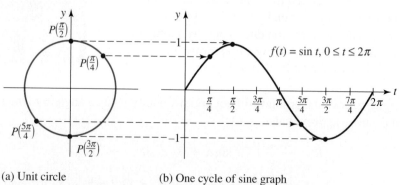

(a) Unit circle          (b) One cycle of sine graph

**FIGURE 4.3.1** Points $P(t)$ on a circle corresponding to points on the graph

is said to be one **cycle** of its graph. In the case of the sine function, the graph over the interval $[0, 2\pi]$ in Figure 4.3.1(b) is one cycle of the graph of $f(t) = \sin t$.

From this point on we will revert to the traditional symbols $x$ and $y$ when graphing trigonometric functions. Thus, $f(t) = \sin t$ will either be written $f(x) = \sin x$ or simply $y = \sin x$.

◀ Note

The graph of a periodic function is easily obtained by repeatedly drawing one cycle of its graph. In other words, the graph of $y = \sin x$ on, say, the intervals $[-2\pi, 0]$ and $[2\pi, 4\pi]$ is the same as that given in Figure 4.3.1(b). Recall from Section 4.2 that the sine function is an odd function since $f(-x) = \sin(-x) = -\sin x = -f(x)$. Thus, as can be seen in **FIGURE 4.3.2**, the graph of $y = \sin x$ is symmetric with respect to the origin.

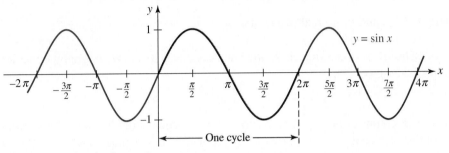

FIGURE 4.3.2 Graph of $y = \sin x$

By working again with the unit circle we can obtain one cycle of the graph of the cosine function $g(x) = \cos x$ on the interval $[0, 2\pi]$. In contrast to the graph of $f(x) = \sin x$ where $f(0) = f(2\pi) = 0$, for the cosine function we have $g(0) = g(2\pi) = 1$. **FIGURE 4.3.3** shows one cycle (in red) of $y = \cos x$ on $[0, 2\pi]$, along with the extension of that cycle (in blue) to the adjacent intervals $[-2\pi, 0]$ and $[2\pi, 4\pi]$. We see from this figure that the graph of the cosine function is symmetric with respect to the $y$-axis. This is a consequence of $g$ being an even function: $g(-x) = \cos(-x) = \cos x = g(x)$.

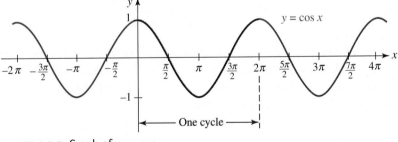

FIGURE 4.3.3 Graph of $y = \cos x$

☐ **Intercepts** In this and subsequent courses in mathematics it is important that you know the $x$-coordinates of the $x$-intercepts of the sine and cosine graphs—in other words, the zeros of $f(x) = \sin x$ and $g(x) = \cos x$. From the sine graph in Figure 4.3.2, we see that the zeros of the sine function, or the numbers for which $\sin x = 0$, are $x = 0, \pm\pi$, $\pm 2\pi, \pm 3\pi, \ldots$. These numbers are integer multiples of $\pi$. From the cosine graph in Figure 4.3.3, we see that $\cos x = 0$ when $x = \pm\pi/2, \pm 3\pi/2, \pm 5\pi/2, \ldots$. These numbers are odd-integer multiples of $\pi/2$.

If $n$ represents an integer, then $2n + 1$ is an odd integer. Therefore the zeros of $f(x) = \sin x$ and $g(x) = \cos x$ can be written in a compact form.

$$\sin x = 0 \quad \text{for} \quad x = n\pi, \, n \text{ an integer.} \qquad (1)$$

$$\cos x = 0 \quad \text{for} \quad x = (2n + 1)\frac{\pi}{2}, \, n \text{ an integer.} \qquad (2)$$

Using the distributive law, the result in (2) is often written as $x = \pi/2 + n\pi$.

As we did in Chapters 2 and 3, we can obtain variations of the basic sine and cosine graphs through rigid and nonrigid transformations. For the remainder of the discussion we will consider graphs of functions of the form

$$y = A\sin(Bx + C) + D \qquad \text{or} \qquad y = A\cos(Bx + C) + D, \qquad (3)$$

where $A$, $B$, $C$, and $D$ are real constants.

□ **Graphs of $y = A\sin x + D$ and $y = A\cos x + D$** We begin by considering the special cases of (3):

$$y = A\sin x \qquad \text{and} \qquad y = A\cos x.$$

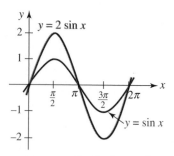

FIGURE 4.3.4 Vertical stretch of $y = \sin x$

For $A > 0$, graphs of these functions are either a vertical stretch or a vertical compression of the graphs of $y = \sin x$ or $y = \cos x$. For $A < 0$, the graphs are also reflected in the $x$-axis. For example, as **FIGURE 4.3.4** shows, we obtain the graph of $y = 2\sin x$ by stretching the graph of $y = \sin x$ vertically by a factor of 2. Note that the maximum and minimum values of $y = 2\sin x$ occur at the same $x$-values as the maximum and minimum values of $y = \sin x$. In general, the maximum distance from any point on the graph of $y = A\sin x$ or $y = A\cos x$ to the $x$-axis is $|A|$. The number $|A|$ is called the **amplitude** of the functions or of their graphs. The amplitude of the basic functions $y = \sin x$ and $y = \cos x$ is $|A| = 1$. In general, if a periodic function $f$ is continuous, then over a closed interval of length equal to its period, $f$ has both a maximum value $M$ and a minimum value $m$. The amplitude is defined by

$$\text{amplitude} = \tfrac{1}{2}[M - m]. \qquad (4)$$

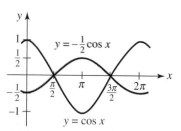

FIGURE 4.3.5 Graph of function in Example 1

## EXAMPLE 1    Vertically Compressed Cosine Graph

Graph $y = -\frac{1}{2}\cos x$.

**Solution** The graph of $y = -\frac{1}{2}\cos x$ is the graph of $y = \cos x$ compressed vertically by a factor of $\frac{1}{2}$ and then reflected in the $x$-axis. With the identification $A = -\frac{1}{2}$, we see that the amplitude of the function is $|A| = |-\frac{1}{2}| = \frac{1}{2}$. The graph of $y = -\frac{1}{2}\cos x$ on the interval $[0, 2\pi]$ is shown in red in **FIGURE 4.3.5**. ∎

The graphs of

$$y = A\sin x + D \qquad \text{and} \qquad y = A\cos x + D$$

are the graphs of $y = A\sin x$ and $y = A\cos x$ shifted vertically, up for $D > 0$ and down for $D < 0$. For example, the graph of $y = 1 + 2\sin x$ is the graph of $y = 2\sin x$ (Figure 4.3.4) shifted up 1 unit. The amplitude of the graph of either $y = A\sin x + D$ or $y = A\cos x + D$ is still $|A|$. Note that in **FIGURE 4.3.6**, the maximum of $y = 1 + 2\sin x$ is $y = 3$ at $x = \pi/2$ and the minimum is $y = -1$ at $x = 3\pi/2$. From (4), the amplitude of $y = 1 + 2\sin x$ is then $\frac{1}{2}[3 - (-1)] = 2$.

By interpreting $x$ as a placeholder in (1) and (2), we can find the $x$-coordinates of the $x$-intercepts of the graphs of sine and cosine functions of the form $y = A\sin Bx$

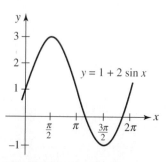

FIGURE 4.3.6 Graph of $y = 2\sin x$ shifted up 1 unit

CHAPTER 4 TRIGONOMETRIC FUNCTIONS

and $y = A\cos Bx$. (We consider this next.) For example, to solve $\sin 2x = 0$, we have from (1)

$$2x = n\pi \quad \text{so that} \quad x = \tfrac{1}{2}n\pi, \, n = 0, \pm 1, \pm 2, \ldots,$$

that is, $x = 0, \pm\tfrac{1}{2}\pi, \pm\tfrac{2}{2}\pi = \pi, \pm\tfrac{3}{2}\pi, \pm\tfrac{4}{2}\pi = 2\pi$, and so on. See **FIGURE 4.3.7**.

☐ **Graphs of $y = A\sin Bx$ and $y = A\cos Bx$** We now consider the graph of $y = \sin Bx$ for $B > 0$. The function has amplitude 1 since $A = 1$. Because the period of $y = \sin x$ is $2\pi$, a cycle of the graph of $y = \sin Bx$ begins at $x = 0$ and will start to repeat its values when $Bx = 2\pi$. In other words, a cycle of the function $y = \sin Bx$ is completed on the interval defined by $0 \le Bx \le 2\pi$. Dividing the last inequality by $B$ shows that the **period** of the function $y = \sin Bx$ is $2\pi/B$ and that the graph over the interval $[0, 2\pi/B]$ is one **cycle** of its graph. For example, the period of $y = \sin 2x$ is $2\pi/2 = \pi$, and therefore one cycle of the graph is completed on the interval $[0, \pi]$. Figure 4.3.7 shows that two cycles of the graph of $y = \sin 2x$ (in red and blue) are completed on the interval $[0, 2\pi]$, whereas the graph of $y = \sin x$ (in green) has completed only one cycle. In terms of transformations, we can characterize the cycle of $y = \sin 2x$ on $[0, \pi]$ as a **horizontal compression** of the cycle of $y = \sin x$ on $[0, 2\pi]$.

◀ Be careful here; $\sin 2x \ne 2\sin x$

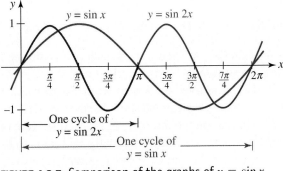

**FIGURE 4.3.7** Comparison of the graphs of $y = \sin x$ and $y = \sin 2x$

In summary, the graphs of

$$y = A\sin Bx \quad \text{and} \quad y = A\cos Bx$$

for $B > 0$ each have amplitude $|A|$ and period $2\pi/B$.

---

**EXAMPLE 2**     **Horizontally Compressed Cosine Graph**

Find the period of $y = \cos 4x$ and graph the function.

**Solution** Since $B = 4$, we see that the period of $y = \cos 4x$ is $2\pi/4 = \pi/2$. We conclude that the graph of $y = \cos 4x$ is the graph of $y = \cos x$ compressed horizontally. To graph the function, we draw one cycle of the cosine graph with amplitude 1 on the interval $[0, \pi/2]$ and then use periodicity to extend the graph. **FIGURE 4.3.8** shows four complete cycles of $y = \cos 4x$ (the basic cycle in red and the extended graph in blue) and one cycle of $y = \cos x$ (in green) on $[0, 2\pi]$. Notice that $y = \cos 4x$ attains its minimum at $x = \pi/4$ since $\cos 4(\pi/4) = \cos \pi = -1$, and attains its maximum at $x = \pi/2$ since $\cos 4(\pi/2) = \cos 2\pi = 1$.  ■

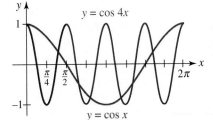

**FIGURE 4.3.8** Graph of function in Example 2

If $B < 0$ in either $y = A\sin Bx$ or $y = A\cos Bx$, we can use the even/odd properties (see (8) of Section 4.2) to rewrite the function with positive $B$. This is illustrated in the next example.

EXAMPLE 3 **Horizontally Stretched Sine Graph**

Find the amplitude and period of $y = \sin\left(-\frac{1}{2}x\right)$. Graph the function.

**Solution** Since we require $B > 0$, we use $\sin(-x) = -\sin x$ to rewrite the function as

$$y = \sin\left(-\tfrac{1}{2}x\right) = -\sin\tfrac{1}{2}x.$$

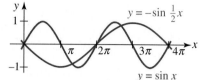

$y = -\sin\frac{1}{2}x$

$y = \sin x$

**FIGURE 4.3.9** Graph of function in Example 3

With the identification $A = -1$, the amplitude is seen to be $|A| = |-1| = 1$. Now with $B = \frac{1}{2}$, we find that the period is $2\pi/\frac{1}{2} = 4\pi$. Hence we can interpret the cycle of $y = -\sin\frac{1}{2}x$ on $[0, 4\pi]$ as a horizontal stretch and a reflection (in the $x$-axis because $A < 0$) of the cycle of $y = \sin x$ on $[0, 2\pi]$. **FIGURE 4.3.9** shows that on the interval $[0, 4\pi]$, the graph of $y = -\sin\frac{1}{2}x$ (in blue) completes one cycle whereas the graph of $y = \sin x$ (in green) completes two cycles. ∎

☐ **Graphs of $y = A\sin(Bx + C)$ and $y = A\cos(Bx + C)$** We have seen that the basic graphs of $y = \sin x$ and $y = \cos x$ can be stretched or compressed vertically ($y = A\sin x$ and $y = A\cos x$), shifted vertically ($y = A\sin x + D$ and $y = A\cos x + D$), and stretched or compressed horizontally ($y = A\sin Bx + D$ and $y = A\cos Bx + D$). The graphs of

$$y = A\sin(Bx + C) + D \qquad \text{and} \qquad y = A\cos(Bx + C) + D$$

are the graphs of $y = A\sin Bx + D$ and $y = A\cos Bx + D$ shifted horizontally.

In the remaining discussion we focus on the graphs of $y = A\sin(Bx + C)$ and $y = A\cos(Bx + C)$. For example, we know from Section 2.2 that the graph of $y = \cos(x - \pi/2)$ is the basic cosine graph shifted to the right. In **FIGURE 4.3.10**, the graph of $y = \cos(x - \pi/2)$ (in red) on the interval $[0, 2\pi]$ is one cycle of $y = \cos x$ on the interval $[-\pi/2, 3\pi/2]$ (in blue) shifted horizontally $\pi/2$ units to the right. Similarly, the graphs of $y = \sin(x + \pi/2)$ and $y = \sin(x - \pi/2)$ are the basic sine graphs shifted $\pi/2$ units to the left and to the right, respectively. See **FIGURES 4.3.11** and **4.3.12**.

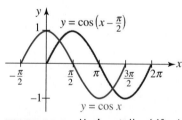

**FIGURE 4.3.10** Horizontally shifted cosine graph

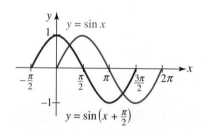

**FIGURE 4.3.11** Horizontally shifted sine graph

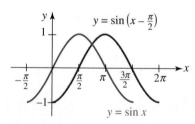

**FIGURE 4.3.12** Horizontally shifted sine graph

By comparing the red graphs in Figures 4.3.10–4.3.12 with the graphs in Figures 4.3.2 and 4.3.3 we see that

- the cosine graph shifted $\pi/2$ units to the right is the sine graph,
- the sine graph shifted $\pi/2$ units to the left is the cosine graph, and
- the sine graph shifted $\pi/2$ units to the right is the cosine graph reflected in the $x$-axis.

In other words, we have graphically verified the identities

$$\cos\left(x - \frac{\pi}{2}\right) = \sin x, \quad \sin\left(x + \frac{\pi}{2}\right) = \cos x, \quad \text{and} \quad \sin\left(x - \frac{\pi}{2}\right) = -\cos x. \quad (5)$$

We now consider the graph of $y = A\sin(Bx + C)$ for $B > 0$. Since the values of $\sin(Bx + C)$ range from $-1$ to $1$, it follows that $A\sin(Bx + C)$ varies between $-A$ and $A$. That is, the **amplitude** of $y = A\sin(Bx + C)$ is $|A|$. Also, as $Bx + C$ varies from $0$ to $2\pi$, the graph will complete one cycle. By solving $Bx + C = 0$ and $Bx + C = 2\pi$, we find that one cycle is completed as $x$ varies from $-C/B$ to $(2\pi - C)/B$. Therefore, the function $y = A\sin(Bx + C)$ has the **period**

$$\frac{2\pi - C}{B} - \left(-\frac{C}{B}\right) = \frac{2\pi}{B}.$$

Moreover, if $f(x) = A\sin Bx$, then

$$f\left(x + \frac{C}{B}\right) = A\sin B\left(x + \frac{C}{B}\right) = A\sin(Bx + C). \qquad (6)$$

The result in (6) shows that the graph of $y = A\sin(Bx + C)$ can be obtained by shifting the graph of $f(x) = A\sin Bx$ horizontally a distance $|C|/B$. If $C < 0$ the shift is to the right, whereas if $C > 0$ the shift is to the left. The number $|C|/B$ is called the **phase shift** of the graph of $y = A\sin(Bx + C)$.

## EXAMPLE 4        Equation of a Shifted Cosine Graph

The graph of $y = 10\cos 4x$ is shifted $\pi/12$ units to the right. Find its equation.
    **Solution**  By writing $f(x) = 10\cos 4x$ and using (6), we find

$$f\left(x - \frac{\pi}{12}\right) = 10\cos 4\left(x - \frac{\pi}{12}\right) \qquad \text{or} \qquad y = 10\cos\left(4x - \frac{\pi}{3}\right).$$

In the last equation we would identify $C = -\pi/3$. The phase shift is $\pi/12$.    ■

As a practical matter the phase shift of $y = A\sin(Bx + C)$ can be obtained by factoring the number $B$ from $Bx + C$:    ◄ Note

$$y = A\sin(Bx + C) = A\sin B\left(x + \frac{C}{B}\right).$$

For convenience we summarize the preceding information.

### SHIFTED SINE AND COSINE GRAPHS

The graphs of

$$y = A\sin(Bx + C) \qquad \text{and} \qquad y = A\cos(Bx + C), \qquad B > 0,$$

are, respectively, the graphs of $y = A\sin Bx$ and $y = A\cos Bx$ shifted horizontally by $|C|/B$. The shift is to the right if $C < 0$ and to the left if $C > 0$. The number $|C|/B$ is called the **phase shift**. The **amplitude** of each graph is $|A|$ and the **period** of each graph is $2\pi/B$.

## EXAMPLE 5        Horizontally Shifted Sine Graph

Graph $y = 3\sin(2x - \pi/3)$.
    **Solution**  For purposes of comparison we will first graph $y = 3\sin 2x$. The amplitude of $y = 3\sin 2x$ is $|A| = 3$ and its period is $2\pi/2 = \pi$. Thus one cycle of $y = 3\sin 2x$ is completed on the interval $[0, \pi]$. Then we extend this graph to the

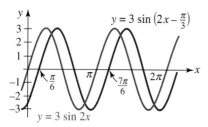

**FIGURE 4.3.13** Graph of function in Example 5

adjacent interval $[\pi, 2\pi]$, as shown in blue in **FIGURE 4.3.13**. Next, we rewrite $y = 3\sin(2x - \pi/3)$ by factoring 2 from $2x - \pi/3$:

$$y = 3\sin\left(2x - \frac{\pi}{3}\right) = 3\sin 2\left(x - \frac{\pi}{6}\right).$$

From the last form we see that the phase shift is $\pi/6$. The graph of the given function, shown in red in Figure 4.3.13, is obtained by shifting the graph of $y = 3\sin 2x$ to the right $\pi/6$ units. Remember, this means that if $(x, y)$ is a point on the blue graph, then $(x + \pi/6, y)$ is the corresponding point on the red graph. For example, $x = 0$ and $x = \pi$ are the $x$-coordinates of two $x$-intercepts of the blue graph. Thus $x = 0 + \pi/6 = \pi/6$ and $x = \pi + \pi/6 = 7\pi/6$ are $x$-coordinates of the $x$-intercepts of the red or shifted graph. These numbers are indicated by the arrows in Figure 4.3.13.

◼

### ▮ EXAMPLE 6　　　Horizontally Shifted Graphs

Determine the amplitude, period, phase shift, and direction of horizontal shift for each of the following functions.

**(a)** $y = 15\cos\left(5x - \dfrac{3\pi}{2}\right)$　　**(b)** $y = -8\sin\left(2x + \dfrac{\pi}{4}\right)$

**Solution**

**(a)** We first make the identifications $A = 15$, $B = 5$, and $C = -3\pi/2$. Thus the amplitude is $|A| = 15$ and the period is $2\pi/B = 2\pi/5$. The phase shift can be computed either by $(|-3\pi|/2)/5 = 3\pi/10$ or by rewriting the function as

$$y = 15\cos 5\left(x - \frac{3\pi}{10}\right).$$

The last form indicates that the graph of $y = 15\cos(5x - 3\pi/2)$ is the graph of $y = 15\cos 5x$ shifted $3\pi/10$ units to the right.

**(b)** Since $A = -8$, the amplitude is $|A| = |-8| = 8$. With $B = 2$ the period is $2\pi/2 = \pi$. By factoring 2 from $2x + \pi/4$, we see from

$$y = -8\sin\left(2x + \frac{\pi}{4}\right) = -8\sin 2\left(x + \frac{\pi}{8}\right)$$

that the phase shift is $\pi/8$. The graph of $y = -8\sin(2x + \pi/4)$ is the graph of $y = -8\sin 2x$ shifted $\pi/8$ units to the left.

◼

### ▮ EXAMPLE 7　　　Horizontally Shifted Cosine Graph

Graph $y = 2\cos(\pi x + \pi)$.

　　**Solution** The amplitude of $y = 2\cos \pi x$ is $|A| = 2$ and the period is $2\pi/\pi = 2$. Thus one cycle of $y = 2\cos \pi x$ is completed on the interval $[0, 2]$. In **FIGURE 4.3.14**, two cycles of the graph of $y = 2\cos \pi x$ (in blue) are shown. The $x$-intercepts of this graph correspond to the values of $x$ for which $\cos \pi x = 0$. By (2), this implies $\pi x = (2n + 1)\pi/2$ or $x = (2n + 1)/2$, $n$ an integer. In other words, for $n = 0, -1, 1, -2, 2, -3, \ldots$, we get $x = \pm\frac{1}{2}, \pm\frac{3}{2}, \pm\frac{5}{2}$, and so on. Now by rewriting the given function as

$$y = 2\cos \pi(x + 1)$$

we see the phase shift is 1. The graph of $y = 2\cos(\pi x + \pi)$ (in red) in Figure 4.3.14 is obtained by shifting the graph of $y = 2\cos \pi x$ to the left 1 unit. This means that the $x$-intercepts are the same for both graphs.

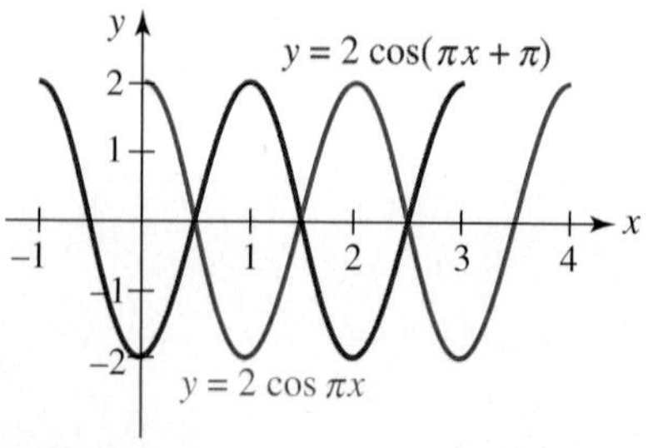

◼

**FIGURE 4.3.14** Graph of function in Example 7

CHAPTER 4 TRIGONOMETRIC FUNCTIONS

## EXAMPLE 8    Alternating Current

The current $I$ (in amperes) in a wire of an alternating-current circuit is given by $I(t) = 30\sin 120\pi t$, where $t$ is time measured in seconds. Sketch one cycle of the graph. What is the maximum value of the current?

**Solution** The graph has amplitude 30 and period $2\pi/120\pi = \frac{1}{60}$. Therefore, we sketch one cycle of the basic sine curve on the interval $\left[0, \frac{1}{60}\right]$, as shown in **FIGURE 4.3.15**. From the figure it is evident that the maximum value of the current is $I = 30$ amperes and occurs at $t = \frac{1}{240}$ second since

$$I\left(\tfrac{1}{240}\right) = 30\sin\left(120\pi \cdot \tfrac{1}{240}\right) = 30\sin\frac{\pi}{2} = 30.$$

■

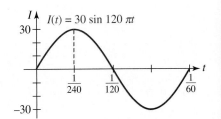

**FIGURE 4.3.15** Graph of current in Example 8

---

| **4.3** | **Exercises** | Answers to selected odd-numbered problems begin on page ANS–12. |

In Problems 1–6, use the techniques of shifting, stretching, compressing, and reflecting to sketch at least one cycle of the graph of the given function.

**1.** $y = \frac{1}{2} + \cos x$  **2.** $y = -1 + \cos x$
**3.** $y = 2 - \sin x$  **4.** $y = 3 + 3\sin x$
**5.** $y = -2 + 4\cos x$  **6.** $y = 1 - 2\sin x$

In Problems 7–10, the given figure shows one cycle of a sine or cosine graph. From the figure, determine $A$ and $D$ and write an equation of the form $y = A\sin x + D$ or $y = A\cos x + D$ for the graph.

**7.**

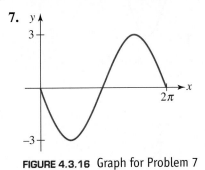

**FIGURE 4.3.16** Graph for Problem 7

**8.**

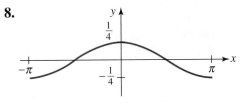

**FIGURE 4.3.17** Graph for Problem 8

**9.**

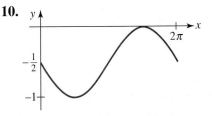

**FIGURE 4.3.18** Graph for Problem 9

**10.**

**FIGURE 4.3.19** Graph for Problem 10

In Problems 11–16, use (1) and (2) of Section 4.3 to find the $x$-intercepts for the graph of the given function. Do not graph.

**11.** $y = \sin \pi x$  **12.** $y = -\cos 2x$

**13.** $y = 10\cos\dfrac{x}{2}$  **14.** $y = 3\sin(-5x)$

**15.** $y = \sin\left(x - \dfrac{\pi}{4}\right)$ $\qquad\qquad\qquad$ **16.** $y = \cos(2x - \pi)$

In Problems 17 and 18, find the $x$-intercepts of the graph of the given function on the interval $[0, 2\pi]$. Then find all intercepts using periodicity.

**17.** $y = -1 + \sin x$ $\qquad\qquad\qquad$ **18.** $y = 1 - 2\cos x$

In Problems 19–24, the given figure shows one cycle of a sine or cosine graph. From the figure, determine $A$ and $B$ and write an equation of the form $y = A\sin Bx$ or $y = A\cos Bx$ for the graph.

**19.**

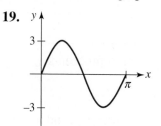

FIGURE 4.3.20 Graph for Problem 19

**20.**

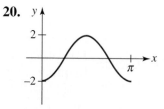

FIGURE 4.3.21 Graph for Problem 20

**21.**

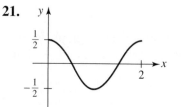

FIGURE 4.3.22 Graph for Problem 21

**22.**

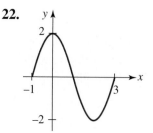

FIGURE 4.3.23 Graph for Problem 22

**23.**

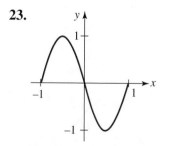

FIGURE 4.3.24 Graph for Problem 23

**24.**

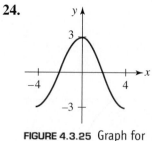

FIGURE 4.3.25 Graph for Problem 24

In Problems 25–32, find the amplitude and period of the given function. Sketch at least one cycle of the graph.

**25.** $y = 4\sin \pi x$ $\qquad\qquad\qquad$ **26.** $y = -5\sin\dfrac{x}{2}$

**27.** $y = -3\cos 2\pi x$ $\qquad\qquad\qquad$ **28.** $y = \dfrac{5}{2}\cos 4x$

**29.** $y = 2 - 4\sin x$

**30.** $y = 2 - 2\sin \pi x$

**31.** $y = 1 + \cos \dfrac{2x}{3}$

**32.** $y = -1 + \sin \dfrac{\pi x}{2}$

In Problems 33–42, find the amplitude, period, and phase shift of the given function. Sketch at least one cycle of the graph.

**33.** $y = \sin\left(x - \dfrac{\pi}{6}\right)$

**34.** $y = \sin\left(3x - \dfrac{\pi}{4}\right)$

**35.** $y = \cos\left(x + \dfrac{\pi}{4}\right)$

**36.** $y = -2\cos\left(2x - \dfrac{\pi}{6}\right)$

**37.** $y = 4\cos\left(2x - \dfrac{3\pi}{2}\right)$

**38.** $y = 3\sin\left(2x + \dfrac{\pi}{4}\right)$

**39.** $y = 3\sin\left(\dfrac{x}{2} - \dfrac{\pi}{3}\right)$

**40.** $y = -\cos\left(\dfrac{x}{2} - \pi\right)$

**41.** $y = -4\sin\left(\dfrac{\pi}{3}x - \dfrac{\pi}{3}\right)$

**42.** $y = 2\cos\left(-2\pi x - \dfrac{4\pi}{3}\right)$

In Problems 43 and 44, write an equation of the function whose graph is described in words.

**43.** The graph of $y = \cos x$ is vertically stretched up by a factor of 3 and shifted down by 5 units. One cycle of $y = \cos x$ on $[0, 2\pi]$ is compressed to $[0, \pi/3]$ and then the compressed cycle is shifted horizontally $\pi/4$ units to the left.

**44.** One cycle of $y = \sin x$ on $[0, 2\pi]$ is stretched to $[0, 8\pi]$ and then the stretched cycle is shifted horizontally $\pi/12$ units to the right. The graph is also compressed vertically by a factor of $\frac{3}{4}$ and then reflected in the $x$-axis.

In Problems 45–48, find horizontally shifted sine and cosine functions so that each function satisfies the given conditions. Graph the functions.

**45.** Amplitude 3, period $2\pi/3$, shifted by $\pi/3$ units to the right
**46.** Amplitude $\frac{2}{3}$, period $\pi$, shifted by $\pi/4$ units to the left
**47.** Amplitude 0.7, period 0.5, shifted by 4 units to the right
**48.** Amplitude $\frac{5}{4}$, period 4, shifted by $1/2\pi$ units to the left

In Problems 49 and 50, graphically verify the given identity.

**49.** $\cos(x + \pi) = -\cos x$

**50.** $\sin(x + \pi) = -\sin x$

## Miscellaneous Applications

**51. Pendulum** The angular displacement $\theta$ of a pendulum from the vertical at time $t$ seconds is given by $\theta(t) = \theta_0 \cos \omega t$, where $\theta_0$ is the initial displacement at $t = 0$ seconds. See **FIGURE 4.3.26**. For $\omega = 2$ rad/s and $\theta_0 = \pi/10$, sketch two cycles of the resulting function.

**52. Current** In a certain kind of electrical circuit, the current $I$ measured in amperes at time $t$ seconds is given by

$$I(t) = 10\cos\left(120\pi t + \dfrac{\pi}{3}\right).$$

Sketch two cycles of the graph of $I$ as a function of time $t$.

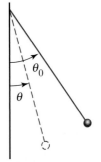

**FIGURE 4.3.26**
Pendulum in
Problem 51

**53. Depth of Water** The depth $d$ of water at the entrance to a small harbor at time $t$ is modeled by a function of the form

$$d(t) = A \sin B\left(t - \frac{\pi}{2}\right) + C,$$

where $A$ is one half the difference between the high- and low-tide depths; $2\pi/B$, $B > 0$, is the tidal period; and $C$ is the average depth. Assume that the tidal period is 12 hours, the depth at high tide is 18 feet, and the depth at low tide is 6 feet. Sketch two cycles of the graph of $d$.

**54. Fahrenheit Temperature** Suppose that

$$T(t) = 50 + 10\sin\frac{\pi}{12}(t - 8),$$

$0 \le t \le 24$, is a mathematical model of the Fahrenheit temperature at $t$ hours after midnight on a certain day of the week.

(a) What is the temperature at 8 A.M.?
(b) At what time(s) does $T(t) = 60$?
(c) Sketch the graph of $T$.
(d) Find the maximum and minimum temperatures and the times at which they occur.

## Calculator Problems

In Problems 55–58, use a calculator to investigate whether the given function is periodic.

**55.** $f(x) = \sin\left(\dfrac{1}{x}\right)$ 

**56.** $f(x) = \dfrac{1}{\sin 2x}$

**57.** $f(x) = 1 + (\cos x)^2$ 

**58.** $f(x) = x \sin x$

## For Discussion

**59.** The function $f(x) = \sin\frac{1}{2}x + \sin 2x$ is periodic. What is the period of $f$?
**60.** Discuss and then sketch the graphs of $y = |\sin x|$ and $y = |\cos x|$.

---

### 4.4 Other Trigonometric Functions

☐ **Introduction** Four additional trigonometric functions are defined in terms of reciprocals and quotients of the sine and cosine functions. In this section we consider the properties and graphs of these new functions.

We begin with a definition.

> ### TANGENT, COTANGENT, SECANT, COSECANT FUNCTIONS
>
> The **tangent**, **cotangent**, **secant**, and **cosecant** functions are denoted by $\tan x$, $\cot x$, $\sec x$, and $\csc x$, respectively, and are defined as follows:
>
> $$\tan x = \frac{\sin x}{\cos x}, \qquad \cot x = \frac{\cos x}{\sin x}, \qquad (1)$$
>
> $$\sec x = \frac{1}{\cos x}, \qquad \csc x = \frac{1}{\sin x}. \qquad (2)$$

Note that the tangent and cotangent functions are related by

$$\cot x = \frac{\cos x}{\sin x} = \frac{1}{\dfrac{\sin x}{\cos x}} = \frac{1}{\tan x}.$$

In view of the definitions in (2) and the foregoing result, $\cot x$, $\sec x$, and $\csc x$ are referred to as the **reciprocal functions**.

☐ **Domain and Range** Because the functions in (1) and (2) are quotients, the **domain** of each function consists of the set of real numbers except those numbers for which the denominator is zero. We have seen in (2) of Section 4.3 that $\cos x = 0$ for $x = (2n + 1)\pi/2, n = 0, \pm1, \pm2, \ldots$, and so

• the domain of $\tan x$ and of $\sec x$ is $\{x \mid x \neq (2n + 1)\pi/2, n = 0, \pm1, \pm2, \ldots\}$.

Similarly, from (1) of Section 4.3, $\sin x = 0$ for $x = n\pi, n = 0, \pm1, \pm2, \ldots$, and so it follows that

• the domain of $\cot x$ and of $\csc x$ is $\{x \mid x \neq n\pi, n = 0, \pm1, \pm2, \ldots\}$.

We know that the values of the sine and cosine are bounded, that is, $|\sin x| \leq 1$ and $|\cos x| \leq 1$. From these last inequalities we have

$$|\sec x| = \left|\frac{1}{\cos x}\right| = \frac{1}{|\cos x|} \geq 1 \tag{3}$$

and

$$|\csc x| = \left|\frac{1}{\sin x}\right| = \frac{1}{|\sin x|} \geq 1. \tag{4}$$

Recall that an inequality such as (3) means that $\sec x \geq 1$ or $\sec x \leq -1$. Hence the range of the secant function is $(-\infty, -1] \cup [1, \infty)$. The inequality in (4) implies that the cosecant function has the same range $(-\infty, -1] \cup [1, \infty)$. When we consider the graphs of the tangent and cotangent functions we will see that they have the same range: $(-\infty, \infty)$.

If we interpret $x$ as an angle, then **FIGURE 4.4.1** illustrates the algebraic signs of the tangent, cotangent, secant, and cosecant functions in each of the four quadrants. This is easily verified using the signs of the sine and cosine functions displayed in Figure 4.2.2.

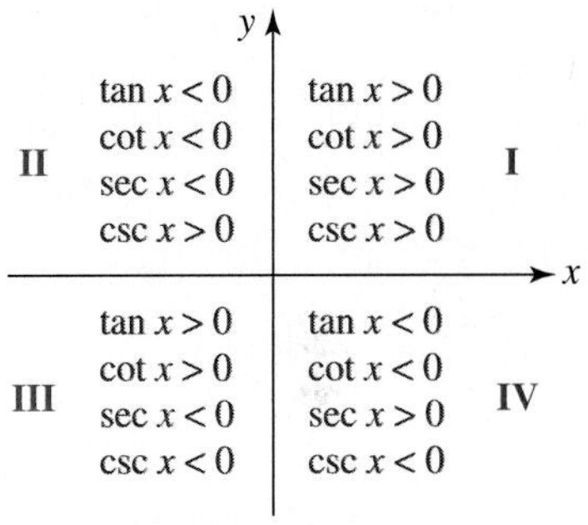

| | |
|---|---|
| II<br>$\tan x < 0$<br>$\cot x < 0$<br>$\sec x < 0$<br>$\csc x > 0$ | $\tan x > 0$<br>$\cot x > 0$<br>$\sec x > 0$<br>$\csc x > 0$   I |
| III<br>$\tan x > 0$<br>$\cot x > 0$<br>$\sec x < 0$<br>$\csc x < 0$ | $\tan x < 0$<br>$\cot x < 0$<br>$\sec x > 0$<br>$\csc x < 0$   IV |

**FIGURE 4.4.1** Signs of $\tan x$, $\cot x$, $\sec x$, and $\csc x$ in the four quadrants

### EXAMPLE 1     Example 5 of Section 4.2 Revisited

Find $\tan x$, $\cot x$, $\sec x$, and $\csc x$ for $x = -\pi/6$.

**Solution** In Example 5 of Section 4.2 we saw that

$$\sin\left(-\frac{\pi}{6}\right) = -\sin\frac{\pi}{6} = -\frac{1}{2} \quad \text{and} \quad \cos\left(-\frac{\pi}{6}\right) = \cos\frac{\pi}{6} = \frac{\sqrt{3}}{2}.$$

Therefore, by the definitions in (1) and (2):

$$\tan\left(-\frac{\pi}{6}\right) = \frac{-1/2}{\sqrt{3}/2} = -\frac{1}{\sqrt{3}}, \quad \cot\left(-\frac{\pi}{6}\right) = \frac{\sqrt{3}/2}{-1/2} = -\sqrt{3}, \quad \leftarrow \text{We could also use } \cot x = 1/\tan x$$

$$\sec\left(-\frac{\pi}{6}\right) = \frac{1}{\sqrt{3}/2} = \frac{2}{\sqrt{3}}, \quad \csc\left(-\frac{\pi}{6}\right) = \frac{1}{-1/2} = -2. \qquad ■$$

## TABLE 1

| $x$ | $0$ | $\frac{\pi}{6}$ | $\frac{\pi}{4}$ | $\frac{\pi}{3}$ | $\frac{\pi}{2}$ |
|---|---|---|---|---|---|
| $\tan x$ | $0$ | $\frac{1}{\sqrt{3}}$ | $1$ | $\sqrt{3}$ | $-$ |
| $\cot x$ | $-$ | $\sqrt{3}$ | $1$ | $\frac{1}{\sqrt{3}}$ | $0$ |
| $\sec x$ | $1$ | $\frac{2}{\sqrt{3}}$ | $\sqrt{2}$ | $2$ | $-$ |
| $\csc x$ | $-$ | $2$ | $\sqrt{2}$ | $\frac{2}{\sqrt{3}}$ | $1$ |

Table 1 summarizes some important values of the tangent, cotangent, secant, and cosecant and was constructed using values of the sine and cosine from Section 4.2. A dash in the table indicates the trigonometric function is not defined at that particular value of $x$.

☐ **Pythagorean Identities** The tangent is related to the secant by a useful identity. If we divide the Pythagorean identity

$$\sin^2 x + \cos^2 x = 1 \tag{5}$$

by $\cos^2 x$, we see that

$$\frac{\sin^2 x}{\cos^2 x} + \frac{\cos^2 x}{\cos^2 x} = \frac{1}{\cos^2 x}. \tag{6}$$

Similarly, dividing (5) by $\sin^2 x$ gives

$$\frac{\sin^2 x}{\sin^2 x} + \frac{\cos^2 x}{\sin^2 x} = \frac{1}{\sin^2 x}. \tag{7}$$

Using the laws of exponents,

$$\frac{\sin^2 x}{\cos^2 x} = \left(\frac{\sin x}{\cos x}\right)^2 = \tan^2 x, \qquad \frac{1}{\cos^2 x} = \left(\frac{1}{\cos x}\right)^2 = \sec^2 x,$$

$$\frac{\cos^2 x}{\sin^2 x} = \left(\frac{\cos x}{\sin x}\right)^2 = \cot^2 x, \qquad \frac{1}{\sin^2 x} = \left(\frac{1}{\sin x}\right)^2 = \csc^2 x,$$

(6) and (7) can be written in a simpler manner. These results summarized next are also known as **Pythagorean identities**.

> ## PYTHAGOREAN IDENTITIES (CONTINUED)
> For every real number $x$ for which the functions are defined,
> $$1 + \tan^2 x = \sec^2 x, \tag{8}$$
> $$1 + \cot^2 x = \csc^2 x. \tag{9}$$

**■ EXAMPLE 2**        **Using a Pythagorean Identity**

Given that $\csc x = -5$ and $3\pi/2 < x < 2\pi$, determine the values of $\tan x$ and $\cot x$.
   **Solution** We first compute $\cot x$. It follows from (9) that

$$\cot^2 x = \csc^2 x - 1.$$

For $3\pi/2 < x < 2\pi$, we see from Figure 4.4.1 that $\cot x$ must be negative and so we take the negative square root:

$$\cot x = -\sqrt{\csc^2 x - 1} = -\sqrt{(-5)^2 - 1} = -\sqrt{24} = -2\sqrt{6}.$$

Using $\cot x = 1/\tan x$, we have

$$\tan x = \frac{1}{\cot x} = \frac{1}{-2\sqrt{6}} = \overset{\text{rationalizing the denominator}}{\overset{\downarrow}{-\frac{\sqrt{6}}{12}}}.$$

In Example 2, given the information $\csc x = -5$ and $3\pi/2 < x < 2\pi$, we could easily find the values of the remaining five trigonometric functions. One way of proceeding would be to use $\csc x = 1/\sin x$ to find $\sin x = 1/\csc x = -\frac{1}{5}$. Then we use $\sin^2 x + \cos^2 x = 1$ to find $\cos x$. After we have found $\cos x$, the remaining three trigonometric functions can be obtained from (1) and (2).

☐ **Periodicity** Because the sine and cosine functions are $2\pi$ periodic, each of the functions in (1) and (2) have a period $2\pi$. But from (10) of Section 4.2 we have

$$\tan(x + \pi) = \frac{\sin(x + \pi)}{\cos(x + \pi)} = \frac{-\sin x}{-\cos x} = \tan x. \tag{10}$$

◄ Also see Problem 49 and 50 in Exercises 4.3.

Thus (10) implies that $\tan x$ and $\cot x$ are periodic with a period $p \le \pi$. In the case of the tangent function, $\tan x = 0$ only if $\sin x = 0$, that is, only if $x = 0, \pm\pi, \pm 2\pi$, and so on. Therefore, the smallest positive number $p$ for which $\tan(x + p) = \tan x$ is $p = \pi$. The cotangent function has the same period since it is the reciprocal of the tangent function.

---

### PERIOD OF THE TANGENT AND COTANGENT

The tangent and cotangent functions are periodic with **period** $\pi$. Therefore,

$$\tan(x + \pi) = \tan x \quad \text{and} \quad \cot(x + \pi) = \cot x \tag{11}$$

for every real number $x$ for which the functions are defined.

---

### PERIOD OF THE SECANT AND COSECANT

The secant and cosecant functions are periodic with **period** $2\pi$. Therefore,

$$\sec(x + 2\pi) = \sec x \quad \text{and} \quad \csc(x + 2\pi) = \csc x \tag{12}$$

for every real number $x$ for which the functions are defined.

---

☐ **Graphs** The numbers that make the denominators of $\tan x$, $\cot x$, $\sec x$, and $\csc x$ equal to zero correspond to vertical asymptotes of their graphs. For example, we encourage you to verify using a calculator that

◄ This is a good time to review (7) of Section 3.5.

$$\tan x \to -\infty \text{ as } x \to -\frac{\pi}{2}^{+} \quad \text{and} \quad \tan x \to \infty \text{ as } x \to \frac{\pi}{2}^{-}.$$

In other words, $x = -\pi/2$ and $x = \pi/2$ are vertical asymptotes. The graph of $y = \tan x$ on the interval $(-\pi/2, \pi/2)$ given in **FIGURE 4.4.2** is one **cycle** of the graph of $y = \tan x$. Using periodicity we extend the cycle in Figure 4.4.2 to adjacent intervals of length $\pi$, as shown in **FIGURE 4.4.3**. The $x$-intercepts of the graph of the tangent function are $(0, 0)$, $(\pm\pi, 0)$, $(\pm 2\pi, 0), \ldots$, and the vertical asymptotes of the graph are $x = \pm\pi/2$, $\pm 3\pi/2, \pm 5\pi/2, \ldots$.

The graph of $y = \cot x$ is similar to the graph of the tangent function and is given in **FIGURE 4.4.4**. In this case, the graph of $y = \cot x$ on the interval $(0, \pi)$ is one **cycle** of the graph of $y = \cot x$. The $x$-intercepts of the graph of the cotangent function are

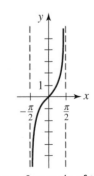

**FIGURE 4.4.2** One cycle of the graph of $y = \tan x$

$(\pm\pi/2, 0), (\pm 3\pi/2, 0), (\pm 5\pi/2, 0), \dots$, and the vertical asymptotes of the graph are $x = 0, \pm\pi, \pm 2\pi, \pm 3\pi, \dots$.

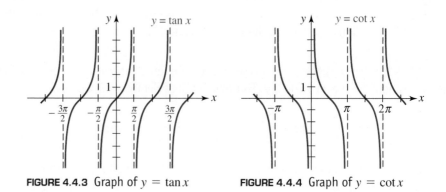

**FIGURE 4.4.3** Graph of $y = \tan x$     **FIGURE 4.4.4** Graph of $y = \cot x$

Note that the graphs of $y = \tan x$ and $y = \cot x$ are symmetric with respect to the origin, since $\tan(-x) = -\tan x$ and $\cot(-x) = -\cot x$.

For both $y = \sec x$ and $y = \csc x$ we know that $|y| \geq 1$, and so no portion of their graphs can appear in the horizontal strip $-1 < y < 1$ of the Cartesian plane. Hence the graphs of $y = \sec x$ and $y = \csc x$ have no $x$-intercepts. Both $y = \sec x$ and $y = \csc x$ have period $2\pi$. The vertical asymptotes for the graph of $y = \sec x$ are the same as $y = \tan x$, namely, $x = \pm\pi/2, \pm 3\pi/2, \pm 5\pi/2, \dots$. Because $y = \cos x$ is an even function, so is $y = \sec x = 1/\cos x$. The graph of $y = \sec x$ is symmetric with respect to the $y$-axis. On the other hand, the vertical asymptotes for the graph of $y = \csc x$ are the same as $y = \cot x$, namely, $x = 0, \pm\pi, \pm 2\pi, \pm 3\pi, \dots$. Because $y = \sin x$ is an odd function, so is $y = \csc x = 1/\sin x$. The graph of $y = \csc x$ is symmetric with respect to the origin. One cycle of the graph of $y = \sec x$ on $[0, 2\pi]$ is extended to the interval $[-2\pi, 0]$ by periodicity (or $y$-axis symmetry) in **FIGURE 4.4.5**. Similarly, in **FIGURE 4.4.6** we extend one cycle of $y = \csc x$ on $(0, 2\pi)$ to the interval $(-2\pi, 0)$ by periodicity (or origin symmetry).

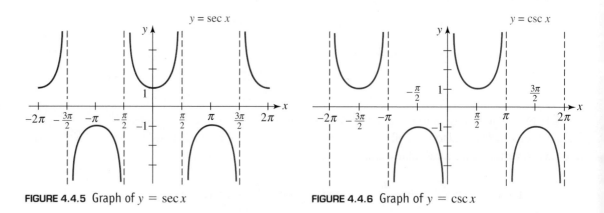

**FIGURE 4.4.5** Graph of $y = \sec x$     **FIGURE 4.4.6** Graph of $y = \csc x$

☐ **Transformations and Graphs** Similar to the sine and cosine graphs, rigid and nonrigid transformations can be applied to the graphs of $y = \tan x, y = \cot x, y = \sec x$, and $y = \csc x$. For example, a function such as $y = A\tan(Bx + C) + D$ can be analyzed in the following manner:

vertical stretch/compression/reflection          vertical shift
↓                                                    ↓

$$y = A\tan(Bx + C) + D \qquad (13)$$

↑      ↑

horizontal stretch/compression by changing period          horizontal shift

If $B > 0$, then the period of

$$y = A\tan(Bx + C) \qquad \text{and} \qquad y = A\cot(Bx + C) \text{ is } \pi/B, \qquad (14)$$

whereas the period of

$$y = A\sec(Bx + C) \qquad \text{and} \qquad y = A\csc(Bx + C) \text{ is } 2\pi/B. \qquad (15)$$

As we see in (13), the number $A$ in each case can be interpreted as either a vertical stretch or a compression of a graph. However, you should be aware of the fact that the functions in (14) and (15) have no amplitude, because none of the functions has a maximum *and* a minimum value.

◀ Of the six trigonometric functions, only the sine and cosine functions have an amplitude.

## ■ EXAMPLE 3　　Comparison of Graphs

Find the period, $x$-intercepts, and vertical asymptotes for the graph of $y = \tan 2x$. Graph the function on $[0, \pi]$.

**Solution** With the identification $B = 2$, we see from (14) that the period is $\pi/2$. Since $\tan 2x = \sin 2x/\cos 2x$, the $x$-intercepts of the graph occur at the zeros of $\sin 2x$. From (1) of Section 4.3, $\sin 2x = 0$ for

$$2x = n\pi \qquad \text{so that} \qquad x = \tfrac{1}{2}n\pi, n = 0, \pm 1, \pm 2, \ldots .$$

That is, $x = 0, \pm\pi/2, \pm 2\pi/2 = \pi, \pm 3\pi/2, \pm 4\pi/2 = 2\pi$, and so on. The $x$-intercepts are $(0, 0), (\pm\pi/2, 0), (\pm\pi, 0), (\pm 3\pi/2, 0), \ldots$. The vertical asymptotes of the graph occur at zeros of $\cos 2x$. From (2) of Section 4.3, the numbers for which $\cos 2x = 0$ are found in the following manner:

$$2x = (2n + 1)\frac{\pi}{2} \qquad \text{so that} \qquad x = (2n + 1)\frac{\pi}{4}, n = 0, \pm 1, \pm 2, \ldots .$$

That is, the vertical asymptotes are $x = \pm\pi/4, \pm 3\pi/4, \pm 5\pi/4, \ldots$. On the interval $[0, \pi]$, the graph of $y = \tan 2x$ has three intercepts $(0, 0), (\pi/2, 0)$, and $(\pi, 0)$ and two vertical asymptotes $x = \pi/4$ and $x = 3\pi/4$. In **FIGURE 4.4.7**, we have compared the graphs of $y = \tan x$ and $y = \tan 2x$ on the interval. The graph of $y = \tan 2x$ is a horizontal compression of the graph of $y = \tan x$.

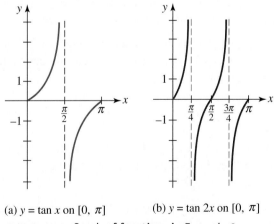

(a) $y = \tan x$ on $[0, \pi]$　　(b) $y = \tan 2x$ on $[0, \pi]$

**FIGURE 4.4.7** Graph of functions in Example 3

## ■ EXAMPLE 4　　Comparison of Graphs

Compare one cycle of the graphs of $y = \tan x$ and $y = \tan(x - \pi/4)$.

**Solution** The graph of $y = \tan(x - \pi/4)$ is the graph of $y = \tan x$ shifted horizontally $\pi/4$ units to the right. The intercept $(0, 0)$ for the graph of $y = \tan x$ is shifted to $(\pi/4, 0)$ on the graph of $y = \tan(x - \pi/4)$. The vertical asymptotes $x = -\pi/2$

and $x = \pi/2$ for the graph of $y = \tan x$ are shifted to $x = -\pi/4$ and $x = 3\pi/4$ for the graph of $y = \tan(x - \pi/4)$. In FIGURES 4.4.8(a) and 4.4.8(b) we see, respectively, that a cycle of the graph of $y = \tan x$ on the interval $(-\pi/2, \pi/2)$ is shifted to the right to yield a cycle of the graph of $y = \tan(x - \pi/4)$ on the interval $(-\pi/4, 3\pi/4)$.

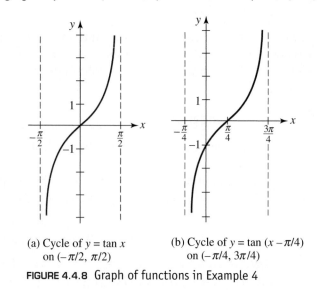

(a) Cycle of $y = \tan x$ on $(-\pi/2, \pi/2)$

(b) Cycle of $y = \tan(x - \pi/4)$ on $(-\pi/4, 3\pi/4)$

FIGURE 4.4.8 Graph of functions in Example 4

As we did in the analysis of the graphs of $y = A\sin(Bx + C)$ and $y = A\cos(Bx + C)$, we can determine the amount of horizontal shift for graphs of functions such as $y = A\tan(Bx + C)$ and $y = A\sec(Bx + C)$ by factoring the number $B > 0$ from $Bx + C$.

### ■ EXAMPLE 5    Two Shifts and Two Compressions

Graph $y = 2 - \frac{1}{2}\sec(3x - \pi/2)$.

**Solution** Let's break down the analysis of the graph into four parts, namely, by transformations.

(i) One cycle of the graph of $y = \sec x$ occurs on $[0, 2\pi]$. Since the period of $y = \sec 3x$ is $2\pi/3$, one cycle of its graph occurs on the interval $[0, 2\pi/3]$.

In other words, the graph of $y = \sec 3x$ is a horizontal compression of the graph of $y = \sec x$. Since $\sec 3x = 1/\cos 3x$, the vertical asymptotes occur at the zeros of $\cos 3x$. Using (2) of Section 4.3, we find

$$3x = (2n + 1)\frac{\pi}{2} \quad \text{or} \quad x = (2n + 1)\frac{\pi}{6}, n = 0, \pm 1, \pm 2, \ldots.$$

FIGURE 4.4.9(a) shows two cycles of the graph $y = \sec 3x$; one cycle on $[-2\pi/3, 0]$ and another on $[0, 2\pi/3]$. Within those intervals the vertical asymptotes are $x = -\pi/2$, $x = -\pi/6$, $x = \pi/6$, and $x = \pi/2$.

(ii) The graph of $y = -\frac{1}{2}\sec 3x$ is the graph of $y = \sec 3x$ compressed vertically by a factor of $\frac{1}{2}$ and then reflected in the $x$-axis. See Figure 4.4.9(b).

(iii) By factoring 3 from $3x - \pi/2$, we see from

$$y = -\frac{1}{2}\sec\left(3x - \frac{\pi}{2}\right) = -\frac{1}{2}\sec 3\left(x - \frac{\pi}{6}\right)$$

that the graph of $y = -\frac{1}{2}\sec(3x - \pi/2)$ is the graph of $y = -\frac{1}{2}\sec 3x$ shifted $\pi/6$ units to the right. By shifting the two intervals $[-2\pi/3, 0]$ and $[0, 2\pi/3]$

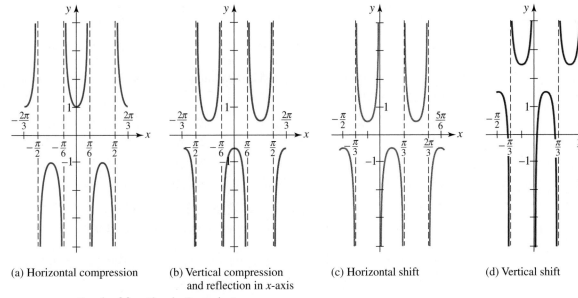

(a) Horizontal compression

(b) Vertical compression and reflection in $x$-axis

(c) Horizontal shift

(d) Vertical shift

**FIGURE 4.4.9** Graph of function in Example 5

in Figure 4.4.9(b) to the right $\pi/6$ units, we see in Figure 4.4.9(c) two cycles of $y = -\frac{1}{2}\sec(3x - \pi/2)$ on the intervals $[-\pi/2, \pi/6]$ and $[\pi/6, 5\pi/6]$. The vertical asymptotes $x = -\pi/2$, $x = -\pi/6$, $x = \pi/6$, and $x = \pi/2$ shown in Figure 4.4.9(b) are shifted to $x = -\pi/3$, $x = 0$, $x = \pi/3$, and $x = 2\pi/3$. Observe that the $y$-intercept $(0, -\frac{1}{2})$ in Figure 4.4.9(b) is now moved to $(\pi/6, -\frac{1}{2})$ in Figure 4.4.9(c).

(*iv*) Finally, we obtain the graph $y = 2 - \frac{1}{2}\sec(3x - \pi/2)$ in Figure 4.4.9(d) by shifting the graph of $y = -\frac{1}{2}\sec(3x - \pi/2)$ in Figure 4.4.9(c) two units upward. ∎

## 4.4 Exercises
Answers to selected odd-numbered problems begin on page ANS–13.

In Problems 1 and 2, complete the given table.

1.

| $x$ | $\frac{2\pi}{3}$ | $\frac{3\pi}{4}$ | $\frac{5\pi}{6}$ | $\pi$ | $\frac{7\pi}{6}$ | $\frac{5\pi}{4}$ | $\frac{4\pi}{3}$ | $\frac{3\pi}{2}$ | $\frac{5\pi}{3}$ | $\frac{7\pi}{4}$ | $\frac{11\pi}{6}$ | $2\pi$ |
|---|---|---|---|---|---|---|---|---|---|---|---|---|
| $\tan x$ | | | | | | | | | | | | |
| $\cot x$ | | | | | | | | | | | | |

2.

| $x$ | $\frac{2\pi}{3}$ | $\frac{3\pi}{4}$ | $\frac{5\pi}{6}$ | $\pi$ | $\frac{7\pi}{6}$ | $\frac{5\pi}{4}$ | $\frac{4\pi}{3}$ | $\frac{3\pi}{2}$ | $\frac{5\pi}{3}$ | $\frac{7\pi}{4}$ | $\frac{11\pi}{6}$ | $2\pi$ |
|---|---|---|---|---|---|---|---|---|---|---|---|---|
| $\sec x$ | | | | | | | | | | | | |
| $\csc x$ | | | | | | | | | | | | |

In Problems 3–18, find the indicated value without the use of a calculator.

**3.** $\cot\dfrac{13\pi}{6}$

**4.** $\csc\left(-\dfrac{3\pi}{2}\right)$

**5.** $\tan\dfrac{9\pi}{2}$

**6.** $\sec 7\pi$

**7.** $\csc\left(-\dfrac{\pi}{3}\right)$

**8.** $\cot\left(-\dfrac{13\pi}{3}\right)$

**9.** $\tan\dfrac{23\pi}{4}$

**10.** $\tan\left(-\dfrac{5\pi}{6}\right)$

**11.** $\sec\dfrac{10\pi}{3}$  **12.** $\cot\dfrac{17\pi}{6}$  **13.** $\csc 5\pi$  **14.** $\sec\dfrac{29\pi}{4}$

**15.** $\sec(-120°)$  **16.** $\tan 405°$  **17.** $\csc 495°$  **18.** $\cot(-720°)$

In Problems 19–26, use the given information to find the values of the remaining five trigonometric functions.

**19.** $\tan x = -2,\ \pi/2 < x < \pi$  **20.** $\cot x = \frac{1}{2},\ \pi < x < 3\pi/2$

**21.** $\csc x = \frac{4}{3},\ 0 < x < \pi/2$  **22.** $\sec x = -5,\ \pi/2 < x < \pi$

**23.** $\sin x = \frac{1}{3},\ \pi/2 < x < \pi$  **24.** $\cos x = -1/\sqrt{5},\ \pi < x < 3\pi/2$

**25.** $\cos x = \frac{12}{13},\ 3\pi/2 < x < 2\pi$  **26.** $\sin x = \frac{4}{5},\ 0 < x < \pi/2$

**27.** If $3\cos x = \sin x$, find all values of $\tan x$, $\cot x$, $\sec x$, and $\csc x$.

**28.** If $\csc x = \sec x$, find all values of $\tan x$, $\cot x$, $\sin x$, and $\cos x$.

In Problems 29–36, find the period, $x$-intercepts, and the vertical asymptotes of the given function. Sketch at least one cycle of the graph.

**29.** $y = \tan \pi x$  **30.** $y = \tan\dfrac{x}{2}$

**31.** $y = \cot 2x$  **32.** $y = -\cot\dfrac{\pi x}{3}$

**33.** $y = \tan\left(\dfrac{x}{2} - \dfrac{\pi}{4}\right)$  **34.** $y = \dfrac{1}{4}\cot\left(x - \dfrac{\pi}{2}\right)$

**35.** $y = -1 + \cot \pi x$  **36.** $y = \tan\left(x + \dfrac{5\pi}{6}\right)$

In Problems 37–44, find the period and the vertical asymptotes of the given function. Sketch at least one cycle of the graph.

**37.** $y = -\sec x$  **38.** $y = 2\sec\dfrac{\pi x}{2}$

**39.** $y = 3\csc \pi x$  **40.** $y = -2\csc\dfrac{x}{3}$

**41.** $y = \sec\left(3x - \dfrac{\pi}{2}\right)$  **42.** $y = \csc x(4x + \pi)$

**43.** $y = 3 + \csc\left(2x + \dfrac{\pi}{2}\right)$  **44.** $y = -1 + \sec(x - 2\pi)$

In Problems 45 and 46, use the graphs of $y = \tan x$ and $y = \sec x$ to find numbers $A$ and $C$ for which the given equality is true.

**45.** $\cot x = A\tan(x + C)$  **46.** $\csc x = A\sec(x + C)$

### Calculator Problems

**47.** Using a calculator in radian mode, compare the values of $\tan 1.57$ and $\tan 1.58$. Explain the difference in these values.

**48.** Using a calculator in radian mode, compare the values of $\cot 3.14$ and $\cot 3.15$.

### For Discussion

**49.** Can $9\csc x = 1$ for any real number $x$?

**50.** Can $7 + 10\sec x = 0$ for any real number $x$?

**51.** For which real numbers $x$ is **(a)** $\sin x \leq \csc x$? **(b)** $\sin x < \csc x$?

**52.** For which real numbers $x$ is **(a)** $\sec x \leq \cos x$? **(b)** $\sec x < \cos x$?

**53.** Discuss and then sketch the graphs of $y = |\sec x|$ and $y = |\csc x|$.

---

## 4.5 Special Identities

☐ **Introduction** In this section we examine trigonometric identities. A **trigonometric identity** is an equation or formula involving trigonometric functions that is valid for all angles or real numbers for which both sides of the equality are defined. There are *numerous* trigonometric identities, but we are going to develop only those of special importance in courses in mathematics and science.

The formulas derived in the discussion that follows apply to a real number $x$ as well as to an angle $x$ measured in degrees or in radians.

☐ **Pythagorean Identities** In Section 4.2, we saw that the sine and cosine are related by the fundamental identity $\sin^2 x + \cos^2 x = 1$. In Section 4.4, we saw that by dividing this identity in turn by $\cos^2 x$, and then by $\sin^2 x$, we obtain two more identities, one relating $\tan^2 x$ to $\sec^2 x$ and the other relating $\cot^2 x$ to $\csc^2 x$. These so-called **Pythagorean identities** are so basic to trigonometry that we give them again for future reference.

---

**PYTHAGOREAN IDENTITIES**

For $x$ a real number for which the functions are defined,

$$\sin^2 x + \cos^2 x = 1, \tag{1}$$
$$1 + \tan^2 x = \sec^2 x, \tag{2}$$
$$1 + \cot^2 x = \csc^2 x. \tag{3}$$

---

☐ **Trigonometric Substitutions** In calculus it is often useful to make use of trigonometric substitution to change the form of certain algebraic expressions involving radicals. Generally, this is done using the Pythagorean identities. The following example illustrates the technique.

**EXAMPLE 1**　　　**Rewriting a Radical**

Rewrite $\sqrt{a^2 - x^2}$ as a trigonometric expression without radicals by means of the substitution $x = a\sin\theta$, $a > 0$, and $-\pi/2 \leq \theta \leq \pi/2$.

**Solution** If $x = a\sin\theta$, then

$$\begin{aligned}
\sqrt{a^2 - x^2} &= \sqrt{a^2 - (a\sin\theta)^2} \\
&= \sqrt{a^2 - a^2\sin^2\theta} \\
&= \sqrt{a^2(1 - \sin^2\theta)} \quad \leftarrow \text{now use (1)} \\
&= \sqrt{a^2\cos^2\theta}.
\end{aligned}$$

Since $a > 0$ and $\cos\theta \geq 0$ for $-\pi/2 \leq \theta \leq \pi/2$, the original radical is the same as

$$\sqrt{a^2 - x^2} = \sqrt{a^2\cos^2\theta} = a\cos\theta. \qquad \blacksquare$$

☐ **Sum and Difference Formulas** The **sum** and **difference formulas** for the cosine and sine functions are identities that reduce $\cos(x_1 + x_2)$, $\cos(x_1 - x_2)$, $\sin(x_1 + x_2)$, and $\sin(x_1 - x_2)$ to expressions that involve $\cos x_1$, $\cos x_2$, $\sin x_1$, and $\sin x_2$. We will derive the formula for $\cos(x_1 - x_2)$ first, and then we will use that result to obtain the others.

For convenience, let us suppose that $x_1$ and $x_2$ represent angles measured in radians. As shown in **FIGURE 4.5.1(a)**, let $d$ denote the distance between $P(x_1)$ and $P(x_2)$. If we place the angle $x_1 - x_2$ in standard position as shown in Figure 4.5.1(b), then $d$ is also the distance between $P(x_1 - x_2)$ and $P(0)$. Equating the squares of these distances gives

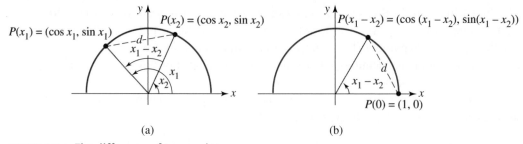

(a)                                    (b)

**FIGURE 4.5.1** The difference of two angles

$$(\cos x_1 - \cos x_2)^2 + (\sin x_1 - \sin x_2)^2 = (\cos(x_1 - x_2) - 1)^2 + \sin^2(x_1 - x_2)$$

or $\quad \cos^2 x_1 - 2\cos x_1 \cos x_2 + \cos^2 x_2 + \sin^2 x_1 - 2\sin x_1 \sin x_2 + \sin^2 x_2$

$$= \cos^2(x_1 - x_2) - 2\cos(x_1 - x_2) + 1 + \sin^2(x_1 - x_2).$$

In view of (1),

$$\cos^2 x_1 + \sin^2 x_1 = 1, \quad \cos^2 x_2 + \sin^2 x_2 = 1, \quad \cos^2(x_1 - x_2) + \sin^2(x_1 - x_2) = 1,$$

and so the preceding equation simplifies to

$$\cos(x_1 - x_2) = \cos x_1 \cos x_2 + \sin x_1 \sin x_2.$$

This last result can be put to work immediately to find the cosine of the sum of two angles. Since $x_1 + x_2$ can be rewritten as the difference $x_1 - (-x_2)$,

$$\cos(x_1 + x_2) = \cos(x_1 - (-x_2))$$
$$= \cos x_1 \cos(-x_2) + \sin x_1 \sin(-x_2).$$

By the even-odd identities, $\cos(-x_2) = \cos x_2$ and $\sin(-x_2) = -\sin x_2$, it follows that the last line is the same as

$$\cos(x_1 + x_2) = \cos x_1 \cos x_2 - \sin x_1 \sin x_2.$$

The two results just obtained are summarized next.

## SUM AND DIFFERENCE FORMULAS FOR THE COSINE

For all real numbers $x_1$ and $x_2$,

$$\cos(x_1 + x_2) = \cos x_1 \cos x_2 - \sin x_1 \sin x_2, \qquad (4)$$
$$\cos(x_1 - x_2) = \cos x_1 \cos x_2 + \sin x_1 \sin x_2. \qquad (5)$$

**EXAMPLE 2**          **Cosine of a Sum**

Evaluate $\cos(7\pi/12)$.

**Solution** We have no way of evaluating $\cos(7\pi/12)$ directly. However, observe that

$$\frac{7\pi}{12} \text{ radians} = 105° = 60° + 45° = \frac{\pi}{3} + \frac{\pi}{4}.$$

Because $7\pi/12$ radians is a second quadrant angle, we know that the value of $\cos(7\pi/12)$ is negative. Proceeding, the sum formula (4) gives

$$\cos\frac{7\pi}{12} = \cos\left(\frac{\pi}{3} + \frac{\pi}{4}\right) = \cos\frac{\pi}{3}\cos\frac{\pi}{4} - \sin\frac{\pi}{3}\sin\frac{\pi}{4}$$

$$= \frac{1}{2}\frac{\sqrt{2}}{2} - \frac{\sqrt{3}}{2}\frac{\sqrt{2}}{2} = \frac{\sqrt{2}}{4}(1 - \sqrt{3}).$$

Using $\sqrt{2}\sqrt{3} = \sqrt{6}$, this result can also be written as $\cos(7\pi/12) = (\sqrt{2} - \sqrt{6})/4$. Since $\sqrt{6} > \sqrt{2}$, we see that $\cos(7\pi/12) < 0$, as expected. ∎

To obtain the corresponding sum/difference identities for the sine function we will make use of two identities:

$$\cos\left(x - \frac{\pi}{2}\right) = \sin x \quad \text{and} \quad \sin\left(x - \frac{\pi}{2}\right) = -\cos x. \qquad (6)$$

◀ See (5) in Section 4.3.

These identities were first discovered in Section 4.3 by shifting the graphs of the cosine and sine. However, both results in (6) can now be proved using (5):

$$\cos\left(x - \frac{\pi}{2}\right) = \cos x\cos\frac{\pi}{2} + \sin x\sin\frac{\pi}{2} = \cos x \cdot 0 + \sin x \cdot 1 = \sin x,$$

$$\underset{\text{zero}}{\downarrow} \qquad \underset{\text{by (5)}}{\downarrow}$$

$$\cos x = \cos\left(\frac{\pi}{2} - \frac{\pi}{2} + x\right) = \cos\left(\frac{\pi}{2} - \left(\frac{\pi}{2} - x\right)\right) = \sin\left(\frac{\pi}{2} - x\right) = -\sin\left(x - \frac{\pi}{2}\right).$$

Now from the first equation in (6), the sine of the sum $x_1 + x_2$ can be written

$$\sin(x_1 + x_2) = \cos\left((x_1 + x_2) - \frac{\pi}{2}\right)$$

$$= \cos\left(x_1 + \left(x_2 - \frac{\pi}{2}\right)\right)$$

$$= \cos x_1\cos\left(x_2 - \frac{\pi}{2}\right) - \sin x_1\sin\left(x_2 - \frac{\pi}{2}\right) \quad \leftarrow \text{by (4)}$$

$$= \cos x_1\sin x_2 - \sin x_1(-\cos x_2). \quad \leftarrow \text{by (6)}$$

The last line is traditionally written as

$$\sin(x_1 + x_2) = \sin x_1\cos x_2 + \cos x_1\sin x_2.$$

To obtain the sine of the difference $x_1 - x_2$, we use again $\cos(-x_2) = \cos x_2$ and $\sin(-x_2) = -\sin x_2$:

$$\sin(x_1 - x_2) = \sin(x_1 + (-x_2)) = \sin x_1\cos(-x_2) + \cos x_1\sin(-x_2)$$

$$= \sin x_1\cos x_2 - \cos x_1\sin x_2.$$

### SUM AND DIFFERENCE FORMULAS FOR THE SINE

For all real numbers $x_1$ and $x_2$,

$$\sin(x_1 + x_2) = \sin x_1\cos x_2 + \cos x_1\sin x_2, \qquad (7)$$
$$\sin(x_1 - x_2) = \sin x_1\cos x_2 - \cos x_1\sin x_2. \qquad (8)$$

EXAMPLE 3          **Sine of a Sum**

Evaluate $\sin(7\pi/12)$.

**Solution** We proceed as in Example 2, except we use the sum formula (7):

$$\sin\frac{7\pi}{12} = \sin\left(\frac{\pi}{3} + \frac{\pi}{4}\right) = \sin\frac{\pi}{3}\cos\frac{\pi}{4} + \cos\frac{\pi}{3}\sin\frac{\pi}{4}$$

$$= \frac{\sqrt{3}}{2}\frac{\sqrt{2}}{2} + \frac{1}{2}\frac{\sqrt{2}}{2} = \frac{\sqrt{2}}{4}(1 + \sqrt{3}).$$

As in Example 2, the result can be rewritten as $\sin(7\pi/12) = (\sqrt{2} + \sqrt{6})/4$. ∎

Since we know the value of $\cos(7\pi/12)$ from Example 2, we can also compute the value of $\sin(7\pi/12)$ using the Pythagorean identity (1):

$$\sin^2\frac{7\pi}{12} + \cos^2\frac{7\pi}{12} = 1.$$

We solve for $\sin(7\pi/12)$ and take the positive square root:

$$\sin\frac{7\pi}{12} = \sqrt{1 - \cos^2\frac{7\pi}{12}} = \sqrt{1 - \left[\frac{\sqrt{2}}{4}(1 - \sqrt{3})\right]^2}$$

$$= \sqrt{\frac{4 + 2\sqrt{3}}{8}} = \frac{\sqrt{2 + \sqrt{3}}}{2}.$$

(9)

Although the number in (9) does not look like the result obtained in Example 3, the values are the same. See Problem 68 in Exercises 4.5.

There are sum and difference formulas for the tangent function as well. We can derive the sum formula using the sum formulas for the sine and cosine as follows:

$$\tan(x_1 + x_2) = \frac{\sin(x_1 + x_2)}{\cos(x_1 + x_2)} = \frac{\sin x_1 \cos x_2 + \cos x_1 \sin x_2}{\cos x_1 \cos x_2 - \sin x_1 \sin x_2}.$$

(10)

We now divide the numerator and denominator of (10) by $\cos x_1 \cos x_2$ (assuming that $x_1$ and $x_2$ are such that $\cos x_1 \cos x_2 \neq 0$),

$$\tan(x_1 + x_2) = \frac{\dfrac{\sin x_1}{\cos x_1}\dfrac{\cos x_2}{\cos x_2} + \dfrac{\cos x_1}{\cos x_1}\dfrac{\sin x_2}{\cos x_2}}{\dfrac{\cos x_1}{\cos x_1}\dfrac{\cos x_2}{\cos x_2} - \dfrac{\sin x_1}{\cos x_1}\dfrac{\sin x_2}{\cos x_2}} = \frac{\tan x_1 + \tan x_2}{1 - \tan x_1 \tan x_2}.$$

(11)

The derivation of the difference formula for $\tan(x_1 - x_2)$ is obtained in a similar manner. We summarize the two results.

### SUM AND DIFFERENCE FORMULAS FOR THE TANGENT

For real numbers $x_1$ and $x_2$ for which the functions are defined,

$$\tan(x_1 + x_2) = \frac{\tan x_1 + \tan x_2}{1 - \tan x_1 \tan x_2},$$

(12)

$$\tan(x_1 - x_2) = \frac{\tan x_1 - \tan x_2}{1 + \tan x_1 \tan x_2}.$$

(13)

**Tangent of a Sum**

Evaluate $\tan(\pi/12)$.

   **Solution** If we think of $\pi/12$ as an angle in radians, then

$$\frac{\pi}{12}\text{ radians } = 15° = 45° - 30° = \frac{\pi}{4} - \frac{\pi}{6}\text{ radians.}$$

It follows from formula (13):

$$\tan\frac{\pi}{12} = \tan\left(\frac{\pi}{4} - \frac{\pi}{6}\right) = \frac{\tan\dfrac{\pi}{4} - \tan\dfrac{\pi}{6}}{1 + \tan\dfrac{\pi}{4}\tan\dfrac{\pi}{6}}$$

◀ You should rework this example using
$$\pi/12 = \pi/3 - \pi/4$$
to see that the result is the same.

$$= \frac{1 - \dfrac{1}{\sqrt{3}}}{1 + 1\cdot\dfrac{1}{\sqrt{3}}} = \frac{\sqrt{3} - 1}{\sqrt{3} + 1}$$

$$= \frac{\sqrt{3} - 1}{\sqrt{3} + 1}\cdot\frac{\sqrt{3} - 1}{\sqrt{3} - 1} \qquad \leftarrow \text{rationalizing the denominator}$$

$$= \frac{(\sqrt{3} - 1)^2}{2} = \frac{4 - 2\sqrt{3}}{2} = 2 - \sqrt{3}.\qquad\blacksquare$$

   Strictly speaking, we really do not need the identities for $\tan(x_1 \pm x_2)$, since we can always compute $\sin(x_1 \pm x_2)$ and $\cos(x_1 \pm x_2)$ using (4)–(8) and then proceed as in (10), that is, form the quotient $\sin(x_1 \pm x_2)/\cos(x_1 \pm x_2)$.

   ☐ **Double-Angle Formulas** Many useful trigonometric formulas can be derived from the sum and difference formulas. The **double-angle formulas** express the cosine and sine of $2x$ in terms of the cosine and sine of $x$.

   If we set $x_1 = x_2 = x$ in (4) and use $\cos(x + x) = \cos 2x$, then

$$\cos 2x = \cos x\cos x - \sin x\sin x = \cos^2 x - \sin^2 x.$$

Similarly, by setting $x_1 = x_2 = x$ in (7) and using $\sin(x + x) = \sin 2x$, then

these two terms are equal
$$\downarrow \qquad\qquad \downarrow$$
$$\sin 2x = \sin x\cos x + \cos x\sin x = 2\sin x\cos x.$$

We summarize the last two results.

---

### DOUBLE-ANGLE FORMULAS FOR THE COSINE AND SINE

For any real number $x$,

$$\cos 2x = \cos^2 x - \sin^2 x, \qquad (14)$$
$$\sin 2x = 2\sin x\cos x. \qquad (15)$$

---

■ EXAMPLE 5 **Using the Double-Angle Formulas**

If $\sin x = -\frac{1}{4}$ and $\pi < x < 3\pi/2$, find the exact values of $\cos 2x$ and $\sin 2x$.

**Solution** First, we compute $\cos x$ using $\sin^2 x + \cos^2 x = 1$. Since $\pi < x < 3\pi/2$, $\cos x < 0$, and so we choose the negative square root:

$$\cos x = -\sqrt{1 - \sin^2 x} = -\sqrt{1 - \left(-\frac{1}{4}\right)^2} = -\frac{\sqrt{15}}{4}.$$

From the double-angle formula (14),

$$\cos 2x = \cos^2 x - \sin^2 x$$
$$= \left(-\frac{\sqrt{15}}{4}\right)^2 - \left(-\frac{1}{4}\right)^2$$
$$= \frac{15}{16} - \frac{1}{16} = \frac{14}{16} = \frac{7}{8}.$$

Finally, from the double-angle formula (15),

$$\sin 2x = 2 \sin x \cos x = 2\left(-\frac{1}{4}\right)\left(-\frac{\sqrt{15}}{4}\right) = \frac{\sqrt{15}}{8}. \qquad \blacksquare$$

The formula in (14) has two useful alternative forms. By (1), we know that $\sin^2 x = 1 - \cos^2 x$. Substituting the last expression into (14) yields $\cos 2x = \cos^2 x - (1 - \cos^2 x)$ or

$$\cos 2x = 2\cos^2 x - 1. \qquad (16)$$

On the other hand, if we substitute $\cos^2 x = 1 - \sin^2 x$ into (14) we get

$$\cos 2x = 1 - 2\sin^2 x. \qquad (17)$$

☐ **Half-Angle Formulas** The alternative forms (16) and (17) of the double-angle formula (14) are the source of two **half-angle formulas**. Solving (16) and (17) for $\cos^2 x$ and $\sin^2 x$ gives, respectively,

$$\cos^2 x = \frac{1}{2}(1 + \cos 2x) \qquad \text{and} \qquad \sin^2 x = \frac{1}{2}(1 - \cos 2x). \qquad (18)$$

By replacing the symbol $x$ in (18) by $x/2$ and using $2(x/2) = x$, we obtain the following formulas.

---

### HALF-ANGLE FORMULAS FOR THE COSINE AND SINE

For any real number $x$,

$$\cos^2 \frac{x}{2} = \frac{1}{2}(1 + \cos x), \qquad (19)$$

$$\sin^2 \frac{x}{2} = \frac{1}{2}(1 - \cos x). \qquad (20)$$

---

**EXAMPLE 6**  **Using the Half-Angle Formulas**

Find the exact values of $\cos(5\pi/8)$ and $\sin(5\pi/8)$.

**Solution** If we let $x = 5\pi/4$, then $x/2 = 5\pi/8$ and formulas (19) and (20) yield, respectively,

$$\cos^2\left(\frac{5\pi}{8}\right) = \frac{1}{2}\left(1 + \cos\frac{5\pi}{4}\right) = \frac{1}{2}\left[1 + \left(-\frac{\sqrt{2}}{2}\right)\right] = \frac{2 - \sqrt{2}}{4},$$

and $\qquad \sin^2\left(\frac{5\pi}{8}\right) = \frac{1}{2}\left(1 - \cos\frac{5\pi}{4}\right) = \frac{1}{2}\left[1 - \left(-\frac{\sqrt{2}}{2}\right)\right] = \frac{2 + \sqrt{2}}{4}.$

Because $5\pi/8$ radians is a second quadrant angle, $\cos(5\pi/8) < 0$ and $\sin(5\pi/8) > 0$. Therefore, we take the negative square root for the value of the cosine,

$$\cos\left(\frac{5\pi}{8}\right) = -\sqrt{\frac{2 - \sqrt{2}}{4}} = -\frac{\sqrt{2 - \sqrt{2}}}{2},$$

and the positive square root for the value of the sine,

$$\sin\left(\frac{5\pi}{8}\right) = \sqrt{\frac{2 + \sqrt{2}}{4}} = \frac{\sqrt{2 + \sqrt{2}}}{2}. \qquad \blacksquare$$

In Section 4.3 we examined the graphs of horizontally shifted sine and cosine graphs. It turns out that any linear combination of a sine and a cosine of the form

$$y = c_1 \cos Bx + c_2 \sin Bx \tag{21}$$

can be expressed either as

$$y = A \sin(Bx + \phi) \tag{22}$$

or as $y = A\cos(Bx + \phi)$. We examine only the reduction of (21) to the form given in (22).

---

### REDUCTION OF (21) TO (22)

For real numbers $c_1$, $c_2$, $B$, and $x$,

$$c_1 \cos Bx + c_2 \sin Bx = A \sin(Bx + \phi), \tag{23}$$

where $A$ and $\phi$ are defined by

$$A = \sqrt{c_1^2 + c_2^2}, \tag{24}$$

and $\qquad\qquad \sin\phi = \dfrac{c_1}{A}, \qquad \cos\phi = \dfrac{c_2}{A}. \tag{25}$

---

To verify this, we use the sum formula (7):

$$\begin{aligned}
A\sin(Bx + \phi) &= A\sin Bx \cos\phi + A\cos Bx \sin\phi \\
&= (A\sin\phi)\cos Bx + (A\cos\phi)\sin Bx \\
&= c_1 \cos Bx + c_2 \sin Bx
\end{aligned}$$

and identify $A\sin\phi = c_1$, $A\cos\phi = c_2$. Thus, $\sin\phi = c_1/A = c_1/\sqrt{c_1^2 + c_2^2}$ and $\cos\phi = c_2/A = c_2/\sqrt{c_1^2 + c_2^2}$.

### EXAMPLE 7      Reduction to (23)

Sketch the graph of $y = \cos 2x - \sqrt{3}\sin 2x$.

**Solution** To start we use (23) to express $y$ as a single sine function. With the identifications $c_1 = 1$, $c_2 = -\sqrt{3}$, and $B = 2$, we have from (24) and (25),

$$A = \sqrt{c_1^2 + c_2^2} = \sqrt{1^2 + (-\sqrt{3})^2} = \sqrt{4} = 2,$$

$$\sin\phi = \frac{1}{2} \quad \text{and} \quad \cos\phi = -\frac{\sqrt{3}}{2}.$$

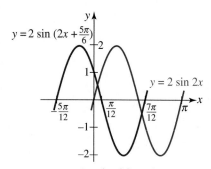

$y = 2 \sin \left(2x + \frac{5\pi}{6}\right)$

$y = 2 \sin 2x$

**FIGURE 4.5.2** Graph of function in Example 7

From the last two equations we conclude that $\phi$ is the second quadrant angle $5\pi/6$ radians. Therefore $y = \cos 2x - \sqrt{3}\sin 2x$ can be rewritten as

$$y = 2\sin\left(2x + \frac{5\pi}{6}\right) \qquad \text{or} \qquad y = 2\sin 2\left(x + \frac{5\pi}{12}\right).$$

Hence the graph of $y = \cos 2x - \sqrt{3}\sin 2x$ is the graph of $y = 2\sin 2x$, which has amplitude 2, period $2\pi/2 = \pi$, and is shifted $5\pi/12$ units to the left. **FIGURE 4.5.2** shows one cycle of $y = 2\sin 2x$ on $[0, \pi]$ in blue, and the shifted cycle on the interval $[-5\pi/12, 7\pi/12]$ in red. ∎

## NOTES FROM THE CLASSROOM

(*i*) Should you memorize all the identities presented in this section? You should consult your instructor about this, but in the opinion of the authors, you should at the very least memorize formulas (1)–(8), (14), (15), and the two formulas in (18).

(*ii*) When you enroll in a calculus course, check the title of your text. If it has the words *Early Transcendentals* in its title, then your knowledge of the graphs and properties of the trigonometric functions will come into play almost immediately.

(*iii*) As discussed in Sections 2.9 and 3.7, the principal topics of study in calculus are *derivatives* and *integrals* of functions. The sum identities (4) and (7) are used to find the derivatives of $\sin x$ and $\cos x$, See Section 4.10. Identities are especially useful in integral calculus. Replacing a radical by a trigonometric function as illustrated in Example 1 in this section is a standard technique for evaluating some types of integrals. Also, to evaluate integrals of $\cos^2 x$ and $\sin^2 x$ you would use the half-angle formulas in the form given in (18):

$$\cos^2 x = \tfrac{1}{2}(1 + \cos 2x) \qquad \text{and} \qquad \sin^2 x = \tfrac{1}{2}(1 - \cos 2x).$$

At some point in your study of integral calculus you may be required to evaluate integrals of products such as

$$\sin 2x \sin 5x \qquad \text{and} \qquad \sin 10x \cos 4x.$$

One way of doing this is to use the sum/difference formulas to devise an identity that converts these products into either a sum of sines or a sum of cosines. See Problems 73–76 in Exercises 4.5.

**4.5** **Exercises** Answers to selected odd-numbered problems begin on page ANS–14.

In Problems 1–8, proceed as in Example 1 and rewrite the given expression as a trigonometric expression without radicals by making the indicated substitution. Assume that $a > 0$.

**1.** $\sqrt{a^2 - x^2}$, $\qquad x = a\cos\theta$, $\qquad 0 \le \theta \le \pi$

**2.** $\sqrt{a^2 + x^2}$, $\qquad x = a\tan\theta$, $\qquad -\pi/2 < \theta < \pi/2$

**3.** $\sqrt{x^2 - a^2}$, $\qquad x = a\sec\theta$, $\qquad 0 \le \theta < \pi/2$

**4.** $\sqrt{16 - 25x^2}, \qquad x = \frac{4}{5}\sin\theta, \qquad -\pi/2 \le \theta \le \pi/2$

**5.** $\dfrac{x}{\sqrt{9 - x^2}}, \qquad x = 3\sin\theta, \qquad -\pi/2 < \theta < \pi/2$

**6.** $\dfrac{\sqrt{x^2 - 3}}{x^2}, \qquad x = \sqrt{3}\sec\theta, \qquad 0 < \theta < \pi/2$

**7.** $\dfrac{1}{\sqrt{7 + x^2}}, \qquad x = \sqrt{7}\tan\theta, \qquad -\pi/2 < \theta < \pi/2$

**8.** $\dfrac{\sqrt{5 - x^2}}{x}, \qquad x = \sqrt{5}\cos\theta, \qquad 0 \le \theta \le \pi$

In Problems 9–30, use a sum or difference formula to find the exact value of the given expression.

**9.** $\cos\dfrac{\pi}{12}$

**10.** $\sin\dfrac{\pi}{12}$

**11.** $\sin 75°$

**12.** $\cos 75°$

**13.** $\sin\dfrac{7\pi}{12}$

**14.** $\cos\dfrac{11\pi}{12}$

**15.** $\tan\dfrac{5\pi}{12}$

**16.** $\cos\left(-\dfrac{5\pi}{12}\right)$

**17.** $\sin\left(-\dfrac{\pi}{12}\right)$

**18.** $\tan\dfrac{11\pi}{12}$

**19.** $\sin\dfrac{11\pi}{12}$

**20.** $\tan\dfrac{7\pi}{12}$

**21.** $\cos 165°$

**22.** $\sin 165°$

**23.** $\tan 165°$

**24.** $\cos 195°$

**25.** $\sin 195°$

**26.** $\tan 195°$

**27.** $\cos 345°$

**28.** $\sin 345°$

**29.** $\cos\dfrac{13\pi}{12}$

**30.** $\tan\dfrac{17\pi}{12}$

In Problems 31–34, use a double-angle formula to write the given expression as a single trigonometric function of twice the angle.

**31.** $2\cos\beta\sin\beta$

**32.** $\cos^2 2t - \sin^2 2t$

**33.** $1 - 2\sin^2\dfrac{\pi}{5}$

**34.** $2\cos^2\left(\dfrac{19}{2}x\right) - 1$

In Problems 35–40, use the given information to find **(a)** $\cos 2x$, **(b)** $\sin 2x$, and **(c)** $\tan 2x$.

**35.** $\sin x = \sqrt{2}/3, \qquad \pi/2 < x < \pi$

**36.** $\cos x = \sqrt{3}/5, \qquad 3\pi/2 < x < 2\pi$

**37.** $\tan x = \frac{1}{2}, \qquad \pi < x < 3\pi/2$

**38.** $\csc x = -3, \qquad \pi < x < 3\pi/2$

**39.** $\sec x = -\frac{13}{5}, \qquad \pi/2 < x < \pi$

**40.** $\cot x = \frac{4}{3}, \qquad 0 < x < \pi/2$

In Problems 41–48, use a half-angle formula to find the exact value of the given expression.

**41.** $\cos(\pi/12)$

**42.** $\sin(\pi/8)$

**43.** $\sin(3\pi/8)$

**44.** $\tan(\pi/12)$

**45.** $\cos 67.5°$

**46.** $\sin 15°$

**47.** $\csc(13\pi/12)$

**48.** $\sec(-3\pi/8)$

In Problems 49–54, use the given information to find **(a)** $\cos(x/2)$, **(b)** $\sin(x/2)$, and **(c)** $\tan(x/2)$.

**49.** $\sin t = \frac{12}{13}, \qquad \pi/2 < t < \pi$

**50.** $\cos t = \frac{4}{5}, \qquad 3\pi/2 < t < 2\pi$

**51.** $\tan x = 2, \qquad \pi < x < 3\pi/2$

**52.** $\csc x = 9, \qquad 0° < x < \pi/2$

**53.** $\sec x = \frac{3}{2}, \qquad 0° < x < 90°$

**54.** $\cot x = -\frac{1}{4}, \qquad 90° < x < 180°$

**55.** If $P(x_1)$ and $P(x_2)$ are points in quadrant II on the terminal side of the angles $x_1$ and $x_2$, respectively, with $\cos x_1 = -\frac{1}{3}$ and $\sin x_2 = \frac{2}{3}$, find **(a)** $\sin(x_1 + x_2)$, **(b)** $\cos(x_1 + x_2)$, **(c)** $\sin(x_1 - x_2)$, and **(d)** $\cos(x_1 - x_2)$.

**56.** If $x_1$ is a quadrant II angle, $x_2$ is a quadrant III angle, $\sin x_1 = \frac{8}{17}$, and $\tan x_2 = \frac{3}{4}$, find **(a)** $\sin(x_1 + x_2)$, **(b)** $\sin(x_1 - x_2)$, **(c)** $\cos(x_1 + x_2)$, and **(d)** $\cos(x_1 - x_2)$.

In Problems 57–62, proceed as in Example 7 and reduce the given expression to the form $y = A\sin(Bx + \phi)$. Sketch the graph and state the amplitude, the period, and the phase shift.

**57.** $y = \cos \pi x - \sin \pi x$

**58.** $y = \sin\frac{\pi}{2}x - \sqrt{3}\cos\frac{\pi}{2}x$

**59.** $y = \sqrt{3}\sin 2x + \cos 2x$

**60.** $y = \sqrt{3}\cos 4x - \sin 4x$

**61.** $y = \frac{\sqrt{2}}{2}(-\sin x - \cos x)$

**62.** $y = \sin x + \cos x$

## Miscellaneous Applications

**63. Mach Number** The ratio of the speed of an airplane to the speed of sound is called the Mach number $M$ of the plane. If $M > 1$, the plane makes sound waves that form a (moving) cone, as shown in **FIGURE 4.5.3**. A sonic boom is heard at the intersection of the cone with the ground. If the vertex angle of the cone is $\theta$, then

$$\sin\frac{\theta}{2} = \frac{1}{M}.$$

If $\theta = \pi/6$, find the exact value of the Mach number.

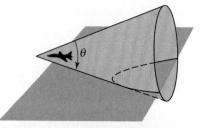

**FIGURE 4.5.3** Airplane in Problem 63

**64. Cardiovascular Branching** A mathematical model for blood flow in a large blood vessel predicts that the optimal values of the angles $\theta_1$ and $\theta_2$, which represent the (positive) angles of the smaller daughter branches (vessels) with respect to the axis of the parent branch, are given by

$$\cos\theta_1 = \frac{A_0^2 + A_1^2 - A_2^2}{2A_0A_1} \quad \text{and} \quad \cos\theta_2 = \frac{A_0^2 - A_1^2 + A_2^2}{2A_0A_2},$$

where $A_0$ is the cross-sectional area of the parent branch and $A_1$ and $A_2$ are the cross-sectional areas of the daughter branches. See **FIGURE 4.5.4**. Let $\psi = \theta_1 + \theta_2$ be the junction angle, as shown in the figure.

**(a)** Show that

$$\cos\psi = \frac{A_0^2 - A_1^2 - A_2^2}{2A_1A_2}.$$

**(b)** Show that for the optimal values of $\theta_1$ and $\theta_2$, the cross-sectional area of the daughter branches, $A_1 + A_2$, is greater than or equal to that of the parent branch. Therefore, the blood must slow down in the daughter branches.

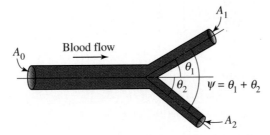

**FIGURE 4.5.4** Branching of a large blood vessel in Problem 64

65. **Area of a Triangle**  Show that the area of an isosceles triangle with equal sides of length $x$ is

$$A = \tfrac{1}{2}x^2\sin\theta,$$

where $\theta$ is the angle formed by the two equal sides. See **FIGURE 4.5.5**. [*Hint:* Consider $\theta/2$ as shown in the figure.]

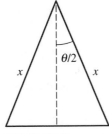

**FIGURE 4.5.5**
Isosceles triangle in
Problem 65

66. **Range of a Projectile**  If a projectile, such as a shot put, is released upward from a height $h$ at an angle $\phi$ with velocity $v_0$, then the range $R$ at which it strikes the ground is given by

$$R = \frac{v_0^2\cos\phi}{g}(\sin\phi + \sqrt{\sin^2\phi + (2gh/v_0^2)}),$$

where $g$ is the acceleration due to gravity. See **FIGURE 4.5.6**. It can be shown that the maximum range $R_{max}$ is achieved if the angle $\phi$ satisfies the equation

$$\cos 2\phi = \frac{gh}{v_0^2 + gh}.$$

Show that

$$R_{max} = \frac{v_0 \sqrt{v_0^2 + 2gh}}{g}$$

by using the expressions for $R$ and $\cos 2\phi$ and the half-angle formulas for the sine and the cosine with $t = 2\phi$.

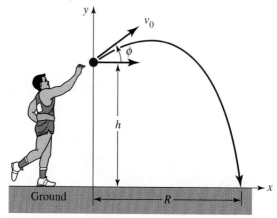

**FIGURE 4.5.6** Projectile in Problem 66

## For Discussion

**67.** Discuss: Why would you expect to get an error message from your calculator when you try to evaluate

$$\frac{\tan 35° + \tan 55°}{1 - \tan 35° \tan 55°}?$$

**68.** In Example 3 we showed that $\sin\dfrac{7\pi}{12} = \dfrac{\sqrt{2} + \sqrt{6}}{4}$. Following the example, we then showed that $\sin\dfrac{7\pi}{12} = \dfrac{\sqrt{2 + \sqrt{3}}}{2}$. Demonstrate that these answers are equivalent.

**69.** Discuss: How would you express $\sin 3\theta$ in terms of $\sin\theta$? Carry out your ideas.

**70.** In Problem 55, in what quadrants do $P(x_1 + x_2)$ and $P(x_1 - x_2)$ lie?

**71.** In Problem 56, in which quadrant does the terminal side of $x_1 + x_2$ lie? The terminal side of $x_1 - x_2$?

**72.** Use the sum/difference formulas (4), (5), (7), and (8) to derive the **product-to-sum formulas**:

$$\sin x_1 \sin x_2 = \tfrac{1}{2}[\cos(x_1 - x_2) - \cos(x_1 + x_2)]$$
$$\cos x_1 \cos x_2 = \tfrac{1}{2}[\cos(x_1 - x_2) + \cos(x_1 + x_2)]$$
$$\sin x_1 \cos x_2 = \tfrac{1}{2}[\sin(x_1 + x_2) + \sin(x_1 - x_2)]$$

In Problems 73–76, use a product-to-sum formula in Problem 72 to write the given product as a sum of sines or a sum of cosines.

**73.** $\cos 4\theta \cos 3\theta$

**74.** $\sin\dfrac{3t}{2}\cos\dfrac{t}{2}$

**75.** $\sin 2x \sin 5x$

**76.** $\sin 10x \cos 4x$

# 4.6 Trigonometric Equations

☐ **Introduction** In Section 4.5 we considered trigonometric identities, which are equations involving trigonometric functions that are satisfied by all values of the variable for which both sides of the equality are defined. In this section we examine **conditional trigonometric equations**, that is, equations that are true for only certain values of the variable. We discuss techniques for finding those values of the variable (if any) that satisfy the equation.

We begin by considering the problem of finding all real numbers $x$ that satisfy $\sin x = \sqrt{2}/2$. As the graph of $y = \sin x$ in **FIGURE 4.6.1** indicates, there exists an infinite number of solutions of this equation:

$$\ldots, -\frac{7\pi}{4}, \quad \frac{\pi}{4}, \quad \frac{9\pi}{4}, \quad \frac{17\pi}{4}, \ldots,$$

and

$$\ldots, -\frac{5\pi}{4}, \quad \frac{3\pi}{4}, \quad \frac{11\pi}{4}, \quad \frac{19\pi}{4}, \ldots. \tag{1}$$

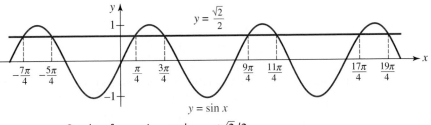

**FIGURE 4.6.1** Graphs of $y = \sin x$ and $y = \sqrt{2}/2$

We observe that in each list in (1), every solution can be obtained by adding $2\pi = 8\pi/4$ to the preceding solution. This is a consequence of the periodicity of the sine function. It is common for trigonometric equations to have an infinite number of solutions because of the periodicity of the trigonometric functions. In general, to obtain solutions of an equation such as $\sin x = \sqrt{2}/2$, it is more convenient to use a unit circle and reference angles rather than a graph of the trigonometric function. We illustrate this approach in the following example.

### EXAMPLE 1 Using the Unit Circle

Find all real numbers $x$ satisfying $\sin x = \sqrt{2}/2$.

**Solution** If $\sin x = \sqrt{2}/2$, the reference angle for $x$ is $\pi/4$ radians. Since the value of $\sin x$ is positive, the terminal side of the angle $x$ lies in either the first or second quadrant. Thus, as shown in **FIGURE 4.6.2**, the only solutions between 0 and $2\pi$ are

$$x = \frac{\pi}{4} \quad \text{or} \quad x = \frac{3\pi}{4}.$$

Since the sine function is periodic with period $2\pi$, all of the remaining solutions can be obtained by adding integer multiples of $2\pi$ to these solutions:

$$x = \frac{\pi}{4} + 2n\pi \quad \text{or} \quad x = \frac{3\pi}{4} + 2n\pi,$$

where $n$ is an integer.

**FIGURE 4.6.2** Unit circle for Example 1

When we are faced with a more complicated equation, such as

$$4\sin^2 x - 8\sin x + 3 = 0,$$

the basic approach is to solve for a single trigonometric function (in this case, it would be $\sin x$) by using methods similar to those for solving algebraic equations. Then the values of the variable $x$ are determined using the unit circle and reference angles. The following example illustrates this technique.

### ■ EXAMPLE 2    Solving a Trigonometric Equation by Factoring

Find all solutions of $4\sin^2 x - 8\sin x + 3 = 0$.

**Solution**  We first observe that this is a quadratic equation in $\sin x$, and that it factors as

$$(2\sin x - 3)(2\sin x - 1) = 0.$$

This implies that either

$$\sin x = \frac{3}{2} \qquad \text{or} \qquad \sin x = \frac{1}{2}.$$

The first equation has no solution, since $|\sin x| \le 1$. As we see in FIGURE 4.6.3, the two angles between $0$ and $2\pi$ for which $\sin x$ equals $\frac{1}{2}$ are

$$x = \frac{\pi}{6} \qquad \text{or} \qquad x = \frac{5\pi}{6}.$$

Therefore, by the periodicity of the sine function, the solutions are

$$x = \frac{\pi}{6} + 2n\pi \qquad \text{or} \qquad x = \frac{5\pi}{6} + 2n\pi,$$

where $n$ is an integer.  ■

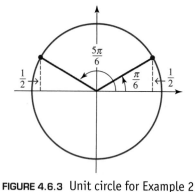

FIGURE 4.6.3  Unit circle for Example 2

### ■ EXAMPLE 3    Checking for Lost Solutions

Find all solutions of $\sin x = \cos x$.  (2)

**Solution**  In order to work with a single trigonometric function, we divide both sides of the equation by $\cos x$ to obtain

$$\tan x = 1.  \qquad (3)$$

Equation (3) is equivalent to (2) *provided* that $\cos x \neq 0$. We observe that if $\cos x = 0$, then by (2) of Section 4.3, $x = (2n + 1)\pi/2 = \pi/2 + n\pi$, for $n$ an integer. By the sum formula for the sine,

$$\underset{\underset{\text{See (7) in Section 4.5.}}{\downarrow}}{\phantom{x}} \quad \underset{\underset{(-1)^n}{\downarrow}}{\phantom{x}} \quad \underset{\underset{0}{\downarrow}}{\phantom{x}}$$

$$\sin\left(\frac{\pi}{2} + n\pi\right) = \sin\frac{\pi}{2}\cos n\pi + \cos\frac{\pi}{2}\sin n\pi = (-1)^n \neq 0,$$

$\cos 0 = 1, \cos\pi = -1, \cos 2\pi = 1,$ ▶ $\cos 3\pi = -1$, and so on. In general, $\cos n\pi = (-1)^n$, where $n$ is an integer.

these values of $x$ do not satisfy the original equation. Thus we will find *all* the solutions to (2) by solving equation (3).

Now $\tan x = 1$ implies that the reference angle for $x$ is $\pi/4$ radian. Since $\tan x = 1 > 0$, the terminal side of the angle of $x$ radians can lie either in the first or in the third quadrant, as shown in FIGURE 4.6.4. Thus the solutions are

$$x = \frac{\pi}{4} + 2n\pi \qquad \text{or} \qquad x = \frac{5\pi}{4} + 2n\pi,$$

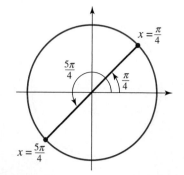

FIGURE 4.6.4  Unit circle in Example 3

CHAPTER 4 TRIGONOMETRIC FUNCTIONS

where $n$ is an integer. We can see from Figure 4.6.4 these two sets of numbers can be written more compactly as

$$x = \frac{\pi}{4} + n\pi,$$

where $n$ is an integer.

◀ This follows from the fact that $\tan x$ is $\pi$-periodic.

■

□ **Losing Solutions** When solving an equation, if you divide by an expression containing a variable, you may lose some solutions of the original equation. For example, in algebra a common mistake in solving equations such as $x^2 = x$ is to divide by $x$ to obtain $x = 1$. But by writing $x^2 = x$ as $x^2 - x = 0$ or $x(x - 1) = 0$, we see that in fact $x = 0$ or $x = 1$. To prevent the loss of a solution you must determine the values that make the expression zero and check to see whether they are solutions of the original equation. Note that in Example 3, when we divided by $\cos x$, we took care to check that no solutions were lost.

Whenever possible, it is preferable to avoid dividing by a variable expression. As illustrated with the algebraic equation $x^2 = x$, this can frequently be accomplished by collecting all nonzero terms on one side of the equation and then factoring (something we could not do in Example 3). Example 4 illustrates this technique.

**■ EXAMPLE 4**        **Solving a Trigonometric Equation by Factoring**

Solve $2 \sin x \cos^2 x = -\dfrac{\sqrt{3}}{2} \cos x.$         (4)

**Solution** To avoid dividing by $\cos x$, we write the equation as

$$2 \sin x \cos^2 x + \frac{\sqrt{3}}{2} \cos x = 0$$

and factor:

$$\cos x \left( 2 \sin x \cos x + \frac{\sqrt{3}}{2} \right) = 0.$$

Thus either

$$\cos x = 0 \qquad \text{or} \qquad 2 \sin x \cos x + \frac{\sqrt{3}}{2} = 0.$$

Since the cosine is zero for all odd multiples of $\pi/2$, the solutions from $\cos x = 0$ are:

$$x = (2n + 1)\frac{\pi}{2} = \frac{\pi}{2} + n\pi,$$

where $n$ is an integer.

In the second equation we replace $2 \sin x \cos x$ by $\sin 2x$ from the double-angle formula for the sine to obtain an equation with a single trigonometric function:

See (15) in Section 4.5.

$$\sin 2x + \frac{\sqrt{3}}{2} = 0 \qquad \text{or} \qquad \sin 2x = -\frac{\sqrt{3}}{2}.$$

Thus the reference angle for $2x$ is $\pi/3$. Since the sine is negative, the angle $2x$ must be in either the third quadrant or in the fourth quadrant. As **FIGURE 4.6.5** illustrates, either

$$2x = \frac{4\pi}{3} + 2n\pi \qquad \text{or} \qquad 2x = \frac{5\pi}{3} + 2n\pi.$$

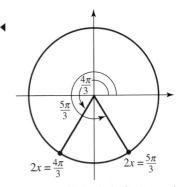

**FIGURE 4.6.5** Unit circle for Example 4

Dividing by 2 gives,

$$x = \frac{2\pi}{3} + n\pi \qquad \text{or} \qquad x = \frac{5\pi}{6} + n\pi.$$

Therefore, all solutions of (4) are

$$x = \frac{\pi}{2} + n\pi, \qquad x = \frac{2\pi}{3} + n\pi, \qquad \text{or} \qquad x = \frac{5\pi}{6} + n\pi,$$

where $n$ is an integer. ∎

In Example 4, had we simplified the equation by dividing by $\cos x$ and not checked to see if the values of $x$ for which $\cos x = 0$ satisfied equation (4), we would have lost the solutions $x = \pi/2 + n\pi$, where $n$ is an integer.

### ■ EXAMPLE 5 — Using a Trigonometric Identity

Solve $3 \cos^2 x - \cos 2x = 1$.

**Solution** We observe that the given equation involves both the cosine of $x$ and the cosine of $2x$. Consequently, we use the double-angle formula for the cosine in the form

$$\cos 2x = 2 \cos^2 x - 1 \qquad \leftarrow \text{See (16) of Section 4.5.}$$

to replace the equation by an equivalent equation that involves $\cos x$ only. We find that

$$3 \cos^2 x - (2 \cos^2 x - 1) = 1 \qquad \text{becomes} \qquad \cos^2 x = 0.$$

Therefore, $\cos x = 0$, and the solutions are

$$x = (2n + 1)\frac{\pi}{2} = \frac{\pi}{2} + n\pi,$$

where $n$ is an integer. ∎

So far in this section we have viewed the variable in the trigonometric equation as representing either a real number or an angle measured in radians. If the variable represents an angle measured in degrees, the technique for solving is the same.

### ■ EXAMPLE 6 — Equation When the Angle Is in Degrees

Solve $\cos 2\theta = -\frac{1}{2}$, where $\theta$ is an angle measured in degrees.

**Solution** Since $\cos 2\theta = -\frac{1}{2}$, the reference angle for $2\theta$ is $60°$ and the angle $2\theta$ must be in either the second or the third quadrant. **FIGURE 4.6.6** illustrates that either $2\theta = 120°$ or $2\theta = 240°$. Any angle that is coterminal with one of these angles will also satisfy $\cos 2\theta = -\frac{1}{2}$. These angles are obtained by adding any integer multiple of $360°$ to $120°$ or to $240°$:

$$2\theta = 120° + 360°n \qquad \text{or} \qquad 2\theta = 240° + 360°n,$$

where $n$ is an integer. Dividing by 2 the last line yields

$$\theta = 60° + 180°n \qquad \text{or} \qquad \theta = 120° + 180°n. \qquad ∎$$

The next example shows that by squaring an equation we may introduce extraneous solutions.

### ■ EXAMPLE 7 — Extraneous Roots

Find all solutions of $1 + \tan \alpha = \sec \alpha$, where $\alpha$ is an angle measured in degrees.

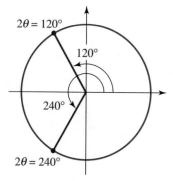

**FIGURE 4.6.6** Unit circle for Example 6

$2\theta = 120°$

$120°$

$240°$

$2\theta = 240°$

CHAPTER 4 TRIGONOMETRIC FUNCTIONS

**Solution** The equation does not factor, but we see that if we square both sides, we can use a fundamental identity to obtain an equation involving a single trigonometric function:

$$(1 + \tan \alpha)^2 = (\sec \alpha)^2$$
$$1 + 2\tan \alpha + \tan^2 \alpha = \sec^2 \alpha \qquad \leftarrow \text{See (8) of Section 4.4.}$$
$$1 + 2\tan \alpha + \tan^2 \alpha = 1 + \tan^2 \alpha$$
$$2\tan \alpha = 0$$
$$\tan \alpha = 0.$$

The values of $\alpha$ in $[0°, 360°)$ for which $\tan \alpha = 0$ are

$$\alpha = 0° \qquad \text{and} \qquad \alpha = 180°.$$

Since we squared each side of the original equation, we may have introduced extraneous solutions. Therefore, it is important that we check all solutions in the original equation. Substituting $\alpha = 0°$ into $1 + \tan \alpha = \sec \alpha$, we obtain the *true* statement $1 + 0 = 1$. But after substituting $\alpha = 180°$, we obtain the *false* statement $1 + 0 = -1$. Therefore, $180°$ is an extraneous solution and $\alpha = 0°$ is the only solution in the interval $[0°, 360°)$. Thus, all the solutions of the equation are given by

$$\alpha = 0° + 360°n = 360°n,$$

where $n$ is an integer. For $n \neq 0$, these are the angles that are coterminal with $0°$. ∎

Recall from Section 2.1 that finding the $x$-intercepts of the graph of a function $y = f(x)$ is equivalent to solving the equation $f(x) = 0$. The following example makes use of this fact.

### EXAMPLE 8    Intercepts of a Graph

Find the first three $x$-intercepts of the graph of $f(x) = \sin 2x \cos x$ on the positive $x$-axis.

**Solution** We must solve $f(x) = 0$, that is, $\sin 2x \cos x = 0$. It follows that either $\sin 2x = 0$ or $\cos x = 0$.

From $\sin 2x = 0$, we obtain $2x = n\pi$, where $n$ is an integer, or $x = n\pi/2$, where $n$ is an integer. From $\cos x = 0$, we find $x = \pi/2 + n\pi$, where $n$ is an integer. Then for $n = 2$, $x = n\pi/2$ gives $x = \pi$, whereas for $n = 0$ and $n = 1$, $x = \pi/2 + n\pi$ gives $x = \pi/2$ and $x = 3\pi/2$. Thus the first three $x$-intercepts on the positive $x$-axis are $(\pi/2, 0), (\pi, 0), (3\pi/2, 0)$. ∎

So far all of the trigonometric equations have had solutions that were related by reference angles to the special angles $0, \pi/6, \pi/4, \pi/3$, or $\pi/2$. If this is not the case, we will see in the next section how to use inverse trigonometric functions and a calculator to find solutions.

---

**4.6** **Exercises** Answers to selected odd-numbered problems begin on page ANS–15.

In Problems 1–6, find all solutions of the given trigonometric equation if $x$ represents an angle measured in radians.

**1.** $\sin x = \sqrt{3}/2$

**2.** $\cos x = -\sqrt{2}/2$

**3.** $\sec x = \sqrt{2}$

**4.** $\tan x = -1$

**5.** $\cot x = -\sqrt{3}$

**6.** $\csc x = 2$

In Problems 7–12, find all solutions of the given trigonometric equation if $x$ represents a real number.

**7.** $\cos x = -1$

**8.** $2 \sin x = -1$

**9.** $\tan x = 0$

**10.** $\sqrt{3} \sec x = 2$

**11.** $-\csc x = 1$

**12.** $\sqrt{3} \cot x = 1$

In Problems 13–18, find all solutions of the given trigonometric equation if $\theta$ represents an angle measured in degrees.

**13.** $\csc \theta = 2\sqrt{3}/3$

**14.** $2 \sin \theta = \sqrt{2}$

**15.** $1 + \cot \theta = 0$

**16.** $\sqrt{3} \sin \theta = \cos \theta$

**17.** $\sec \theta = -2$

**18.** $2 \cos \theta + \sqrt{2} = 0$

In Problems 19–46, find all solutions of the given trigonometric equation if $x$ is a real number and $\theta$ is an angle measured in degrees.

**19.** $\cos^2 x - 1 = 0$

**20.** $2 \sin^2 x - 3 \sin x + 1 = 0$

**21.** $3 \sec^2 x = \sec x$

**22.** $\tan^2 x + (\sqrt{3} - 1) \tan x - \sqrt{3} = 0$

**23.** $2 \cos^2 \theta - 3 \cos \theta - 2 = 0$

**24.** $2 \sin^2 \theta - \sin \theta - 1 = 0$

**25.** $\cot^2 \theta + \cot \theta = 0$

**26.** $2 \sin^2 \theta + (2 - \sqrt{3}) \sin \theta - \sqrt{3} = 0$

**27.** $\cos 2x = -1$

**28.** $\sec 2x = 2$

**29.** $2 \sin 3\theta = 1$

**30.** $\tan 4\theta = -1$

**31.** $\cot(x/2) = 1$

**32.** $\csc(\theta/3) = -1$

**33.** $\sin 2x + \sin x = 0$

**34.** $\cos 2x + \sin^2 x = 1$

**35.** $\cos 2\theta = \sin \theta$

**36.** $\sin 2\theta + 2 \sin \theta - 2 \cos \theta = 2$

**37.** $\sin^4 x - 2 \sin^2 x + 1 = 0$

**38.** $\tan^4 \theta - 2 \sec^2 \theta + 3 = 0$

**39.** $\sec x \sin^2 x = \tan x$

**40.** $\dfrac{1 + \cos \theta}{\cos \theta} = 2$

**41.** $1 + \cot \theta = \csc \theta$

**42.** $\sin x + \cos x = 0$

**43.** $\sqrt{\dfrac{1 + 2 \sin x}{2}} = 1$

**44.** $\sin x + \sqrt{\sin x} = 0$

**45.** $\cos \theta - \sqrt{\cos \theta} = 0$

**46.** $\cos \theta \sqrt{1 + \tan^2 \theta} = 1$

In Problems 47–54, find the first three $x$-intercepts of the graph of the given function on the positive $x$-axis.

**47.** $f(x) = -5 \sin(3x + \pi)$

**48.** $f(x) = 2 \cos\left(x + \dfrac{\pi}{4}\right)$

**49.** $f(x) = 2 - \sec \dfrac{\pi}{2} x$

**50.** $f(x) = 1 + \cos \pi x$

**51.** $f(x) = \sin x + \tan x$

**52.** $f(x) = 1 - 2 \cos\left(x + \dfrac{\pi}{3}\right)$

**53.** $f(x) = \sin x - \sin 2x$

**54.** $f(x) = \cos x + \cos 3x$ [*Hint*: Write $3x = x + 2x$.]

In Problems 55–58, by graphing determine whether the given equation has any solutions.

**55.** $\tan x = x$ [*Hint*: Graph $y = \tan x$ and $y = x$ on the same set of axes.]

**56.** $\sin x = x$

**57.** $\cot x - x = 0$

**58.** $\cos x + x + 1 = 0$

## Miscellaneous Applications

**59. Isosceles Triangle** From Problem 65 in Exercises 4.5, the area of the isosceles triangle with vertex angle $\theta$ as shown in Figure 4.5.5 is given by $A = \frac{1}{2}x^2 \sin\theta$. If the length $x$ is 4, what value of $\theta$ will give a triangle with area 4?

**60. Circular Motion** An object travels in a circular path centered at the origin with constant angular speed. The $y$-coordinate of the object at any time $t$ seconds is given by $y = 8\cos(\pi t - \pi/12)$. At what time(s) does the object cross the $x$-axis?

**61. Mach Number** Use Problem 63 in Exercises 4.5 to find the vertex angle of the cone of sound waves made by an airplane flying at Mach 2.

**62. Alternating Current** An electric generator produces a 60-cycle alternating current given by $I(t) = 30\sin 120\pi\left(t - \frac{7}{36}\right)$, where $I(t)$ is the current in amperes at $t$ seconds. Find the smallest positive value of $t$ for which the current is 15 amperes.

**63. Electrical Circuits** If the voltage given by $V = V_0\sin(\omega t + \alpha)$ is impressed on a series circuit, an alternating current is produced. If $V_0 = 110$ volts, $\omega = 120\pi$ radians per second, and $\alpha = -\pi/6$, when is the voltage equal to zero?

**64. Refraction of Light** Consider a ray of light passing from one medium (such as air) into another medium (such as a crystal). Let $\phi$ be the angle of incidence and $\theta$ the angle of refraction. As shown in **FIGURE 4.6.7**, these angles are measured from a vertical line. According to **Snell's law**, there is a constant $c$, that depends on the two mediums, such that $\dfrac{\sin\phi}{\sin\theta} = c$. Assume that for light passing from air into a crystal, $c = 1.437$. Find $\phi$ and $\theta$ such that the angle of incidence is twice the angle of refraction.

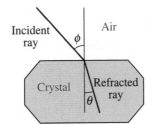

**FIGURE 4.6.7** Light rays in Problem 64

**65. Snow Cover** On the basis of data collected from 1966 to 1980, the extent of snow cover $S$ in the northern hemisphere, measured in millions of square kilometers, can be modeled by the function

$$S(w) = 25 + 21\cos\left(\tfrac{1}{26}\pi(w - 5)\right),$$

where $w$ is the number of weeks past January 1.
**(a)** How much snow cover does this formula predict for April Fool's Day? (Round $w$ to the nearest integer.)
**(b)** In which week does the formula predict the least amount of snow cover?
**(c)** What month does this fall in?

---

## **4.7** Inverse Trigonometric Functions

☐ **Introduction** Although we can find the values of the trigonometric functions of real numbers or angles, in many applications we must do the reverse: Given the value of a trigonometric function, find a corresponding angle or number. This suggests we consider inverse trigonometric functions. Before we define the inverse trigono-

Recall that a function $f$ is one-to-one if ▶
every $y$ in its range corresponds to exactly
one $x$ in its domain.

metric functions, let's recall from Section 2.7 some of the properties of a one-to-one function $f$ and its inverse $f^{-1}$.

☐ **Properties of Inverse Functions** If $y = f(x)$ is a one-to-one function then there is a unique inverse function $f^{-1}$ with the following properties:

- The domain of $f^{-1}$ = range of $f$.
- The range of $f^{-1}$ = domain of $f$.
- $y = f(x)$ is equivalent to $x = f^{-1}(y)$.
- The graphs of $f$ and $f^{-1}$ are reflections in the line $y = x$.
- $f(f^{-1}(x)) = x$ for $x$ in the domain of $f^{-1}$.
- $f^{-1}(f(x)) = x$ for $x$ in the domain of $f$.

See Example 7 in Section 2.7. ▶

Inspection of the graphs of the various trigonometric functions clearly shows that *none* of these functions is one-to-one. In Section 2.7 we discussed the fact that if a function $f$ is not one-to-one, it may be possible to restrict the function to a portion of its domain where it is one-to-one. Then we can define an inverse for $f$ on that restricted domain. Normally, when we restrict the domain, we make sure to preserve the entire range of the original function.

☐ **Arcsine Function** From FIGURE 4.7.1(a) we see that the function $y = \sin x$ on the closed interval $[-\pi/2, \pi/2]$ takes on all values in its range $[-1, 1]$. Notice that any horizontal line drawn to intersect the red portion of the graph can do so at most once. Thus the sine function on this restricted domain is one-to-one and has an inverse. There are two commonly used notations to denote the inverse of the function shown in Figure 4.7.1(b):

$$\arcsin x \qquad \text{or} \qquad \sin^{-1} x,$$

and are read **arcsine of $x$** and **inverse sine of $x$**, respectively.

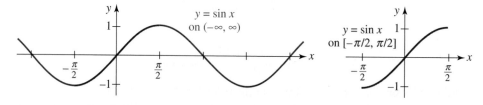

(a) Not a one-to-one function   (b) A one-to-one function

FIGURE 4.7.1  Restricting the domain of $y = \sin x$ to produce a one-to-one function

In FIGURE 4.7.2(a) we have reflected the portion of the graph of $y = \sin x$ on the interval $[-\pi/2, \pi/2]$ (the red graph in Figure 4.7.1(b)) about the line $y = x$ to obtain the graph of $y = \arcsin x$ (in blue). For clarity, we have reproduced this blue graph in

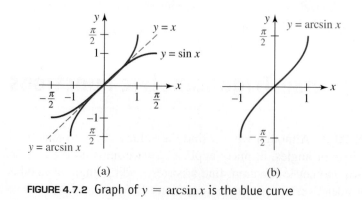

(a)   (b)

FIGURE 4.7.2  Graph of $y = \arcsin x$ is the blue curve

Figure 4.7.2(b). As this curve shows, the domain of the arcsine function is $[-1, 1]$ and the range is $[-\pi/2, \pi/2]$.

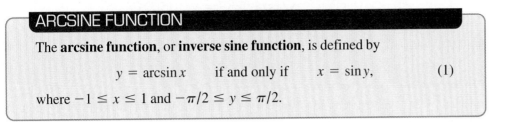

**ARCSINE FUNCTION**

The **arcsine function**, or **inverse sine function**, is defined by

$$y = \arcsin x \quad \text{if and only if} \quad x = \sin y, \qquad (1)$$

where $-1 \le x \le 1$ and $-\pi/2 \le y \le \pi/2$.

In other words:

> *The arcsine of the number x is that number y (or radian-measured angle) between $-\pi/2$ and $\pi/2$ whose sine is x.*

When using the notation $\sin^{-1} x$ it is important to realize that "$-1$" is not an exponent; rather, it denotes an inverse function. The notation $\arcsin x$ has an advantage over the notation $\sin^{-1} x$ in that there is no "$-1$" and hence no potential for misinterpretation; moreover, the prefix "arc" refers to an angle—*the* angle whose sine is $x$. But since $y = \arcsin x$ and $y = \sin^{-1} x$ are used interchangeably in calculus and in applications, we will continue to alternate their use so that you become comfortable with both notations.

◀ Note of Caution:

$$(\sin x)^{-1} = \frac{1}{\sin x} \ne \sin^{-1} x$$

▮ **EXAMPLE 1**    **Evaluating the Inverse Sine Function**

Find **(a)** $\arcsin\frac{1}{2}$, **(b)** $\sin^{-1}\left(-\frac{1}{2}\right)$, and **(c)** $\sin^{-1}(-1)$.

**Solution**

**(a)** If we let $y = \arcsin\frac{1}{2}$, then by (1) we must find the number $y$ (or radian measured angle) that satisfies $\sin y = \frac{1}{2}$ *and* $-\pi/2 \le y \le \pi/2$. Since $\sin(\pi/6) = \frac{1}{2}$, and $\pi/6$ satisfies the inequality $-\pi/2 \le y \le \pi/2$ it follows that $y = \pi/6$.

**(b)** If we let $y = \sin^{-1}\left(-\frac{1}{2}\right)$, then $\sin y = -\frac{1}{2}$. Since we must choose $y$ such that $-\pi/2 \le y \le \pi/2$, we find that $y = -\pi/6$.

**(c)** Letting $y = \sin^{-1}(-1)$, we have that $\sin y = -1$ and $-\pi/2 \le y \le \pi/2$. Hence $y = -\pi/2$. ▮

In parts (b) and (c) of Example 1 we were careful to choose $y$ so that $-\pi/2 \le y \le \pi/2$. For example, it is a common error to think that because $\sin(3\pi/2) = -1$, then necessarily $\sin^{-1}(-1)$ can be taken to be $3\pi/2$. Remember: If $y = \sin^{-1} x$, then $y$ is subject to the restriction $-\pi/2 \le y \le \pi/2$, and $3\pi/2$ does not satisfy this inequality.

◀ Read this paragraph several times.

▮ **EXAMPLE 2**    **Evaluating a Composition**

Without using a calculator, find $\tan\left(\sin^{-1}\frac{1}{4}\right)$.

**Solution** We must find the tangent of the angle of $t$ radians with sine equal to $\frac{1}{4}$, that is, $\tan t$, where $t = \sin^{-1}\frac{1}{4}$. The angle $t$ is shown in **FIGURE 4.7.3**. Since

$$\tan t = \frac{\sin t}{\cos t} = \frac{1/4}{\cos t},$$

we want to determine the value of $\cos t$. From Figure 4.7.3 and the Pythagorean identity $\sin^2 t + \cos^2 t = 1$, we see that

$$\left(\frac{1}{4}\right)^2 + \cos^2 t = 1 \quad \text{or} \quad \cos t = \frac{\sqrt{15}}{4}.$$

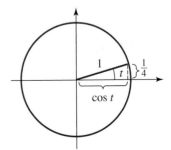

**FIGURE 4.7.3** The angle $t = \sin^{-1}\frac{1}{4}$ in Example 2

Hence we have

$$\tan t = \frac{1/4}{\sqrt{15}/4} = \frac{1}{\sqrt{15}} = \frac{\sqrt{15}}{15},$$

and so

$$\tan\left(\sin^{-1}\frac{1}{4}\right) = \tan t = \frac{\sqrt{15}}{15}.$$

□ **Arccosine Function** If we restrict the domain of the cosine function to the closed interval $[0, \pi]$, the resulting function is one-to-one and thus has an inverse. We denote this inverse by

$$\arccos x \qquad \text{or} \qquad \cos^{-1} x,$$

which gives us the following definition.

---

**ARCCOSINE FUNCTION**

The **arccosine function**, or **inverse cosine function**, is defined by

$$y = \arccos x \qquad \text{if and only if} \qquad x = \cos y, \qquad (2)$$

where $-1 \leq x \leq 1$ and $0 \leq y \leq \pi$.

---

The graphs shown in **FIGURE 4.7.4** illustrate how the function $y = \cos x$ restricted to the interval $[0, \pi]$ becomes a one-to-one function. The inverse of the function shown in Figure 4.7.4(b) is $y = \arccos x$.

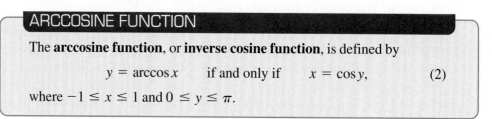

(a) Not a one-to-one function   (b) A one-to-one function

**FIGURE 4.7.4** Restricting the domain of $y = \cos x$ to produce a one-to-one function

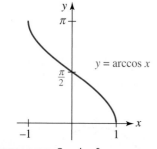

**FIGURE 4.7.5** Graph of $y = \arccos x$

By reflecting the graph of the one-to-one function in Figure 4.7.4(b) in the line $y = x$ we obtain the graph of $y = \arccos x$ shown in **FIGURE 4.7.5**.

Note that the figure clearly shows that the domain and range of $y = \arccos x$ are $[-1, 1]$ and $[0, \pi]$, respectively.

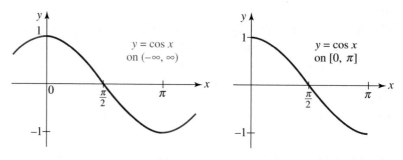

**EXAMPLE 3**   **Evaluating the Inverse Cosine Function**

Find **(a)** $\arccos(\sqrt{2}/2)$ and **(b)** $\cos^{-1}(-\sqrt{3}/2)$.

**Solution**

**(a)** If we let $y = \arccos(\sqrt{2}/2)$, then $\cos y = \sqrt{2}/2$, and $0 \leq y \leq \pi$. Thus $y = \pi/4$.

**(b)** Letting $y = \cos^{-1}(-\sqrt{3}/2)$, we have that $\cos y = -\sqrt{3}/2$, and we must find $y$ such that $0 \leq y \leq \pi$. Therefore, $y = 5\pi/6$, since $\cos(5\pi/6) = -\sqrt{3}/2$.

## EXAMPLE 4    Evaluating the Compositions of Functions

Write $\sin(\cos^{-1}x)$ as an algebraic expression in $x$.

**Solution**  In **FIGURE 4.7.6** we have constructed an angle of $t$ radians with cosine equal to $x$. Then $t = \cos^{-1}x$, or $x = \cos t$, where $0 \le t \le \pi$. Now to find $\sin(\cos^{-1}x) = \sin t$, we use the identity $\sin^2 t + \cos^2 t = 1$. Thus

$$\sin^2 t + x^2 = 1$$
$$\sin^2 t = 1 - x^2$$
$$\sin t = \sqrt{1 - x^2}.$$

We use the nonnegative square root of $1 - x^2$, since the range of $\cos^{-1}x$ is $[0, \pi]$, and the sine of an angle $t$ in the first or second quadrant is positive.  ■

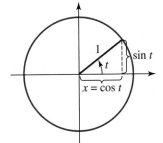

**FIGURE 4.7.6**  The angle $t = \cos^{-1}x$ in Example 4

☐ **Arctangent Function**  If we restrict the domain of $\tan x$ to the open interval $(-\pi/2, \pi/2)$, then the resulting function is one-to-one and thus has an inverse. This inverse is denoted by

$$\arctan x \qquad \text{or} \qquad \tan^{-1}x.$$

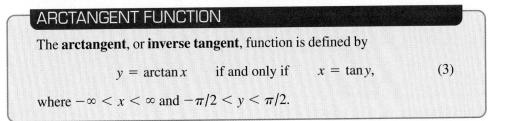

**ARCTANGENT FUNCTION**

The **arctangent**, or **inverse tangent**, function is defined by

$$y = \arctan x \qquad \text{if and only if} \qquad x = \tan y, \qquad (3)$$

where $-\infty < x < \infty$ and $-\pi/2 < y < \pi/2$.

The graphs shown in **FIGURE 4.7.7** illustrate how the function $y = \tan x$ restricted to the open interval $(-\pi/2, \pi/2)$ becomes a one-to-one function.

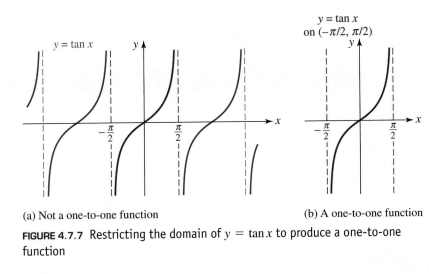

(a) Not a one-to-one function

(b) A one-to-one function

**FIGURE 4.7.7**  Restricting the domain of $y = \tan x$ to produce a one-to-one function

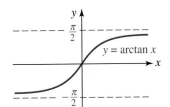

**FIGURE 4.7.8**  Graph of $y = \arctan x$

By reflecting the graph of the one-to-one function in Figure 4.7.7(b) in the line $y = x$ we obtain the graph of $y = \arctan x$ shown in **FIGURE 4.7.8**. We see in the figure that the domain and range of $y = \arctan x$ are, in turn, the intervals $(-\infty, \infty)$ and $(-\pi/2, \pi/2)$.

EXAMPLE 5                    Evaluating the Inverse Tangent

Find $\tan^{-1}(-1)$.

**Solution** If $\tan^{-1}(-1) = y$, then $\tan y = -1$, where $-\pi/2 < y < \pi/2$. It follows that $\tan^{-1}(-1) = y = -\pi/4$. ∎

EXAMPLE 6                    Evaluating Compositions of Functions

Without using a calculator, find $\sin\left(\arctan\left(-\frac{5}{3}\right)\right)$.

**Solution** If we let $t = \arctan\left(-\frac{5}{3}\right)$, then $\tan t = -\frac{5}{3}$. The Pythagorean identity $1 + \tan^2 t = \sec^2 t$ can be used to find $\sec t$:

$$1 + \left(-\frac{5}{3}\right)^2 = \sec^2 t$$

$$\sec t = \sqrt{\frac{25}{9} + 1} = \sqrt{\frac{34}{9}} = \frac{\sqrt{34}}{3}.$$

In the preceding line we take the positive square root because $t = \arctan\left(-\frac{5}{3}\right)$ is in the interval $(-\pi/2, \pi/2)$ (the range of the arctangent function) and the secant of an angle $t$ in the first or fourth quadrant is positive. Also, from $\sec t = \sqrt{34}/3$ we find the value of $\cos t$ from the reciprocal identity:

$$\cos t = \frac{1}{\sec t} = \frac{1}{\sqrt{34}/3} = \frac{3}{\sqrt{34}}.$$

Finally we can use the identity $\tan t = \sin t/\cos t$ in the form $\sin t = \tan t \cos t$ to compute $\sin\left(\arctan\left(-\frac{5}{3}\right)\right)$. It follows that

$$\sin t = \tan t \cos t = \left(-\frac{5}{3}\right)\left(\frac{3}{\sqrt{34}}\right) = -\frac{5}{\sqrt{34}}.$$ ∎

☐ **Properties of the Inverses** Recall from Section 2.7 that $f^{-1}(f(x)) = x$ and $f(f^{-1}(x)) = x$ hold for any function $f$ and its inverse under suitable restrictions on $x$. Thus for the inverse trigonometric functions, we have the following properties.

PROPERTIES OF INVERSE TRIGONOMETRIC FUNCTIONS

$(i)$ $\arcsin(\sin x) = \sin^{-1}(\sin x) = x$   if   $-\pi/2 \leq x \leq \pi/2$
$(ii)$ $\sin(\arcsin x) = \sin(\sin^{-1} x) = x$   if   $-1 \leq x \leq 1$
$(iii)$ $\arccos(\cos x) = \cos^{-1}(\cos x) = x$   if   $0 \leq x \leq \pi$
$(iv)$ $\cos(\arccos x) = \cos(\cos^{-1} x) = x$   if   $-1 \leq x \leq 1$
$(v)$ $\arctan(\tan x) = \tan^{-1}(\tan x) = x$   if   $-\pi/2 < x < \pi/2$
$(vi)$ $\tan(\arctan x) = \tan(\tan^{-1} x) = x$   if   $-\infty < x < \infty$

EXAMPLE 7                    Applying the Inverse Properties

Without using a calculator, evaluate:

**(a)** $\sin^{-1}\left(\sin\frac{\pi}{12}\right)$   **(b)** $\cos\left(\cos^{-1}\frac{1}{3}\right)$   **(c)** $\tan^{-1}\left(\tan\frac{3\pi}{4}\right)$.

**Solution**

**(a)** By (*i*) of the properties of the inverse trigonometric functions,

$$\sin^{-1}\left(\sin\frac{\pi}{12}\right) = \frac{\pi}{12}.$$

**(b)** By property (*iv*), $\cos\left(\cos^{-1}\frac{1}{3}\right) = \frac{1}{3}$.

**(c)** In this case we *cannot* apply property (*v*), since the number $3\pi/4$ is not in the interval $(-\pi/2, \pi/2)$. If we first evaluate $\tan(3\pi/4) = -1$, then we have

See Example 5
↓

$$\tan^{-1}\left(\tan\frac{3\pi}{4}\right) = \tan^{-1}(-1) = -\frac{\pi}{4}. \qquad \blacksquare$$

The next example shows how inverse trigonometric functions can be used to solve trigonometric equations.

### ■ EXAMPLE 8    Solving Equations Using Inverse Functions

Find the solutions of $4\cos^2 x - 3\cos x - 2 = 0$ in the interval $[0, \pi]$.

**Solution** We recognize that this is a quadratic equation in $\cos x$. Since it does not readily factor, we apply the quadratic formula to obtain

$$\cos x = \frac{3 \pm \sqrt{41}}{8}.$$

At this point we can discard the value $(3 + \sqrt{41})/8 \approx 1.18$, since $\cos x$ cannot be greater than 1. We then use the inverse cosine function (and the aid of a calculator) to solve the remaining equation:

$$\cos x = \frac{3 - \sqrt{41}}{8} \qquad \text{which implies} \qquad x = \cos^{-1}\left(\frac{3 - \sqrt{41}}{8}\right) \approx 2.01.$$

Of course, had we attempted to compute $\cos^{-1}[(3 + \sqrt{41})/8]$ with a calculator, we would have received an error message. ■

☐ **Simple Harmonic Motion** Many physical objects vibrate or oscillate in a regular manner, repeatedly moving back and forth over a definite time interval. Some examples are clock pendulums, a mass on a spring, sound waves, strings on a guitar when plucked, the human heart, tides, and alternating current. Consider a mass on a spring shown in **FIGURE 4.7.9**. In the absence of frictional or damping forces, a model for the displacement (or directed distance) of the mass measured from a position called the **equilibrium position** is given by the function

$$y(t) = y_0\cos\omega t + \frac{v_0}{\omega}\sin\omega t. \qquad (4)$$

In (4), $\omega = \sqrt{k/m}$, where $k$ is the spring constant (an indicator of the stiffness of the spring), $m$ is the mass attached to the spring (measured in slugs or kilograms), $y_0$ is the initial displacement of the mass (measured above or below the equilibrium position), $v_0$ is the initial velocity of the mass, and $t$ is time measured in seconds. Oscillatory motion modeled by the function (4) is said to be **simple harmonic motion**. The **period**

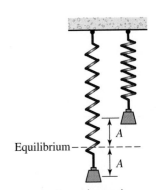

**FIGURE 4.7.9** An undamped spring/mass system exhibits simple harmonic motion

$p$ of motion is $p = 2\pi/\omega$ seconds. The number $f = 1/p = 1/(2\pi/\omega) = \omega/2\pi$ is called the **frequency** of motion. The frequency indicates the number of cycles completed by the graph per unit time. For example, if the period of (4) is, say, $p = 2$ seconds, then one cycle of the function is complete in 2 seconds. The frequency $f = 1/p = \frac{1}{2}$ means one-half of a cycle is complete in 1 second. In the study of simple harmonic motion it is convenient to recast the linear combination (4) as a single expression involving only the sine function:

$$y(t) = A\sin(\omega t + \phi). \qquad (5)$$

The reduction of (4) to (5) can be done in exactly the same manner as illustrated in Example 7 of Section 4.5. In this situation we make the following identifications in (23)–(25) of Section 4.5:

$$c_1 = y_0, \qquad c_2 = v_0/\omega, \qquad A = \sqrt{c_1^2 + c_2^2}, \qquad \text{and} \qquad B = \omega.$$

To find the $\phi$ (in radians) we use

$$\left. \begin{aligned} \sin\phi &= \frac{c_1}{A} \\ \cos\phi &= \frac{c_2}{A} \end{aligned} \right\} \tan\phi = \frac{c_1}{c_2}. \qquad (6)$$

### ■ EXAMPLE 9　　　Equation of Motion

Find the equation of simple harmonic motion (5) for a spring mass system if $m = \frac{1}{16}$ slug, $y_0 = \frac{2}{3}$ ft, $k = 4$ lb/ft, and $v_0 = -\frac{4}{3}$ ft/s.

**Solution** We begin with the simple harmonic motion equation (4). Since $k/m = 4/(\frac{1}{16}) = 64$, $\omega = \sqrt{k/m} = 8$, and $v_0/\omega = (-\frac{4}{3})/8 = -\frac{1}{6}$; therefore (4) becomes

$$y(t) = \tfrac{2}{3}\cos 8t - \tfrac{1}{6}\sin 8t.$$

The period of motion is $2\pi/8 = \pi/4$ second; the frequency is $4/\pi \approx 1.27$ cycles per second. With $c_1 = \frac{2}{3}$, $c_2 = -\frac{1}{6}$, we find the amplitude of motion is $A = \sqrt{(\frac{2}{3})^2 + (-\frac{1}{6})^2} = \frac{1}{6}\sqrt{17}$ ft. Although $\tan\phi = \frac{2}{3}/(-\frac{1}{6}) = -4$, we cannot blindly assume that $\phi = \tan^{-1}(-4)$. The angle that we take for $\phi$ must be consistent with the algebraic signs of $\sin\phi$ and $\cos\phi$ defined in (6). Because $c_1 > 0$ and $c_2 < 0$, $\sin\phi > 0$ and $\cos\phi < 0$ indicate that the angle $\phi$ lies in the second quadrant. However, because the range of the inverse tangent function is $(-\pi/2, \pi/2)$, $\tan^{-1}(-4) = -1.3258$ is a fourth quadrant angle. The correct angle is found by using the reference angle 1.3258 for $\tan^{-1}(-4)$ to find the second quadrant angle

$$\phi = \pi - 1.3258 = 1.8158 \text{ radians.}$$

The equation of motion is then $y(t) = \frac{1}{2}\sqrt{17}\sin(8t + 1.8158)$. ■

Only in the two cases, $c_1 > 0$, $c_2 > 0$ or $c_1 < 0$, $c_2 > 0$, can we use $\tan\phi$ in (6) to write $\phi = \tan^{-1}(c_1/c_2)$. (Why?) Correspondingly, $\phi$ is a first or a fourth quadrant angle.

□ **Postscript—The Other Inverse Trig Functions** The functions $\cot x$, $\sec x$, and $\csc x$ also have inverses when their domains are suitably restricted. See Problems 57–59 in Exercises 4.7. Because these functions are not used as often as arctan, arccos, and arcsin, most scientific calculators do not have keys for them. However, any calculator that computes arcsin, arccos, and arctan can be used to obtain values for **arccsc**, **arcsec**, and **arccot**. Unlike the fact that $\sec x = 1/\cos x$, we note that $\sec^{-1} x \neq 1/\cos^{-1} x$; rather, $\sec^{-1} x = \cos^{-1}(1/x)$ for $|x| \geq 1$. Similar relationships hold for $\csc^{-1} x$ and $\cot^{-1} x$. See Problems 64–66 in Exercises 4.7.

---

**4.7** | Exercises Answers to selected odd-numbered problems begin on page ANS–15.

In Problems 1–14, find the indicated value without using a calculator.

**1.** $\sin^{-1} 0$

**2.** $\tan^{-1} \sqrt{3}$

**3.** $\arccos(-1)$

**4.** $\arcsin \dfrac{\sqrt{3}}{2}$

**5.** $\arccos \frac{1}{2}$

**6.** $\arctan(-\sqrt{3})$

**7.** $\sin^{-1}\left(-\dfrac{\sqrt{3}}{2}\right)$

**8.** $\cos^{-1} \dfrac{\sqrt{3}}{2}$

**9.** $\tan^{-1} 1$

**10.** $\sin^{-1} \dfrac{\sqrt{2}}{2}$

**11.** $\arctan\left(-\dfrac{\sqrt{3}}{3}\right)$

**12.** $\arccos\left(-\frac{1}{2}\right)$

**13.** $\sin^{-1}\left(-\dfrac{\sqrt{2}}{2}\right)$

**14.** $\arctan 0$

In Problems 15–32, find the indicated value without using a calculator.

**15.** $\sin\left(\cos^{-1} \frac{3}{5}\right)$

**16.** $\cos\left(\sin^{-1} \frac{1}{3}\right)$

**17.** $\tan\left(\arccos\left(-\frac{2}{3}\right)\right)$

**18.** $\sin\left(\arctan \frac{1}{4}\right)$

**19.** $\cos(\arctan(-2))$

**20.** $\tan\left(\sin^{-1}\left(-\frac{1}{6}\right)\right)$

**21.** $\csc\left(\sin^{-1} \frac{3}{5}\right)$

**22.** $\sec(\tan^{-1} 4)$

**23.** $\sin\left(\sin^{-1} \frac{1}{5}\right)$

**24.** $\cos\left(\cos^{-1}\left(-\frac{4}{5}\right)\right)$

**25.** $\tan(\tan^{-1} 1.2)$

**26.** $\sin(\arcsin 0.75)$

**27.** $\arcsin\left(\sin \dfrac{\pi}{16}\right)$

**28.** $\arccos\left(\cos \dfrac{2\pi}{3}\right)$

**29.** $\tan^{-1}(\tan \pi)$

**30.** $\sin^{-1}\left(\sin \dfrac{5\pi}{6}\right)$

**31.** $\cos^{-1}\left(\cos\left(-\dfrac{\pi}{4}\right)\right)$

**32.** $\arctan\left(\tan \dfrac{\pi}{7}\right)$

In Problems 33–40, write the given expression as an algebraic expression in $x$.

**33.** $\sin(\tan^{-1} x)$

**34.** $\cos(\tan^{-1} x)$

**35.** $\tan(\arcsin x)$

**36.** $\sec(\arccos x)$

**37.** $\cot(\sin^{-1} x)$

**38.** $\cos(\sin^{-1} x)$

**39.** $\csc(\arctan x)$

**40.** $\tan(\arccos x)$

In Problems 41–48, sketch the graph of the given function.

**41.** $y = \arctan|x|$

**42.** $y = \dfrac{\pi}{2} - \arctan x$

**43.** $y = |\arcsin x|$

**44.** $y = \sin^{-1}(x + 1)$

**45.** $y = 2\cos^{-1}x$

**46.** $y = \cos^{-1}2x$

**47.** $y = \arccos(x - 1)$

**48.** $y = \cos(\arcsin x)$

In Problems 49–54, find the solutions of the given equation in the indicated interval. (Give your answers to two decimal places.)

**49.** $20\cos^2 x + \cos x - 1 = 0,\quad [0, \pi]$
**50.** $3\sin^2 x - 8\sin x + 4 = 0,\quad [-\pi/2, \pi/2]$
**51.** $\tan^2 x + \tan x - 1 = 0,\quad (-\pi/2, \pi/2)$
**52.** $3\sin 2x + \cos x = 0,\quad [-\pi/2, \pi/2]$
**53.** $5\cos^3 x - 3\cos^2 x - \cos x = 0,\quad [0, \pi]$
**54.** $\tan^4 x - 3\tan^2 x + 1 = 0,\quad (-\pi/2, \pi/2).$

In Problems 55 and 56, proceed as in Example 9 and use the given information to find the equation of simple harmonic motion (5) for a spring/mass system. Find the period, frequency, and amplitude of motion.

**55.** $m = \frac{1}{4}$ slug, $y_0 = \frac{1}{2}$ ft, $k = 1$ lb/ft, and $v_0 = \frac{3}{2}$ ft/s
**56.** $m = 1.6$ slug, $y_0 = -\frac{1}{3}$ ft, $k = 40$ lb/ft, and $v_0 = -\frac{5}{4}$ ft/s

**57.** The **arccotangent** function can be defined by $y = \text{arccot } x$ (or $y = \cot^{-1}x$) if and only if $x = \cot y$, where $0 < y < \pi$. Graph $y = \text{arccot } x$, and give the domain and the range of this function.
**58.** The **arccosecant** function can be defined by $y = \text{arccsc } x$ (or $y = \csc^{-1}x$) if and only if $x = \csc y$, where $-\pi/2 \leq y \leq \pi/2$ and $y \neq 0$. Graph $y = \text{arccsc } x$ and give the domain and the range of this function.
**59.** One definition of the **arcsecant** function is $y = \text{arcsec } x$ (or $y = \sec^{-1}x$) if and only if $x = \sec y$, where $0 \leq y \leq \pi$ and $y \neq \pi/2$. (See Problem 60 for an alternative definition.) Graph $y = \text{arcsec } x$ and give the domain and the range of this function.
**60.** An alternative definition of the arcsecant function can be made by restricting the domain of the secant function to $[0, \pi/2) \cup [\pi, 3\pi/2)$. Under this restriction, define the arcsecant function. Graph $y = \text{arcsec } x$, and give the domain and the range of this function.
**61.** Using the definition of the arccotangent function from Problem 57, for what values of $x$ is it true that **(a)** $\cot(\text{arccot } x) = x$ and **(b)** $\text{arccot}(\cot x) = x$?
**62.** Using the definition of the arccosecant function from Problem 58, for what values of $x$ is it true that **(a)** $\csc(\text{arccsc } x) = x$ and **(b)** $\text{arccsc}(\csc x) = x$?
**63.** Using the definition of the arcsecant function from Problem 59, for what values of $x$ is it true that **(a)** $\sec(\text{arcsec } x) = x$ and **(b)** $\text{arcsec}(\sec x) = x$?

**64.** Verify that $\text{arccot } x = \dfrac{\pi}{2} - \arctan x$, for all real numbers $x$.

**65.** Verify that $\text{arccsc } x = \arcsin(1/x)$ for $|x| \geq 1$.
**66.** Verify that $\text{arcsec } x = \arccos(1/x)$ for $|x| \geq 1$.

In Problems 67–72, use the results of Problems 64–66 and a calculator to find the indicated value.

**67.** $\cot^{-1}0.75$

**68.** $\csc^{-1}(-1.3)$

**69.** $\operatorname{arccsc}(-1.5)$

**70.** $\operatorname{arccot}(-0.3)$

**71.** $\operatorname{arcsec}(-1.2)$

**72.** $\sec^{-1} 2.5$

## Miscellaneous Applications

**73. Projectile Motion** The departure angle $\theta$ for a bullet to hit a target at a distance $R$ (assuming that the target and the gun are at the same height) satisfies

$$R = \frac{v_0^2 \sin 2\theta}{g},$$

where $v_0$ is the muzzle velocity and $g$ is the acceleration due to gravity. If the target is 800 ft from the gun and the muzzle velocity is 200 ft/s, find the departure angle. Use $g = 32$ ft/s$^2$. [*Hint:* There are two solutions.]

**74. Olympic Sports** For the Olympic event, the hammer throw, it can be shown that the maximum distance is achieved for the release angle $\theta$ (measured from the horizontal) that satisfies

$$\cos 2\theta = \frac{gh}{v_0^2 + gh},$$

where $h$ is the height of the hammer above the ground at release, $v_0$ is the initial velocity, and $g$ is the acceleration due to gravity. For $v_0 = 13.7$ m/s and $h = 2.25$ m, find the optimal release angle. Use $g = 9.81$ m/s$^2$.

**75. Highway Design** In the design of highways and railroads, curves are banked to provide centripetal force for safety. The optimal banking angle $\theta$ is given by $\tan\theta = v^2/Rg$, where $v$ is the speed of the vehicle, $R$ is the radius of the curve, and $g$ is the acceleration due to gravity. See **FIGURE 4.7.10**. As the formula indicates, for a given radius there is no one correct angle for all speeds. Consequently, curves are banked for the average speed of the traffic over them. Find the correct banking angle for a curve of radius 600 ft on a country road where speeds average 30 mph. Use $g = 32$ ft/s$^2$. [*Hint:* Use consistent units.]

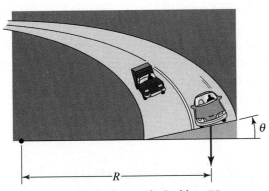

**FIGURE 4.7.10** Banked curve in Problem 75

**76. Highway Design–Continued** If $\mu$ is the coefficient of friction between the car and the road, then the maximum velocity $v_m$ that a car can travel around a curve without slipping is given by $v_m^2 = gR \tan(\theta + \tan^{-1}\mu)$, where $\theta$ is the banking angle of the curve. Find $v_m$ for the country road in Problem 75 if $\mu = 0.26$.

**77. Geology** Viewed from the side, a volcanic cinder cone usually looks like an isosceles trapezoid. See FIGURE 4.7.11. Studies of cinder cones less than 50,000 years old indicate that cone height $H_{co}$ and crater width $W_{cr}$ are related to the cone width $W_{co}$ by the equations $H_{co} = 0.18W_{co}$ and $W_{cr} = 0.40W_{co}$. If $W_{co} = 1.00$, use these equations to determine the base angle $\phi$ of the trapezoid in Figure 4.7.11.

volcanic cone

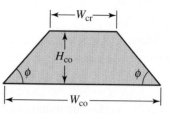

**FIGURE 4.7.11** Volcanic cinder cone in Problem 77

**78. Electrical Circuits** Under certain conditions the current $I(t)$ in an electrical circuit at time $t$ is given by $I(t) = I_0[\sin(\omega t + \theta)\cos\phi + \cos(\omega t + \theta)\sin\phi]$. Solve for $t$. [*Hint*: Use the sum formula for the sine to write the expression in brackets as a single sine function. See (7) of Section 4.5.]

## For Discussion

**79.** Using a calculator set in radian mode, evaluate $\arctan(\tan 1.8)$, $\arccos(\cos 1.8)$, and $\arcsin(\sin 1.8)$. Explain the results.

**80.** Using a calculator set in radian mode, evaluate $\tan^{-1}(\tan(-1))$, $\cos^{-1}(\cos(-1))$, and $\sin^{-1}(\sin(-1))$. Explain the results.

**81.** In Section 4.3 we saw that the graphs of $y = \sin x$ and $y = \cos x$ are related by shifting and reflecting. Justify the identity

$$\arcsin x + \arccos x = \frac{\pi}{2},$$

for all $x$ in $[-1, 1]$, by finding a similar relationship between the graphs of $y = \arcsin x$ and $y = \arccos x$.

**82.** With a calculator set in radian mode determine which of the following inverse trigonometric evaluations result in an error message: **(a)** $\sin^{-1}(-2)$, **(b)** $\cos^{-1}(-2)$, **(c)** $\tan^{-1}(-2)$. Explain.

**83.** Discuss: Can any periodic function be one-to-one?

## 4.8 Right Triangle Trigonometry

☐ **Introduction** The word *trigonometry* (from the Greek *trigonon* meaning triangle and *metria* meaning measurement) refers to the measurement of triangles. In Section 4.2 we defined the trigonometric functions using coordinates of points on the unit circle and by using radian measure we were able to define the trigonometric functions of any angle. In this section we will show that the trigonometric functions of an acute angle in a right triangle have an equivalent definition in terms of the lengths of the sides of the triangle.

☐ **Opposite, Adjacent, and Hypotenuse** In **FIGURE 4.8.1(a)** we have drawn a right triangle with sides labeled $a$, $b$, and $c$ (indicating their respective lengths) and one of the acute angles denoted by $\theta$. From the Pythagorean theorem we know that $a^2 + b^2 = c^2$. The side opposite the right angle is called the **hypotenuse**; the remaining sides are referred to as the **legs** of the triangle. The legs labeled $a$ and $b$ are, in turn, said to be the side **adjacent** to the angle $\theta$ and the side **opposite** the angle $\theta$. We will also use the abbreviations **hyp**, **adj**, and **opp** to denote the lengths of these sides.

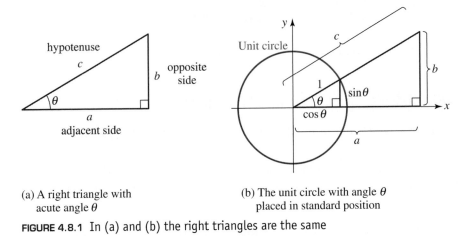

(a) A right triangle with acute angle $\theta$

(b) The unit circle with angle $\theta$ placed in standard position

**FIGURE 4.8.1** In (a) and (b) the right triangles are the same

If we place $\theta$ in standard position and draw a unit circle centered at the origin, we see from Figure 4.8.1(b) that there are two similar right triangles containing the same angle $\theta$. Since corresponding sides of similar triangles are proportional, it follows that

$$\frac{\sin\theta}{1} = \frac{b}{c} = \frac{\text{opp}}{\text{hyp}} \quad \text{and} \quad \frac{\cos\theta}{1} = \frac{a}{c} = \frac{\text{adj}}{\text{hyp}}.$$

Also, we have

$$\frac{\tan\theta}{1} = \frac{\sin\theta}{\cos\theta} = \frac{b/c}{a/c} = \frac{b}{a} = \frac{\text{opp}}{\text{adj}}.$$

Then, applying the reciprocal identities in Section 4.5, each trigonometric function of $\theta$ can be written as the ratio of the lengths of the sides of a right triangle as follows. See **FIGURE 4.8.2**.

---

**TRIGONOMETRIC FUNCTIONS OF $\theta$ IN A RIGHT TRIANGLE**

For an acute angle $\theta$ in a right triangle as shown in Figure 4.8.2,

$$\sin\theta = \frac{\text{opp}}{\text{hyp}} \qquad \cos\theta = \frac{\text{adj}}{\text{hyp}}$$

$$\tan\theta = \frac{\text{opp}}{\text{adj}} \qquad \cot\theta = \frac{\text{adj}}{\text{opp}} \qquad (1)$$

$$\sec\theta = \frac{\text{hyp}}{\text{adj}} \qquad \csc\theta = \frac{\text{hyp}}{\text{opp}}.$$

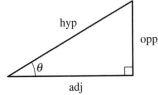

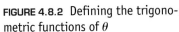

**FIGURE 4.8.2** Defining the trigonometric functions of $\theta$

---

**EXAMPLE 1**     **Values of the Six Trigonometric Functions**

Find the exact values of the six trigonometric functions of the angle $\theta$ in the right triangle shown in **FIGURE 4.8.3**.

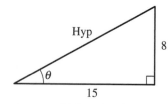

**FIGURE 4.8.3** Right triangle in Example 1

**Solution** From Figure 4.8.3 we see that the side opposite $\theta$ has length 8 and the side adjacent has length 15. From the Pythagorean theorem the hypotenuse $c$ is

$$c^2 = 8^2 + 15^2 = 289 \qquad \text{and so} \qquad c = \sqrt{289} = 17.$$

Thus from (1) the values of the six trigonometric functions are:

$$\sin\theta = \frac{\text{opp}}{\text{hyp}} = \frac{8}{17}, \qquad \cos\theta = \frac{\text{adj}}{\text{hyp}} = \frac{15}{17},$$

$$\tan\theta = \frac{\text{opp}}{\text{adj}} = \frac{8}{15}, \qquad \cot\theta = \frac{\text{adj}}{\text{opp}} = \frac{15}{8},$$

$$\sec\theta = \frac{\text{hyp}}{\text{adj}} = \frac{17}{15}, \qquad \csc\theta = \frac{\text{hyp}}{\text{opp}} = \frac{17}{8}.$$

■

### EXAMPLE 2     Using a Triangle Sketch

If $\theta$ is an acute angle and $\sin\theta = \frac{2}{7}$, find the values of the other trigonometric functions of $\theta$.

**Solution** We sketch a right triangle with an acute angle $\theta$ satisfying $\sin\theta = \frac{2}{7}$, by making opp = 2 and hyp = 7 as shown in **FIGURE 4.8.4**. From the Pythagorean theorem we have

$$2^2 + (\text{adj})^2 = 7^2 \qquad \text{so that} \qquad (\text{adj})^2 = 7^2 - 2^2 = 45.$$

Thus, $$\text{adj} = \sqrt{45} = 3\sqrt{5}.$$

The values of the remaining five trigonometric functions are obtained from the definitions in (1):

$$\cos\theta = \frac{\text{adj}}{\text{hyp}} = \frac{3\sqrt{5}}{7}, \qquad\qquad \sec\theta = \frac{\text{hyp}}{\text{adj}} = \frac{7}{3\sqrt{5}} = \frac{7\sqrt{5}}{15},$$

$$\tan\theta = \frac{\text{opp}}{\text{adj}} = \frac{2}{3\sqrt{5}} = \frac{2\sqrt{5}}{15}, \qquad \cot\theta = \frac{\text{adj}}{\text{opp}} = \frac{3\sqrt{5}}{2},$$

$$\csc\theta = \frac{\text{hyp}}{\text{opp}} = \frac{7}{2}.$$

■

**FIGURE 4.8.4** Right triangle for Example 2

☐ **Solving Right Triangles** Applications of right triangle trigonometry in fields such as surveying and navigation involve **solving right triangles**. The expression "to solve a triangle" means that we wish to find the length of each side and the measure of each angle in the triangle. We can solve any right triangle if we know either two sides or one acute angle and one side. As the following examples will show, sketching and labeling the triangle is an essential part of the solution process. It will be our general practice to label a right triangle as shown in **FIGURE 4.8.5**. The three vertices will be denoted by $A$, $B$, and $C$ with $C$ at the vertex of the right angle. We denote the angles at $A$ and $B$ by $\alpha$ and $\beta$ and the lengths of the sides opposite these angles by $a$ and $b$, respectively. The length of the side opposite the right angle at $C$ is denoted by $c$.

**FIGURE 4.8.5** Standard labeling for a right triangle

### EXAMPLE 3     Solving a Right Triangle

Solve the right triangle having a hypotenuse of length $4\sqrt{3}$ and one $60°$ angle.

**Solution** First we make a sketch of the triangle and label it as shown in **FIGURE 4.8.6**. We wish to find $a$, $b$, and $\beta$. Since $\alpha$ and $\beta$ are complementary angles, $\alpha + \beta = 90°$ yields

$$\beta = 90° - \alpha = 90° - 60° = 30°.$$

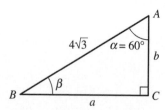

**FIGURE 4.8.6** Right triangle for Example 3

    CHAPTER 4 TRIGONOMETRIC FUNCTIONS

We are given the length of the hypotenuse, namely, hyp $= 4\sqrt{3}$. To find $a$, the length of the side opposite the angle $\alpha = 60°$, we select the sine function. From $\sin\alpha = $ opp/hyp, we obtain

$$\sin 60° = \frac{a}{4\sqrt{3}} \qquad \text{or} \qquad a = 4\sqrt{3}\sin 60°.$$

Since $\sin 60° = \sqrt{3}/2$, we have

$$a = 4\sqrt{3}\sin 60° = 4\sqrt{3}\left(\frac{\sqrt{3}}{2}\right) = 6.$$

Finally, the length $b$ of the side adjacent to the $60°$ angle, we select the cosine function. From $\cos\alpha = $ adj/hyp, we obtain

$$\cos 60° = \frac{b}{4\sqrt{3}}, \qquad \text{or} \qquad b = 4\sqrt{3}\cos 60°.$$

Because $\cos 60° = \frac{1}{2}$, we find

$$b = 4\sqrt{3}\cos 60° = 4\sqrt{3}\left(\tfrac{1}{2}\right) = 2\sqrt{3}. \qquad \blacksquare$$

In Example 3 once we determined $a$, we could have found $b$ by using either the Pythagorean theorem or the tangent function. In general, there are usually several ways to solve a triangle.

□ **Use of a Calculator** If angles other than $30°$, $45°$, or $60°$ are involved in a problem, we obtain approximations of the desired trigonometric function values with a calculator. For the remainder of this chapter, whenever an approximation is used, we will round the final results to the nearest hundredth unless the problem specifies otherwise. To take full advantage of the calculator's accuracy, store the computed values of the trigonometric functions in the calculator for subsequent calculations. If, instead, a rounded version of a displayed value is written down and then later keyed back into the calculator, the accuracy of the final result may be diminished.

### ■ EXAMPLE 4    Solving a Right Triangle

Solve the right triangle with legs of length 4 and 5.

**Solution** After sketching and labeling the triangle as shown in **FIGURE 4.8.7**, we see that we need to find $c$, $\alpha$, and $\beta$. From the Pythagorean theorem, the hypotenuse $c$ is given by

$$c = \sqrt{5^2 + 4^2} = \sqrt{41} \approx 6.40.$$

To find $\beta$, we use $\tan\beta = $ opp/adj. (By choosing to work with the given quantities, we avoid error due to previous approximations.) Thus we have

$$\tan\beta = \tfrac{4}{5} = 0.8.$$

From a calculator set in degree mode, we find $\beta \approx 38.66°$. Since, $\alpha = 90° - \beta$ we obtain $\alpha \approx 51.34°$. $\qquad \blacksquare$

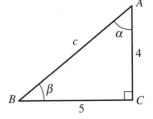

**FIGURE 4.8.7** Right triangle for Example 4

### ■ EXAMPLE 5    Finding the Height of a Tree

A kite is caught in the top branches of a tree. If the 90-ft kite string makes an angle of $22°$ with the ground, estimate the height of the tree by finding the distance from the kite to the ground.

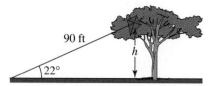

FIGURE 4.8.8 Tree in Example 5

**Solution** Let $h$ denote the height of the kite. From **FIGURE 4.8.8** we see that

$$\frac{h}{90} = \sin 22° \qquad \text{and so} \qquad h = 90\sin 22° \approx 33.71\text{ft.}$$

■

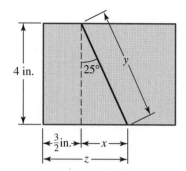

FIGURE 4.8.9 Saw cut in Example 6

**EXAMPLE 6**  **Finding Dimensions of a Saw Cut**

A carpenter cuts the end of a 4-in.-wide board on a 25° bevel from the vertical, starting at a point $1\frac{1}{2}$ in. from the end of the board. Find the lengths of the diagonal cut and the remaining side. See **FIGURE 4.8.9**.

**Solution** Let $x$, $y$, and $z$ be the (unknown) dimensions, as labeled in Figure 4.8.9. It follows from the definition of the tangent function that

$$\tan 25° = \frac{x}{4} \qquad \text{so therefore} \qquad x = 4\tan 25° \approx 1.87 \text{ in.}$$

To find $y$ we observe that

$$\cos 25° = \frac{4}{y} \qquad \text{so} \qquad y = \frac{4}{\cos 25°} \approx 4.41 \text{ in.}$$

Since $z = \frac{3}{2} + x$ and $x \approx 1.87$ in., we see that $z \approx 1.5 + 1.87 \approx 3.37$ in.

■

☐ **Angles of Elevation and Depression** The angle between an observer's line of sight to an object and the horizontal is given a special name. As **FIGURE 4.8.10** illustrates, if the line of sight is to an object above the horizontal, the angle is called an **angle of elevation**, whereas if the line of sight is to an object below the horizontal, the angle is called an **angle of depression**.

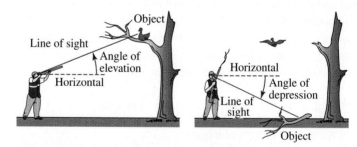

FIGURE 4.8.10 Angles of elevation and depression

**EXAMPLE 7**  **Using Angles of Elevation**

A surveyor uses an instrument called a theodolite to measure the angle of elevation between ground level and the top of a mountain. At one point the angle of elevation is measured to be 41°. A half kilometer farther from the base of the mountain, the angle of elevation is measured to be 37°. How high is the mountain?

**Solution** Let $h$ represent the height of the mountain. **FIGURE 4.8.11** shows that there are two right triangles sharing the common side $h$, so we obtain two equations in two unknowns $z$ and $h$:

$$\frac{h}{z + 0.5} = \tan 37° \qquad \text{and} \qquad \frac{h}{z} = \tan 41°.$$

We can solve each of these for $h$, obtaining

$$h = (z + 0.5)\tan 37° \qquad \text{and} \qquad h = z\tan 41°.$$

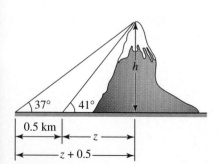

FIGURE 4.8.11 Mountain in Example 7

Equating the last two results gives an equation from which we can determine the distance $z$:

$$(z + 0.5)\tan 37° = z\tan 41°.$$

Solving for $z$ gives us

$$z = \frac{-0.5\tan 37°}{\tan 37° - \tan 41°}.$$

Now $h$ can be found using $h = z\tan 41°$:

$$h = \frac{-0.5\tan 37°\tan 41°}{\tan 37° - \tan 41°} \approx 2.83 \text{ km.}\qquad\blacksquare$$

Section 2.8 was devoted to setting up or constructing functions that were described or expressed in words. As emphasized in that section, this is a task that you will surely face in a course in calculus. Our last example illustrates a recommended procedure of sketching a figure and labeling quantities of interest with appropriate variables.

## ■ EXAMPLE 8    Defining Functions That Involve Trigonometry

A plane flying horizontally at an altitude of 2 miles approaches a radar station as shown in **FIGURE 4.8.12**.

(a) Express the distance $d$ between the plane and the radar station as a function of the angle of elevation $\theta$.

(b) Express the angle of elevation $\theta$ of the plane as a function of the horizontal separation $x$ between the plane and the radar station.

**Solution** As shown in Figure 4.8.12, $\theta$ is an acute angle in a right triangle.

(a) We can relate the distance $d$ and the angle $\theta$ by $\sin\theta = 2/d$. Solving for $d$ gives

$$d(\theta) = \frac{2}{\sin\theta} \quad \text{or} \quad d(\theta) = 2\csc\theta,$$

where $0° < \theta \le 90°$.

(b) The horizontal separation $x$ and $\theta$ are related by $\tan\theta = 2/x$. We make use of the inverse tangent function to solve for $\theta$:

$$\theta(x) = \tan^{-1}\frac{2}{x},$$

where $0 < x < \infty$.  $\blacksquare$

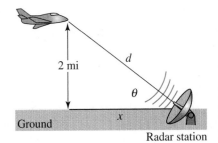

**FIGURE 4.8.12** Plane in Example 8

---

**4.8 ▌ Exercises**    Answers to selected odd-numbered problems begin on page ANS–15.

In Problems 1–10, find the values of the six trigonometric functions of the angle $\theta$ in the given triangle.

**1.**

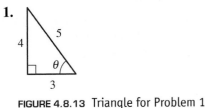

**FIGURE 4.8.13** Triangle for Problem 1

**2.**

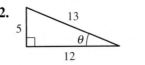

**FIGURE 4.8.14** Triangle for Problem 2

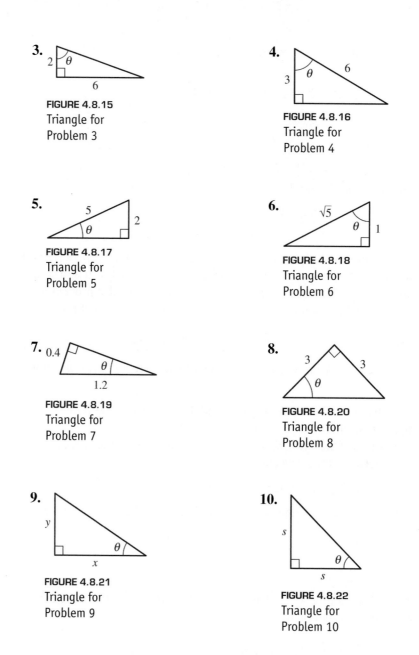

**3.**

**FIGURE 4.8.15**
Triangle for
Problem 3

**4.**

**FIGURE 4.8.16**
Triangle for
Problem 4

**5.**

**FIGURE 4.8.17**
Triangle for
Problem 5

**6.**

**FIGURE 4.8.18**
Triangle for
Problem 6

**7.**

**FIGURE 4.8.19**
Triangle for
Problem 7

**8.**

**FIGURE 4.8.20**
Triangle for
Problem 8

**9.**

**FIGURE 4.8.21**
Triangle for
Problem 9

**10.**

**FIGURE 4.8.22**
Triangle for
Problem 10

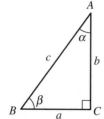

**FIGURE 4.8.23**
Triangle for
Problems 11-22

In Problems 11–22, find the indicated unknowns. Each problem refers to the triangle shown in **FIGURE 4.8.23**.

**11.** $a = 4, \beta = 27°; b, c$

**12.** $c = 10, \beta = 49°; a, b$

**13.** $b = 8, \beta = 34.33°; a, c$

**14.** $c = 25, \alpha = 50°; a, b$

**15.** $b = 1.5, c = 3; \alpha, \beta, a$

**16.** $a = 5, b = 2; \alpha, \beta, c$

**17.** $a = 4, b = 10; \alpha, \beta, c$

**18.** $b = 4, \alpha = 58°; a, c$

**19.** $a = 9, c = 12; \alpha, \beta, b$

**20.** $b = 3, c = 6; \alpha, \beta, a$

**21.** $b = 20, \alpha = 23°; a, c$

**22.** $a = 11, \alpha = 33.5°; b, c$

**23.** A building casts a shadow 20 m long. If the angle from the tip of the shadow to a point on top of the building is 69°, how high is the building?

**24.** Two trees are on opposite sides of a river, as shown in **FIGURE 4.8.24**. A base line of 100 ft is measured from tree $T_1$, and from that position the angle $\beta$ to $T_2$ is measured to be 29.7°. If the base line is perpendicular to the line segment between $T_1$ and $T_2$, find the distance between the two trees.

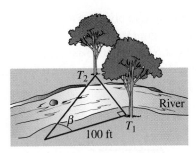

**FIGURE 4.8.24** Trees and river in Problem 24

25. A 50-ft tower is located on the edge of a river. The angle of elevation between the opposite bank and the top of the tower is $37°$. How wide is the river?

26. A surveyor uses a geodometer to measure the straight-line distance from a point on the ground to a point on top of a mountain. Use the information given in **FIGURE 4.8.25** to find the height of the mountain.

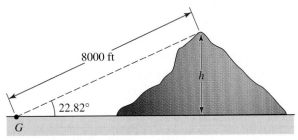

**FIGURE 4.8.25** Mountain in Problem 26

27. An observer on the roof of building $A$ measures a $27°$ angle of depression between the horizontal and the base of building $B$. The angle of elevation from the same point to the roof of the second building is $41.42°$. What is the height of building $B$ if the height of building $A$ is 150 ft? Assume buildings $A$ and $B$ are on the same horizontal plane.

28. Find the height $h$ of a mountain using the information given in **FIGURE 4.8.26**.

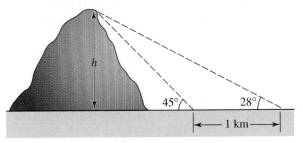

**FIGURE 4.8.26** Mountain in Problem 28

29. The top of a 20-ft ladder is leaning against the edge of the roof of a house. If the angle of inclination of the ladder from the horizontal is $51°$, what is the approximate height of the house and how far is the bottom of the ladder from the base of the house?

30. An airplane flying horizontally at an altitude of 25,000 ft approaches a radar station located on a 2000-ft-high hill. At one instant in time, the angle between the radar dish pointed at the plane and the horizontal is $57°$. What is the straight-line distance in miles between the airplane and the radar station at that particular instant?

**31.** A 5-mi straight segment of a road climbs a 4000-ft hill. Determine the angle that the road makes with the horizontal.

**32.** A box has dimensions as shown in **FIGURE 4.8.27**. Find the length of the diagonal between the corners $P$ and $Q$. What is the angle $\theta$ formed between the diagonal and the bottom edge of the box?

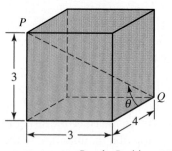

**FIGURE 4.8.27** Box in Problem 32

**33.** Observers in two towns $A$ and $B$ on either side of a 12,000-ft mountain measure the angles of elevation between the ground and the top of the mountain. See **FIGURE 4.8.28**. Assuming that the towns and the mountain top lie in the same vertical plane, find the horizontal distance between them.

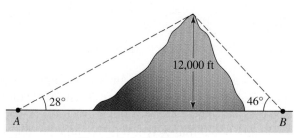

**FIGURE 4.8.28** Mountain in Problem 33

**34.** A drawbridge* measures 7.5 m from shore to shore, and when completely open it makes an angle of $43°$ with the horizontal. See **FIGURE 4.8.29(a)**. When the bridge is

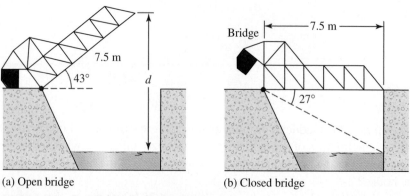

(a) Open bridge    (b) Closed bridge

**FIGURE 4.8.29** Drawbridge in Problem 34

---

*The drawbridge shown in Figure 4.8.29, where the span is continuously balanced by a counterweight, is called a *bascule* bridge.

closed, the angle of depression from the shore to a point on the surface of the water below the opposite end is 27°. See Figure 4.8.29(b). When the bridge is fully open, what is the distance $d$ between the highest point of the bridge and the water below?

**35.** A flagpole is located at the edge of a sheer 50-ft cliff at the bank of a river of width 40 feet. See **FIGURE 4.8.30**. An observer on the opposite side of the river measures an angle of 9° between her line of sight to the top of the flagpole and her line of sight to the top of the cliff. Find the height of the flagpole.

**36.** From an observation sight 1000 ft from the base of Mt. Rushmore, the angle of elevation to the top of the sculpted head of George Washington is measured to be 80.05°, whereas the angle of elevation to the bottom of his head is 79.946°. Determine the height of George Washington's head.

Bust of George Washington on Mt. Rushmore

**37.** The length of a Boeing 747 airplane is 231 ft. What is the plane's altitude if it subtends an angle of 2° when it is directly above an observer on the ground? See **FIGURE 4.8.31**.

**38.** The height of a gnomon (pin) of a sundial is 4 in. If it casts a 6-in. shadow, what is the angle of elevation of the sun?

**39.** Weather radar is capable of measuring both the angle of elevation to the top of a thunderstorm and its range (the horizontal distance to the storm). If the range of a storm is 90 km and the angle of elevation is 4°, can a passenger plane that is able to climb to 10 km fly over the storm?

Sundial

**40.** Cloud ceiling is the lowest altitude at which solid cloud is present. The cloud ceiling at airports must be sufficiently high for safe takeoffs and landings. At night the cloud ceiling can be determined by illuminating the base of the clouds with a searchlight pointed vertically upward. If an observer is 1 km from the searchlight and the angle of elevation to the base of the illuminated cloud is 8°, find the cloud ceiling. See **FIGURE 4.8.32**. (During the day cloud ceilings are generally estimated by sight. However, if an accurate reading is required, a balloon is inflated so that it will rise at a known constant rate. Then it is released and timed until it disappears into the cloud. The cloud ceiling is determined by multiplying the rate by the time of the ascent; trigonometry is not required for this calculation.)

**FIGURE 4.8.30** Flagpole in Problem 35

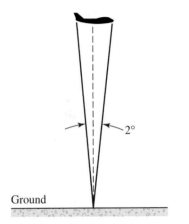

**FIGURE 4.8.31** Airplane in Problem 37

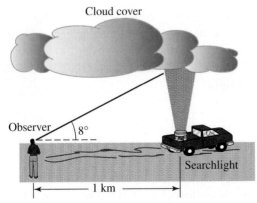

**FIGURE 4.8.32** Searchlight in Problem 40

**41.** Assuming that the Earth is a sphere, show that $C_\theta = C_e \cos\theta$, where $C_\theta$ is the circumference of the parallel of latitude at the latitude angle $\theta$ and $C_e$ is the Earth's circumference at the equator. See **FIGURE 4.8.33**. [*Hint*: $R\cos\theta = r$.]

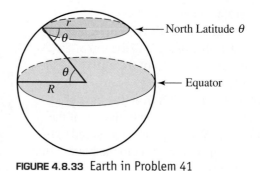

**FIGURE 4.8.33** Earth in Problem 41

**42.** Use Problem 41 and the fact that the radius $R$ of the Earth is 6400 km to find:
  **(a)** the circumference of the Arctic Circle, which lies at $66°33'$ N $(66.55°$ N$)$ latitude,
  **(b)** the distance "around the world" at the $58°40'$ N $(58.67°$ N$)$ latitude.

**43.** The distance between the Earth and the Moon varies as the Moon revolves around the Earth. At a particular time the **geocentric parallax** angle shown in **FIGURE 4.8.34** is measured to be $1°$. Calculate to the nearest hundred miles the distance between the center of the Earth and the center of the Moon at this instant. Assume that the radius of the Earth is 3963 miles.

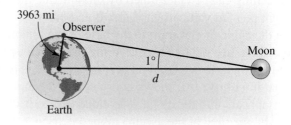

**FIGURE 4.8.34** Angle in Problem 43

**44.** The final length of a volcanic lava flow seems to decrease as the elevation of the lava vent from which it originates increases. An empirical study of Mt. Etna gives the final lava flow length $L$ in terms of elevation $h$ by the formula

$$L = 23 - 0.0053h,$$

where $L$ is measured in kilometers and $h$ is measured in meters. Suppose that a Sicilian village at elevation 750 m is on a $10°$ slope directly below a lava vent at 2500 m. See **FIGURE 4.8.35**. According to the formula, how close will the lava flow get to the village?

Mount Etna

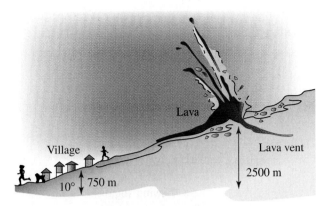

**FIGURE 4.8.35** Lava flow in Problem 44

**45.** As shown in **FIGURE 4.8.36**, two tracking stations $S_1$ and $S_2$ sight a weather balloon between them at elevation angles $\alpha$ and $\beta$, respectively. Express the height $h$ of the balloon in terms of $\alpha$ and $\beta$, and the distance $c$ between the tracking stations. Asssume the tracking stations and the balloon lie in the same vertical plane

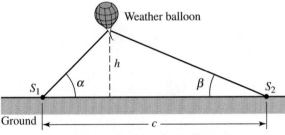

**FIGURE 4.8.36** Weather balloon in Problem 45

**46.** An entry in a soapbox derby rolls down a hill. Using the information given in **FIGURE 4.8.37**, find the total distance $d_1 + d_2$ that the soapbox travels.

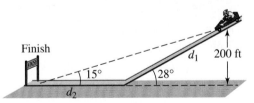

**FIGURE 4.8.37** Soapbox in Problem 46

In Problems 47–50, proceed as in Example 7 and translate the words into an appropriate function.

**47.** A tracking telescope, located 1.25 km from the point of a rocket launch, follows a vertically ascending rocket. Express the height $h$ of the rocket as a function of the angle of elevation $\theta$.

**48.** A searchlight one-half mile offshore illuminates a point $P$ on the shore. Express the distance $d$ from the searchlight to the point of illumination $P$ as a function of the angle $\theta$ shown in **FIGURE 4.8.38**.

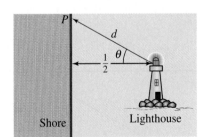

**FIGURE 4.8.38** Searchlight in Problem 48

**49.** A statue is placed on a pedestal as shown in **FIGURE 4.8.39**. Express the viewing angle $\theta$ as a function of the distance $x$ from the pedestal.

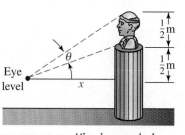

**FIGURE 4.8.39** Viewing angle in Problem 49

**50.** A woman on an island wishes to reach a point $R$ on a straight shore on the mainland from a point $P$ on the island. The point $P$ is 9 mi from the shore and 15 mi from point $R$. See **FIGURE 4.8.40**. If the woman rows a boat at a rate of 3 mi/h to a point $Q$ on land, then walks the rest of the way at a rate of 5 mi/h, express the total time it takes the woman to reach point $R$ as a function of the indicated angle $\theta$. [*Hint*: Distance = rate $\times$ time.]

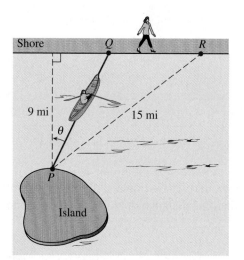

**FIGURE 4.8.40** Woman rowing to shore in Problem 50

## 4.9 Law of Sines and Law of Cosines

☐ **Introduction** In Section 4.8 we saw how to solve right triangles. In this section we consider two techniques for solving general triangles.

☐ **Law of Sines** Consider the triangle $ABC$, shown in **FIGURE 4.9.1**, with angles $\alpha$, $\beta$, and $\gamma$, and corresponding opposite sides $BC$, $AC$, and $AB$. If we know the length of one side and two other parts of the triangle, we can then find the remaining three parts. One way of doing this is by the **Law of Sines**.

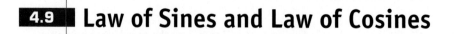

**FIGURE 4.9.1** General triangle

## THE LAW OF SINES

Suppose angles $\alpha$, $\beta$, and $\gamma$, and opposite sides of length $a$, $b$, and $c$ are as shown in Figure 4.9.1. Then

$$\frac{\sin\alpha}{a} = \frac{\sin\beta}{b} = \frac{\sin\gamma}{c}. \qquad (1)$$

Although the Law of Sines is valid for any triangle, we will derive it only for acute triangles—that is, a triangle in which all three angles $\alpha$, $\beta$, and $\gamma$ are less than 90°. As shown in **FIGURE 4.9.2**, let $h$ be the length of the altitude from vertex $A$ to side $BC$. Since the altitude is perpendicular to the base $BC$ it determines two right triangles. Consequently, we can write

$$\frac{h}{c} = \sin\beta \qquad \text{and} \qquad \frac{h}{b} = \sin\gamma. \qquad (2)$$

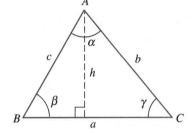

**FIGURE 4.9.2** Acute triangle

Thus (2) gives

$$h = c\sin\beta \qquad \text{and} \qquad h = b\sin\gamma. \qquad (3)$$

Equating the two expressions in (3) gives $c\sin\beta = b\sin\gamma$ so that

$$\frac{\sin\beta}{b} = \frac{\sin\gamma}{c}. \qquad (4)$$

If we use the altitude from the vertex $C$ to the side $AB$, it follows in the same manner that

$$\frac{\sin\alpha}{a} = \frac{\sin\beta}{b}. \qquad (5)$$

Combining (4) and (5) yields the result in (1).

### EXAMPLE 1     Determining the Parts of a Triangle

Find the remaining parts of the triangle shown in **FIGURE 4.9.3**.

**Solution** Let $\beta = 20°$, $\alpha = 130°$, and $b = 6$. It follows then that $\gamma = 180° - 20° - 130° = 30°$. From (1) we see that

$$\frac{\sin 130°}{a} = \frac{\sin 20°}{6} = \frac{\sin 30°}{c}. \qquad (6)$$

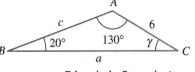

**FIGURE 4.9.3** Triangle in Example 1

We use the first equality in (6) to solve for $a$:

$$a = 6\,\frac{\sin 130°}{\sin 20°} \approx 13.44.$$

The second equality in (6) gives $c$:

$$c = 6\,\frac{\sin 30°}{\sin 20°} \approx 8.77. \qquad \blacksquare$$

### EXAMPLE 2     Height of a Building

A building is situated on the side of a hill that slopes downward at an angle of 15°. The Sun is uphill from the building at an angle of elevation of 42°. Find the building's height if it casts a shadow 36 ft long.

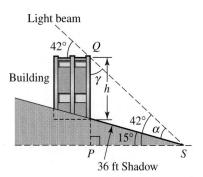

**FIGURE 4.9.4** Triangle $QPS$ in Example 2

**Solution** Denote the height of the building on the downward slope by $h$ and construct a right triangle $QPS$ as shown in **FIGURE 4.9.4**. Now $\alpha + 15° = 42°$ so that $\alpha = 27°$. Since $\triangle QPS$ is a right triangle, $\gamma = 90° - 42° = 48°$. From the Law of Sines (1),

$$\frac{\sin 27°}{h} = \frac{\sin 48°}{36} \qquad \text{so} \qquad h = 36\frac{\sin 27°}{\sin 48°} \approx 21.99 \text{ ft.} \qquad \blacksquare$$

In Examples 1 and 2, where we were given *two angles and a side*, each triangle had a unique solution. However, this may not always be true for triangles where we know *two sides and an angle opposite one of these sides*. The next example illustrates the latter situation.

**■ EXAMPLE 3**  **Given Parts Determine Two Triangles**

Find the remaining parts of the triangle with $\beta = 50°$, $b = 5$, and $c = 6$.
   **Solution** From the Law of Sines, we have

$$\frac{\sin 50°}{5} = \frac{\sin \gamma}{6} \qquad \text{or} \qquad \sin \gamma = \frac{6}{5}\sin 50° \approx 0.9193.$$

From a calculator set in degree mode, we obtain $\gamma \approx 66.82°$. At this point it is essential to recall that the sine function is also positive for second quadrant angles. In other words, there is another angle satisfying $0° \leq \gamma \leq 180°$ for which $\sin \gamma \approx 0.9193$. Using $66.82°$ as a reference angle we find the second quadrant angle to be $180° - 66.82° = 113.18°$. Therefore, the two possibilities for $\gamma$ are $\gamma_1 \approx 66.82°$ and $\gamma_2 \approx 113.18°$. Thus, as shown in **FIGURE 4.9.5**, there are two possible triangles $ABC_1$ and $ABC_2$ satisfying the given three conditions.
   To complete the solution of triangle $ABC_1$ (Figure 4.9.5(a)), we first find $\alpha_1 = 180° - \gamma_1 - \beta \approx 63.18°$. To find the side opposite this angle we use

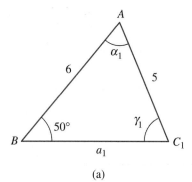

(a)

$$\frac{\sin 63.18°}{a_1} = \frac{\sin 50°}{5} \qquad \text{which gives} \qquad a_1 = 5\left(\frac{\sin 63.18°}{\sin 50°}\right) \approx 5.83.$$

To complete the solution of triangle $ABC_2$ (Figure 4.9.5(b)), we find $\alpha_2 = 180° - \gamma_2 - \beta \approx 16.82°$. Then from

$$\frac{\sin 16.82°}{a_2} = \frac{\sin 50°}{5} \qquad \text{we find} \qquad a_2 = 5\left(\frac{\sin 16.82°}{\sin 50°}\right) \approx 1.89. \qquad \blacksquare$$

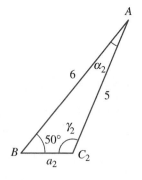

(b)

**FIGURE 4.9.5** Triangles in Example 3

☐ **Ambiguous Case** When solving triangles, the situation where two sides and an angle opposite one of these sides are given is called the **ambiguous case**. We have just seen in Example 3 that the given information may determine two different triangles. In the ambiguous case other complications can arise. For instance, suppose that the length of sides $AB$ and $AC$ (that is, $c$ and $b$) and the angle $\beta$ in triangle $ABC$ are specified. As shown in **FIGURE 4.9.6**, we draw the angle $\beta$ and mark off side $AB$ with length $c$ to locate the vertices $A$ and $B$. The third vertex $C$ is located on the base by drawing an arc of a circle of radius $b$ (the length of $AC$) with center $A$. As shown in **FIGURE 4.9.7**, there are four possible outcomes of this construction:

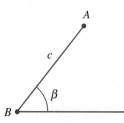

**FIGURE 4.9.6** Horizontal base, the angle $\beta$, and side $AB$

- The arc does not intersect the base and no triangle is formed.
- The arc intersects the base in two distinct points $C_1$ and $C_2$ and two triangles are formed (as in Example 3).
- The arc intersects the base in one point and one triangle is formed.
- The arc is tangent to the base and a single right triangle is formed.

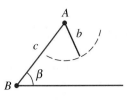
(a) No triangle

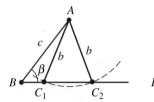
(b) Two triangles

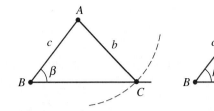
(c) Single triangle

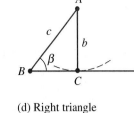
(d) Right triangle

**FIGURE 4.9.7** Solution possibilities for the ambiguous case in the Law of Sines

## ▌EXAMPLE 4    Determining the Parts of a Triangle

Find the remaining parts of the triangle with $\beta = 40°$, $b = 5$, and $c = 9$.

**Solution** From the Law of Sines (1), we have

$$\frac{\sin 40°}{5} = \frac{\sin \gamma}{9} \qquad \text{and so} \qquad \sin \gamma = \frac{9}{5}\sin 40° \approx 1.1570.$$

Since the sine of any angle must be between $-1$ and $1$, $\sin \gamma \approx 1.1570$ is impossible. This means the triangle has no solution; the side with length $b$ is not long enough to reach the base. This is the case illustrated in Figure 4.9.7(a). ▘

Triangles for which we know either *three sides* or *two sides and the included angle* (that is, the angle formed by the given sides) cannot be solved directly using the Law of Sines. The method we consider next can be used to solve triangles in these two cases.

☐ **Pythagorean Theorem** In a right triangle, such as the one shown in **FIGURE 4.9.8**, the length $c$ of the hypotenuse is related to the lengths $a$ and $b$ of the other two sides by the Pythagorean theorem

$$c^2 = a^2 + b^2. \qquad (7)$$

This last equation is a special case of a general formula that relates the lengths of the sides of *any* triangle.

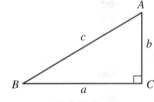

**FIGURE 4.9.8** Right triangle

☐ **Law of Cosines** Suppose again that the triangle $ABC$, shown in Figure 4.9.1, represents a general triangle. The generalization of (7) is called the **Law of Cosines**.

### THE LAW OF COSINES

Suppose angles $\alpha$, $\beta$, and $\gamma$ and opposite sides $a$, $b$, and $c$ are as shown in Figure 4.9.1. Then

$$\begin{aligned} a^2 &= b^2 + c^2 - 2bc\cos\alpha, \\ b^2 &= a^2 + c^2 - 2ac\cos\beta, \\ c^2 &= a^2 + b^2 - 2ab\cos\gamma. \end{aligned} \qquad (8)$$

Like (1), the Law of Cosines is valid for any triangle. But for convenience, we will derive the first two equations in (8) using the same acute triangle given in Figure 4.9.2. However, this time let $P$ denote the point where the altitude from the vertex $A$ intersects side $BC$. Then, since both $\triangle BPA$ and $\triangle CPA$ in **FIGURE 4.9.9** are right triangles we have from (7),

$$c^2 = h^2 + (c\cos\beta)^2 \qquad (9)$$

and

$$b^2 = h^2 + (b\cos\gamma)^2. \qquad (10)$$

Now the length of $BC$ is $a = c\cos\beta + b\cos\gamma$ so that

$$c\cos\beta = a - b\cos\gamma. \qquad (11)$$

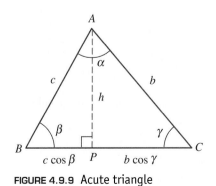

**FIGURE 4.9.9** Acute triangle

Moreover, from (10),

$$h^2 = b^2 - (b\cos\gamma)^2. \qquad (12)$$

Substituting (11) and (12) into (9) and simplifying yields the third equation in (8):

$$c^2 = b^2 - (b\cos\gamma)^2 + (a - b\cos\gamma)^2$$
$$= b^2 - b^2\cos^2\gamma + a^2 - 2ab\cos\gamma + b^2\cos^2\gamma$$

or
$$c^2 = a^2 + b^2 - 2ab\cos\gamma. \qquad (13)$$

Note that equation (13) reduces to the Pythagorean theorem (7) when $\gamma = 90°$.

Similarly, if we use $b\cos\gamma = a - c\cos\beta$ and $h^2 = c^2 - (c\cos\beta)^2$ to eliminate $b\cos\gamma$ and $h^2$ in (10) we obtain the second equation in (8).

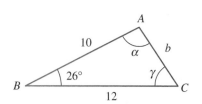

FIGURE 4.9.10 Triangle in Example 5

### ▮ EXAMPLE 5        Determining the Parts of a Triangle

Find the remaining parts of the triangle shown in FIGURE 4.9.10.

**Solution** First, if we call the unknown side $b$ and identify $a = 12$, $c = 10$, and $\beta = 26°$, then from the second equation in (8) we can write

$$b^2 = (12)^2 + (10)^2 - 2(12)(10)\cos 26°.$$

Therefore, $b^2 \approx 28.2894$ and $b \approx 5.32$.

Next, we use the Law of Cosines to determine the remaining angles in the triangle in Figure 4.9.10. If $\gamma$ is the angle at the vertex $C$, then the third equation in (8) gives

$$10^2 = 12^2 + (5.32)^2 - 2(12)(5.32)\cos\gamma \qquad \text{or} \qquad \cos\gamma \approx 0.5663.$$

With the aid of a calculator we find $\gamma \approx 55.51°$. Note that since the cosine of an angle between $90°$ and $180°$ is negative, there is no need to consider two possibilities as we did in Example 2. Finally, the angle at the vertex $A$ is $\alpha = 180° - \beta - \gamma$ or $\alpha \approx 98.89°$.    ▪

In Example 5, observe that after $b$ is found, we know two sides and an angle opposite one of these sides. Hence we could have used the Law of Sines to find the angle $\gamma$.

In the next example we consider the case in which the lengths of the three sides of a triangle are given.

### ▮ EXAMPLE 6        Determining the Angles in a Triangle

Find the angles $\alpha$, $\beta$, and $\gamma$ in the triangle shown in FIGURE 4.9.11.

**Solution** We use the Law of Cosines to find the angle opposite the longest side:

$$9^2 = 6^2 + 7^2 - 2(6)(7)\cos\gamma \qquad \text{or} \qquad \cos\gamma = \tfrac{1}{21}.$$

A calculator then gives $\gamma \approx 87.27°$. Although we could use the Law of Cosines, we choose to find $\beta$ by the Law of Sines:

$$\frac{\sin\beta}{6} = \frac{\sin 87.27°}{9} \qquad \text{or} \qquad \sin\beta = \frac{6}{9}\sin 87.27° \approx 0.6659.$$

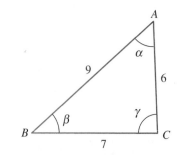

FIGURE 4.9.11 Triangle in Example 6

Since $\gamma$ is the angle opposite the longest side it is the largest angle in the triangle, so $\beta$ must be an acute angle. Thus, $\sin\beta \approx 0.6659$ yields $\beta \approx 41.75°$. Finally, from $\alpha = 180° - \beta - \gamma$ we find $\alpha \approx 50.98°$.    ▪

☐ **Bearing** In navigation directions are given using bearings. A **bearing** designates the acute angle that a line makes with the north–south line. For example, FIGURE 4.9.12(a) illustrates a bearing of S40°W, meaning 40 degrees west of south. The bearings in Figures 4.9.12(b) and 4.9.12(c) are N65°E and S80°E, respectively.

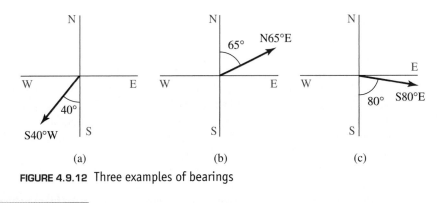

**FIGURE 4.9.12** Three examples of bearings

## EXAMPLE 7    Bearings of Two Ships

Two ships leave a port at 7:00 A.M., one traveling at 12 knots (nautical miles per hour) and the other at 10 knots. If the faster ship maintains a bearing of N47°W and the other ship maintains a bearing of S20°W, what is their separation (to the nearest nautical mile) at 11:00 A.M. that day?

**Solution** Since the elapsed time is 4 hours, the faster ship has traveled $4 \cdot 12 = 48$ nautical miles from port and the slower ship $4 \cdot 10 = 40$ nautical miles. Using these distances and the given bearings, we can sketch the triangle (valid at 11:00 A.M.) shown in **FIGURE 4.9.13**. In the triangle $c$ denotes the distance separating the ships and $\gamma$ is the angle opposite that side. Since $47° + \gamma + 20° = 180°$ we find $\gamma = 113°$. Finally, the Law of Cosines

$$c^2 = 48^2 + 40^2 - 2(48)(40)\cos 113°,$$

gives $c^2 \approx 5404.41$ or $c \approx 73.51$. Thus the distance between the ships (to the nearest nautical mile) is 74 nautical miles. ∎

**FIGURE 4.9.13** Ships in Example 7

## NOTES FROM THE CLASSROOM

(*i*) An important first step in solving a triangle is determining which of the three approaches we have discussed to use: right triangle trigonometry, the Law of Sines, or the Law of Cosines. The following table describes the various types of problems and gives the most appropriate approach for each. The term *oblique* refers to any triangle that is not a right triangle.

| Type of Triangle | Information Given | Technique |
|---|---|---|
| Right | Two sides or angle and a side | Basic definitions of sine, cosine, and tangent; the Pythagorean theorem |
| Oblique | Three sides | The Law of Cosines |
| Oblique | Two sides and the included angle | The Law of Cosines |
| Oblique | Two angles and a side | The Law of Sines |
| Oblique | Two sides and an angle opposite one of the sides | The Law of Sines (if the given angle is acute; it is an ambiguous case.) |

(ii) Here are some additional bits of advice for solving triangles.
- Students will frequently use the Law of Sines when a right triangle trigonometric function could have been used. A right triangle approach is the simplest and most efficient.
- When three sides are given, check first to see whether the length of the longest side is greater than or equal to the sum of the lengths of the other two sides. If it is, there can be no solution (even though the given information indicates a Law of Cosines approach). This is because the shortest distance between two points is the length of the line segment joining them.
- In applying the Law of Sines, if you obtain a value greater than 1 for the sine of an angle, there is no solution.
- In the ambiguous case of the Law of Sines, when solving for the first unknown angle, you must consider *both the acute angle found from your calculator and its supplement as possible solutions*. The supplement will be a solution if the sum of the supplement and the angle given in the triangle is less than 180°.

| **4.9** | Exercises | Answers to selected odd-numbered problems begin on page ANS–16. |

In Problems 1–32, refer to Figure 4.9.1. ▶

In Problems 1–10, use the Law of Sines to solve the triangle.

**1.** $\alpha = 80°, \beta = 20°, b = 7$      **2.** $\alpha = 60°, \beta = 15°, c = 30$

**3.** $\beta = 37°, \gamma = 51°, a = 5$      **4.** $\alpha = 30°, \gamma = 75°, a = 6$

**5.** $\beta = 72°, b = 12, c = 6$      **6.** $\alpha = 120°, a = 9, c = 4$

**7.** $\gamma = 62°, b = 7, c = 4$      **8.** $\beta = 110°, \gamma = 25°, a = 14$

**9.** $\gamma = 15°, a = 8, c = 5$      **10.** $\alpha = 55°, a = 20, c = 18$

In Problems 11–20, use the Law of Cosines to solve the triangle.

**11.** $\gamma = 65°, a = 5, b = 8$      **12.** $\beta = 48°, a = 7, c = 6$

**13.** $a = 8, b = 10, c = 7$      **14.** $\gamma = 31.5°, a = 4, b = 8$

**15.** $\gamma = 97.33°, a = 3, b = 6$      **16.** $a = 7, b = 9, c = 4$

**17.** $a = 11, b = 9.5, c = 8.2$      **18.** $\alpha = 162°, b = 11, c = 8$

**19.** $a = 5, b = 7, c = 10$      **20.** $a = 6, b = 5, c = 7$

In Problems 21–32, use either the Law of Sines or the Law of Cosines as appropriate to solve the triangle.

**21.** $\gamma = 150°, b = 7, c = 5$      **22.** $a = 5, b = 12, c = 13$

**23.** $a = 3, b = 4, c = 5$      **24.** $\alpha = 35°, a = 9, b = 12$

**25.** $\beta = 30°, a = 10, b = 7$      **26.** $\alpha = 140°, \gamma = 20, c = 12$

**27.** $a = 6, b = 8, c = 12$      **28.** $\beta = 130°, a = 4, c = 7$

**29.** $\alpha = 22°, b = 3, c = 9$      **30.** $\alpha = 75°, \gamma = 45°, b = 8$

**31.** $\alpha = 20°, a = 8, c = 27$      **32.** $\beta = 100°, a = 22.3, b = 16.1$

### Miscellaneous Applications

In Problems 33–44, use either the Law of Sines or the Law of Cosines as appropriate.

**33. Length of a Pool** A 10-ft rope that is available to measure the length between two points $A$ and $B$ at opposite ends of a kidney-shaped swimming pool is not long enough. A third point $C$ is found such that the distance from $A$ to $C$ is 10 ft. It is determined that angle $ACB$ is 115° and angle $ABC$ is 35°. Find the distance from $A$ to $B$. See **FIGURE 4.9.14**.

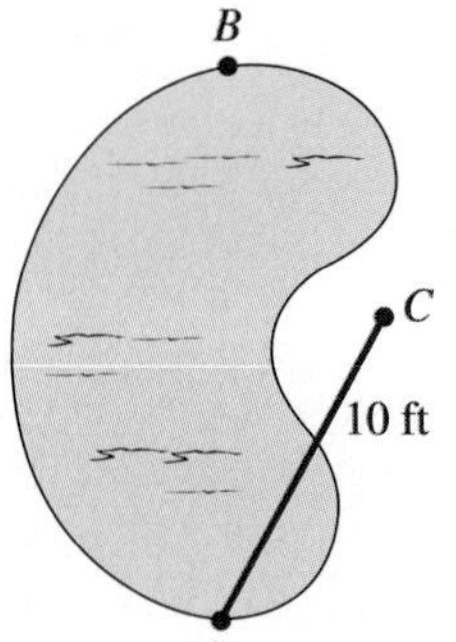

**FIGURE 4.9.14** Pool in Problem 33

CHAPTER 4 TRIGONOMETRIC FUNCTIONS

272

**34. Width of a River**  Two points $A$ and $B$ lie on opposite sides of a river. Another point $C$ is located on the same side of the river as $B$ at a distance of 230 ft from $B$. If angle $ABC$ is 105° and angle $ACB$ is 20°, find the distance across the river from $A$ to $B$.

**35. Length of a Telephone Pole**  A telephone pole makes an angle of 82° with the level ground. As shown in **FIGURE 4.9.15**, the angle of elevation of the Sun is 76°. Find the length of the telephone pole if its shadow is 3.5 m. (Assume that the tilt of the pole is away from the sun and in the same plane as the pole and the sun.)

**36. Not on the Level**  A man 5 ft 9 in. tall stands on a sidewalk that slopes down at a constant angle. A vertical street lamp directly behind him causes his shadow to be 25 ft long. The angle of depression from the top of the man to the tip of his shadow is 31°. Find the angle $\alpha$, as shown in **FIGURE 4.9.16**, that the sidewalk makes with the horizontal.

**FIGURE 4.9.15**
Telephone pole
in Problem 35

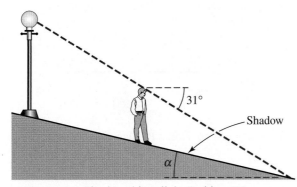

**FIGURE 4.9.16**  Sloping sidewalk in Problem 36

**37. How High?**  If the man in Problem 36 is 20 ft down the sidewalk from the lamppost, find the height of the light above the sidewalk.

**38. Plane with an Altitude**  Angles of elevation to an airplane are measured from the top and the base of a building that is 20 m tall. The angle from the top of the building is 38°, and the angle from the base of the building is 40°. Find the altitude of the airplane.

**39. How Far?**  A ship sails due west from a harbor for 22 nautical miles. It then sails S62°W for another 15 nautical miles. How far is the ship from the harbor?

**40. How Far Apart?**  Two hikers leave their camp simultaneously, taking bearings of N42°W and S20°E, respectively. If they each average a rate of 5 km/h, how far apart are they after 1 hr?

**41. Bearings**  On a hiker's map point $A$ is 2.5 in. due west of point $B$ and point $C$ is 3.5 in. from $B$ and 4.2 in. from $A$, respectively. See **FIGURE 4.9.17**. Find **(a)** the bearing of $A$ from $C$, **(b)** the bearing of $B$ from $C$.

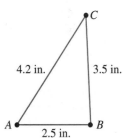

**FIGURE 4.9.17**  Triangle in Problem 41

**42. How Long Will it Take?**  Two ships leave port simultaneously, one traveling at 15 knots and the other at 12 knots. They maintain bearings of S42°W and S10°E, respectively. After 3 hrs the first ship runs aground and the second ship immediately goes to its aid.
**(a)** How long will it take the second ship to reach the first ship if it travels at 14 knots?
**(b)** What bearing should it take?

**43. A Robotic Arm**  A two-dimensional robot arm "knows" where it is by keeping track of a "shoulder" angle $\alpha$ and an "elbow" angle $\beta$. As shown in **FIGURE 4.9.18**, this arm has a fixed point of rotation at the origin. The shoulder angle is measured counterclockwise from the $x$-axis, and the elbow angle is measured counterclockwise from the upper to the lower arm. Suppose that the upper and lower

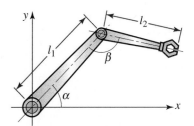

**FIGURE 4.9.18**  Robotic arm in Problem 43

arms are both of length 2 and that the elbow angle $\beta$ is prevented from "hyper-extending" beyond $180°$. Find the angles $\alpha$ and $\beta$ that will position the robot's hand at the point $(1, 2)$.

44. **Which Way?** Two lookout towers are situated on mountain tops $A$ and $B$, 4 mi from each other. A helicopter firefighting team is located in a valley at point $C$, 3 mi from $A$ and 2 mi from $B$. Using the line between $A$ and $B$ as a reference, a lookout spots a fire at an angle of $40°$ from tower $A$ and $82°$ from tower $B$. See **FIGURE 4.9.19**. At what angle, measured from $CB$, should the helicopter fly in order to head directly for the fire?

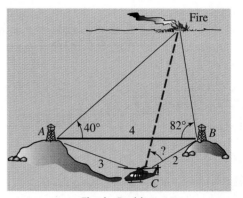

**FIGURE 4.9.19** Fire in Problem 44

# 4.10 The Limit Concept Revisited

∫**Calculus PREVIEW**  □ **Introduction** As we saw in Section 2.9, the fundamental motivating problem of differential calculus, *find a tangent line to the graph of the function*, is answered by the concept of a *limit*. In that section we purposely kept the discussion about limits at an intuitive level; our emphasis was on reviewing the appropriate algebra, such as factoring and rationalization, necessary to be able to compute a limit analytically. In the study of the calculus of the trigonometric functions you will, of course, be expected to compute limits involving trigonometric functions. As the examples in this section will illustrate, computation of trigonometric limits entail both algebraic manipulations and knowledge of basic trigonometric identities.

We begin with a fundamental limit result for the sine function.

□ **An Important Trigonometric Limit** To do the calculus of the trigonometric functions, $\sin x$, $\cos x$, $\tan x$, and so on, it is important to realize that the variable $x$ is a real number or an angle $x$ measured in radians. With that in mind, consider the numerical values of $\sin x$ as $x$ approaches 0 from the right $(x \to 0^+)$ given in the table that follows.

| $x \to 0^+$ | 0.1 | 0.01 | 0.001 | 0.0001 |
|---|---|---|---|---|
| $\dfrac{\sin x}{x}$ | 0.99833416 | 0.99998333 | 0.99999983 | 0.99999999 |

It is easy to see that the same results given in the table hold as $x \to 0^-$. Because $\sin x$ is an odd function, for $x > 0$ and $-x < 0$ we have $\sin(-x) = -\sin x$ and as a consequence $\dfrac{\sin(-x)}{-x} = \dfrac{\sin x}{x}$. In other words, when the value of $x$ is small in absolute value

$$\frac{\sin x}{x} \approx 1.$$

While numerical calculations such as this do not constitute a proof, they do suggest that $\dfrac{\sin x}{x} \to 1$ as $x \to 0$. Using the limit symbol, we have motivated the following result

$$\lim_{x \to 0} \frac{\sin x}{x} = 1. \tag{1}$$

Problem 40 in Exercises 4.10 gives a guided tour through the basic steps of a proof of (1) that is usually presented in calculus.

In this discussion we make the same assumption that we did in Sections 1.5 and 2.9, namely, that all limits under consideration actually exist. Everything that we do, algebraic manipulations, taking limits of products and quotients in the examples in this section is predicated on this assumption. ◀ Important

Other limits of importance are

$$\lim_{x \to a} \sin x = \sin a, \tag{2}$$

$$\lim_{x \to a} \cos x = \cos a. \tag{3}$$

The results (2) and (3) are immediate consequences of the fact that $f(x) = \sin x$ and $g(x) = \cos x$ are continuous functions for all $x$. As we have seen in Section 4.3 the graphs of $\sin x$ and $\cos x$ are smooth and unbroken. For example, from (2),

$$\lim_{x \to \pi/6} \sin x = \sin\frac{\pi}{6} = \frac{1}{2} \tag{4}$$

and

$$\lim_{x \to 0} \sin x = \sin 0 = 0.$$

Also, from (3),

$$\lim_{x \to 0} \cos x = \cos 0 = 1. \tag{5}$$

The results in (1), (2), and (3) are used often to compute other limits. As in Section 1.5 many of the limits considered in this section are limits of fractional expressions where *both* the numerator and the denominator are approaching 0. Recall, these kinds of limits are said to have the **indeterminate form** $0/0$. Note that the limit (1) is of this indeterminate form.

## ■ EXAMPLE 1　　　Using (1)

Find $\displaystyle\lim_{x \to 0} \frac{10x - 3\sin x}{x}$.

**Solution** We rewrite the fractional expression as two fractions with the same denominator $x$:

$$\lim_{x \to 0} \frac{10x - 3\sin x}{x} = \lim_{x \to 0} \left[ \frac{10x}{x} - \frac{3\sin x}{x} \right]$$

$$= \lim_{x \to 0} \frac{10x}{x} - 3\lim_{x \to 0} \frac{\sin x}{x} \quad \leftarrow \text{cancel the } x \text{ in the first expression}$$

$$= \lim_{x \to 0} 10 - 3\lim_{x \to 0} \frac{\sin x}{x} \quad \leftarrow \text{now use (1)}$$

$$= 10 - 3 \cdot 1$$

$$= 7. \quad \blacksquare$$

### EXAMPLE 2　　Using the Double-Angle Formula

Find $\displaystyle\lim_{x \to 0} \frac{\sin 2x}{x}$.

**Solution** To evaluate the given limit, we make use of the double-angle formula $\sin 2x = 2\sin x \cos x$ of Section 4.5 and the results in (1) and (5):

$$\overset{\text{from (5)} \quad \text{from (1)}}{\downarrow \qquad \downarrow}$$

$$\lim_{x \to 0} \frac{\sin 2x}{x} = \lim_{x \to 0} \frac{2\cos x \sin x}{x} = 2\lim_{x \to 0} \cos x \cdot \frac{\sin x}{x} = 2 \cdot 1 \cdot 1 = 2.$$

Thus, $$\lim_{x \to 0} \frac{\sin 2x}{x} = 2. \qquad (6) \quad \blacksquare$$

☐ **Using a Substitution** We are often interested in limits similar to that considered in Example 2. But if we wish to find, say, $\displaystyle\lim_{x \to 0} \frac{\sin 5x}{x}$ the procedure employed in Example 2 breaks down at a practical level since we have not developed a trigonometric identity for $\sin 5x$. There is an alternative procedure that allows us to quickly find $\displaystyle\lim_{x \to 0} \frac{\sin kx}{x}$, where $k \neq 0$ is any real constant, by simply changing the variable by means of a **substitution**. If we let $t = kx$, then $x = t/k$. Notice that as $x \to 0$ then necessarily $t \to 0$. Thus we can write

$$\overset{\text{this limit is 1 from (1)}}{\downarrow}$$

$$\lim_{x \to 0} \frac{\sin kx}{x} = \lim_{t \to 0} \frac{\sin t}{t/k} = \lim_{t \to 0} \frac{\sin t}{1} \cdot \frac{k}{t} = k\lim_{t \to 0} \frac{\sin t}{t} = k.$$

Thus we have proved the general result

$$\lim_{x \to 0} \frac{\sin kx}{x} = k. \qquad (7)$$

Hence $\displaystyle\lim_{x \to 0} \frac{\sin 5x}{x} = 5$. See Problem 25 in Exercises 4.10.

### EXAMPLE 3　　Trigonometric Limit

Find $\displaystyle\lim_{x \to 0} \frac{\tan x}{x}$.

**Solution** Using the definition $\tan x = \sin x/\cos x$ we can write

$$\lim_{x \to 0} \frac{\tan x}{x} = \lim_{x \to 0} \frac{\dfrac{\sin x}{\cos x}}{x} = \lim_{x \to 0} \frac{1}{\cos x} \cdot \frac{\sin x}{x} = \lim_{x \to 0} \frac{1}{\cos x} \cdot \frac{\sin x}{x}.$$

From (5) and (1) we know that $\cos x \to 1$ and $(\sin x)/x \to 1$ as $x \to 0$, and so the preceding line becomes

$$\lim_{x \to 0} \frac{\tan x}{x} = \frac{1}{1} \cdot 1 = 1. \qquad \blacksquare$$

### EXAMPLE 4      Using a Pythagorean Identity

Find $\displaystyle\lim_{x \to 0} \frac{1 - \cos x}{x}$.

**Solution** To compute this limit we start with a bit of algebraic cleverness by multiplying the numerator and denominator by the conjugate factor of the numerator. Next we use the fundamental Pythagorean identity $\sin^2 x + \cos^2 x = 1$ in the form $1 - \cos^2 x = \sin^2 x$:

$$\lim_{x \to 0} \frac{1 - \cos x}{x} = \lim_{x \to 0} \frac{1 - \cos x}{x} \cdot \frac{1 + \cos x}{1 + \cos x}$$

$$= \lim_{x \to 0} \frac{1 - \cos^2 x}{x(1 + \cos x)}$$

$$= \lim_{x \to 0} \frac{\sin^2 x}{x(1 + \cos x)}.$$

For the next step we resort back to algebra to rewrite the fractional expression as a product, then use the results in (1), (4), and (5):

$$\lim_{x \to 0} \frac{1 - \cos x}{x} = \lim_{x \to 0} \frac{\sin^2 x}{x(1 + \cos x)}$$

$$= \lim_{x \to 0} \frac{\sin x}{x} \cdot \frac{\sin x}{1 + \cos x}$$

$$= 1 \cdot \frac{0}{1} \qquad (8)$$

$$= 0.$$

That is, $$\lim_{x \to 0} \frac{1 - \cos x}{x} = 0. \qquad \blacksquare$$

From (8) we obtain a limit result that is used in calculus to find the derivatives of the sine and cosine functions. Since the limit in (8) is equal to 0, we can write

$$\lim_{x \to 0} \frac{1 - \cos x}{x} = \lim_{x \to 0} \frac{-(\cos x - 1)}{x} = (-1)\lim_{x \to 0} \frac{\cos x - 1}{x} = 0.$$

Dividing by $-1$ then gives

$$\lim_{x \to 0} \frac{\cos x - 1}{x} = 0. \qquad (9)$$

☐ **The Calculus Connection** In Section 2.9 we saw that the derivative of a function $y = f(x)$ is the function $f'(x)$ defined by a limit of a difference quotient:

$$f'(x) = \lim_{h \to 0} \frac{f(x + h) - f(x)}{h}. \tag{10}$$

In computing this limit we shrink $h$ to zero but $x$ is held fixed. Recall too, if a number $x = a$ is in the domains of $f$ and $f'$, then $f(a)$ is the $y$-coordinate of the point of tangency $(a, f(a))$ and $f'(a)$ is the slope of the tangent line at that point.

☐ **Derivatives of** $f(x) = \sin x$ **and** $f(x) = \cos x$ To find the derivative of $f(x) = \sin x$ we use the four-step process illustrated in Example 3 of Section 2.9. In the first step we use from Section 4.5 the sum formula for the sine function:

$$\sin(x_1 + x_2) = \sin x_1 \cos x_2 + \cos x_1 \sin x_2. \tag{11}$$

(*i*) With $x$ and $h$ playing the parts of $x_1$ and $x_2$, we have from (11):

$$f(x + h) = \sin(x + h) = \sin x \cos h + \cos x \sin h.$$

(*ii*) $f(x + h) - f(x) = \sin x \cos h + \cos x \sin h - \sin x$
$$= \sin x (\cos h - 1) + \cos x \sin h$$

As we see in the next line, we cannot cancel the $h$'s in the difference quotient but we can rewrite the expression to make use of the limit results in (1) and (9).

(*iii*) $\dfrac{f(x + h) - f(x)}{h} = \dfrac{\sin x (\cos h - 1) + \cos x \sin h}{h}$

$$= \sin x \frac{\cos h - 1}{h} + \cos x \frac{\sin h}{h}$$

(*iv*) In this line, the symbol $h$ plays the part of the symbol $x$ in (1) and (9):

$$f'(x) = \lim_{h \to 0} \frac{f(x + h) - f(x)}{h} = \sin x \lim_{h \to 0} \frac{\cos h - 1}{h} + \cos x \lim_{h \to 0} \frac{\sin h}{h}.$$

From the limit results in (1) and (9), the last line is the same as

$$f'(x) = \lim_{h \to 0} \frac{f(x + h) - f(x)}{h} = \sin x \cdot 0 + \cos x \cdot 1 = \cos x.$$

In summary:

- the derivative of $f(x) = \sin x$ is $f'(x) = \cos x$. $\qquad$ (12)

It is left to you the student to show that:

- the derivative of $f(x) = \cos x$ is $f'(x) = -\sin x$. $\qquad$ (13)

See Problems 23 and 24 in Exercises 4.10.

---

■ EXAMPLE 5 $\qquad\qquad$ **Equation of a Tangent Line**
_____

Find an equation of the tangent line to the graph of $f(x) = \sin x$ at $x = 4\pi/3$.

**Solution** We start by finding the point of tangency. From

$$f\left(\frac{4\pi}{3}\right) = \sin\frac{4\pi}{3} = -\frac{\sqrt{3}}{2}$$

we see that the point of tangency is $(4\pi/3, -\sqrt{3}/2)$. The slope of the tangent line at that point is the derivative of $f(x) = \sin x$ evaluated at the $x$-coordinate. From (12) we know that $f'(x) = \cos x$ and so the slope at $(4\pi/3, -\sqrt{3}/2)$ is

$$f'\left(\frac{4\pi}{3}\right) = \cos\frac{4\pi}{3} = -\frac{1}{2}.$$

From the point-slope form of a line, an equation of the tangent line is

$$y + \frac{\sqrt{3}}{2} = -\frac{1}{2}\left(x - \frac{4\pi}{3}\right) \qquad \text{or} \qquad y = -\frac{1}{2}x + \frac{2\pi}{3} - \frac{\sqrt{3}}{2}.$$

See **FIGURE 4.10.1**. ■

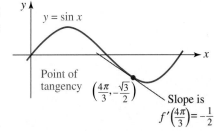

FIGURE 4.10.1 Tangent line in Example 5

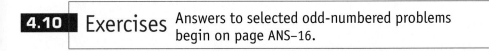

## 4.10 Exercises Answers to selected odd-numbered problems begin on page ANS–16.

In Problems 1–18, use the results in (1), (2), (3), (7), and (9) to find the indicated limit.

**1.** $\displaystyle\lim_{x\to 0}\frac{\sin\frac{1}{2}x}{x}$

**2.** $\displaystyle\lim_{x\to 0}\frac{\sin\pi x}{x}$

**3.** $\displaystyle\lim_{\theta\to 0}\frac{\sin(-\theta)}{\theta}$

**4.** $\displaystyle\lim_{t\to 0}\frac{\sin 3t}{4t}$

**5.** $\displaystyle\lim_{x\to 5\pi/6}\cos x$

**6.** $\displaystyle\lim_{x\to \pi/4}\sin x$

**7.** $\displaystyle\lim_{x\to \pi/2}(\cos x + 5\sin x)$

**8.** $\displaystyle\lim_{x\to \pi/6}\cos x\sin x$

**9.** $\displaystyle\lim_{x\to 0}\frac{\cos x - 1}{10x}$

**10.** $\displaystyle\lim_{\theta\to 0}\frac{8(1 - \cos\theta)}{\theta}$

**11.** $\displaystyle\lim_{x\to 0}\frac{4x^2 - 2\sin x}{x}$

**12.** $\displaystyle\lim_{x\to 0}\frac{2\sin 4x + 1 - \cos x}{x}$

**13.** $\displaystyle\lim_{x\to 0}\frac{\sin^2 x}{x}$

**14.** $\displaystyle\lim_{x\to 0}\frac{\sin^2 x}{x^2}$

**15.** $\displaystyle\lim_{x\to \pi/2}\frac{\cos x}{\cot x}$

**16.** $\displaystyle\lim_{x\to 0}\frac{\cos x\tan x}{x}$

**17.** $\displaystyle\lim_{x\to 0}x\cot x$

**18.** $\displaystyle\lim_{x\to \pi/4}\frac{\cos 2x}{\cos x - \sin x}$

In Problems 19–22, proceed as in Example 5 to find an equation of the tangent line to the graph of $f(x) = \sin x$ at the indicated value of $x$.

**19.** $x = 0$

**20.** $x = \pi/2$

**21.** $x = \pi/6$

**22.** $x = 2\pi/3$

**23.** Proceed as on page 280 and find the derivative of $f(x) = \cos x$.

**24.** Use the result of Problem 23 to find an equation of the tangent line to the graph of $f(x) = \cos x$ at $x = \pi/3$.

**25.** Use the facts that

$$\lim_{x\to 0}\frac{\cos 5x - 1}{x} = 0 \qquad \text{and} \qquad \lim_{x\to 0}\frac{\sin 5x}{x} = 5$$

to find the derivative of $f(x) = \sin 5x$.

**26.** Use the result of Problem 25, to find an equation of the tangent line to the graph of $f(x) = \sin 5x$ at $x = \pi$.

## Calculator/Computer Problems

In Problems 27 and 28, use a calculator or computer to estimate the given limit by completing each table. Round the entries in each table to eight decimal places.

**27.** $\displaystyle\lim_{x \to 0} \frac{1 - \cos x}{x^2}$

| $x \to 0^+$ | 0.1 | 0.01 | 0.001 | 0.0001 | 0.00001 |
|---|---|---|---|---|---|
| $\dfrac{1 - \cos x}{x^2}$ | | | | | |

Explain why we do not have to consider $x \to 0^-$.

**28.** $\displaystyle\lim_{x \to 2} \frac{x^2 - 4}{\sin(x - 2)}$

| $x \to 2^+$ | 2.1 | 2.01 | 2.001 | 2.0001 | 2.00001 |
|---|---|---|---|---|---|
| $\dfrac{x^2 - 4}{\sin(x - 2)}$ | | | | | |

| $x \to 2^-$ | 1.9 | 1.99 | 1.999 | 1.9999 | 1.99999 |
|---|---|---|---|---|---|
| $\dfrac{x^2 - 4}{\sin(x - 2)}$ | | | | | |

## For Discussion

In Problems 29–36, discuss how to use the result in (1) along with some clever algebra, trigonometry, or a substitution to find the given limit.

**29.** $\displaystyle\lim_{x \to 0} \frac{x}{\sin 3x}$

**30.** $\displaystyle\lim_{x \to 0} \frac{\sin 4x}{\sin 5x}$

**31.** $\displaystyle\lim_{x \to 0} \frac{\sin x^2}{x^2}$

**32.** $\displaystyle\lim_{x \to \pi} \frac{\sin x}{\pi - x}$

**33.** $\displaystyle\lim_{x \to 0} \frac{x^2}{1 - \cos x}$

**34.** $\displaystyle\lim_{x \to 0} \frac{\cos\left(x + \frac{1}{2}\pi\right)}{x}$

**35.** $\displaystyle\lim_{x \to 0^+} \frac{\sin x}{\sqrt{x}}$

**36.** $\displaystyle\lim_{x \to 1} \frac{\sin(x - 1)}{x^2 + 2x - 3}$

**37.** Using what you have learned in Problems 29 and 36, find the limit

$$\lim_{x \to 2} \frac{x^2 - 4}{\sin(x - 2)}$$

without the aid of the numerical table in Problem 28.

**38. (a)** Use a calculator to complete the following table.

| $x \to 0^+$ | 0.1 | 0.01 | 0.001 | 0.0001 | 0.00001 |
|---|---|---|---|---|---|
| $\dfrac{1 - \cos x^2}{x^4}$ | | | | | |

**(b)** Find the limit $\displaystyle\lim_{x\to 0} \dfrac{1 - \cos x^2}{x^4}$ using the method given in Example 4.

**(c)** Discuss any differences that you observe between parts **(a)** and **(b)**.

**39. (a)** A regular $n$-gon is an $n$-sided polygon inscribed in a circle; the polygon is formed by $n$ equally spaced points on the circle. Suppose the polygon shown in **FIGURE 4.10.2** represents a regular $n$-gon inscribed in a circle of radius $r$. Use trigonometry to show that the area $A(n)$ of the $n$-gon is given by

$$A(n) = \frac{n}{2}r^2 \sin\left(\frac{2\pi}{n}\right).$$

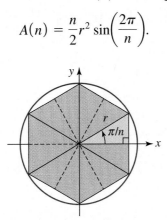

**FIGURE 4.10.2** Inscribed $n$-gon for Problem 39

**(b)** It stands to reason that the area $A(n)$ approaches the area of the circle as the number of sides of the $n$-gon increases. Compute $A_{100}$ and $A_{1000}$.

**(c)** Let $x = 2\pi/n$ in $A(n)$ and note that as $n \to \infty$ then $x \to 0$. Use (1) of this section to show that $\displaystyle\lim_{n\to\infty} A(n) = \pi r^2$.

**40.** Consider a circle centered at the origin $O$ with radius 1. As shown in **FIGURE 4.10.3(a)**, let the shaded region $OPR$ be a sector of the circle with central angle $t$ such that $0 < t < \pi/2$. We see from Figures 4.10.3(a)–4.10.3(d) that

$$\text{area of } \triangle OPR < \text{area of sector } OPR < \text{area of } \triangle OQR. \qquad (12)$$

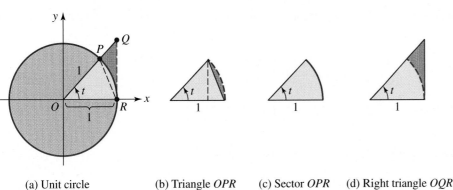

(a) Unit circle      (b) Triangle $OPR$    (c) Sector $OPR$    (d) Right triangle $OQR$

**FIGURE 4.10.3** Unit circle in Problem 40

(a) Show that the area of $\triangle OPR$ is $\frac{1}{2}\sin t$ and that the area of $\triangle OQR$ is $\frac{1}{2}\tan t$.

(b) Since the area of a sector of a circle is $\frac{1}{2}r^2\theta$, where $r$ is its radius and $\theta$ is measured in radians, it follows that the area of sector $OPR$ is $\frac{1}{2}t$. Use this result, along with the areas in part (a), to show that the inequality in (12) yields

$$\cos t < \frac{\sin t}{t} < 1.$$

(c) Discuss how the preceding inequality proves (1) when we let $t \to 0^+$.

---

| CHAPTER 4 | Review Exercises | Answers to selected odd-numbered problems begin on page ANS–16. |

In Problems 1–20, fill in the blanks.

1. $\pi/5$ radians = _____ degrees.

2. 10 degrees = _____ radians.

3. The exact values of the coordinates of the point $P(t)$ on the unit circle corresponding to $t = 5\pi/6$ are _____.

4. The reference angle for $4\pi/3$ radians is _____ radians.

5. $\tan\dfrac{\pi}{3} = $ _____.

6. In standard position, the terminal side of the angle $\dfrac{8\pi}{5}$ radians lies in the _____ quadrant.

7. If $\sin\theta = -\frac{1}{3}$ and $\theta$ is in quadrant IV, then $\sec\theta = $ _____.

8. If $\tan t = 2$ and $t$ is in quadrant III, then $\cos t = $ _____.

9. The $y$-intercept for the graph of the function $y = 2\sec(x + \pi)$ is _____.

10. The values of $t$ in the interval $[0, 2\pi]$ that satisfy $\sin 2t = \frac{1}{2}$ are _____.

11. If $\sin u = \frac{3}{5}, 0 < u < \pi/2$, and $\cos v = 1/\sqrt{5}, 3\pi/2 < v < 2\pi$, then $\cos(u + v) = $ _____.

12. If $\cos t = -\frac{2}{3}, \pi < t < 3\pi/2$, then $\cos\frac{1}{2}t = $ _____.

13. A sine function with period 4 and amplitude 6 is given by _____.

14. The first vertical asymptote for the graph of $y = \tan\left(x - \dfrac{\pi}{4}\right)$ to the right of the $y$-axis is _____.

15. $\sin t + \cos t = $ _____ $\sin\left(t + \dfrac{\pi}{4}\right)$.

16. If $\sin t = \frac{1}{6}$, then $\cos\left(t - \dfrac{\pi}{2}\right) = $ _____.

17. The amplitude of $y = -10\cos\left(\dfrac{\pi}{3}x\right)$ is _____.

18. $\cos\left(\dfrac{\pi}{6} - \dfrac{5\pi}{4}\right) = $ _____

**19.** The exact value of $\arccos\left(\cos\dfrac{9\pi}{5}\right) = $ _____.

**20.** The period of the function $y = 2\sin\dfrac{\pi}{3}t$ is _____.

In Problems 21–40, answer true or false.

**21.** If $\tan t = \frac{3}{4}$, then $\sin t = 3$ and $\cos t = 4$. _____

**22.** In a right triangle, If $\sin\theta = \frac{11}{61}$, then $\cot\theta = \frac{60}{11}$. _____

**23.** $\sec(-\pi) = \csc\left(\dfrac{3\pi}{2}\right)$ _____

**24.** There is no angle $t$ such that $\sec t = \frac{1}{2}$. _____

**25.** $\sin(2\pi - t) = -\sin t$ _____

**26.** $1 + \sec^2\theta = \tan^2\theta$ _____

**27.** $(5, 0)$ is an $x$-intercept of the graph of $y = 3\sin\pi x$. _____

**28.** $(2\pi/3, -1/\sqrt{3})$ is a point on the graph of $y = \cot x$. _____

**29.** The range of the function $y = \csc x$ is $(-\infty, -1] \cup [1, \infty)$. _____

**30.** The graph of $y = \csc x$ does not intersect the $y$-axis. _____

**31.** The line $x = \pi/2$ is a vertical asymptote for the graph of $y = \tan x$. _____

**32.** If $\tan(x + \pi) = 0.3$, then $\tan x = 0.3$. _____

**33.** For the sine function $y = -2\sin x$ we have $-2 \le y \le 2$. _____

**34.** $\sin 6x = 2\sin 3x \cos 3x$ _____

**35.** The graph of $y = \sin\left(2x - \dfrac{\pi}{3}\right)$ is the graph of $y = \sin 2x$ shifted $\pi/3$ units to

the right. _____

**36.** Since $\tan(5\pi/4) = 1$, then $\arctan(1) = 5\pi/4$. _____

**37.** $\arcsin\left(\frac{1}{2}\right) = 30°$ _____

**38.** $f(x) = \arcsin x$ is not periodic. _____

**39.** $f(x) = x\sin x$ is $2\pi$ periodic. _____

**40.** $f(x) = \sin(\cos x)$ is an even function. _____

In Problems 41–46, find all $t$ in the interval $[0, 2\pi]$ that satisfy the given equation.

**41.** $\cos t \sin t - \cos t + \sin t - 1 = 0$    **42.** $\cos t - \sin t = 0$

**43.** $4\sin^2 t - 1 = 0$    **44.** $\sin t = 2\tan t$

**45.** $\sin t + \cos t = 1$    **46.** $\tan t - 3\cot t = 2$

In Problems 47–50, solve the triangle satisfying the given conditions.

**47.** $\alpha = 30°, \beta = 70°, b = 10$    **48.** $\gamma = 145°, a = 25, c = 20$

**49.** $\alpha = 51°, b = 20, c = 10$    **50.** $a = 4, b = 6, c = 3$

In Problems 51–58, find the indicated value without using a calculator.

**51.** $\cos^{-1}\left(-\frac{1}{2}\right)$    **52.** $\arcsin(-1)$

**53.** $\cot\left(\cos^{-1}\frac{3}{4}\right)$    **54.** $\cos\left(\arcsin\frac{2}{5}\right)$

**55.** $\sin^{-1}(\sin\pi)$    **56.** $\cos(\arccos 0.42)$

**57.** $\sin\left(\arccos\left(\frac{5}{13}\right)\right)$    **58.** $\arctan(\cos\pi)$

In Problems 59 and 60, write the given expression as an algebraic expression in $x$.

**59.** $\sin(\arccos x)$

**60.** $\sec(\tan^{-1} x)$

In Problems 61–64, the given graph can be interpreted as a rigid/nonrigid transformation of the graph of $y = \sin x$ and of the graph of $y = \cos x$. Find an equation of the graph using the sine function. Then find an equation of the same graph using the cosine function.

**61.**

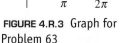

FIGURE 4.R.1 Graph for Problem 61

**62.**

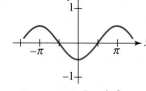

FIGURE 4.R.2 Graph for Problem 62

**63.**

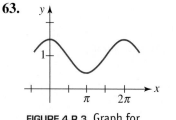

FIGURE 4.R.3 Graph for Problem 63

**64.**

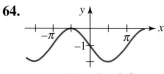

FIGURE 4.R.4 Graph for Problem 64

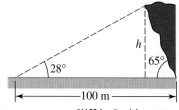

FIGURE 4.R.5 Cliff in Problem 65

**65.** A surveyor 100 m from the base of an overhanging cliff measures a 28° angle of elevation from that point to the top of the cliff. See FIGURE 4.R.5. If the cliff makes an angle of 65° with the horizontal ground, determine its height $h$.

**66.** A rocket is launched from ground level at an angle of elevation of 43°. If the rocket hits a drone target plane flying at 20,000 ft, find the horizontal distance between the rocket launch site and the point directly beneath the plane. What is the straight-line distance between the rocket launch site and the target plane?

**67.** A competition water-skier leaves a ramp at point $R$ and lands at $S$. See FIGURE 4.R.6. A judge at point $J$ measures an $\angle RJS$ as 47°. If the distance from the ramp to the judge is 110 ft, find the length of the jump. Assume that $\angle SRJ$ is 90°.

**68.** The angle between two sides of a parallelogram is 40°. If the lengths of the sides are 5 and 10 cm, find the lengths of the two diagonals.

**69.** A weather satellite orbiting the equator of the Earth at a height of $H = 36{,}000$ km spots a thunderstorm to the north at $P$ at an angle of $\theta = 6.5°$ from its vertical. See FIGURE 4.R.7.

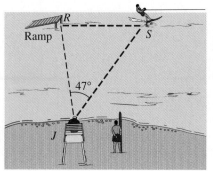

FIGURE 4.R.6 Water-skier in Problem 67

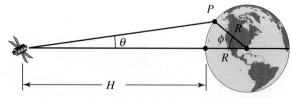

FIGURE 4.R.7 Satellite in Problem 69

**(a)** Given that the Earth's radius is approximately $R = 6370$ km, find the latitude $\phi$ of the thunderstorm.

**(b)** Show that angles $\theta$ and $\phi$ are related by

$$\tan\theta = \frac{R\sin\phi}{H + R(1 - \cos\phi)}.$$

**70.** It can be shown that a basketball of diameter $d$ approaching the basket from an angle $\theta$ to the horizontal will pass through a hoop of diameter $D$ if $D\sin\theta > d$, where $0° \leq \theta \leq 90°$. See **FIGURE 4.R.8**. If the basketball has diameter 24.6 cm and the hoop has diameter 45 cm, what range of approach angles $\theta$ will result in a basket?

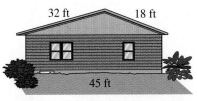

**FIGURE 4.R.8** Basketball in Problem 70

**71.** Each of the 24 NAVSTAR Global Positioning System (GPS) satellites orbits the Earth at an altitude of $h = 20{,}200$ km. Using this network of satellites, an inexpensive hand-held GPS receiver can determine its position on the surface of the Earth to within 10 m. Find the greatest distance $s$ (in km) on the surface of the Earth that can be observed from a single GPS satellite. See **FIGURE 4.R.9**. Take the radius of the Earth to be 6370 km. [*Hint*: Find the central angle $\theta$ subtended by $s$.]

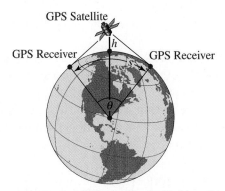

**FIGURE 4.R.9** GPS satellite in Problem 71

**72.** An airplane flying horizontally at a speed of 400 miles per hour is climbing at an angle of 6° from the horizontal. When it passes directly over a car traveling 60 miles per hour, it is 2 miles above the car. Assuming that the airplane and the car remain in the same vertical plane, find the angle of elevation from the car to the airplane after 30 minutes.

**73.** A house measures 45 ft from front to back. The roof measures 32 ft from the front of the house to the peak and 18 ft from the peak to the back of the house. See **FIGURE 4.R.10**. Find the angles of elevation of the front and back parts of the roof.

32 ft      18 ft

45 ft

**FIGURE 4.R.10** House in Problem 73

**74.** A regular five-sided polygon is called a regular pentagon. See Figure 4.10.2. Using radian measure, determine the sum of the vertex angles in a regular pentagon.

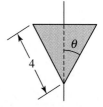

FIGURE 4.R.11 End of water trough in Problem 75

In Problems 75–82, translate the words into an appropriate function.

**75.** A 20-ft-long water trough has ends in the form of isosceles triangles with sides that are 4 ft long. See Figure 2.8.21 in Exercises 2.8. As shown in FIGURE 4.R.11, let $\theta$ denote the angle between the vertical and one of the sides of a triangular end. Express the volume of the trough as a function of $2\theta$.

**76.** A person driving a car approaches a freeway sign as shown in FIGURE 4.R.12. Let $\theta$ be her viewing angle of the sign and let $x$ represent her horizontal distance (measured in feet) to that sign. Express $\theta$ as a function of $x$.

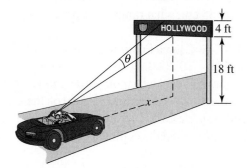

FIGURE 4.R.12 Freeway sign in Problem 76

**77.** As shown in FIGURE 4.R.13, a plank is supported by a sawhorse so that one end rests on the ground and the other end rests against a building. Express the length of the plank as a function of the indicated angle $\theta$.

**78.** A farmer wishes to enclose a pasture in the form of a right triangle using 2000 ft of fencing on hand. See FIGURE 4.R.14. Show that the area of the pasture as a function of the indicated angle $\theta$ is

$$A(\theta) = \cot\theta \cdot \left(\frac{2000}{1 + \cot\theta + \csc\theta}\right)^2.$$

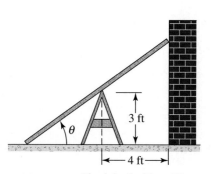

FIGURE 4.R.13 Plank in Problem 77

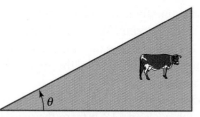

FIGURE 4.R.14 Pasture in Problem 78

**79.** Express the volume of the box shown in FIGURE 4.R.15 as a function of the indicated angle $\theta$.

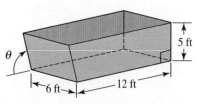

FIGURE 4.R.15 Box in Problem 79

80. A corner of an 8.5 in. × 11 in. piece of paper is folded over to the other edge of the paper as shown in FIGURE 4.R.16. Express the length $L$ of the crease as a function of the angle $\theta$ shown in the figure.

81. A gutter is to be made from a sheet of metal 30 cm wide by turning up the edges of width 10 cm along each side so that the sides make equal angles $\phi$ with the vertical. See FIGURE 4.R.17. Express the cross-sectional area of the gutter as a function of the angle $\phi$.

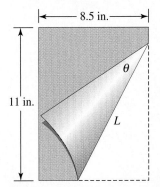

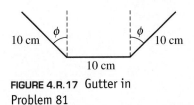

FIGURE 4.R.17  Gutter in Problem 81

FIGURE 4.R.16  Folded paper in Problem 80

82. A metal pipe is to be carried horizontally around a right-angled corner from a hallway 8 feet wide into a hallway that is 6 feet wide. See FIGURE 4.R.18. Express the length L of the pipe as a function of the angle $\theta$ shown in the figure.

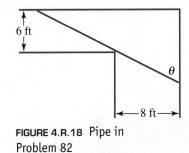

FIGURE 4.R.18  Pipe in Problem 82

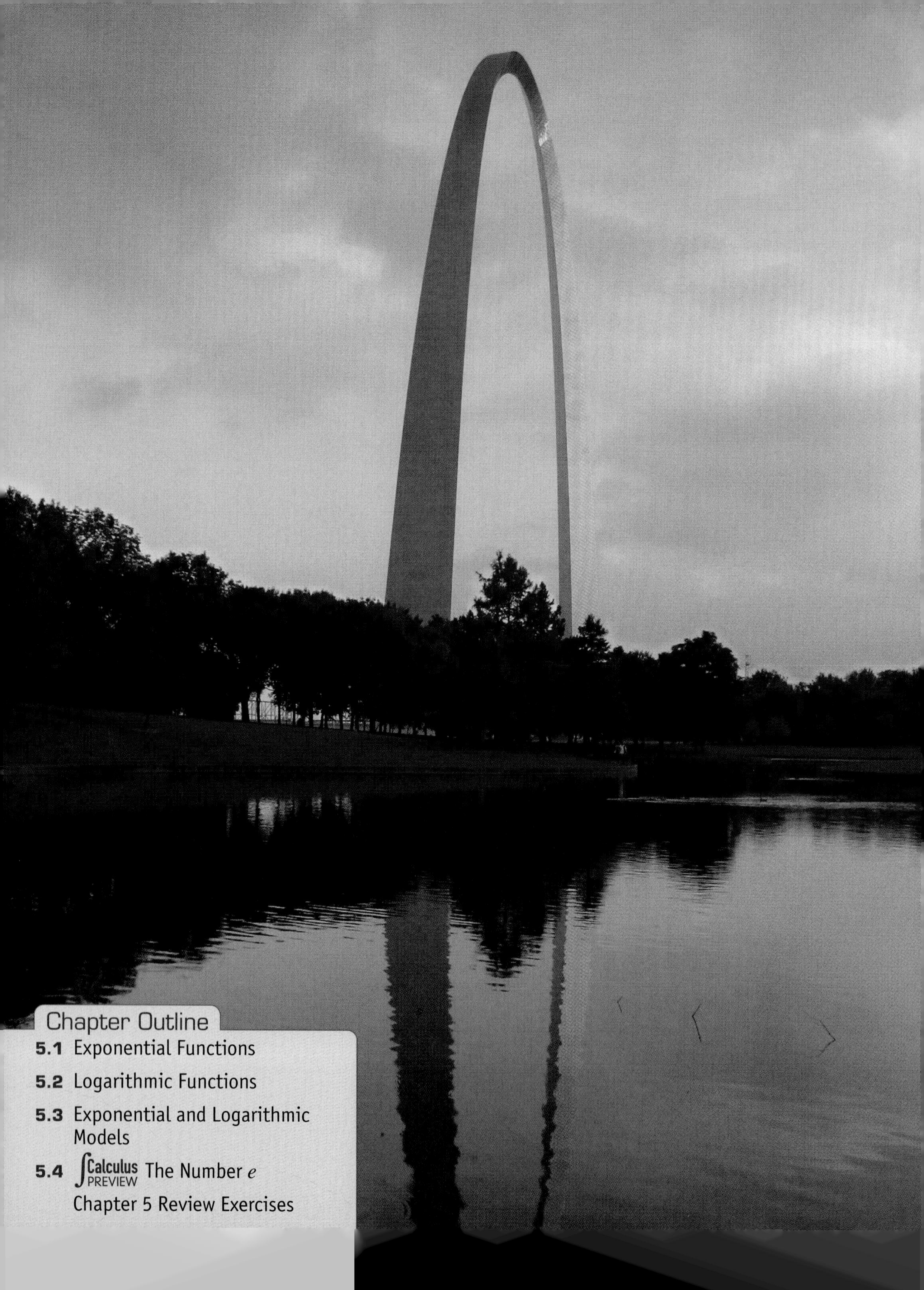

## Chapter Outline

# Exponential and Logarithmic Functions

# 5

## 5.1 Exponential Functions

☐ **Introduction** In the preceding chapters we considered functions such as $f(x) = x^2$, that is, a function with a variable base $x$ and constant power or exponent 2. We now examine functions such as $f(x) = 2^x$ having a constant base 2 and a variable exponent $x$.

---

**EXPONENTIAL FUNCTION**

If $b > 0$ and $b \neq 1$, then an **exponential function** $y = f(x)$ is function of the form

$$f(x) = b^x. \tag{1}$$

The number $b$ is called the **base** and $x$ is called the **exponent**.

---

The **domain** of an exponential $f$ defined in (1) is the set of all real numbers $(-\infty, \infty)$.

In (1) the base $b$ is restricted to positive numbers in order to guarantee that $b^x$ is always a real number. For example, with this restriction we avoid complex numbers such as $(-4)^{1/2}$. Also, the base $b = 1$ is of little interest to us since (1) is the constant function $f(x) = 1^x = 1$.

☐ **Exponents** As just mentioned, the domain of an exponential function (1) is the set of all real numbers. This means that the exponent $x$ can be either a rational or an irrational number. For example, if the base $b = 3$ and the exponent $x$ is a *rational number*—for example, $x = \frac{1}{5}$ and $x = 1.4$—then

$$3^{1/5} = \sqrt[5]{3} \quad \text{and} \quad 3^{1.4} = 3^{14/10} = 3^{7/5} = \sqrt[5]{3^7}.$$

For an exponent $x$ that is an *irrational number*, $b^x$ is defined, but its precise definition is beyond the scope of this text. We can, however, suggest a procedure for defining a number such as $3^{\sqrt{2}}$. From the decimal representation $\sqrt{2} = 1.414213562\ldots$ we see that the rational numbers

$$1, 1.4, 1.41, 1.414, 1.4142, 1.41421, \ldots,$$

are successively better approximations to $\sqrt{2}$. By using these rational numbers as exponents, we would expect that the numbers

$$3^1, 3^{1.4}, 3^{1.41}, 3^{1.414}, 3^{1.4142}, 3^{1.41421}, \ldots,$$

are then successively better approximations to $3^{\sqrt{2}}$. In fact, this can be shown to be true with a precise definition of $b^x$ for an irrational value of $x$. But on a practical level, we can use the $\boxed{y^x}$ key on a calculator to obtain the approximation 4.728804388 to $3^{\sqrt{2}}$.

☐ **Laws of Exponents** In most algebra texts the laws of exponents are stated first for integer exponents and then for rational exponents. Since $b^x$ can be defined for all real numbers $x$ when $b > 0$, it can be proved that these same **laws of exponents** hold for all real-number exponents. If $a > 0$, $b > 0$ and $x$, $x_1$, and $x_2$ denote real numbers, then

$(i)$ $b^{x_1} \cdot b^{x_2} = b^{x_1 + x_2}$ $\qquad\qquad$ $(ii)$ $\dfrac{b^{x_1}}{b^{x_2}} = b^{x_1 - x_2}$

$(iii)$ $\dfrac{1}{b^x} = b^{-x}$ $\qquad\qquad$ $(iv)$ $(b^{x_1})^{x_2} = b^{x_1 x_2}$

$(v)$ $(ab)^x = a^x b^x$ $\qquad\qquad$ $(vi)$ $\left(\dfrac{a}{b}\right)^x = \dfrac{a^x}{b^x}.$

---

**■ EXAMPLE 1** $\qquad\qquad$ **Rewriting a Function**

At times, we will use the laws of exponents to rewrite a function in a different form. For example, neither $f(x) = 2^{3x}$ nor $g(x) = 4^{-2x}$ have the precise form of the exponential function defined in (1). However, by the laws of exponents $f$ can be rewritten as $f(x) = 8^x$ ($b = 8$ in (1)), and $g$ can be recast as $g(x) = \left(\frac{1}{16}\right)^x$ ($b = \frac{1}{16}$ in (1)). The details are shown below:

$$f(x) = 2^{3x} \overset{\underset{\text{by }(iv)}{\downarrow}}{=} (2^3)^x \overset{\underset{\text{form is now }b^x}{\downarrow}}{=} 8^x,$$

$$g(x) = 4^{-2x} \overset{\underset{\text{by }(iv)}{\downarrow}}{=} (4^{-2})^x \overset{\underset{\text{by }(iii)}{\downarrow}}{=} \left(\frac{1}{4^2}\right)^x \overset{\underset{\text{form is now }b^x}{\downarrow}}{=} \left(\frac{1}{16}\right)^x.$$ ■

☐ **Graphs** We distinguish two types of graphs for (1) depending on whether the base $b$ satisfies $b > 1$ or $0 < b < 1$. The next two examples illustrate, in turn, the graphs of $f(x) = 3^x$ and $f(x) = \left(\frac{1}{3}\right)^x$. Before graphing, we can make some intuitive observations about both functions. Since the bases $b = 3$ and $b = \frac{1}{3}$ are positive, the values of $3^x$ and $\left(\frac{1}{3}\right)^x$ are positive for every real number $x$. Moreover, neither $3^x$ nor $\left(\frac{1}{3}\right)^x$ can be zero for any $x$ and so the graphs of $f(x) = 3^x$ and $f(x) = \left(\frac{1}{3}\right)^x$ have no $x$-intercepts. Also, $3^0 = 1$ and $\left(\frac{1}{3}\right)^0 = 1$, and so $f(0) = 1$ in each case. This means that the graphs of $f(x) = 3^x$ and $f(x) = \left(\frac{1}{3}\right)^x$ have the same $y$-intercept $(0, 1)$.

**■ EXAMPLE 2** $\qquad\qquad$ **Graph for $b > 1$**

Graph the function $f(x) = 3^x$.

$\qquad$ **Solution** We first construct a table of some function values corresponding to preselected values of $x$. As shown in **FIGURE 5.1.1**, we plot the corresponding points obtained from the table and connect them with a continuous curve. The graph shows that $f$ is an increasing function on the interval $(-\infty, \infty)$.

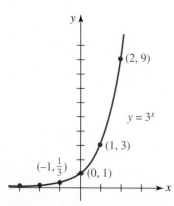

**FIGURE 5.1.1** Graph of function in Example 2

| $x$ | $-3$ | $-2$ | $-1$ | $0$ | $1$ | $2$ |
|---|---|---|---|---|---|---|
| $f(x)$ | $\frac{1}{27}$ | $\frac{1}{9}$ | $\frac{1}{3}$ | $1$ | $3$ | $9$ |

■

CHAPTER 5 EXPONENTIAL AND LOGARITHMIC FUNCTIONS

EXAMPLE 3 **Graph for $0 < b < 1$**

Graph the function $f(x) = \left(\frac{1}{3}\right)^x$.

**Solution** Proceeding as in Example 2, we construct a table of some function values corresponding to pre-selected values of $x$. Note, for example, by the laws of exponents

$$f(-2) = \left(\tfrac{1}{3}\right)^{-2} = (3^{-1})^{-2} = 3^2 = 9.$$

As shown in FIGURE 5.1.2, we plot the corresponding points obtained from the table and connect them with a continuous curve. In this case the graph shows that $f$ is a decreasing function on the interval $(-\infty, \infty)$.

| $x$ | $-3$ | $-2$ | $-1$ | $0$ | $1$ | $2$ |
|-----|------|------|------|-----|-----|-----|
| $f(x)$ | $27$ | $9$ | $3$ | $1$ | $\frac{1}{3}$ | $\frac{1}{9}$ |

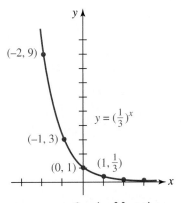

FIGURE 5.1.2 Graph of function in Example 3

Exponential functions with bases satisfying $0 < b < 1$, such as $b = \frac{1}{3}$, are frequently written in an alternative manner. We note that $y = \left(\frac{1}{3}\right)^x$ is the same as $y = 3^{-x}$. From this last result we see that the graph of $y = 3^{-x}$ is simply the graph of $y = 3^x$ reflected in the $y$-axis.

◀ For reflections in a coordinate axis, see (*ii*) on page 62 of Section 2.2

☐ **Horizontal Asymptote** FIGURE 5.1.3 illustrates the two general shapes that the graph of an exponential function $f(x) = b^x$ can have. There is, however, one more important aspect of all such graphs. Observe in Figure 5.1.3 that for $b > 1$,

$$f(x) = b^x \to 0 \quad \text{as} \quad x \to -\infty, \quad \leftarrow \text{blue graph}$$

whereas for $0 < b < 1$,

$$f(x) = b^x \to 0 \quad \text{as} \quad x \to \infty. \quad \leftarrow \text{red graph}$$

In other words, the line $y = 0$ (the $x$-axis) is a **horizontal asymptote** for both types of exponential graphs.

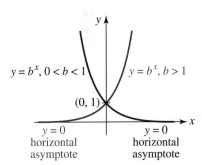

FIGURE 5.1.3 $f$ increasing for $b > 1$; $f$ decreasing for $0 < b < 1$

☐ **Properties of an Exponential Function** The following list summarizes some of the important properties of the exponential function $f$ with base $b$. Reexamine the graphs in Figures 5.1.1–5.1.3 as you read this list.

- The domain of $f$ is the set of real numbers, that is, $(-\infty, \infty)$.
- The range of $f$ is the set of positive real numbers, that is, $(0, \infty)$.
- The $y$-intercept of $f$ is $(0, 1)$. The graph of $f$ has no $x$-intercept.
- The function $f$ is increasing for $b > 1$ and decreasing for $0 < b < 1$.     (2)
- The $x$-axis, that is, $y = 0$, is a horizontal asymptote for the graph of $f$.
- The function $f$ is continuous on $(-\infty, \infty)$.
- The function $f$ is one-to-one.

Although the graphs $y = b^x$ in the case, say, when $b > 1$, all share the same basic shape and all pass through the same point $(0, 1)$, there are subtle differences. The larger the base $b$ the more steeply the graph rises as $x$ increases. In FIGURE 5.1.4 we compare the graphs of $y = 5^x$, $y = 3^x$, $y = 2^x$, and $y = (1.2)^x$ in green, blue, gold, and red, respectively, on the same coordinate axes. We see from its graph that the values of $y = (1.2)^x$ increase slowly as $x$ increases. For example, for $y = (1.2)^x$, $f(3) = (1.2)^3 = 1.728$, whereas for $y = 5^x$, $f(3) = 5^3 = 125$.

The fact that (1) is a one-to-one function follows from the horizontal line test discussed in Section 2.7. Note in Figures 5.1.1–5.1.4 a horizontal line can cross or intersect an exponential graph in at most one point.

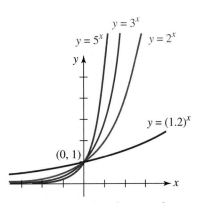

FIGURE 5.1.4 Graphs of $y = b^x$ for $b = 1.2, 2, 3, 5$

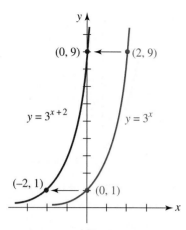

**FIGURE 5.1.5** Shifted graph in Example 4

Of course, we can obtain other kinds of graphs by rigid and nonrigid transformations, or when an exponential function is combined with other functions by either an arithmetic operation or by function composition. In the next several examples we examine variations of the exponential graph.

**EXAMPLE 4**                    **Horizontally Shifted Graph**

Graph the function $f(x) = 3^{x+2}$.

   **Solution**  From the discussion in Section 2.2 you should recognize that the graph of $f(x) = 3^{x+2}$ is the graph of $y = 3^x$ shifted 2 units to the left. Recall that since the shift is a rigid transformation to the left, the points on the graph of $f(x) = 3^{x+2}$ are the points on the graph of $y = 3^x$ moved horizontally 2 units to the left. This means that the $y$-coordinates of points $(x, y)$ on the graph of $y = 3^x$ remain unchanged, but 2 is subtracted from all the $x$-coordinates of the points. Thus we see from **FIGURE 5.1.5** that the points $(0, 1)$ and $(2, 9)$ on the graph of $y = 3^x$ are moved, in turn, to the points $(-2, 1)$ and $(0, 9)$ on the graph of $f(x) = 3^{x+2}$.                                        ▪

   The function $f(x) = 3^{x+2}$ in Example 4 can be rewritten, if desired, as $f(x) = 9 \cdot 3^x$. By (*i*) of the laws of exponents, $3^{x+2} = 3^2 3^x = 9 \cdot 3^x$. In this manner we can reinterpret the graph of $f(x) = 3^{x+2}$ as a vertical stretch of the graph of $y = 3^x$ by a factor of 9. For example, $(1, 3)$ is on the graph of $y = 3^x$, whereas, $(1, 9 \cdot 3) = (1, 27)$ is on the graph of $f(x) = 3^{x+2}$.

☐ **The Number** $e$  Most every student of mathematics has heard of, and has likely worked with, the famous irrational number $\pi = 3.141592654\ldots$. Recall that an irrational number is a nonrepeating and nonterminating decimal. In calculus and applied mathematics the irrational number

$$e = 2.718281828459\ldots$$

arguably plays a role more important than the number $\pi$. The usual definition of the number $e$ is the number that the function $f(x) = (1 + 1/x)^x$ approaches as we let $x$ become large without bound in the positive direction, that is, $f(x) \to e$ as $x \to \infty$. Using the limit notation introduced in Sections 1.5 and 2.9, we write

$$e = \lim_{x \to \infty}\left(1 + \frac{1}{x}\right)^x. \tag{3}$$

See Problems 55 and 57 in Exercises 5.1. You will often see an alternative definition of the number $e$. If we let $h = 1/x$ in (3), then as $x \to \infty$ we have simultaneously $h \to 0$. Hence an equivalent form of (3) is

$$e = \lim_{h \to 0}(1 + h)^{1/h}. \tag{4}$$

See Problems 56 and 58 in Exercises 5.1. Of course, advancing (3) and (4) as *definitions* of the number $e$ raises the obvious question: Where do these strange limits come from? An unsatisfying partial answer is: Definitions (3) and (4) come from calculus. While we cannot prove in this course that the limits in (3) and (4) exist, we will, however, discuss the origins of $e$ in Section 5.4.

☐ **The Natural Exponential Function**  When the base in (1) is chosen to be $b = e$, the function

$$f(x) = e^x \tag{5}$$

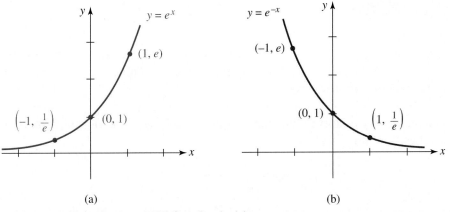

(a)

(b)

**FIGURE 5.1.6** Natural exponential function in (a)

is called the **natural exponential function**. Since $b = e > 1$ and $b = 1/e < 1$, the graphs of $y = e^x$ and $y = e^{-x}$ (or $y = (1/e)^x$) are given in **FIGURE 5.1.6**.

On the face of it, the exponential function (5) possesses no noticeable graphical characteristic that distinguishes it from, say, the function $f(x) = 3^x$, and has no special properties other than the ones given in the bulleted list (2). As mentioned, questions as to why (5) is a "natural" and frankly, the most important exponential function, can only be answered fully in courses in calculus and beyond. We will explore some of the importance of the number $e$ in Sections 5.3 and 5.4.

**EXAMPLE 5**                    **Vertically Shifted Graph**

Graph the function $f(x) = 2 - e^{-x}$. State the range.

**Solution** We first draw the graph of $y = e^{-x}$ as shown in part (a) of **FIGURE 5.1.7**. Then we reflect the first graph in the $x$-axis to obtain the graph of $y = -e^{-x}$ in part (b) of Figure 5.1.7. Finally, the graph in part (c) of Figure 5.1.7 is obtained by shifting the graph in part (b) upward 2 units.

The $y$-intercept $(0, -1)$ of $y = -e^{-x}$ when shifted upward 2 units returns us to the original $y$-intercept in part (a) of Figure 5.1.7. Finally, the horizontal asymptote $y = 0$ in parts (a) and (b) of the figure is shifted to $y = 2$ in part (c) of Figure 5.1.7. From the last graph we can conclude that the range of the function $f(x) = 2 - e^{-x}$ is the set of real numbers defined by $y < 2$, that is, the interval $(-\infty, 2)$ on the $y$-axis.

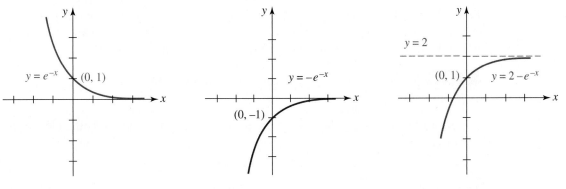

(a) Start with graph of $y = e^{-x}$       (b) Graph in (a) reflected in $x$-axis       (c) Graph in (b) shifted upward 2 units

**FIGURE 5.1.7** Graph of the function in Example 5

In the next example we graph the function composition of the natural exponential function $y = e^x$ with the simple quadratic polynomial function $y = -x^2$.

## EXAMPLE 6  A Function Composition

Graph the function $f(x) = e^{-x^2}$.

**Solution** Because $f(0) = e^{-0^2} = e^0 = 1$, the $y$-intercept of the graph is $(0, 1)$. Also, $f(x) \neq 0$ since $e^{-x^2} \neq 0$ for every real number $x$. This means that the graph of $f$ has no $x$-intercepts. Then from

$$f(-x) = e^{-(-x)^2} = e^{-x^2} = f(x)$$

we conclude that $f$ is an even function, and so its graph is symmetric with respect to the $y$-axis. Lastly, observe that

$$f(x) = \frac{1}{e^{x^2}} \to 0 \text{ as } x \to \infty.$$

By symmetry we can also conclude that $f(x) \to 0$ as $x \to -\infty$. This shows that $y = 0$ is a horizontal asymptote for the graph of $f$. The graph of $f$ is given in **FIGURE 5.1.8**. ■

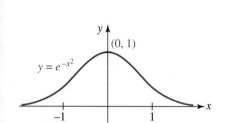

**FIGURE 5.1.8** Graph of the function in Example 6

Bell-shaped graphs such as that given in Figure 5.1.8 are very important in the study of probability and statistics.

☐ **One-to-One Property** Recall from (1) of Section 2.7 that a one-to-one function $f$ possesses the property that if $f(x_1) = f(x_2)$, then $x_1 = x_2$. Because $f(x) = b^x$ is one-to-one, we have:

$$\text{If } b^{x_1} = b^{x_2}, \quad \text{then} \quad x_1 = x_2. \tag{6}$$

As the next example shows, the property in (6) enables us to solve certain kinds of exponential equations.

## EXAMPLE 7  An Exponential Equation

Solve $2^{x-3} = 8^{x+1}$ for $x$.

**Solution** Observe on the right-hand side of the given equality that 8 can be written as a power of 2, that is, $8 = 2^3$. Furthermore, by the laws of exponents

$$\overset{\text{by } (iv) \text{ we multiply exponents}}{\overset{\downarrow \quad \downarrow}{8^{x+1} = (2^3)^{x+1} = 2^{3x+3}}}.$$

Thus, the equation is the same as

$$2^{x-3} = 2^{3x+3}.$$

From the one-to-one property (6) it follows that the exponents are equal, that is, $x - 3 = 3x + 3$. Solving for $x$ then gives $2x = -6$ or $x = -3$. You are encouraged to check this answer by substituting $-3$ for $x$ in the original equation. ■

## EXAMPLE 8  An Exponential Equation

Solve $7^{2(x+1)} = 343$ for $x$.

**Solution** By noting that $343 = 7^3$, we have the same base on both sides of the equality:

$$7^{2(x+1)} = 7^3.$$

Thus by (6) we can equate exponents and solve for $x$:

$$2(x + 1) = 3$$
$$2x + 2 = 3$$
$$2x = 1$$
$$x = \tfrac{1}{2}.$$

■

In Problems 1–12, sketch the graph of the given function $f$. Find the $y$-intercept and the horizontal asymptote of the graph. State whether the function is increasing or decreasing.

**1.** $f(x) = \left(\frac{3}{4}\right)^x$  
**3.** $f(x) = -2^x$  
**5.** $f(x) = 2^{x+1}$  
**7.** $f(x) = -5 + 3^x$  
**9.** $f(x) = 3 - \left(\frac{1}{5}\right)^x$  
**11.** $f(x) = -1 + e^{x-3}$  

**2.** $f(x) = \left(\frac{4}{3}\right)^x$  
**4.** $f(x) = -2^{-x}$  
**6.** $f(x) = 2^{2-x}$  
**8.** $f(x) = 2 + 3^{-x}$  
**10.** $f(x) = 9 - e^x$  
**12.** $f(x) = -3 - e^{x+5}$  

In Problems 13–16, find an exponential function $f(x) = b^x$ such that the graph of $f$ passes through the given point.

**13.** $(3, 216)$  
**15.** $(-1, e^2)$  

**14.** $(-1, 5)$  
**16.** $(2, e)$  

In Problems 17 and 18, determine the range of the given function.

**17.** $f(x) = 5 + e^{-x}$  

**18.** $f(x) = 4 - 2^{-x}$  

In Problems 19–24, find the $x$- and $y$-intercepts of the graph of the given function. Do not graph.

**19.** $f(x) = 2^x - 4$  
**21.** $f(x) = xe^x + 10e^x$  

**23.** $f(x) = x^3 8^x + 5x^2 8^x + 6x 8^x$  

**20.** $f(x) = -3^{2x} + 9$  
**22.** $f(x) = x^2 2^x - 2^x$  

**24.** $f(x) = \dfrac{2^x - 6 + 2^{3-x}}{x + 2}$  

In Problems 25–28, use a graph to solve the given inequality.

**25.** $2^x > 16$  
**27.** $e^{x-2} < 1$  

**26.** $e^x \leq 1$  
**28.** $\left(\frac{1}{2}\right)^x \geq 8$  

In Problems 29 and 30, use the graph in Figure 5.1.8 to sketch the graph of the given function $f$.

**29.** $f(x) = e^{-(x-3)^2}$  

**30.** $f(x) = 3 - e^{-(x+1)^2}$  

In Problems 31 and 32, use $f(-x) = f(x)$ to demonstrate that the given function is even. Sketch the graph of $f$.

**31.** $f(x) = e^{x^2}$  

**32.** $f(x) = e^{-|x|}$  

In Problems 33–36, use the graphs obtained in Problems 31 and 32 as an aid in sketching the graph of the given function $f$.

**33.** $f(x) = 1 - e^{x^2}$  
**35.** $f(x) = -e^{|x-3|}$  

**34.** $f(x) = 2 + 3e^{|x|}$  
**36.** $f(x) = e^{(x+2)^2}$  

**37.** Show that $f(x) = 2^x + 2^{-x}$ is an even function. Sketch the graph of $f$.  
**38.** Show that $f(x) = 2^x - 2^{-x}$ is an odd function. Sketch the graph of $f$.

In Problems 39–46, use the one-to-one property (6) to solve the given exponential equation.

**39.** $10^{-2x} = \dfrac{1}{1000}$

**40.** $\left(\dfrac{1}{e}\right)^x = e^3$

**41.** $8^{x-7} - 1 = 0$

**42.** $3^x - 27(\frac{1}{3})^x = 0$

**43.** $2^x \cdot 3^x = 36$

**44.** $\dfrac{4^x}{3^x} = \dfrac{9}{16}$

**45.** $3^x = 27^{x^2}$

**46.** $5^{x^2-3} = 25^x$

In Problems 47–50, either factor or use the quadratic formula to solve the given equation.

**47.** $(5^x)^2 - 26(5^x) + 25 = 0$
**48.** $64^x - 10(8^x) + 16 = 0$
**49.** $2^x + 2^{-x} = 2$
**50.** $(10^x)^2 + 10(10^x) - 1000(10^x) - 10{,}000 = 0$

In Problems 51 and 52, find the $x$-intercept of the graph of the given function.

**51.** $f(x) = e^{x+4} - e$

**52.** $f(x) = 2 - \frac{1}{5}(0.1)^x$

In Problems 53 and 54, sketch the graph of the given piecewise-defined function $f$.

**53.** $f(x) = \begin{cases} -e^x, & x < 0 \\ -e^{-x}, & x \geq 0 \end{cases}$

**54.** $f(x) = \begin{cases} e^{-x}, & x \leq 0 \\ -e^x, & x > 0 \end{cases}$

## Calculator Problems

In Problems 55 and 56, use a calculator to fill out the given table.

**55.**

| $x$ | 10 | 100 | 1000 | 10,000 | 100,000 | 1,000,000 |
|---|---|---|---|---|---|---|
| $(1 + 1/x)^x$ | | | | | | |

**56.**

| $h$ | 0.1 | 0.01 | 0.001 | 0.0001 | 0.00001 | 0.000001 |
|---|---|---|---|---|---|---|
| $(1 + h)^{1/h}$ | | | | | | |

**57. (a)** Use a graphing utility to graph the functions $f(x) = (1 + 1/x)^x$ and $g(x) = e$ on the same set of coordinate axes. Use the intervals $(0, 10]$, $(0, 100]$, $(0, 1000]$. Describe the behavior of $f$ for large values of $x$. In graphical terms, what is $g(x) = e$?

**(b)** Graph the function $f$ in part (a) on the interval $[-10, 0)$. Superimpose that graph with the graph of $f$ on $(0, 10]$ obtained in part (a). Is $f$ a continuous function?

**58.** Use a graphing utility to graph the function $f(x) = (1 + x)^{1/x}$ on the intervals $[0.1, 1]$, $[0.01, 1]$, and $[0.001, 1]$. Describe the behavior of $f$ near $x = 0$.

In Problems 59 and 60, use a graphing utility as an aid in determining the $x$-coordinates of the points of intersection of the graphs of the functions $f$ and $g$.

**59.** $f(x) = x^2, g(x) = 2^x$

**60.** $f(x) = x^3, g(x) = 3^x$

## For Discussion

**61.** Suppose $2^t = 5$ and $6^t = 2$. Answer the following questions using the laws of exponents given in this section.

**(a)** What does $12^t$ equal?      **(b)** What does $3^t$ equal?
**(c)** What does $6^{-t}$ equal?      **(d)** What does $6^{3t}$ equal?
**(e)** What does $2^{-3t}2^{7t}$ equal?      **(f)** What does $6^{t^2}$ equal?

**62.** Discuss: What does the graph of $y = e^{e^x}$ look like? Do not use a graphing utility.

In Problems 63 and 64, the given fractional expression can be decomposed into partial fractions. See Section 3.6. Discuss how this can be done and carry out your ideas.

**63.** $\dfrac{e^t}{(e^t + 1)^2(e^t - 2)}$            **64.** $\dfrac{e^{2t}}{(e^t + 1)^3}$

## 5.2 Logarithmic Functions

☐ **Introduction** Since an exponential function $y = b^x$ is one-to-one, we know that it has an inverse function. To find this inverse, we interchange the variables $x$ and $y$ to obtain $x = b^y$. This last formula defines $y$ as a function of $x$:

> *y is that exponent of the base b that produces x.*

By replacing the word *exponent* with the word *logarithm*, we can rephrase the preceding line as

> *y is that logarithm of the base b that produces x.*

This last line is abbreviated by the notation $y = \log_b x$ and is called the logarithmic function.

> **LOGARITHMIC FUNCTION**
>
> The **logarithmic function** with base $b > 0$, $b \neq 1$, is defined by
>
> $$y = \log_b x \quad \text{if and only if} \quad x = b^y. \tag{1}$$

For $b > 0$ there is no real number $y$ for which $b^y$ can either be 0 or negative. It then follows from $x = b^y$ that $x > 0$. In other words, the **domain** of a logarithmic function $y = \log_b x$ is the set of positive real numbers $(0, \infty)$.

For emphasis, all that is being said in the preceding sentences is:

> *The logarithmic expression $y = \log_b x$ and the exponential expression $x = b^y$ are equivalent,*

that is, they mean the same thing. As a consequence, within a specific context such as solving a problem, we can use whichever form happens to be more convenient. The following table lists several examples of equivalent logarithmic and exponential statements.

| Logarithmic Form | Exponential Form |
|---|---|
| $\log_3 9 = 2$ | $9 = 3^2$ |
| $\log_8 2 = \frac{1}{3}$ | $2 = 8^{1/3}$ |
| $\log_{10} 0.001 = -3$ | $0.001 = 10^{-3}$ |
| $\log_b 5 = -1$ | $5 = b^{-1}$ |

☐ **Graphs** Recall from Section 2.7 that the graph of an inverse function can be obtained by reflecting the graph of the original function in the line $y = x$. This technique was used to obtain the red graphs from the blue graphs in **FIGURE 5.2.1**. As you inspect the two graphs in Figure 5.2.1(a) and in Figure 5.2.1(b), remember that the domain $(-\infty, \infty)$ and range $(0, \infty)$ of $y = b^x$ become, in turn, the range $(-\infty, \infty)$ and domain $(0, \infty)$ of $y = \log_b x$. Also note that the $y$-intercept $(0, 1)$ for the exponential function (blue graphs) becomes the $x$-intercept $(1, 0)$ for the logarithmic function (red graphs).

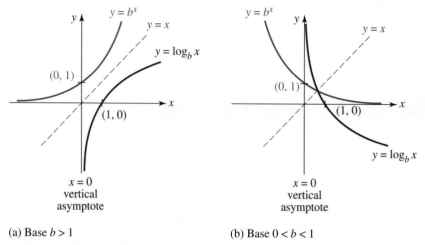

(a) Base $b > 1$                               (b) Base $0 < b < 1$

**FIGURE 5.2.1** Graphs of logarithmic functions

☐ **Vertical Asymptote** When the exponential function is reflected in the line $y = x$, the horizontal asymptote $y = 0$ for the graph of $y = b^x$ becomes a vertical asymptote for the graph of $y = \log_b x$. In Figure 5.2.1 we see that for $b > 1$,

$$\log_b x \to -\infty \quad \text{as} \quad x \to 0^+, \quad \leftarrow \text{red graph in (a)}$$

whereas for $0 < b < 1$,

$$\log_b x \to \infty \quad \text{as} \quad x \to 0^+. \quad \leftarrow \text{red graph in (b)}$$

From (7) of Section 3.5 we conclude that $x = 0$, which is the equation of the $y$-axis, is a **vertical asymptote** for the graph of $y = \log_b x$.

☐ **Properties of a Logarithmic Function** The following list summarizes some of the important properties of the logarithmic function $f(x) = \log_b x$.

- The domain of $f$ is the set of positive real numbers, that is, $(0, \infty)$.
- The range of $f$ is the set of real numbers, that is, $(-\infty, \infty)$.
- The $x$-intercept of $f$ is $(1, 0)$. The graph of $f$ has no $y$-intercept.      (2)
- The function $f$ is increasing for $b > 1$ and decreasing for $0 < b < 1$.
- The $y$-axis, that is, $x = 0$, is a vertical asymptote for the graph of $f$.
- The function $f$ is continuous on $(0, \infty)$.
- The function $f$ is one-to-one.

We would like to call attention to the third entry in the foregoing list for special emphasis:

$$\log_b 1 = 0 \quad \text{since} \quad b^0 = 1. \qquad (3)$$

Also,                  $$\log_b b = 1 \quad \text{since} \quad b^1 = b. \qquad (4)$$

Thus, in addition to $(1, 0)$, the graph of any logarithmic function (1) with base $b$ also contains the point $(b, 1)$. The equivalence of $y = \log_b x$ and $x = b^y$ also yields two sometimes-useful identities. By substituting $y = \log_b x$ into $x = b^y$ and then $x = b^y$ into $y = \log_b x$ we get:

$$x = b^{\log_b x} \quad \text{and} \quad y = \log_b b^y. \tag{5}$$

For example, from (5), $8^{\log_8 10} = 10$ and $\log_{10} 10^5 = 5$.

<h2>EXAMPLE 1      Logarithmic Graph for $b > 1$</h2>

Graph $f(x) = \log_{10}(x + 10)$.

    **Solution** This is the graph of $y = \log_{10} x$, which has the shape shown in Figure 5.2.1(a) shifted 10 units to the left. To reinforce the fact that the domain of a logarithmic function is the set of positive real numbers, we can obtain the domain of $f(x) = \log_{10}(x + 10)$ by the requirement that we must have $x + 10 > 0$ or $x > -10$. In interval notation, the domain of $f$ is $(-10, \infty)$. In the accompanying table, we have chosen convenient values of $x$ in order to plot a few points.

| $x$ | $-9$ | $0$ | $90$ |
|---|---|---|---|
| $f(x)$ | $0$ | $1$ | $2$ |

Notice,

$$f(-9) = \log_{10} 1 = 0 \quad \leftarrow \text{by (3)}$$
$$f(0) = \log_{10} 10 = 1 \quad \leftarrow \text{by (4)}$$

The vertical asymptote $x = 0$ for the graph of $y = \log_{10} x$ becomes $x = -10$ for the shifted graph. This asymptote is the dashed vertical line in FIGURE 5.2.2. ∎

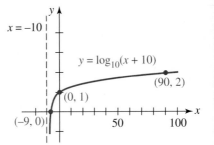

**FIGURE 5.2.2** Graph of function in Example 1

□ **Natural Logarithm** Logarithms with base $b = 10$ are called **common logarithms** and logarithms with base $b = e$ are called **natural logarithms**. Furthermore, it is customary to write the natural logarithm

$$\log_e x \quad \text{as} \quad \ln x.$$

The symbol "$\ln x$" is usually read phonetically as "ell-en of $x$." Since $b = e > 1$, the graph of $y = \ln x$ has the characteristic logarithmic shape shown in Figure 5.2.1(a). See FIGURE 5.2.3. For base $b = e$, (1) becomes

$$y = \ln x \quad \text{if and only if} \quad x = e^y. \tag{6}$$

The analogues of (3) and (4) for the natural logarithm are:

$$\ln 1 = 0 \quad \text{since} \quad e^0 = 1. \tag{7}$$
$$\ln e = 1 \quad \text{since} \quad e^1 = e. \tag{8}$$

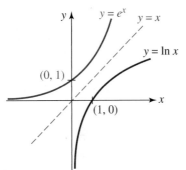

**FIGURE 5.2.3** Graph of the natural logarithm

The identities in (5) become

$$x = e^{\ln x} \quad \text{and} \quad y = \ln e^y. \tag{9}$$

For example, from (9), $e^{\ln 13} = 13$.

    Common and natural logarithms can be found on all calculators.

□ **Laws of Logarithms** The laws of exponents given in Section 5.1 can be restated equivalently as the laws of logarithms. To see this, suppose we write $M = b^{x_1}$ and $N = b^{x_2}$. Then by (1), $x_1 = \log_b M$ and $x_2 = \log_b N$.

    *Product*: By (*i*) of Section 5.1, $MN = b^{x_1 + x_2}$. Expressed as a logarithm this is $x_1 + x_2 = \log_b MN$. Substituting for $x_1$ and $x_2$ gives

$$\log_b M + \log_b N = \log_b MN.$$

    *Quotient*: By (*ii*) of Section 5.1, $M/N = b^{x_1 - x_2}$. Expressed as a logarithm this is $x_1 - x_2 = \log_b(M/N)$. Substituting for $x_1$ and $x_2$ gives

$$\log_b M - \log_b N = \log_b(M/N).$$

*Power*: By (*iv*) of Section 5.1, $M^c = b^{cx_1}$. Expressed as a logarithm this is $cx_1 = \log_b M^c$. Substituting for $x_1$ gives

$$c\log_b M = \log_b M^c.$$

For convenience and future reference, we summarize these product, quotient, and power laws of logarithms next.

---

**LAWS OF LOGARITHMS**

For any base $b > 0$, $b \neq 1$, and positive numbers $M$ and $N$:

(*i*) $\log_b MN = \log_b M + \log_b N$

(*ii*) $\log_b \left(\dfrac{M}{N}\right) = \log_b M - \log_b N$

(*iii*) $\log_b M^c = c\log_b M$, for $c$ any real number.

---

### EXAMPLE 2      Laws of Logarithms

Simplify and write as a single logarithm

$$\tfrac{1}{2}\ln 36 + 2\ln 4 - \ln 4.$$

**Solution** There are several ways to approach this problem. Note, for example, that the second and third terms can be combined arithmetically as

$$2\ln 4 - \ln 4 = \ln 4. \quad \leftarrow \text{analogous to } 2x - x = x$$

Alternatively, we can use law (*iii*) followed by law (*ii*) to combine these terms:

$$\begin{aligned}
2\ln 4 - \ln 4 &= \ln 4^2 - \ln 4 \\
&= \ln 16 - \ln 4 \\
&= \ln \tfrac{16}{4} \\
&= \ln 4.
\end{aligned}$$

Hence, 

$$\begin{aligned}
\tfrac{1}{2}\ln 36 + 2\ln 4 - \ln 4 &= \ln (36)^{1/2} + \ln 4 \quad \leftarrow \text{by }(iii) \\
&= \ln 6 + \ln 4 \\
&= \ln 24. \quad \leftarrow \text{by }(i)
\end{aligned}$$ ■

### EXAMPLE 3      Rewriting Logarithmic Expressions

Use the laws of logarithms to rewrite each expression and evaluate.

**(a)** $\ln \sqrt{e}$     **(b)** $\ln 5e$     **(c)** $\ln \dfrac{1}{e}$

**Solution**

**(a)** Since $\sqrt{e} = e^{1/2}$, we have from (*iii*) of the laws of logarithms:

$$\ln \sqrt{e} = \ln e^{1/2} = \tfrac{1}{2}\ln e = \tfrac{1}{2} \quad \leftarrow \text{from (8), } \ln e = 1$$

**(b)** From (*i*) of the laws of logarithms and a calculator:

$$\ln 5e = \ln 5 + \ln e = \ln 5 + 1 \approx 2.6094.$$

**(c)** From (*ii*) of the laws of logarithms:

$$\ln \frac{1}{e} = \ln 1 - \ln e = 0 - 1 = -1. \quad \leftarrow \text{from (7) and (8)}$$

Note that (*iii*) of the laws of logarithms can also be used here:

$$\ln\frac{1}{e} = \ln e^{-1} = (-1)\ln e = -1. \quad \leftarrow \ln e = 1 \qquad \blacksquare$$

☐ **One-to-One Property** The logarithmic analogue of the one-to-one property of exponential functions, (6) of Section 5.1, is:

$$\text{If} \quad \log_b x_1 = \log_b x_2, \quad \text{then} \quad x_1 = x_2. \qquad (10)$$

As illustrated in Section 5.1 for the exponential function $y = b^x$, the one-to-one property for the logarithmic function can be used to solve certain types of equations.

**EXAMPLE 4**        **Using the One-to-One Property**

Solve $\ln 2 + \ln(4x - 1) = \ln(2x + 5)$ for $x$.

    **Solution** By (*i*) of the laws of logarithms, the left-hand side of the equation can be written

$$\ln 2 + \ln(4x - 1) = \ln 2(4x - 1) = \ln(8x - 2).$$

The original equation is then

$$\ln(8x - 2) = \ln(2x + 5).$$

Since two logarithms with the same base are equal, it follows immediately from the one-to-one property (10) that

$$8x - 2 = 2x + 5 \quad \text{or} \quad 6x = 7 \quad \text{or} \quad x = \frac{7}{6}. \qquad \blacksquare$$

☐ **Solving Equations** Example 4 illustrates just one of several procedures that we can now use to solve a variety of exponential and logarithmic equations. Here is a brief list of equation-solving strategies:

- Use the one-to-one properties of $b^x$ and $\log_b x$.
- Rewrite an exponential expression as a logarithmic expression.
- Rewrite a logarithmic expression as an exponential expression.
- For equations $a^{x_1} = b^{x_2}$, where $a \neq b$, take the natural logarithm of both sides and simplify using (*iii*) of the laws of logarithms.

**EXAMPLE 5**        **Rewriting an Exponential Expression**

Solve $e^{10k} = 7$ for $k$.

    **Solution** We use (6) to rewrite the given exponential expression as a logarithmic expression:

$$e^{10k} = 7 \quad \text{means} \quad 10k = \ln 7.$$

Therefore, with the aid of a calculator,

$$k = \frac{1}{10}\ln 7 \approx 0.1946. \qquad \blacksquare$$

**EXAMPLE 6**        **Rewriting an Exponential Expression**

Solve $\log_2 x = 5$ for $x$.

    **Solution** We use (1) to rewrite the logarithm statement in its equivalent exponential form:

$$x = 2^5 = 32. \qquad \blacksquare$$

**Taking the ln of Both Sides**

Solve $e^{2x} = 3^{x-4}$ for $x$.

**Solution** Since the bases of the exponential expression on each side of the equality are different, one way to proceed is to take the natural logarithm (the common logarithm could also be used) of both sides. From the equality

$$\ln e^{2x} = \ln 3^{x-4}$$

and (*iii*) of the laws of logarithms, we get

$$2x \ln e = (x - 4) \ln 3.$$

Now using $\ln e = 1$ and the distributive law, the last equation becomes

$$2x = x \ln 3 - 4 \ln 3.$$

Gathering the terms involving the symbol $x$ to one side of the equality then gives

factor $x$ out of these terms

$$2x - x \ln 3 = -4 \ln 3 \quad \text{or} \quad (2 - \ln 3)x = -4 \ln 3 \quad \text{or} \quad x = \frac{-4 \ln 3}{2 - \ln 3}.$$

You are encouraged to verify the calculation that $x \approx -4.8752$.    ■

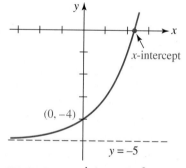

**FIGURE 5.2.4** $x$-intercept of $y = 2^x - 5$

☐ **Change of Base** Suppose we want the $x$-intercept of the graph of $y = 2^x - 5$. See **FIGURE 5.2.4**. By setting $y = 0$, we see that $x$ is the solution of the equation $2^x - 5 = 0$ or $2^x = 5$. Now a perfectly valid solution is $x = \log_2 5$. But from a computational viewpoint (that is, expressing $x$ as a number), the last answer is not desirable since no calculator has a logarithmic function with base 2. We can compute the answer by changing $\log_2 5$ to the natural logarithm by simply taking the natural log of both sides of the exponential equation $2^x = 5$:

$$\ln 2^x = \ln 5$$

*Note:* We actually divide the logarithms here →     $x \ln 2 = \ln 5$

$$x = \frac{\ln 5}{\ln 2} \approx 2.3219.$$

By the way, since we started with $x = \log_2 5$, the last result also proves the equality $\log_2 5 = \frac{\ln 5}{\ln 2}$.

■ EXAMPLE 8                    **Changing the Base**

Find the $x$ in the domain of $f(x) = 8^x$ for which $f(x) = 73$.

**Solution** We must find a solution of the equation $8^x = 73$. Taking the natural logarithm of both sides of the last equation and solving for $x$ yields

$$\ln 8^x = \ln 73 \quad \text{and so} \quad x = \frac{\ln 73}{\ln 8}.$$

With the aid of a calculator, we find $x = \frac{\ln 73}{\ln 8} \approx 2.0633$. As in the discussion that precedes this example, we have actually changed bases from $x = \log_8 73$ to $x = \frac{\ln 73}{\ln 8}$.    ■

To convert a logarithm with any base $b > 0$ to the natural logarithm, we first rewrite the logarithmic expression $x = \log_b N$ as an equivalent exponential expression $b^x = N$. We then take the natural logarithm of both sides of the last equality and solve the resulting equation $x \ln b = \ln N$ for $x$. This yields the general formula:

$$\log_b N = \frac{\ln N}{\ln b}. \tag{11}$$

# NOTES FROM THE CLASSROOM

($i$) Students often struggle with the concept of a *logarithm*. It may help if you repeat to yourself a few dozen times, "A logarithm is an exponent." It may also help if you begin reading a statement such as $3 = \log_{10} 1000$ as "3 is the exponent of 10 that. . . ."

($ii$) For logarithmic equations of the kind in Example 4, you should get accustomed to checking your answer by substituting it back into the original equation. It is possible for a logarithmic equation to have an extraneous solution. Now work Problems 55 and 56 in Exercises 5.2.

($iii$) Be *very* careful applying the laws logarithms. The logarithm does *not* distribute over addition. In other words,

$$\log_b (M + N) \neq \log_b M + \log_b N.$$

Also, $\dfrac{\log_b M}{\log_b N} \neq \log_b M - \log_b N.$

In general, there is no way that we can rewrite either

$$\log_b (M + N) \quad \text{or} \quad \frac{\log_b M}{\log_b N}.$$

($iv$) In calculus, the first step in a procedure known as *logarithmic differentiation* requires the student to take the natural logarithm of both sides of a complicated function such as $y = \dfrac{x^{10}\sqrt{x^2 + 5}}{\sqrt[3]{8x^3 + 2}}$. The idea is to use the laws of logarithms to transform powers into constant multiples, products into sums, and quotients into differences. See Problems 49–52 in Exercises 5.2.

($v$) As you advance through higher courses in mathematics, science, and engineering, you may see different notations for the natural exponential function and for the natural logarithm. For example, on some calculators you may see $y = \exp x$ instead of $y = e^x$. In the computer algebra system *Mathematica* the natural exponential function is written $\text{Exp}[x]$ and the natural logarithm is written $\text{Log}[x]$.

---

**5.2** | Exercises — Answers to selected odd-numbered problems begin on page ANS–17.

In Problems 1–6, rewrite the given exponential expression as an equivalent logarithmic expression.

**1.** $4^{-1/2} = \frac{1}{2}$

**2.** $9^0 = 1$

**3.** $10^4 = 10{,}000$

**4.** $10^{0.3010} = 2$

**5.** $t^{-s} = v$

**6.** $(a + b)^2 = a^2 + 2ab + b^2$

In Problems 7–12, rewrite the given logarithmic expression as an equivalent exponential expression.

**7.** $\log_2 128 = 7$

**8.** $\log_5 \frac{1}{25} = -2$

**9.** $\log_{\sqrt{3}} 81 = 8$

**10.** $\log_{16} 2 = \frac{1}{4}$

**11.** $\log_b u = v$

**12.** $\log_b b^2 = 2$

In Problems 13–18, find the exact value of the given logarithm.

**13.** $\log_{10}(0.0000001)$

**14.** $\log_4 64$

**15.** $\log_2(2^2 + 2^2)$

**16.** $\log_9 \frac{1}{3}$

**17.** $\ln e^e$

**18.** $\ln(e^4 e^9)$

In Problems 19–22, find the exact value of the given expression.

**19.** $10^{\log_{10} 6^2}$

**20.** $25^{\log_5 8}$

**21.** $e^{-\ln 7}$

**22.** $e^{\frac{1}{2}\ln \pi}$

In Problems 23 and 24, find a logarithmic function $f(x) = \log_b x$ such that the graph of $f$ passes through the given point.

**23.** $(49, 2)$

**24.** $(4, \frac{1}{3})$

In Problems 25–32, find the domain of the given function $f$. Find the $x$-intercept and the vertical asymptote of the graph. Sketch the graph of $f$.

**25.** $f(x) = -\log_2 x$

**26.** $f(x) = -\log_2(x + 1)$

**27.** $f(x) = \log_2(-x)$

**28.** $f(x) = \log_2(3 - x)$

**29.** $f(x) = 3 - \log_2(x + 3)$

**30.** $f(x) = 1 - 2\log_4(x - 4)$

**31.** $f(x) = -1 + \ln x$

**32.** $f(x) = 1 + \ln(x - 2)$

In Problems 33 and 34, use a graph to solve the given inequality.

**33.** $\ln(x + 1) < 0$

**34.** $\log_{10}(x + 3) > 1$

**35.** Show that $f(x) = \ln|x|$ is an even function. Sketch the graph of $f$. Find the $x$-intercepts and the vertical asymptote of the graph.

**36.** Use the graph obtained in Problem 35 to sketch the graph of $y = \ln|x - 2|$. Find the $x$-intercept and the vertical asymptote of the graph.

In Problems 37 and 38, sketch the graph of the given function $f$.

**37.** $f(x) = |\ln x|$

**38.** $f(x) = |\ln(x + 1)|$

In Problems 39–42, find the domain of the given function $f$.

**39.** $f(x) = \ln(2x - 3)$

**40.** $f(x) = \ln(3 - x)$

**41.** $f(x) = \ln(9 - x^2)$

**42.** $f(x) = \ln(x^2 - 2x)$

In Problems 43–48, use the laws of logarithms to rewrite the given expression as one logarithm.

**43.** $\log_{10} 2 + 2\log_{10} 5$

**44.** $\frac{1}{2}\log_5 49 - \frac{1}{3}\log_5 8 + 13\log_5 1$

**45.** $\ln(x^4 - 4) - \ln(x^2 + 2)$

**46.** $\ln\left(\dfrac{x}{y}\right) - 2\ln x^3 - 4\ln y$

**47.** $\ln 5 + \ln 5^2 + \ln 5^3 - \ln 5^6$

**48.** $5\ln 2 + 2\ln 3 - 3\ln 4$

In Problems 49–52, use the laws of logarithms so that $\ln y$ contains no products, quotients, or powers.

**49.** $y = \dfrac{x^{10}\sqrt{x^2 + 5}}{\sqrt[3]{8x^3 + 2}}$

**50.** $y = \sqrt{\dfrac{(2x + 1)(3x + 2)}{4x + 3}}$

**51.** $y = \dfrac{(x^3 - 3)^5(x^4 + 3x^2 + 1)^8}{\sqrt{x}(7x + 5)^9}$

**52.** $y = 64x^6\sqrt{x + 1}\sqrt[3]{x^2 + 2}$

In Problems 53–58, use the one-to-one property (10) to solve the given logarithmic equation.

**53.** $\log_2 x - \log_2 10 = \log_2 9.3$
**54.** $\ln 3 + \ln(2x - 1) = \ln 4 + \ln(x + 1)$
**55.** $\ln x + \ln(x - 2) = \ln 3$
**56.** $\log_2(x + 3) + \ln(x - 4) - \ln x = \ln 3$
**57.** $\log_2(x - 3) - \log_2(2x + 1) = -\log_2 4$
**58.** $\log_6 3x - \log_6(x + 1) = \log_6 1$

In Problems 59–62, either factor or use the quadratic formula to solve the given equation.

**59.** $(5^x)^2 - 2(5^x) - 1 = 0$
**60.** $2^{2x} - 12(2^x) + 35 = 0$
**61.** $(\ln x)^2 + \ln x = 2$
**62.** $(\log_{10} 2x)^2 = \log_{10}(2x)^2$

In Problems 63–70, use the properties of logarithms to solve the given equation.

**63.** $\log_{10} \dfrac{1}{x} = 2$
**64.** $2 - \log_3 \sqrt{x^2 + 17} = 0$
**65.** $\log_2(\log_3 x) = 2$
**66.** $\log_5 |1 - x| = 1$
**67.** $\log_2(10x - x^2) = 4$
**68.** $\log_3(x^2 + 1) = 4$
**69.** $\log_3 81^x - \log_3 3^{2x} = 3$
**70.** $\dfrac{\log_2 8^x}{\log_2 \frac{1}{4}} = \dfrac{1}{2}$

In Problems 71 and 72, use the natural logarithm to find $x$ in the domain of the given function for which $f$ takes on the indicated value.

**71.** $f(x) = 6^x;\ f(x) = 51$
**72.** $f(x) = \left(\tfrac{1}{2}\right)^x;\ f(x) = 7$

In Problems 73–76, use the natural logarithm to find $x$.

**73.** $2^{x+5} = 9$
**74.** $4 \cdot 7^{2x} = 9$
**75.** $5^x = 2e^{x+1}$
**76.** $3^{2(x-1)} = 2^{x-3}$

**77.** In science it is sometimes useful to display data using logarithmic coordinates. Which of the following equations determines the graph shown in **FIGURE 5.2.5**?
  (*i*) $y = 2x + 1$    (*ii*) $y = e + x^2$
  (*iii*) $y = ex^2$    (*iv*) $x^2 y = e$

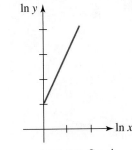

**FIGURE 5.2.5** Graph for Problem 77

## For Discussion

**78.** If $a > 0$ and $b > 0$, $a \neq b$, then $\log_a x$ is a constant multiple of $\log_b x$. That is, $\log_a x = k \log_b x$. Find $k$.
**79.** Show that $(\log_{10} e)(\log_e 10) = 1$. Can you generalize this result?
**80.** Discuss: How can the graphs of the given function be obtained from the graph of $f(x) = \ln x$ by means of a rigid transformation (a shift or a reflection)?

  **(a)** $y = \ln 5x$    **(b)** $y = \ln \dfrac{x}{4}$    **(c)** $y = \ln x^{-1}$    **(d)** $y = \ln(-x)$

# 5.3 Exponential and Logarithmic Models

□ **Introduction** In this section we consider some **mathematical models**. Roughly speaking, a mathematical model is a mathematical description of something that we will call a *system*. To construct a mathematical model we start with a set of reasonable assumptions about the system that we are trying to describe. These assumptions include any empirical laws that are applicable to the system. The end result could be a description as simple as a single function.

□ **Exponential Models** In the physical sciences, the exponential expression $Ce^{kt}$, where $C$ and $k$ are constants, frequently appears in mathematical models of systems that change with time $t$. As a consequence, mathematical models are often used to predict a future state of a system. For example, extremely complicated mathematical models are used to predict the weather over various regions of the country for, say, the next week.

□ **Population Growth** In one model of a growing population, it is assumed that the *rate* of growth of the population is proportional to the *number present* at time $t$. If $P(t)$ denotes the population or number present at time $t$, then with the aid of calculus it can be shown that this assumption gives rise to

$$P(t) = P_0 e^{kt}, \; k > 0, \tag{1}$$

where $t$ is time, and $P_0$ and $k$ are constants. The function (1) is used to describe the growth of populations of bacteria, small animals, and, in some rare circumstances, humans. Setting $t = 0$ gives $P(0) = P_0$, and so $P_0$ is called the **initial population**. The constant $k > 0$ is called the **growth constant** or **growth rate**. Since $e^{kt}$, $k > 0$, is an increasing function on the interval $[0,\infty)$ the model in (1) describes uninhibited growth.

### EXAMPLE 1      Bacterial Growth

It is known that the doubling time* of *E. coli* bacteria, which reside in the large intestine of healthy people, is just 20 minutes. Use the exponential growth model (1) to find the number of *E. coli* bacteria in a culture after 6 hours.

**Solution** Let us use hours as our unit of time, so that 20 min $= \frac{1}{3}$h. Because the initial number of *E. coli* in the culture is not specified, we will simply denote the initial size of the culture as $P_0$. Now using (1), a function interpretation of the first sentence in this example is $P(\frac{1}{3}) = 2P_0$. This means $P_0 e^{k/3} = 2P_0$ or $e^{k/3} = 2$. Solving this last equation for $k$ gives the growth constant

$$\frac{k}{3} = \ln 2 \quad \text{or} \quad k = 3\ln 2 \approx 2.0794.$$

A model for the size of the culture after $t$ hours is then $P(t) = P_0 e^{2.0794t}$. Setting $t = 6$ gives $P(6) = P_0 e^{2.0794(6)} \approx 262{,}144 P_0$. Put another way, if the culture consists of only *one* bacterium at $t = 0$, then (with $P_0 = 1$) the model predicts that there will be 262,144 cells 6 hours later. ■

In the early nineteenth century the English clergyman and economist Thomas R. Malthus used the growth model (1) to predict the world population. For specific values

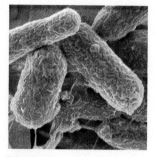

*E. coli* bacteria

When working problems such as this, be ▶ sure to store the value of $k$ in the memory of your calculator.

---

* In biology the doubling time is sometimes referred to as the **generation time**.

of $P_0$ and $k$, the function values $P(t)$ were actually reasonable approximations to the world population for a period of time during the nineteenth century. Since $P(t)$ is an increasing function, Malthus predicted that the future population growth would surpass the world's ability to produce food. As a consequence he also predicted wars and worldwide famine. More a doomsayer than a seer, Malthus failed to foresee that the food supply would keep pace with the increased population through simultaneous advances in science and technology.

In 1840 a more realistic model for predicting human populations in small countries was advanced by the Belgian mathematician/biologist P.F. Verhulst. The so-called **logistic function**

$$P(t) = \frac{K}{1 + ce^{rt}}, \quad r < 0, \tag{2}$$

Thomas R. Malthus (1776–1834)

where $K$, $c$, and $r$ are constants, has over the years proved to be an accurate growth model for populations of protozoa, bacteria, fruit flies, water fleas, and animals confined to limited spaces. In contrast to uninhibited growth of the Malthusian model (1), (2) exhibits bounded growth. More specifically, the population predicted by (2) will not increase beyond the number $K$, called the **carrying capacity** of the system. For $r < 0$, $e^{rt} \to 0$ and $P(t) \to K$ as $t \to \infty$. You are asked to graph a special case of (2) in Problems 7 and 8 in Exercises 5.3.

☐ **Radioactive Decay** Element 88, better known as **radium**, is radioactive. This means that a radium atom spontaneously **decays**, or disintegrates, by emitting radiation in the form of alpha particles, beta particles, and gamma rays. When an atom disintegrates in this manner, its nucleus is transmuted into a nucleus of another element. The nucleus of the radium atom is transmuted into the nucleus of a radon atom. Radon is an odorless and colorless, but highly dangerous, radioactive gas that usually originates in the ground. Since it can penetrate a sealed concrete floor, radon frequently accumulates in basements of some new and highly insulated homes. Some medical organizations have claimed that radon is the second leading cause of lung cancer.

If it is assumed that the rate of decay of a radioactive substance is proportional to the amount remaining or present at time $t$, then we arrive at basically the same model as in (1). The important difference is that $k < 0$. If $A(t)$ represents the amount of the decaying substance that remains at time $t$, then

Madame Curie (1867–1934), discoverer of radium

$$A(t) = A_0 e^{kt}, k < 0, \tag{3}$$

where $A_0$ is the initial amount of the substance present, that is, $A(0) = A_0$. The constant $k < 0$ in (3) is called the **decay constant** or **decay rate**.

## EXAMPLE 2    Decay of Radium

Suppose there are 20 grams of radium on hand initially. After $t$ years the amount remaining is modeled by the function $A(t) = 20e^{-0.000418t}$. Find the amount of radium remaining after 100 years. What percent of the original 20 grams has decayed after 100 years?

**Solution** Using a calculator, we find that after 100 years there remains

$$A(100) = 20e^{-0.000418(100)} \approx 19.18 \text{ g.}$$

Thus, only

$$\frac{20 - 19.18}{20} \times 100\% = 4.1\%$$

of the initial 20 grams has decayed. ∎

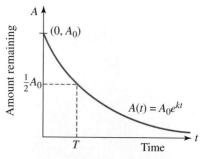

FIGURE 5.3.1 Time $T$ is the half-life

□ **Half-Life** The half-life of a radioactive substance is the time $T$ it takes for one-half of a given amount of that element to disintegrate and change into a new element. See FIGURE 5.3.1. Half-life is a measure of the stability of an element, that is, the shorter the half-life, the more unstable the element. For example, the half-life of the highly radioactive strontium 90, Sr-90, produced in nuclear explosions, is 29 days, whereas the half-life of the uranium isotope U-238 is 4,560,000 years. The half-life of californium, Cf-244, first discovered in 1950, is only 45 minutes. Polonium, Po-213, has a half-life of 0.000001 second.

**EXAMPLE 3**      **Half-life of Radium**

Use the exponential model in Example 2 to determine the half-life of radium.

**Solution** If $A(t) = 20e^{-0.000418t}$, then we must find the time $T$ for which

$$A(T) = \tfrac{1}{2}(20) = 10.$$

(one-half the initial amount)

From $20e^{-0.000418T} = 10$ we get $e^{-0.000418T} = \tfrac{1}{2}$. By rewriting the last expression in the logarithmic form $-0.000418T = \ln\tfrac{1}{2}$ we can solve for $T$:

$$T = \frac{\ln\tfrac{1}{2}}{-0.000418} \approx 1660 \text{ years.}$$ ∎

A careful reading of Example 3 reveals that the initial amount present plays no part in the actual calculation of the half-life. Since the solution of $A(T) = A_0 e^{-0.000418T} = \tfrac{1}{2}A_0$ leads to $e^{-0.000418T} = \tfrac{1}{2}$, we see that $T$ is independent of $A_0$. Thus the half-life of 1 gram, 20 grams, or 10,000 grams of radium is the same. It takes about 1660 years for one-half of *any* given quantity of radium to transmute into radon.

Medications also have half-lives. In this case, the half-life of a drug is the time $T$ that it takes for the body to eliminate, by metabolism or excretion, one-half of the amount of the drug taken. For example, the most popular NSAIDs (nonsteroidal anti-inflammatory drugs, such as aspirin and ibuprofen) taken for the relief of continuing pain have relatively short half-lives of a few hours, and as a consequence must be taken several times a day. The NSAID naproxen has a longer half-life and is usually taken once every 12 hours. See Problem 31 in Exercises 5.3.

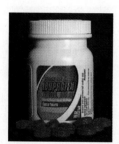

Ibuprofen is an NSAID

□ **Carbon Dating** The approximate age of fossils of once-living matter can be determined by a method known as **carbon dating**. The radioactive isotope of carbon, carbon 14 or C-14, is formed presumably at a constant rate in the atmosphere by the interaction of cosmic rays on nitrogen 14. The carbon dating method, invented by the chemist Willard Libby around 1950, is based on the fact that a plant or an animal absorbs C-14 through the process of breathing and eating, and ceases to absorb C-14 when it dies. As the next example shows, the carbon dating procedure is based on the knowledge that the half-life of C-14 is about 5730 years. Carbon 14 decays back to the original nitrogen 14.

Willard Libby (1908–1980)

The Psalms scroll

Libby won the 1960 Nobel prize in chemistry for his work. Libby's method has been used to date wooden furniture found in Egyptian tombs, the Dead Sea scrolls written on papyrus and animal skin, the famous linen Shroud of Turin, and a recently discovered copy of the Gnostic Gospel of Judas written on papyrus.

## EXAMPLE 4      Carbon Dating a Fossil

A fossilized bone is found to contain $\frac{1}{1000}$ of the initial amount of C-14 that the organism contained while it was alive. Determine the approximate age of the fossil.

**Solution** If there was an initial amount of $A_0$ grams of C-14 in the organism, then $t$ years after its death there are $A(t) = A_0 e^{kt}$ grams remaining. When $t = 5730$, $A(5730) = \frac{1}{2} A_0$, and so $\frac{1}{2} A_0 = A_0 e^{5730k}$. Solving this last equation for the decay constant $k$ gives

$$e^{5740k} = \frac{1}{2} \quad \text{and so} \quad k = \frac{\ln\frac{1}{2}}{5730} \approx -0.00012097.$$

Hence a model for the amount of C-14 remaining is $A(t) = A_0 e^{-0.00012097t}$. Using this model, we now solve $A(t) = \frac{1}{1000} A_0$ for $t$:

$$A_0 e^{-0.00012097t} = \frac{1}{1000} A_0 \quad \text{implies} \quad t = \frac{\ln\frac{1}{1000}}{-0.00012097} \approx 57{,}100 \text{ years.} \quad \blacksquare$$

The age determined in the last example is actually beyond the border of accuracy for the carbon 14 dating method. After 9 half-lives of the isotope, or about 52,000 years, about 99.7% of carbon 14 has decayed, making its measurement in a fossil nearly impossible.

☐ **Newton's Law of Cooling/Warming** Suppose an object or body is placed in a medium (air, water, etc.) that is held at constant temperature $T_m$, called the **ambient temperature**. If the initial temperature $T_0$ of the body or object at the moment it is placed into the medium is greater than the ambient temperature $T_m$, then the body will cool. On the other hand, if $T_0$ is less than $T_m$, then it will warm up. For example, in an office kept at, say, 70°F, a steaming cup of coffee will cool off, whereas a glass of ice water will warm up. The usual cooling/warming assumption is that the rate at which an object cools/warms is proportional to the difference $T(t) - T_m$, where $T(t)$ represents the temperature of the object at time $t$. In either case, cooling or warming, this assumption leads to $T(t) - T_m = (T_0 - T_m)e^{kt}$, where $k$ is a negative constant. Observe that since $e^{kt} \to 0$ for $k < 0$, the last expression is consistent with one's intuitive expectation that $T(t) - T_m \to 0$, or equivalently $T(t) \to T_m$, as $t \to \infty$ (the coffee cools to room temperature; the ice water warms to room temperature). Solving for $T(t)$ we obtain a function for the temperature of the object,

$$T(t) = T_m + (T_0 - T_m)e^{kt}, \, k < 0. \tag{4}$$

The mathematical model in (4), named after its discoverer, is called **Newton's law of cooling/warming**.

## EXAMPLE 5      Cooling of a Cake

A cake is removed from an oven where the temperature was 350°F into a kitchen where the temperature is 75°F. One minute later the temperature of the cake is measured to be 300°F.

   (a) What is the temperature of the cake after 6 minutes?
   (b) At what time is the temperature of the cake 80°F?
   (c) Graph $T(t)$.

Cake will cool

**Solution**

**(a)** When the cake is removed from the oven its temperature is also 350°F, that is, $T_0 = 350$. The ambient temperature is the temperature of the kitchen, $T_m = 75$. Thus (4) becomes $T(t) = 75 + 275e^{kt}$. The measurement that $T(1) = 300$ is the condition that determines $k$. From $T(1) = 75 + 275e^k = 300$ we find

$$e^k = \frac{225}{275} = \frac{9}{11} \quad \text{or} \quad k = \ln\frac{9}{11} \approx -0.2007.$$

From the model $T(t) = 75 + 275e^{-0.2007t}$ we then find

$$T(6) = 75 + 275e^{-0.2007(6)} \approx 157.5°\text{F}. \tag{5}$$

**(b)** To determine when the temperature of the cake will be 80°F, we solve the equation $T(t) = 80$ for $t$. Rewriting $T(t) = 75 + 275e^{-0.2007t} = 80$ as

$$e^{-0.2007t} = \frac{5}{275} = \frac{1}{55} \quad \text{we find} \quad t = \frac{\ln\frac{1}{55}}{-0.2007} \approx 20 \text{ min}.$$

**(c)** With the aid of a graphing utility we obtain the graph of $T(t)$ shown in blue in **FIGURE 5.3.2**. Since $T(t) = 75 + 275e^{-0.2007t} \rightarrow 75$ as $t \rightarrow \infty$, $T = 75$, shown in red in Figure 5.3.2, is a horizontal asymptote for the graph of $T(t) = 75 + 275e^{-0.2007t}$. ∎

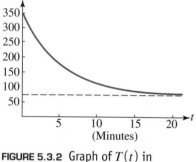

**FIGURE 5.3.2** Graph of $T(t)$ in Example 5

☐ **Compound Interest** Investments such as savings accounts pay an annual rate of interest that can be compounded annually, quarterly, monthly, weekly, daily, and so on. In general, if a principal of $P$ dollars is invested at an annual rate $r$ of interest that is compounded $n$ times a year, then the amount $S$ accrued at the end of $t$ years is given by

$$S = P\left(1 + \frac{r}{n}\right)^{nt}. \tag{6}$$

$S$ is called the **future value** of the principal $P$. If the number $n$ is increased without bound, then interest is said to be **compounded continuously**. To find the future value of $P$ in this case, we let $m = n/r$. Then $n = mr$ and

$$\left(1 + \frac{r}{n}\right)^{nt} = \left(1 + \frac{1}{m}\right)^{mrt} = \left[\left(1 + \frac{1}{m}\right)^m\right]^{rt}.$$

Since $n \rightarrow \infty$ implies that $m \rightarrow \infty$, we see from (3) of Section 5.1 that $(1 + 1/m)^m \rightarrow e$. The right-hand side of (6) becomes

$$P\left[\left(1 + \frac{1}{m}\right)^m\right]^{rt} \rightarrow P[e]^{rt} \quad \text{as} \quad m \rightarrow \infty.$$

Thus, if an annual rate $r$ of interest is compounded continuously, the future value $S$ of a principal $P$ in $t$ years is

$$S = Pe^{rt}. \tag{7}$$

EXAMPLE 6 **Comparing Future Values**

Suppose that $1000 is deposited in a savings account whose annual rate of interest is 3%. Compare the future value of this principal in 10 years (**a**) if interest is compounded monthly and (**b**) if interest is compounded continuously.

**Solution**

(**a**) Since there are 12 months in a year, we identify $n = 12$. Furthermore, with $P = 1000, r = 0.03$, and $t = 10$, (6) becomes

$$S = 1000\left(1 + \frac{0.03}{12}\right)^{12(10)} = 1000(1.0025)^{120} \approx \$1,349.35.$$

(**b**) From (7),

$$S = 1000e^{(0.03)(10)} = 1000e^{0.3} \approx \$1,349.86.$$

Thus over 10 years we have gained $0.51 by compounding continuously rather than monthly. ∎

☐ **Logarithmic Models** Probably the most famous application of the base 10 logarithm, or common logarithm, is the **Richter scale**. In 1935 the American seismologist Charles F. Richter devised a logarithmic scale for comparing the energies of different earthquakes. The magnitude $M$ of an earthquake is defined by

$$M = \log_{10}\frac{A}{A_0}, \tag{8}$$

where $A$ is the amplitude of the largest seismic wave of the earthquake and $A_0$ is a reference amplitude that corresponds to the magnitude $M = 0$. The number $M$ is calculated to one decimal place. Earthquakes of magnitude 6 or greater are considered potentially destructive.

Charles F. Richter (1900–1985)

EXAMPLE 7 **Comparing Intensities**

The earthquake on December 26, 2004, off of the west coast of Northern Sumatra, which spawned a tsunami causing over 200,000 deaths, was initially classified a 9.3 on the Richter scale. On March 28, 2005, an aftershock in the same area was classified as an 8.7 on the Richter scale. How many times more intense was the 2004 earthquake?

**Solution** From (8) we have

$$9.3 = \log_{10}\left(\frac{A}{A_0}\right)_{2004} \quad \text{and} \quad 8.7 = \log_{10}\left(\frac{A}{A_0}\right)_{2005}.$$

This means, in turn, that

$$\left(\frac{A}{A_0}\right)_{2004} = 10^{9.3} \quad \text{and} \quad \left(\frac{A}{A_0}\right)_{2005} = 10^{8.7}.$$

Now, since $9.3 = 8.7 + 0.6$, it follows from the laws of exponents that

$$\left(\frac{A}{A_0}\right)_{2004} = 10^{9.3} = 10^{0.6}10^{8.7} = 10^{0.6}\left(\frac{A}{A_0}\right)_{2005} \approx 3.98\left(\frac{A}{A_0}\right)_{2005}.$$

Thus the original earthquake in 2004 was approximately 4 times as intense as the after-shock in 2004. ∎

You can see from Example 7 that if, say, one earthquake is a 6.0 and another is a 4.0 on the Richter scale, then the 6.0 earthquake is $10^2 = 100$ times more intense than the 4.0 earthquake.

☐ **pH of a Solution**  In chemistry, the hydrogen potential, or **pH**, of a solution is defined as

$$pH = -\log_{10}[H^+], \tag{9}$$

where the symbol $[H^+]$ denotes the concentration of hydrogen ions in solution meas-ured in moles per liter. The pH scale was invented in 1909 by the Danish biochemist Soren Sorensen. Solutions are classified according to their pH value as *acidic*, *base*, or *neu-tral*. A solution with a pH in the range $0 < pH < 7$ is said to be acid; when $pH > 7$, the solution is base (or alkaline). In the case when $pH = 7$, the solution is neutral. Water, if uncontaminated by other solutions or by acid rain, is an example of a neutral solution, whereas undiluted lemon juice is highly acid and has a pH in the range $pH \le 3$. A solu-tion with $pH = 6$ is ten times more acidic than a neutral solution. See Problems 47–50 in Exercises 5.3.

As the next example illustrates, pH values are usually calculated to one decimal place.

**EXAMPLE 8**        **pH of Human Blood**

The concentration of hydrogen ions in the blood of a healthy person is found to be $[H^+] = 3.98 \times 10^{-8}$ moles/liter. Find the pH of blood.

**Solution**  From (9) and the laws of logarithms,

$$
\begin{aligned}
pH &= -\log_{10}[3.98 \times 10^{-8}] \\
&= -[\log_{10}3.98 + \log_{10}10^{-8}] \\
&= -[\log_{10}3.98 - 8\log_{10}10] \quad \leftarrow \log_{10}10 = 1 \\
&= -[\log_{10}3.98 - 8].
\end{aligned}
$$

With the help of the base-10 log key on a calculator, we find that

$$pH \approx -[0.5999 - 8] \approx 7.4. \quad ∎$$

Human blood in usually a base solution. The pH values of blood usually fall within the rather narrow range $7.2 < pH < 7.6$. A person with a blood pH outside these lim-its can suffer illness and even death.

**5.3**    Exercises    Answers to selected odd-numbered problems begin on page ANS–17.

**Population Growth**

1. After 2 hours the number of bacteria in a culture is observed to have doubled.
   (a) Find an exponential model (1) for the number of bacteria in the culture at time $t$.
   (b) Find the number of bacteria present in the culture after 5 hours.
   (c) Find the time that it takes the culture to grow to 20 times its initial size.

**2.** A model for the number of bacteria in a culture after $t$ hours is given by (1).
   **(a)** Find the growth constant $k$ if it is known that after 1 hour the colony has expanded to 1.5 times its initial population.
   **(b)** Find the time that it takes for the culture to quadruple in size.

**3.** A model for the population in a small community is given by $P(t) = 1500e^{kt}$. If the initial population increases by 25% in 10 years, what will the population be in 20 years?

**4.** A model for the population in a small community after $t$ years is given by (1).
   **(a)** If the initial population has doubled in 5 years, how long will it take to triple? To quadruple?
   **(b)** If the population of the community in part (a) is 10,000 after 3 years, what was the initial population?

**5.** A model for the number of bacteria in a culture after $t$ hours is given by $P(t) = P_0 e^{kt}$. After 3 hours it is observed that 400 bacteria are present. After 10 hours 2000 bacteria are present. What was the initial number of bacteria?

**6.** In genetic research a small colony of *drosophila* (small two-winged fruit flies) is grown in a laboratory environment. After 2 days it is observed that the population of flies in the colony has increased to 200. After 5 days the colony has 400 flies.
   **(a)** Find a model $P(t) = P_0 e^{kt}$ for the population of the fruit-fly colony after $t$ days.
   **(b)** What will be the population of the colony in 10 days?
   **(c)** When will the population of the colony be 5000 fruit flies?

**7.** A student sick with a flu virus returns to an isolated college campus of 2000 students. The number of students infected with the flu $t$ days after the student's return is predicted by the logistic function

$$P(t) = \frac{2000}{1 + 1999e^{-0.8905t}}.$$

   **(a)** According to this model, how many students will be infected with the flu after 5 days?
   **(b)** How long will it take for one-half of the student population to become infected?
   **(c)** How many students does the model predict will become infected after a very long period of time?
   **(d)** Sketch a graph of $P(t)$.

**8.** In 1920, Pearl and Reed proposed a logistic model for the population of the United States based on the years 1790, 1850, and 1910. The logistic function they proposed was

$$P(t) = \frac{2930.3009}{0.014854 + e^{-0.0313395t}},$$

   where $P$ is measured in thousands and $t$ represents the number of years past 1780.
   **(a)** The model agrees quite well with the census figures between 1790 and 1910. Determine the population figures for 1790, 1850, and 1910.
   **(b)** What does this model predict for the population of the United States after a very long time? How does this prediction compare with the 2000 census population of 281 million?

### Radioactive Decay and Half-Life

**9.** Initially 200 milligrams of a radioactive substance was present. After 6 hours the mass had decreased by 3%. Construct an exponential model $A(t) = A_0 e^{kt}$ for the amount remaining of the decaying substance after $t$ hours. Find the amount remaining after 24 hours.

**10.** Determine the half-life of the substance in Problem 9.

**11.** Do this problem without using the exponential model (2). Initially there are 400 grams of a radioactive substance on hand. If the half-life of the substance is 8 hours, give an educated guess of how much remains (approximately) after 17 hours. After 23 hours. After 33 hours.

**12.** Construct an exponential model $A(t) = A_0 e^{kt}$ for the amount remaining of the decaying substance in Problem 11. Compare the predicted values $A(17)$, $A(23)$, and $A(33)$ with your guesses.

**13.** Iodine 131, used in nuclear medicine procedures, is radioactive and has a half-life of 8 days. Find the decay constant $k$ for iodine 131. If the amount remaining of an initial sample after $t$ days is given by the exponential model $A(t) = A_0 e^{kt}$, how long will it take for 95% of the sample to decay?

**14.** The amount remaining of a radioactive substance after $t$ hours is given by $A(t) = 100e^{kt}$. After 12 hours, the initial amount has decreased by 7%. How much remains after 48 hours? What is the half-life of the substance?

**15.** The half-life of polonium 210, Po-210, is 140 days. If $A(t) = A_0 e^{kt}$ represents the amount of Po-210 remaining after $t$ days, what is the amount remaining after 80 days? After 300 days?

**16.** Strontium 90 is a dangerous radioactive substance found in acid rain. As such it can make its way into the food chain by polluting the grass in a pasture on which milk cows graze. The half-life of strontium 90 is 29 years.

   **(a)** Find an exponential model (2) for the amount remaining after $t$ years.

   **(b)** Suppose a pasture is found to contain Str-90 that is 3 times a safe level $A_0$. How long will it be before the pasture can be used again for grazing cows?

## Carbon Dating

**17.** Charcoal drawings were discovered on walls and ceilings in a cave in Lascaux, France. Determine the approximate age of the drawings if it was found that 86% of C-14 in a piece of charcoal found in the cave had decayed through radioactivity.

Charcoal drawing in Problem 17

**18.** Analysis on an animal bone fossil at an archeological site reveals that the bone has lost between 90% and 95% of C-14. Give an interval for the possible ages of the bone.

**19.** The shroud of Turin shows the negative image of the body of a man who appears to have been crucified. It is believed by many to be the burial shroud of Jesus of Nazareth. In 1988 the Vatican granted permission to have the shroud carbon dated. Several independent scientific laboratories analyzed the cloth and the consensus opinion was that the shroud is approximately 660 years old, an age consistent with its historical appearance. This age has been disputed by many scholars. Using this age, determine what percentage of the original amount of C-14 remained in the cloth as of 1988.

Shroud image in Problem 19

CHAPTER 5 EXPONENTIAL AND LOGARITHMIC FUNCTIONS

**20.** In 1991 hikers found a preserved body of a man partially frozen in a glacier in the Austrian Alps. Through carbon dating techniques it was found that the body of Ötzi—as the iceman came to be called—contained 53% as much C-14 as found in a living person. What is the approximate date of his death?

The iceman in Problem 20

## Newton's Law of Cooling/Warming

**21.** Suppose a pizza is removed from an oven at 400°F into a kitchen whose temperature is a constant 80°F. Three minutes later the temperature of the pizza is found to be 275°F.
   **(a)** What is the temperature $T(t)$ of the pizza after 5 minutes?
   **(b)** Determine the time when the temperature of the pizza is 150°F.
   **(c)** After a very long period of time, what is the approximate temperature of the pizza?

**22.** A glass of cold water is removed from a refrigerator whose interior temperature is 39°F into a room maintained at 72°F. One minute later the temperature of the water is 43°F. What is the temperature of the water after 10 minutes? After 25 minutes?

**23.** A thermometer is brought from the outside, where the air temperature is −20°F, into a room where the air temperature is a constant 70°F. After one minute inside the room the thermometer reads 0°F. How long will it take for the thermometer to read 60°F?

**24.** A thermometer is taken from inside a house to the outside, where the air temperature is 5°F. After one minute outside the thermometer reads 59°F, and after 5 minutes it reads 32°F. What is the temperature inside the house?

Thermometer in Problem 24

**25.** A dead body was found within a closed room of a house where the temperature was a constant 70°F. At the time of discovery, the core temperature of the body was determined to be 85°F. One hour later a second measurement showed that the core temperature of the body was 80°F. Assume that the time of death corresponds to $t = 0$ and that the core temperature at that time was 98.6°F. Determine how many hours elapsed before the body was found.

**26.** Repeat Problem 25 if evidence indicated that the dead person was running a fever of 102°F at the time of death.

## Compound Interest

**27.** Suppose that 1¢ is deposited in a savings account paying 1% annual interest compounded continuously. How much money will have accrued in the account after 2000 years? What is the future value of 1¢ in 2000 years if the account pays 2% annual interest compounded continuously?

**28.** Suppose that $100,000 is invested at an annual rate of interest of 5%. Use (6) and (7) to compare the future values of that amount in 1 year by completing the following table.

| Interest Compounded | $n$ | Future Value $S$ |
|---|---|---|
| Annually | 1 | |
| Semiannually | 2 | |
| Quarterly | 4 | |
| Monthly | 12 | |
| Weekly | 52 | |
| Daily | 365 | |
| Hourly | 8760 | |
| Continuously | $n \to \infty$ | |

**29.** Suppose that $5000 is deposited in a savings account paying 6% annual interest compounded continuously. How much interest will be earned in 8 years?

**30.** If (7) is solved for $P$, that is, $P = Se^{-rt}$, we obtain the amount that should be invested now at an annual rate $r$ of interest in order to be worth $S$ dollars after $t$ years. We say that $P$ is the **present value** of the amount $S$. What is the present value of $100,000 at an annual rate of 3% compounded continuously for 30 years?

## Miscellaneous Exponential Models

**31. Effective Half-life** Radioactive substances are removed from living organisms by two processes: natural physical decay and biological metabolism. Each process contributes to an effective half-life $E$ that is defined by

$$1/E = 1/P + 1/B,$$

where $P$ is the physical half-life of the radioactive substance and $B$ is the biological half-life.

(a) Radioactive iodine, I-131, is used to treat hyperthyroidism (overactive thyroid). It is known that for human thyroids, $P = 8$ days and $B = 24$ days. Find the effective half-life of I-131.

(b) Suppose the amount of I-131 in the human thyroid after $t$ days is modeled by $A(t) = A_0 e^{kt}$, $k < 0$. Use the effective half-life found in part (a) to determine the percentage of radioactive iodine remaining in the human thyroid gland two weeks after its ingestion.

**32. Newton's Law of Cooling Revisited** The rate at which a body cools also depends on its exposed surface area $S$. If $S$ is a constant, then a modification of (4) is

$$T(t) = T_m + (T_0 - T_m)e^{kSt}, \quad k < 0.$$

Suppose two cups $A$ and $B$ are filled with coffee at the same time. Initially the temperature of the coffee is 150°F. The exposed surface area of the coffee in cup $B$ is twice the surface area of the coffee in cup $A$. After 30 min, the temperature of the coffee in cup $A$ is 100°F. If $T_m = 70$°F, what is the temperature of the coffee in cup $B$ after 30 min?

33. **Series Circuit** In a simple series circuit consisting of a constant voltage $E$, an inductance of $L$ henries, and a resistance of $R$ ohms, it can be shown that the current $I(t)$ is given by

$$I(t) = \frac{E}{R}(1 - e^{-(R/L)t}).$$

Solve for $t$ in terms of the other symbols.

34. **Drug Concentration** Under some conditions the concentration of a drug at time $t$ after injection is given by

$$C(t) = \frac{a}{b} + \left(C_0 - \frac{a}{b}\right)e^{-bt}.$$

Here $a$ and $b$ are positive constants and $C_0$ is the concentration of the drug at $t = 0$. Determine the steady-state concentration of a drug, that is, the limiting value of $C(t)$ as $t \to \infty$. Determine the time $t$ at which $C(t)$ is one-half the steady-state concentration.

## Richter Scale

35. Two of the most devastating earthquakes in the San Francisco Bay area occurred in 1906 along the San Andreas fault and in 1989 in the Santa Cruz Mountains near Loma Prieta peak. The 1906 and 1989 earthquakes measured 8.5 and 7.1 on the Richter scale, respectively. How much greater was the intensity of the 1906 earthquake compared to the 1989 earthquake?

36. How much greater was the intensity of the 2004 Northern Sumatra earthquake (Example 7) compared to the 1964 Alaskan earthquake of magnitude 8.9?

37. If an earthquake has a magnitude 4.2 on the Richter scale, what is the magnitude on the Richter scale of an earthquake that has an intensity 20 times greater? [*Hint*: First solve the equation $10^x = 20$.]

38. Show that the Richter scale defined in (8) of this section can be written

$$M = \frac{\ln A - \ln A_0}{\ln 10}.$$

Marina district in San Francisco, 1989

## pH of a Solution

In Problems 39–42, determine the pH of a solution with the given hydrogen-ion concentration $[H^+]$.

39. $10^{-6}$      40. $4 \times 10^{-7}$

41. $2.8 \times 10^{-8}$      42. $5.1 \times 10^{-5}$

In Problems 43–46, determine the hydrogen-ion concentration $[H^+]$ of a solution with the given pH.

43. 3.3      44. 7.3

45. 6.6      46. 8.1

In Problems 47–50, determine how many more times acidic the first substance is compared to the second substance.

47. lemon juice: pH = 2.3,    vinegar: pH = 3.3

48. battery acid: pH = 1,      lye: pH = 13

**49.** clean rain: pH = 5.6,     acid rain: pH = 3.8

**50.** NaOH: $[H^+] = 10^{-14}$,     HCl: $[H^+] = 1$

## Miscellaneous Logarithmic Models

**51. Richter Scale and Energy** Charles Richter, working with Beno Gutenberg developed the model

$$M = \tfrac{2}{3}\left[\log_{10}E - 11.8\right]$$

that relates the Richter magnitude $M$ of an earthquake and its seismic energy $E$ (measured in ergs). Calculate the seismic energy $E$ of the 2004 Northern Sumatra earthquake where $M = 9.3$.

**52. Intensity Level** The **intensity level $b$** of a sound measured in decibels (dB) is defined by

$$b = 10\log_{10}\frac{I}{I_0}, \tag{10}$$

where $I$ is the **intensity of the sound** measured in watts/cm$^2$ and $I_0 = 10^{-16}$ watts/cm$^2$ is the intensity of the faintest sound that can be heard (0 dB). Use (10) and complete the following table.

| Sound | Intensity $I$ (watts/cm$^2$) | Intensity Level $b$ (dB) |
|---|---|---|
| Whisper | $10^{-14}$ | |
| Conversation | $10^{-11}$ | |
| TV commercials | $10^{-10}$ | |
| Smoke alarm | $10^{-9}$ | |
| Jet taking off | $10^{-7}$ | |
| Rock band | $10^{-4}$ | |

**53.** The threshold of pain is generally taken to be around 140 dB. Find the intensity of sound $I$ corresponding to 140 dB.

**54.** The intensity of sound $I$ is inversely proportional to the square of the distance $d$ from its source, that is,

$$I = \frac{k}{d^2}, \tag{11}$$

where $k$ is the constant of proportionality. Suppose $d_1$ and $d_2$ are distances from a source of sound, and that the corresponding intensity levels of the sounds are $b_1$ and $b_2$. Use (11) in (10) to show that $b_1$ and $b_2$ are related by

$$b_2 = b_1 + 20\log_{10}\frac{d_1}{d_2}. \tag{12}$$

**55.** When a plane $P_1$, flying at an altitude of 1500 ft, passed over a point on the ground its intensity level $b_1$ was measured as 70 dB. Use (12) to find the intensity level $b_2$ of a second plane $P_2$, flying at an altitude of 2600 ft, when it passed over the same point.

**56.** At a distance of 4 ft, the intensity level of an animated conversation is 50 dB. Use (12) to find the intensity level 14 ft from the conversation.

**57. Pupil of the Eye** An empirical model devised by DeGroot and Gebhard relates the diameter $d$ of the pupil of the eye (measured in millimeters, mm) to the luminance $B$ of light source (measured in millilambert's, mL):

$$\log_{10} d = 0.8558 - 0.000401(8.1 + \log_{10} B)^3.$$

See **FIGURE 5.3.3**.

**(a)** The average luminance of clear sky is approximately $B = 255$ mL. Find the corresponding pupil diameter.

**(b)** The luminance of the sun varies from approximately $B = 190{,}000$ mL at sunrise to $B = 51{,}000{,}000$ mL at noon. Find the corresponding pupil diameters.

**(c)** Find the luminance $B$ corresponding to a pupil diameter of 7 mm.

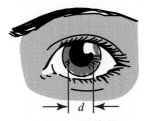

**FIGURE 5.3.3** Pupil diameter in Problem 57

---

## 5.4 ■ The Number $e$

∫**Calculus PREVIEW** ☐ Logarithms were invented in the late sixteenth century by the Scottish lord—and nonmathematician—John Napier. It was he who coined the word "logarithm" from the two Greek words *logos*, meaning ratio, and *arithmos*, meaning number or power. But it took almost two centuries and the genius of the Swiss mathematician Leonhard Euler before the mathematical community became fully aware of the irrational number $e$ and its importance. It is his work that we emulate below in showing why the number $e$ is the natural choice of base for the exponential and logarithmic functions.

☐ **Difference Quotient Revisited** We return to the difference quotient concept first introduced in Section 2.9. Recall that we compute

$$\frac{f(x + h) - f(x)}{h} \tag{1}$$

in three steps. For the exponential function $f(x) = b^x$, we have

$(i)$ $f(x + h) = b^{x+h} = b^x b^h$      ← laws of exponents

$(ii)$ $f(x + h) - f(x) = b^{x+h} - b^x$
            $= b^x b^h - b^x = b^x(b^h - 1)$      ← law of exponents and factoring

$(iii)$ $\dfrac{f(x + h) - f(x)}{h} = \dfrac{b^x(b^h - 1)}{h} = b^x \dfrac{b^h - 1}{h}$

In the fourth step, the calculus step, we let $h \to 0$ but, unlike all the problems given in Exercises 2.9, there is no apparent way of canceling the $h$ in $(iii)$. Nonetheless, the derivative of $f(x) = b^x$ is

$$f'(x) = \lim_{h \to 0} b^x \cdot \frac{b^h - 1}{h}. \tag{2}$$

Since $b^x$ does not depend on the variable $h$, we can rewrite (2) as

$$f'(x) = b^x \cdot \lim_{h \to 0} \frac{b^h - 1}{h}. \tag{3}$$

Now here are the amazing results. The limit in (3),

$$\lim_{h \to 0} \frac{b^h - 1}{h}, \tag{4}$$

can be shown to exist for every positive base $b$. However, as one might expect, we will get a different answer for each base $b$. So let's denote the expression in (4) by the symbol $m(b)$. The derivative of $f(x) = b^x$ is then

$$f'(x) = b^x m(b). \tag{5}$$

You are asked to approximate the value of $m(b)$ in the four cases $b = 1.5, 2, 3$, and 5 in Problems 21–24 of Exercises 5.4. For example, it can be shown that $m(10) \approx 2.302585 \ldots$, and as a consequence the derivative of $f(x) = 10^x$ is

$$f'(x) = (2.302585 \ldots) 10^x. \tag{6}$$

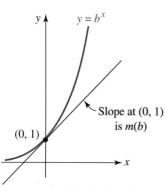

**FIGURE 5.4.1** Find a base $b$ so that the slope $m(b)$ of tangent line at $(0, 1)$ is 1

We can get a better understanding of what $m(b)$ is by evaluating (5) at $x = 0$. Since $b^0 = 1$, we have $f'(0) = m(b)$. In other words, $m(b)$ is the slope of the tangent line to the graph of $f(x) = b^x$ at $x = 0$, that is, at the $y$-intercept $(0, 1)$. See **FIGURE 5.4.1**. Given that we have to calculate a different $m(b)$ for each base $b$, and that $m(b)$ is likely to be an "ugly" number as in (6), over time the following question arose naturally:

*Is there a base b for which $m(b) = 1$?* (7)

☐ **The Answer** To answer the question posed in (7), we must return to the definitions of $e$ given in Section 5.1. Specifically, (4) of Section 5.1,

$$e = \lim_{h \to 0} (1 + h)^{1/h}, \tag{8}$$

provides the means for answering the question posed in (7). If you have studied Sections 1.5, 2.9, and 4.10 you should have an intuitive understanding that the equality in (8) means that as $h$ gets closer and closer to 0 then $(1 + h)^{1/h}$ can be made arbitrarily close to the number $e$. Thus for values of $h$ near 0, we have the approximation $(1 + h)^{1/h} \approx e$, and so it follows that $1 + h \approx e^h$. By rewriting the last expression in the form

$$\frac{e^h - 1}{h} \approx 1 \tag{9}$$

we can conclude that

$$1 = \lim_{h \to 0} \frac{e^h - 1}{h}. \tag{10}$$

Since the right-hand side of (10) is $m(e)$, we have the answer to the question in (7):

*The base b for which $m(b) = 1$ is $b = e$.* (11)

In addition, from (3) we have discovered a wonderfully simple result: The derivative of $f(x) = e^x$ is

$$f'(x) = e^x. \tag{12}$$

The result in (12) is the same as

$$f'(x) = f(x).$$

Moreover, the only other nonzero function $f$ in calculus whose derivative is equal to itself is $f(x) = ce^x$, where $c \neq 0$ is a constant.

☐ **What's Next?** Since $y = \log_b x$ and $y = b^x$ are inverse functions, one would expect that since the simplest derivative of $y = b^x$ is obtained when $b = e$ that the simplest derivative of $y = \log_b x$ also occurs for that base. That is indeed the case. You are encouraged to reexamine (3) of Section 5.1 and then work Problems 1–4 in Exercises 5.4.

☐ **Hyperbolic Functions** We have already seen in Section 5.3 the usefulness of the exponential function $e^x$ in various mathematical models. As a further application, consider a long rope or a flexible wire, such as a telephone wire hanging only under its own weight between two fixed supports. It can be shown that under certain conditions the hanging wire assumes the shape of the graph of the function

$$f(x) = c\frac{e^{x/c} + e^{-x/c}}{2}. \tag{13}$$

The symbol $c$ stands for a positive constant that depends on the physical characteristics of the wire. Functions such as (13), consisting of certain combinations of $e^x$ and $e^{-x}$, appear in so many applications that mathematicians have given them names. In particular, when $c = 1$ in (13), the resulting function $f(x) = \dfrac{e^x + e^{-x}}{2}$ is called the hyperbolic cosine.

---

**THE HYPERBOLIC SINE AND COSINE**

For any real number $x$, the **hyperbolic sine** of $x$, denoted $\sinh x$, is

$$\sinh x = \frac{e^x - e^{-x}}{2}, \tag{14}$$

and the **hyperbolic cosine** of $x$, denoted $\cosh x$, is

$$\cosh x = \frac{e^x + e^{-x}}{2}. \tag{15}$$

---

Analogous to the trigonometric functions $\tan x$, $\cot x$, $\sec x$, and $\csc x$ that are defined in terms of $\sin x$ and $\cos x$, there are four additional hyperbolic functions $\tanh x$, $\coth x$, $\operatorname{sech} x$, and $\operatorname{csch} x$ that are defined in terms of $\sinh x$ and $\cosh x$. For example, the hyperbolic tangent and hyperbolic secant functions are defined as

$$\tanh x = \frac{\sinh x}{\cosh x} \quad \text{and} \quad \operatorname{sech} x = \frac{1}{\cosh x}.$$

☐ **Graphs** The graph of the hyperbolic cosine, shown in **FIGURE 5.4.2**, is called a **catenary**. The word *catenary* derives from the Latin word for a chain, *catena*. The shape of the famous 630-foot tall Gateway arch in St. Louis, Missouri, is an inverted

Gateway arch in St. Louis, MO

catenary. Compare the shape in Figure 5.4.2 with that in the accompanying photo. The graph of $y = \sinh x$ is given in **FIGURE 5.4.3**.

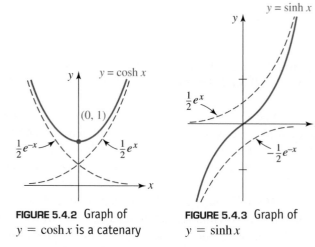

**FIGURE 5.4.2** Graph of $y = \cosh x$ is a catenary

**FIGURE 5.4.3** Graph of $y = \sinh x$

☐ **Identities** Although the hyperbolic functions are not periodic, they possess identities that are similar to trigonometric identities. Analogous to the basic Pythagorean identity of trigonometry $\sin^2 x + \cos^2 x = 1$, for the hyperbolic sine and cosine we have

$$\cosh^2 x - \sinh^2 x = 1.$$

See Problems 9–14 in Exercises 5.4.

**5.4** | Exercises  Answers to selected odd-numbered problems begin on page ANS–17.

**1.** Use the laws of logarithms to show that for $f(x) = \log_b x$,

$$\frac{f(x + h) - f(x)}{h} = \frac{1}{h}\log_b\left(1 + \frac{h}{x}\right) = \frac{1}{x}\log_b\left(1 + \frac{h}{x}\right)^{x/h}.$$

**2.** From Problem 1, the derivative of $f(x) = \log_b x$ is

$$f'(x) = \lim_{h \to 0}\frac{f(x + h) - f(x)}{h} = \frac{1}{x}\lim_{h \to 0}\log_b\left(1 + \frac{h}{x}\right)^{x/h}.$$

Let us assume that the limiting process can be taken inside the logarithm:

$$f'(x) = \frac{1}{x}\log_b\left[\lim_{h \to 0}\left(1 + \frac{h}{x}\right)^{x/h}\right].$$

Rewrite the foregoing result using the substitution $n = x/h$. Notice that since $x$ is held fixed, as $h \to 0$ we must have $n \to \infty$. Give the precise value of $f'(x)$.

In Problems 3 and 4, use the result of Problem 2 to find $f'(x)$ for the given function.

**3.** $f(x) = \log_{10} x$          **4.** $f(x) = \ln x$

In Problems 5 and 6, use the result of Problems 1 and 2 to find $f'(x)$ for the given function. Before using the difference quotient, use the laws of logarithms to rewrite the function.

**5.** $f(x) = \ln\dfrac{x}{9}$            **6.** $f(x) = \log_{10} 6x$

In Problems 7 and 8, compute $\dfrac{f(x+h) - f(x)}{h}$ for the given function.

**7.** $f(x) = e^{5x}$            **8.** $f(x) = e^{-x+4}$

In Problems 9–14, use the definitions of $\cosh x$ and $\sinh x$ in (11) and (12) to verify the given identity.

**9.** $\cosh^2 x - \sinh^2 x = 1$          **10.** $1 - \tanh^2 x = \text{sech}^2 x$.
**11.** $\cosh(-x) = \cosh x$             **12.** $\sinh(-x) = -\sinh x$
**13.** $\sinh 2x = 2\sinh x \cosh x$       **14.** $\cosh 2x = \cosh^2 x + \sinh^2 x$

**15.** If $\cosh x = \frac{3}{2}$, use the identity given in Problem 9 to find the value of $\sinh x$.
**16.** If $\tanh x = \frac{1}{2}$, use the identity given in Problem 10 to find the value of $\cosh x$.
**17.** As can be seen in Figure 5.4.3, the function $y = \sinh x$ is one-to-one. Use the definition of the hyperbolic sine in the form $e^x - 2y - e^{-x} = 0$ to find the inverse of $y = \sinh x$. [*Hint*: See Problems 59 and 60 in Exercises 5.2.]
**18.** The function $y = \cosh x$ on the restricted domain $[0, \infty)$ is one-to-one. Proceed as in Problem 17 to find the inverse of $y = \cosh x$, $x \geq 0$.

## Calculator Problems

**19.** Use a calculator to investigate $\lim\limits_{h \to 0} \dfrac{f(x+h) - f(x)}{h}$ for the function in Problem 7. Determine $f'(x)$.

**20.** Use a calculator to investigate $\lim\limits_{h \to 0} \dfrac{f(x+h) - f(x)}{h}$ for the function in Problem 8. Determine $f'(x)$.

In Problems 21–24, use a calculator to estimate the value $m(b) = \lim\limits_{h \to 0} \dfrac{b^h - 1}{h}$ for $b = 1.5, b = 2, b = 3$, and $b = 5$ by filling out the given table.

**21.**

| $h \to 0$ | 0.1 | 0.01 | 0.001 | 0.0001 | 0.00001 | 0.000001 |
|---|---|---|---|---|---|---|
| $\dfrac{(1.5)^h - 1}{h}$ | | | | | | |

**22.**

| $h \to 0$ | 0.1 | 0.01 | 0.001 | 0.0001 | 0.00001 | 0.000001 |
|---|---|---|---|---|---|---|
| $\dfrac{2^h - 1}{h}$ | | | | | | |

**23.**

| $h \to 0$ | 0.1 | 0.01 | 0.001 | 0.0001 | 0.00001 | 0.000001 |
|---|---|---|---|---|---|---|
| $\dfrac{3^h - 1}{h}$ | | | | | | |

**24.**

| $h \to 0$ | 0.1 | 0.01 | 0.001 | 0.0001 | 0.00001 | 0.000001 |
|---|---|---|---|---|---|---|
| $\dfrac{5^h - 1}{h}$ | | | | | | |

**25.** Fill out a table of the kind in Problems 21–24, but this time use $\dfrac{e^h - 1}{h}$.

**26. A Curiosity** The logarithm developed by John Napier (see page 322) was actually

$$10^7 \log_{1/e}\left(\frac{x}{10^7}\right).$$

Use (11) of Section 5.2 to express this logarithm in terms of the natural logarithm.

---

| CHAPTER 5 | Review Exercises | Answers to selected odd-numbered problems begin on page ANS–17. |
|---|---|---|

In Problems 1–22, fill in the blanks.

**1.** The graph of $y = 6 - e^{-x}$ has the $y$-intercept _____ and horizontal asymptote $y =$ _____.

**2.** The $x$-intercept of the graph of $y = -10 + 10^{5x}$ is _____.

**3.** The graph of $y = \ln(x + 4)$ has the $x$-intercept _____ and vertical asymptote $x =$ _____.

**4.** The $y$-intercept of the graph of $y = \log_8(x + 2)$ is _____.

**5.** $\log_5 2 - \log_5 10 =$ _____

**6.** $6 \ln e + 3 \ln \dfrac{1}{e} =$ _____

**7.** $e^{3 \ln 10} =$ _____

**8.** $10^{\log_{10} 4.89} =$ _____

**9.** $\log_4(4 \cdot 4^2 \cdot 4^3) =$ _____

**10.** $\dfrac{\log_5 625}{\log_5 125} =$ _____

**11.** If $\log_3 N = -2$, then $N =$ _____.

**12.** If $\log_b 6 = \frac{1}{2}$, then $b =$ _____.

**13.** If $\ln e^3 = y$, then $y =$ _____.

**14.** If $\ln 3 + \ln(x - 1) = \ln 2 + \ln x$, then $x =$ _____.

**15.** If $-1 + \ln(x - 3) = 0$, then $x =$ _____.

**16.** If $\ln(\ln x) = 1$, then $x =$ _____.

**17.** If $100 - 20e^{-0.15t} = 35$, then to four rounded decimals $t =$ _____.

**18.** If $3^x = 5$, then $3^{-2x} =$ _____.

**19.** $f(x) = 4^{3x} = (\underline{\quad})^x$

**20.** $f(x) = (e^2)^{x/6} = (\underline{\quad})^x$

**21.** If the graph of $y = e^{x-2} + C$ passes through $(2, 9)$, then $C =$ _____.

**22.** By rigid transformations, the point $(0, 1)$ on the graph of $y = e^x$ is moved to the point _____ on the graph of $y = 4 + e^{x-3}$.

In Problems 23–36, answer true or false.

**23.** $y = \ln x$ and $y = e^x$ are inverse functions. _____

**24.** The point $(b, 1)$ is on the graph of $f(x) = \log_b x$. _____

**25.** $y = 10^{-x}$ and $y = (0.1)^x$ are the same function. _____

**26.** If $f(x) = e^{x^2} - 1$, then $f(x) = 1$ when $x = \pm\ln\sqrt{2}$. _____

**27.** $4^{x/2} = 2^x$ _____

**28.** $\dfrac{2^{x^2}}{2^x} = 2^x$ _____

**29.** $2^x + 2^{-x} = (2 + 2^{-1})^x$ _____

**30.** $2^{3+3x} = 8^{1+x}$ _____

**31.** $-\ln 2 = \ln\left(\frac{1}{2}\right)$ _____

**32.** $\ln\dfrac{e^a}{e^b} = a - b$ _____

**33.** $\ln(\ln e) = 1$ _____

**34.** $\ln\sqrt{43} = \dfrac{\ln 43}{2}$ _____

**35.** $\ln(e + e) = 1 + \ln 2$ _____

**36.** $\log_6(36)^{-1} = -2$ _____

In Problems 37 and 38, rewrite the given exponential expression as an equivalent logarithmic expression.

**37.** $5^{-1} = 0.2$

**38.** $\sqrt[3]{512} = 8$

In Problems 39 and 40, rewrite the given logarithmic expression as an equivalent exponential expression.

**39.** $\log_9 27 = 1.5$

**40.** $\log_6(36)^{-2} = -4$

In Problems 41–48, solve for $x$.

**41.** $2^{1-x} = 8$

**42.** $3^{2x} = 81$

**43.** $e^{1-2x} = e^2$

**44.** $e^{x^2} - e^5 e^{x-1} = 0$

**45.** $2^{1-x} = 7$

**46.** $3^x = 7^{x-1}$

**47.** $e^{x+2} = 6$

**48.** $3e^x = 4e^{-3x}$

In Problems 49 and 50, graph the given functions on the same coordinate axes.

**49.** $y = 4^x$, $y = \log_4 x$

**50.** $y = \left(\frac{1}{2}\right)^x$, $y = \log_{1/2} x$

**51.** Match the letter of the graph in FIGURE 5.R.1 with the appropriate function.

    (i) $f(x) = b^x, b > 2$

    (ii) $f(x) = b^x, 1 < b < 2$

    (iii) $f(x) = b^x, \frac{1}{2} < b < 1$

    (iv) $f(x) = b^x, 0 < b < \frac{1}{2}$

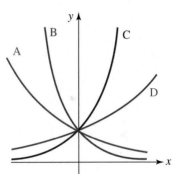

**FIGURE 5.R.1** Graphs for Problem 51

**52.** In FIGURE 5.R.2, fill in the blanks for the coordinates of the points on each graph.

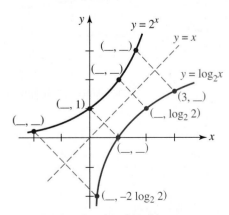

**FIGURE 5.R.2** Graphs of Problem 52

In Problems 53 and 54, find the slope of the line $L$ given in each figure.

**53.** $f(x) = 3^{-(x+1)}$

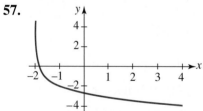

**FIGURE 5.R.3** Graph for Problem 53

**54.**

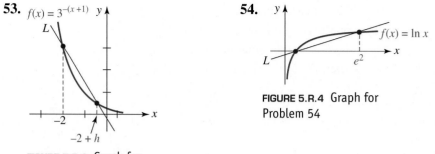

**FIGURE 5.R.4** Graph for Problem 54

In Problems 55–60, match each of the following functions with one of the given graphs.

$(i)$ $y = \ln(x - 2)$    $(ii)$ $y = 2 - \ln x$
$(iii)$ $y = 2 + \ln(x + 2)$    $(iv)$ $y = -2 - \ln(x + 2)$
$(v)$ $y = -\ln(2x)$    $(vi)$ $y = 2 + \ln(-x + 2)$

**55.**

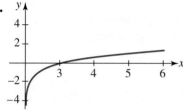

**FIGURE 5.R.5** Graph for Problem 55

**56.**

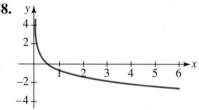

**FIGURE 5.R.6** Graph for Problem 56

**57.**

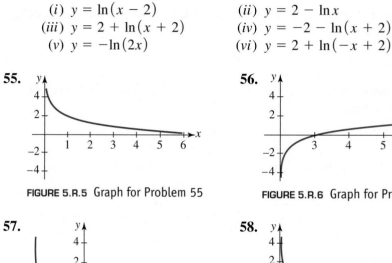

**FIGURE 5.R.7** Graph for Problem 57

**58.**

**FIGURE 5.R.8** Graph for Problem 58

**59.**

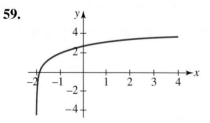

**FIGURE 5.R.9** Graph for Problem 59

**60.**

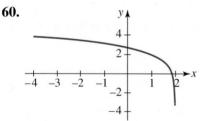

**FIGURE 5.R.10** Graph for Problem 60

In Problems 61 and 62, in words describe the graph of the function $f$ in terms of a transformation of the graph of $y = \ln x$.

**61.** $f(x) = \ln\dfrac{1}{x}$           **62.** $f(x) = \ln x^3$

**63.** Find a function $f(x) = Ae^{kx}$ if $(0, 5)$ and $(6, 1)$ are points on the graph of $f$.

**64.** Find a function $f(x) = A10^{kx}$ if $f(3) = 8$ and $f(0) = \frac{1}{2}$.

**65.** Find a function $f(x) = a + b^x, 0 < b < 1$, if $f(1) = 5.5$ and the graph of $f$ has a horizontal asymptote $y = 5$.

**66.** Find a function $f(x) = a + \log_3(x - c)$ if $f(11) = 10$ and the graph of $f$ has a vertical asymptote $x = 2$.

**67.** If the initial number of bacteria present in a culture doubles after 9 hours, how long will it take for the number of bacteria in the culture to double again?

**68.** A commercial fishing lake is stocked with 10,000 fingerlings. Find a model $P(t) = P_0 e^{kt}$ for the fish population of the lake at time $t$ if the owner of the lake estimates that there will be 5000 fish left after six months. After how many months does the model predict that there will be 1000 fish left?

**69.** Tritium, an isotope of hydrogen, has a half-life of 12.5 years. How much of an initial quantity of this element remains after 50 years?

**70.** It is found that 97% of C-14 has been lost in a human skeleton found at an archeological site. What is the approximate age of the skeleton?

**71.** A person facing retirement invests \$650,000 in a savings account. She wants the account to be worth \$1,000,000 in 10 years. What annual rate $r$ of interest compounded continuously will achieve this dream?

**72.** According to the **Bouguer-Lambert law**, the intensity $I$ (measured in lumens) of a vertical beam of light passing through a transparent substance decreases according to the exponential function $I(x) = I_0 e^{kx}, k < 0$, where $I_0$ is the intensity of the incident beam and $x$ is the depth measured in meters. If the intensity of light 1 meter below the surface of water is 30% of $I_0$, what is the intensity 3 meters below the surface?

**73.** The graph of the function $y = ae^{-be^{-cx}}$ is called a Gompertz curve. Solve for $x$ in terms of the other symbols.

## Chapter Outline

# Conic Sections

**6**

## 6.1 The Parabola

☐ **Introduction** Hypatia is the first woman in the history of mathematics about whom we have considerable knowledge. Born in 370 C.E. in Alexandria, she was renowned as a mathematician and philosopher. Among her writings is *On the Conics of Apollonius*, which popularized Apollonius' (200 B.C.E.) work on curves that can be obtained by intersecting a double-napped cone with a plane: the circle, parabola, ellipse, and hyperbola. See FIGURE 6.1.1. With the close of the Greek period, interest in conics sections waned; after Hypatia the study of these curves was neglected for over 1000 years.

Hypatia

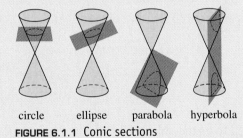

circle    ellipse    parabola    hyperbola

**FIGURE 6.1.1** Conic sections

In the seventeenth century, Galileo showed that in the absence of air resistance the path of a projectile follows a parabolic arc. At about the same time, Johannes Kepler hypothesized that the orbits of planets about the Sun are ellipses with the Sun at one focus. This was later verified by Isaac Newton, using the methods of the newly developed calculus. Kepler also experimented with the reflecting properties of parabolic mirrors; these investigations sped the development of the reflecting telescope. The Greeks had known little of these practical applications. They had studied the conics for their beauty and fascinating properties. In the first three sections of this chapter, we will examine both the ancient properties and the modern applications of these curves. Rather than using a cone, we shall see how the parabola, ellipse, and hyperbola are defined by means of distance. Using a rectangular coordinate system and the distance formula, we obtain equations for the conics. Each of these equations will be in the form of a quadratic equation in variables $x$ and $y$:

$$Ax^2 + Bxy + Cy^2 + Dx + Ey + F = 0,$$

where $A, B, C, D, E$, and $F$ are constants. We have already studied the special case $y = ax^2 + bx + c$ of the foregoing equation in Section 2.4.

Solar system

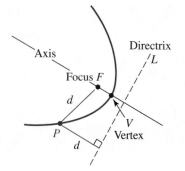

FIGURE 6.1.2 A parabola

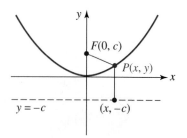

FIGURE 6.1.3 Parabola with vertex $(0, 0)$ and focus on the y-axis

## PARABOLA

A **parabola** is the set of points $P(x, y)$ in the plane that are equidistant from a fixed line $L$, called the **directrix**, and a fixed point $F$, called the **focus**.

A parabola is shown in FIGURE 6.1.2. The line through the focus perpendicular to the directrix is called the **axis** of the parabola. The point of intersection of the parabola and the axis is called the **vertex**, denoted by $V$ in Figure 6.1.2.

☐ **Parabola with Vertex $(0, 0)$** To describe a parabola analytically, we use a rectangular coordinate system where the directrix is a horizontal line $y = -c$, where $c > 0$, and the focus is the point $F(0, c)$. Then we see that the axis of the parabola is along the y-axis, as FIGURE 6.1.3 shows. The origin is necessarily the vertex, since it lies on the axis $c$ units from both the focus and the directrix. The distance from a point $P(x, y)$ to the directrix is

$$y - (-c) = y + c.$$

Using the distance formula, the distance from $P$ to the focus $F$ is

$$d(P, F) = \sqrt{(x - 0)^2 + (y - c)^2}.$$

From the definition of the parabola it follows that $d(P, F) = y + c$, or

$$\sqrt{(x - 0)^2 + (y - c)^2} = y + c.$$

By squaring both sides and simplifying, we obtain

$$x^2 + (y - c)^2 = (y + c)^2$$
$$x^2 + y^2 - 2cy + c^2 = y^2 + 2cy + c^2 \tag{1}$$

or
$$x^2 = 4cy.$$

Equation (1) is referred to as the **standard form** of the equation of a parabola with focus $(0, c)$, directrix $y = -c, c > 0$, and vertex $(0, 0)$. The graph of any parabola with standard form (1) is symmetric with respect to the y-axis

Equation (1) does not depend on the assumption that $c > 0$. However, the direction in which the parabola opens does depend on the sign of $c$. Specifically, if $c > 0$ the parabola opens *upward* as in Figure 6.1.3; if $c < 0$, the parabola opens *downward*.

If the focus of a parabola is assumed to lie on the x-axis at $F(c, 0)$ and the directrix is $x = -c$, then the x-axis is the axis of the parabola and the vertex is $(0, 0)$. If $c > 0$ the parabola opens to the right; if $c < 0$, it opens to the left. In either case, the **standard form** of the equation is

$$y^2 = 4cx. \tag{2}$$

The graph of any parabola with standard form (2) is symmetric with respect to the x-axis.

A summary of all this information for equations (1) and (2) is given in FIGURES 6.1.4 and 6.1.5, respectively. You may be surprised to see in Figure 6.1.4(b) that the directrix above the x-axis is labeled $y = -c$ and the focus on the negative y-axis has coordinates $F(0, c)$. Bear in mind that in this case the assumption is that $c < 0$ and so $-c > 0$. A similar remark holds for Figure 6.1.5(b).

CHAPTER 6 CONIC SECTIONS

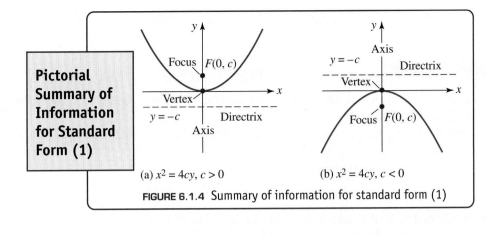

**Pictorial Summary of Information for Standard Form (1)**

(a) $x^2 = 4cy$, $c > 0$     (b) $x^2 = 4cy$, $c < 0$

**FIGURE 6.1.4** Summary of information for standard form (1)

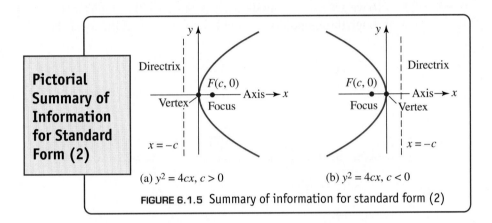

**Pictorial Summary of Information for Standard Form (2)**

(a) $y^2 = 4cx$, $c > 0$     (b) $y^2 = 4cx$, $c < 0$

**FIGURE 6.1.5** Summary of information for standard form (2)

---

### EXAMPLE 1 — The Simplest Parabola

We first encountered the graph of $y = x^2$ in Section 2.2. By comparing this equation with (1) we see

$$\overset{4c}{\underset{\downarrow}{\phantom{x}}}$$
$$x^2 = 1 \cdot y$$

and so $4c = 1$ or $c = \frac{1}{4}$. Therefore the graph of $y = x^2$ is a parabola with vertex at the origin, focus at $\left(0, \frac{1}{4}\right)$, and directrix $y = -\frac{1}{4}$. These details are indicated in the graph in **FIGURE 6.1.6**. ∎

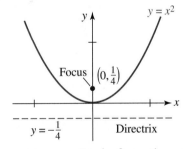

**FIGURE 6.1.6** Graph of equation in Example 1

Knowing the basic parabolic shape, all we need to know to sketch a *rough* graph of either equations (1) and (2) is the fact that the graph passes through its vertex $(0, 0)$ and the direction in which the parabola opens. To add more accuracy to the graph it is convenient to use the number $c$ determined by the standard form equation to plot two additional points. Note that if we choose $y = c$ in (1), then $x^2 = 4c^2$ implies $x = \pm 2c$. Thus $(2c, c)$ and $(-2c, c)$ lie on the graph of $x^2 = 4cy$. Similarly, the choice $x = c$ in (2) implies $y = \pm 2c$, and so $(c, 2c)$ and $(c, -2c)$ are points on the graph of $y^2 = 4cx$. The *line segment* through the focus with endpoints $(2c, c)$, $(-2c, c)$ for equations with standard form (1), and $(c, 2c)$, $(c, -2c)$ for equations with standard form (2) is called the **focal chord**. For example, in Figure 6.1.6, if we choose $y = \frac{1}{4}$, then

◀ Graphing tip for equations (1) and (2).

$x^2 = \frac{1}{4}$ implies $x = \pm\frac{1}{2}$. Endpoints of the horizontal focal chord for $y = x^2$ are $\left(-\frac{1}{2}, \frac{1}{4}\right)$ and $\left(\frac{1}{2}, \frac{1}{4}\right)$.

■ **EXAMPLE 2**     **Finding an Equation of a Parabola**

Find the equation in standard form of the parabola with directrix $x = 2$ and focus $(-2, 0)$. Graph.

   **Solution**  In FIGURE 6.1.7 we have graphed the directrix and the focus. We see from their placement that the equation we seek is of the form $y^2 = 4cx$. Since $c = -2$, the parabola opens to the left and so

$$y^2 = 4(-2)x \qquad \text{or} \qquad y^2 = -8x.$$

As mentioned in the discussion preceding this example, if we substitute $x = c$, or in this case $x = -2$, into the equation $y^2 = -8x$ we can find two points on its graph. From $y^2 = -8(-2) = 16$ we get $y = \pm 4$. As shown in FIGURE 6.1.8, the graph passes through $(0, 0)$ as well as through the endpoints $(-2, -4)$ and $(-2, 4)$ of the focal chord.

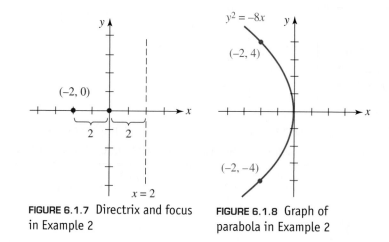

FIGURE 6.1.7 Directrix and focus in Example 2

FIGURE 6.1.8 Graph of parabola in Example 2                  ■

☐ **Parabola with Vertex $(h, k)$**  Suppose that a parabola is shifted both horizontally and vertically so that its vertex is at the point $(h, k)$ and its axis is the vertical line $x = h$. The **standard form** of the equation of the parabola is then

$$(x - h)^2 = 4c(y - k). \tag{3}$$

Similarly, if its axis is the horizontal line $y = k$, the standard form of the equation of the parabola with vertex $(h, k)$ is

$$(y - k)^2 = 4c(x - h). \tag{4}$$

   The parabolas defined by these equations are identical in shape to the parabolas defined by equations (1) and (2) because equations (3) and (4) represent rigid transformations (shifts up, down, left, and right) of the graphs of (1) and (2). For example, the parabola

$$(x + 1)^2 = 8(y - 5)$$

has vertex $(-1, 5)$. Its graph is the graph of $x^2 = 8y$ shifted horizontally one unit to the left followed by an upward vertical shift of five units.

   For each of the equations, (1) and (2) or (3) and (4), the *distance* from the vertex to the focus, as well as the distance from the vertex to the directrix, is $|c|$.

## EXAMPLE 3     Finding an Equation of a Parabola

Find the equation in standard form of the parabola with vertex $(-3, -1)$ and directrix $y = 3$. Find the focus.

**Solution** We begin by graphing the vertex at $(-3, -1)$ and the directrix $y = 3$. From FIGURE 6.1.9 we can see that the parabola must open downward, and so its standard form is (3). This fact, plus the observation that the vertex lies 4 units below the directrix, indicates that the appropriate solution of $|c| = 4$ is $c = -4$. Substituting $h = -3$, $k = -1$, and $c = -4$ into (3) gives

$$[x - (-3)]^2 = 4(-4)[y - (-1)]^2 \quad \text{or} \quad (x + 3)^2 = -16(y + 1)^2. \quad \blacksquare$$

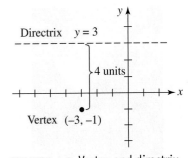

FIGURE 6.1.9 Vertex and directrix in Example 3

## EXAMPLE 4     Find Everything

Find the vertex, focus, axis, directrix, and graph of the parabola

$$y^2 - 4y - 8x - 28 = 0. \tag{5}$$

**Solution** In order to write the equation in one of the standard forms we complete the square in $y$:

$$y^2 - 4y + 4 = 8x + 28 + 4 \quad \leftarrow \text{Add 4 to both sides.}$$

$$(y - 2)^2 = 8x + 32.$$

Thus the standard form of equation (5) is $(y - 2)^2 = 8(x + 4)$. Comparing this equation with (4) we conclude that the vertex is $(-4, 2)$ and that $4c = 8$ or $c = 2$. Thus the parabola opens to the right. From $c = 2 > 0$, the focus is 2 units to the right of the vertex at $(-4 + 2, 2)$ or $(-2, 2)$. The directrix is the vertical line 2 units to the left of the vertex, $x = -4 - 2$ or $x = -6$. Knowing the parabola opens to the right from the point $(-4, 2)$ also tells us that the graph has intercepts. To find the $x$-intecept we set $y = 0$ in (5) and find immediately that $x = -\frac{28}{8} = -\frac{7}{2}$. The $x$-intercept is $(-\frac{7}{2}, 0)$. To find the $y$-intercepts we set $x = 0$ in (5) and find from the quadratic formula that $y = 2 \pm 4\sqrt{2}$ or $y \approx 7.66$ and $y \approx -3.66$. The $y$-intercepts are $(0, 2 - 4\sqrt{2})$ and $(0, 2 + 4\sqrt{2})$. Putting all this information together we get the graph in FIGURE 6.1.10. $\quad \blacksquare$

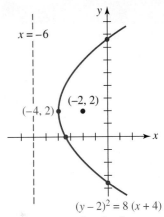

FIGURE 6.1.10 Graph of equation in Example 4

☐ **Applications of the Parabola** The parabola has many interesting properties that make it suitable for certain applications. Reflecting surfaces are often designed to take advantage of a reflection property of parabolas. Such surfaces, called **paraboloids**, are three-dimensional and are formed by rotating a parabola about its axis. As illustrated in FIGURE 6.1.11(a), rays of light (or electronic signals) from a point source located

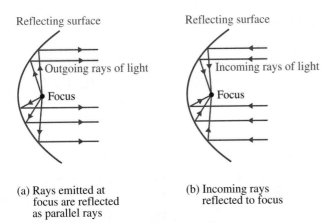

(a) Rays emitted at focus are reflected as parallel rays

(b) Incoming rays reflected to focus

FIGURE 6.1.11 Parabolic reflecting surface

Searchlight

TV satellite dish

at the focus of a parabolic reflecting surface will be reflected along lines parallel to the axis. This is the idea behind the design of searchlights, some flashlights, and on-location satellite dishes. Conversely, if the incoming rays of light are parallel to the axis of a parabola, they will be reflected off the surface along lines passing through the focus. See Figure 6.1.11(b). Beams of light from a distant object such as a galaxy are essentially parallel, and so when these beams enter a reflecting telescope they are reflected by the parabolic mirror to the focus, where a camera is usually placed to capture the image over time. A parabolic home satellite dish operates on the same principle as the reflecting telescope; the digital signal from a TV satellite is captured at the focus of the dish by a receiver.

Parabolas are also important in the design of suspension bridges. It can be shown that if the weight of the bridge is distributed uniformly along its length, then a support cable in the shape of a parabola will bear the load evenly.

The trajectory of an obliquely launched projectile—say, a basketball thrown from the free throw line—will travel in a parabolic arc.

Tuna, which prey on smaller fish, have been observed swimming in schools of 10 to 20 fish arrayed approximately in a parabolic shape. One possible explanation for this is that the smaller fish caught in the school of tuna will try to escape by "reflecting" off the parabola. As a result, they are concentrated at the focus and become easy prey for the tuna.

The Brooklyn bridge is a suspension bridge

The ball travels in a parabolic arc

**6.1** Exercises Answers to selected odd-numbered problems begin on page ANS–18.

In Problems 1–24, find the vertex, focus, directrix, and axis of the given parabola. Graph the parabola.

1. $y^2 = 4x$
2. $y^2 = \frac{7}{2}x$
3. $y^2 = -\frac{4}{3}x$
4. $y^2 = -10x$
5. $x^2 = -16y$
6. $x^2 = \frac{1}{10}y$
7. $x^2 = 28y$
8. $x^2 = -64y$
9. $(y - 1)^2 = 16x$
10. $(y + 3)^2 = -8(x + 2)$
11. $(x + 5)^2 = -4(y + 1)$
12. $(x - 2)^2 + y = 0$
13. $y^2 + 12y - 4x + 16 = 0$
14. $x^2 + 6x + y + 11 = 0$
15. $x^2 + 5x - \frac{1}{4}y + 6 = 0$
16. $x^2 - 2x - 4y + 17 = 0$
17. $y^2 - 8y + 2x + 10 = 0$
18. $y^2 - 4y - 4x + 3 = 0$

**19.** $4x^2 = 2y$

**20.** $3(y-1)^2 = 9x$

**21.** $-2x^2 + 12x - 8y - 18 = 0$

**22.** $4y^2 + 16y - 6x - 2 = 0$

**23.** $6y^2 - 12y - 24x - 42 = 0$

**24.** $3x^2 + 30x - 8y + 75 = 0$

In Problems 25–44, find an equation of the parabola that satisfies the given conditions.

**25.** Focus $(0, 7)$, directrix $y = -7$

**26.** Focus $(0, -5)$, directrix $y = 5$

**27.** Focus $(-4, 0)$, directrix $x = 4$

**28.** Focus $\left(\frac{3}{2}, 0\right)$, directrix $x = -\frac{3}{2}$

**29.** Focus $\left(\frac{5}{2}, 0\right)$, vertex $(0, 0)$

**30.** Focus $(0, -10)$, vertex $(0, 0)$

**31.** Focus $(2, 3)$, directrix $y = -3$

**32.** Focus $(1, -7)$, directrix $x = -5$

**33.** Focus $(-1, 4)$, directrix $x = 5$

**34.** Focus $(-2, 0)$, directrix $y = \frac{3}{2}$

**35.** Focus $(1, 5)$, vertex $(1, -3)$

**36.** Focus $(-2, 3)$, vertex $(-2, 5)$

**37.** Focus $(8, -3)$, vertex $(0, -3)$

**38.** Focus $(1, 2)$, vertex $(7, 2)$

**39.** Vertex $(0, 0)$, directrix $y = -\frac{7}{4}$

**40.** Vertex $(0, 0)$, directrix $x = 6$

**41.** Vertex $(5, 1)$, directrix $y = 7$

**42.** Vertex $(-1, 4)$, directrix $x = 0$

**43.** Vertex $(0, 0)$, through $(-2, 8)$, axis along the $y$-axis

**44.** Vertex $(0, 0)$, through $\left(1, \frac{1}{4}\right)$, axis along the $x$-axis

In Problems 45–48, find the $x$- and $y$-intercepts of the given parabola.

**45.** $(y + 4)^2 = 4(x + 1)$

**46.** $(x - 1)^2 = -2(y - 1)$

**47.** $x^2 + 2y - 18 = 0$

**48.** $y^2 - 8y - x + 15 = 0$

## Miscellaneous Applications

**49. Spotlight** A large spotlight is designed so that a cross section through its axis is a parabola and the light source is at the focus. Find the position of the light source if the spotlight is 4 ft across at the opening and 2 ft deep.

**50. Reflecting Telescope** A reflecting telescope has a parabolic mirror that is 20 ft across at the top and 4 ft deep at the center. Where should the eyepiece be located?

**51.** Suppose that a light ray emanating from the focus of the parabola $y^2 = 4x$ strikes the parabola at $(1, -2)$. What is the equation of the reflected ray?

**52. Suspension Bridge** Suppose that two towers of a suspension bridge are 350 ft apart and the vertex of the parabolic cable is tangent to the road midway between the towers. If the cable is 1 ft above the road at a point 20 ft from the vertex, find the height of the towers above the road.

**53. Another Suspension Bridge** Two 75–ft towers of a suspension bridge with a parabolic cable are 250 ft apart. The vertex of the parabola is tangent to the road midway between the towers. Find the height of the cable above the roadway at a point 50 ft from one of the towers.

**54. Drainpipe** Assume that the water gushing from the end of a horizontal pipe follows a parabolic arc with vertex at the end of the pipe. The pipe is 20 m above the ground. At a point 2 m below the end of the pipe, the horizontal distance from the water to a vertical line through the end of the pipe is 4 m. See FIGURE 6.1.12. Where does the water strike the ground?

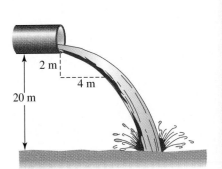

FIGURE 6.1.12 Pipe in Problem 54

**55. A Bull's–Eye** A dart thrower releases a dart 5 ft above the ground. The dart is thrown horizontally and follows a parabolic path. It hits the ground $10\sqrt{10}$ ft from the dart thrower. At a distance of 10 ft from the dart thrower, how high should a bull's–eye be placed in order for the dart to hit it?

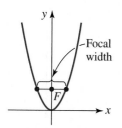

**FIGURE 6.1.13** Focal width in Problem 57

**56. Path of a Projectile** The vertical position of a projectile is given by the equation $y = -16t^2$ and the horizontal position by $x = 40t$ for $t \geq 0$. By eliminating $t$ between the two equations, show that the path of the projectile is a parabolic arc. Graph the path of the projectile.

**57. Focal Width** The focal width of a parabola is the length of the focal chord, that is, the line segment through the focus perpendicular to the axis, with endpoints on the parabola. See **FIGURE 6.1.13**.

(a) Find the focal width of the parabola $x^2 = 8y$.

(b) Show that the focal width of the parabola $x^2 = 4cy$ and $y^2 = 4cx$ is $4|c|$.

**58. Parabolic Orbit** The orbit of a comet is a parabola with the Sun at the focus. When the comet is 50,000,000 km from the Sun, the line from the comet to the Sun is perpendicular to the axis of the parabola. Use the result of Problem 57(b) to write an equation of the comet's path. (A comet with a parabolic path will not return to the solar system.)

## For Discussion

**59.** Suppose that two parabolic reflecting surfaces face one another (with foci on a common axis). Any sound emitted at one focus will be reflected off the parabolas and concentrated at the other focus. **FIGURE 6.1.14** shows the paths of two typical sound waves. Using the definition of a parabola on page 330, show that all waves will travel the same distance. [*Note*: This result is important for the following reason: If the sound waves traveled paths of different lengths, then the waves would arrive at the second focus at different times. The result would be interference rather than clear sound.]

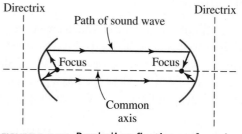

**FIGURE 6.1.14** Parabolic reflecting surfaces in Problem 59

**60.** The point closest to the focus is the vertex. How would you go about proving this? Carry out your ideas.

**61.** For the comet in Problem 58, use the result of Problem 60 to determine the shortest distance between the Sun and the comet.

---

**6.2** **The Ellipse**

☐ **Introduction** The ellipse occurs frequently in astronomy. For example, the paths of the planets around the Sun are elliptical with the Sun located at one focus. Similarly, communication satellites, the Hubble space telescope, and the international space station revolve around the Earth in elliptical orbits with the Earth at one focus. In this section we define the ellipse and study some of its properties and applications.

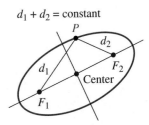

$d_1 + d_2 = $ constant

FIGURE 6.2.1  An ellipse

## ELLIPSE

An **ellipse** is the set of points $P(x, y)$ in the plane such that the sum of the distances between $P$ and two fixed points $F_1$ and $F_2$ is constant. The fixed points $F_1$ and $F_2$ are called **foci** (plural for **focus**). The midpoint of the line segment joining points $F_1$ and $F_2$ is called the **center** of the ellipse.

As shown in **FIGURE 6.2.1**, if $P$ is a point on the ellipse and if $d_1 = d(F_1, P)$ and $d_2 = d(F_2, P)$ are the distances from the foci to $P$, then the preceding definition asserts that

$$d_1 + d_2 = k, \tag{1}$$

where $k > 0$ is some constant.

On a practical level, equation (1) suggests a way of generating an ellipse. **FIGURE 6.2.2** shows that if a string of length $k$ is attached to a piece of paper by two tacks, then an ellipse can be traced out by inserting a pencil against the string and moving it in such a manner that the string remains taut.

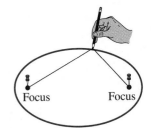

FIGURE 6.2.2  A way to draw an ellipse

□ **Ellipse with Center (0, 0)** We now derive an equation of the ellipse. For algebraic convenience, let us choose $k = 2a > 0$ and put the foci on the $x$-axis with coordinates $F_1(-c, 0)$ and $F_2(c, 0)$ as shown in **FIGURE 6.2.3**. It follows from (1) that

$$\sqrt{(x + c)^2 + y^2} + \sqrt{(x - c)^2 + y^2} = 2a$$

or

$$\sqrt{(x + c)^2 + y^2} = 2a - \sqrt{(x - c)^2 + y^2}. \tag{2}$$

We square both sides of the second equation in (2) and simplify,

$$(x + c)^2 + y^2 = 4a^2 - 4a\sqrt{(x - c)^2 + y^2} + (x - c)^2 + y^2$$

$$a\sqrt{(x - c)^2 + y^2} = a^2 - cx.$$

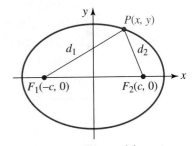

FIGURE 6.2.3  Ellipse with center $(0, 0)$ and foci on the $x$-axis

Squaring a second time gives,

$$a^2[(x - c)^2 + y^2] = a^4 - 2a^2cx + c^2x^2$$

or

$$(a^2 - c^2)x^2 + a^2y^2 = a^2(a^2 - c^2). \tag{3}$$

Referring to Figure 6.2.3, we see that the points $F_1$, $F_2$, and $P$ form a triangle. Because the sum of the lengths of any two sides of a triangle is greater than the remaining side, we must have $2a > 2c$ or $a > c$. Hence, $a^2 - c^2 > 0$. When we let $b^2 = a^2 - c^2$, then (3) becomes $b^2x^2 + a^2y^2 = a^2b^2$. Dividing this last equation by $b^2$ gives

$$\frac{x^2}{a^2} + \frac{y^2}{b^2} = 1. \tag{4}$$

Equation (4) is called the **standard form** of the equation of an ellipse centered at $(0, 0)$ with foci $(-c, 0)$ and $(c, 0)$, where $c$ is defined by $b^2 = a^2 - c^2$, and $a > b > 0$.

If the foci are placed on the $y$-axis, then a repetition of the above analysis leads to

$$\frac{x^2}{b^2} + \frac{y^2}{a^2} = 1. \tag{5}$$

Equation (5) is called the **standard form** of the equation of an ellipse centered at $(0, 0)$ with foci $(0, -c)$ and $(0, c)$, where $c$ is defined by $b^2 = a^2 - c^2$ and $a > b > 0$.

☐ **Major and Minor Axes** The **major axis** of an ellipse is the line segment through its center, containing the foci, and with endpoints on the ellipse. For an ellipse with standard equation (4) the major axis is horizontal, whereas for (5) the major axis is vertical. The line segment through the center, perpendicular to the major axis, and with endpoints on the ellipse, is called the **minor axis**. The two endpoints of the major axis are called the **vertices** of the ellipse. For (4) the vertices are the $x$-intercepts. Setting $y = 0$ in (4) gives $x = \pm a$. The vertices are then $(-a, 0)$ and $(a, 0)$. For (5) the vertices are the $y$-intercepts $(0, -a)$ and $(0, a)$. For equation (4), the endpoints of the minor axis are $(0, -b)$ and $(0, b)$; for (5) the endpoints are $(-b, 0)$ and $(b, 0)$. For either (4) or (5), the **length of the major axis** is $a - (-a) = 2a$; the length of the minor axis is $2b$. Since $a > b$, the major axis of an ellipse is always longer than its minor axis.

A summary of all this information for equations (4) and (5) is given in FIGURE 6.2.4.

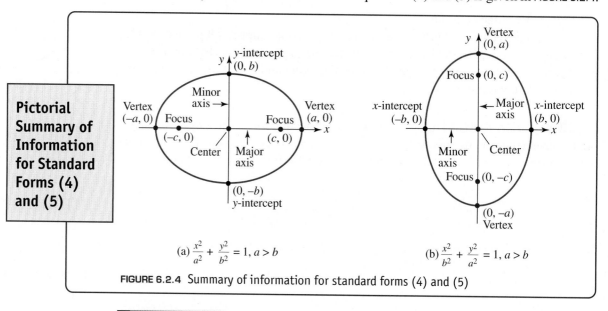

**Pictorial Summary of Information for Standard Forms (4) and (5)**

(a) $\dfrac{x^2}{a^2} + \dfrac{y^2}{b^2} = 1, a > b$

(b) $\dfrac{x^2}{b^2} + \dfrac{y^2}{a^2} = 1, a > b$

FIGURE 6.2.4 Summary of information for standard forms (4) and (5)

### EXAMPLE 1    Vertices and Foci

Find the vertices and foci of the ellipse whose equation is $9x^2 + 3y^2 = 27$. Graph.

   **Solution** By dividing both sides of the equality by 27 the standard form of the equation is

$$\frac{x^2}{3} + \frac{y^2}{9} = 1.$$

We see that $9 > 3$ and so we identify the equation with (5). From $a^2 = 9$ and $b^2 = 3$, we see that $a = 3$ and $b = \sqrt{3}$. The major axis is vertical with endpoints $(0, -3)$ and $(0, 3)$. The minor axis is horizontal with endpoints $(-\sqrt{3}, 0)$ and $(\sqrt{3}, 0)$. Of course, the vertices are also the $y$-intercepts and the endpoints of the minor axis are the $x$-intercepts. Now, to find the foci we use $b^2 = a^2 - c^2$ or $c^2 = a^2 - b^2$ to write $c = \sqrt{a^2 - b^2}$. With $a = 3, b = \sqrt{3}$, we get $c = \sqrt{9 - 3} = \sqrt{6}$. Hence, the foci are on the $y$-axis at $(0, -\sqrt{6})$ and $(0, \sqrt{6})$. The graph is given in FIGURE 6.2.5. ∎

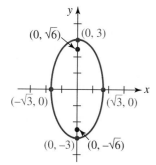

FIGURE 6.2.5 Ellipse in Example 1

### EXAMPLE 2    Finding an Equation of an Ellipse

Find an equation of the ellipse with a focus $(2, 0)$ and an $x$-intercept $(5, 0)$.

   **Solution** Since the given focus is on the $x$-axis, we can find an equation in standard form (4). Consequently, $c = 2, a = 5, a^2 = 25$, and $b^2 = a^2 - c^2$ or $b^2 = 5^2 - 2^2 = 21$. The desired equation is

$$\frac{x^2}{25} + \frac{y^2}{21} = 1.$$ ∎

☐ **Ellipse with Center $(h, k)$**  When the center is at $(h, k)$, the **standard form** for the equation of an ellipse is either

$$\frac{(x - h)^2}{a^2} + \frac{(y - k)^2}{b^2} = 1 \tag{6}$$

or

$$\frac{(x - h)^2}{b^2} + \frac{(y - k)^2}{a^2} = 1. \tag{7}$$

The ellipses defined by these equations are identical in shape to the ellipses defined by equations (4) and (5) since equations (6) and (7) represent rigid transformations of the graphs of (4) and (5). For example, the ellipse

$$\frac{(x - 1)^2}{9} + \frac{(y + 3)^2}{16} = 1$$

has center $(1, -3)$. Its graph is the graph of $x^2/9 + y^2/16 = 1$ shifted horizontally one unit to the right followed by a downward vertical shift of three units.

It is not a good idea to memorize formulas for the vertices and foci of an ellipse with center $(h, k)$. Everything is the same as before: $a$, $b$, and $c$ are positive and $a > b$, $a > c$. You can locate vertices, foci, and endpoints of the minor axis using the fact that $a$ is the distance from the center to a vertex, $b$ is the distance from the center to an endpoint on the minor axis, and $c$ is the distance from the center to a focus. Also, we still have $c^2 = a^2 - b^2$.

### EXAMPLE 3    Ellipse Centered at $(h, k)$

Find the vertices and foci of the ellipse $4x^2 + 16y^2 - 8x - 96y + 84 = 0$. Graph.

**Solution**  To write the given equation in one of the standard forms (6) or (7) we must complete the square in $x$ and in $y$. Recall that in order to complete the square we want the coefficients of the quadratic terms $x^2$ and $y^2$ to be 1. To do this we factor 4 from both $x^2$ and $x$ and factor 16 from both $y^2$ and $y$:

$$4(x^2 - 2x \quad) + 16(y^2 - 6y \quad) = -84.$$

Then from

4 · 1 and 16 · 9 are added to both sides

$$4(x^2 - 2x + 1) + 16(y^2 - 6y + 9) = -84 + 4 \cdot 1 + 16 \cdot 9$$

we obtain

$$4(x - 1)^2 + 16(y - 3)^2 = 64$$

or

$$\frac{(x - 1)^2}{16} + \frac{(y - 3)^2}{4} = 1. \tag{8}$$

From (8) we see that the center of the ellipse is $(1, 3)$. Since the last equation has the standard form (6), we identify $a^2 = 16$ or $a = 4$ and $b^2 = 4$ or $b = 2$. The major axis is horizontal and lies on the horizontal line $y = 3$ passing through $(1, 3)$. This is the red horizontal dashed line segment in **FIGURE 6.2.6**. By measuring $a = 4$ units to the left and then to the right of the center along the line $y = 3$, we arrive at the vertices $(-3, 3)$ and $(5, 3)$. By measuring $b = 2$ units both down and up the vertical line $x = 1$ through the center, we arrive at the endpoints of the minor axis $(1, 1)$ and $(1, 5)$. The minor axis is the black dashed vertical line segment in Figure 6.2.6. Because, $c^2 = a^2 - b^2 = 16 - 4 = 12$, $c = 2\sqrt{3}$. Finally, by measuring $c = 2\sqrt{3}$ units to the left and right of the center along $y = 3$, we obtain the foci $(1 - 2\sqrt{3}, 3)$ and $(1 + 2\sqrt{3}, 3)$.

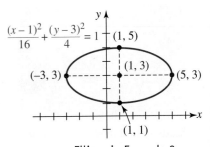

**FIGURE 6.2.6** Ellipse in Example 3

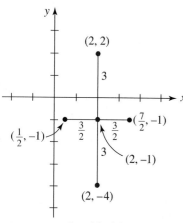

FIGURE 6.2.7 Graphical interpretation of data in Example 4

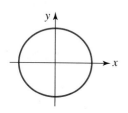

(a) *e* close to zero

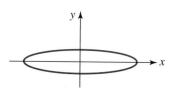

(b) *e* close to 1

FIGURE 6.2.8 Effect of eccentricity on the shape of an ellipse

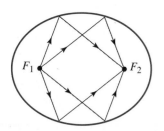

FIGURE 6.2.9 Reflection property of an ellipse

Statuary Hall in Washington, DC

EXAMPLE 4     **Finding an Equation of an Ellipse**

Find an equation of the ellipse with center $(2, -1)$, vertical major axis of length 6, and minor axis of length 3.

**Solution** The length of the major axis is $2a = 6$ hence $a = 3$. Similarly, the length of the minor axis is $2b = 3$, so $b = \frac{3}{2}$. By sketching the center and the axes, we see from FIGURE 6.2.7 that the vertices are $(2, 2)$ and $(2, -4)$ and the endpoints of the minor axis are $(\frac{1}{2}, -1)$ and $(\frac{7}{2}, -1)$. Because the major axis is vertical, the standard equation of this ellipse is

$$\frac{(x-2)^2}{(\frac{3}{2})^2} + \frac{(y-(-1))^2}{3^2} = 1 \quad \text{or} \quad \frac{(x-2)^2}{\frac{9}{4}} + \frac{(y+1)^2}{9} = 1. \quad \blacksquare$$

□ **Eccentricity** Associated with each conic section is a number $e$ called its **eccentricity**. The eccentricity of an ellipse is defined to be

$$e = \frac{c}{a},$$

where $c = \sqrt{a^2 - b^2}$. Since $0 < \sqrt{a^2 - b^2} < a$, the eccentricity of an ellipse satisfies $0 < e < 1$.

EXAMPLE 5     **Example 3 Revisited**

Determine the eccentricity of the ellipse in Example 3.

**Solution** In the solution of Example 3 we found that $a = 4$ and $c = 2\sqrt{3}$. Hence, the eccentricity of the ellipse is $e = (2\sqrt{3})/4 = \sqrt{3}/2 \approx 0.87$. $\quad \blacksquare$

Eccentricity is an indicator of the shape of an ellipse. When $e \approx 0$, that is, $e$ is close to zero, the ellipse is nearly circular, and when $e \approx 1$ the ellipse is flattened or elongated. To see this, observe that if $e$ is close to 0, it follows from $e = \sqrt{a^2 - b^2}/a$ that $c = \sqrt{a^2 - b^2} \approx 0$ and consequently $a \approx b$. As you can see from the standard equations in (4) and (5), this means that the shape of the ellipse is close to circular. Also, because $c$ is the distance from the center of the ellipse to a focus, the two foci are close together near the center. See FIGURE 6.2.8(a). On the other hand, if $e \approx 1$ or $\sqrt{a^2 - b^2}/a \approx 1$, then $c = \sqrt{a^2 - b^2} \approx a$ and so $b \approx 0$. Also, $c \approx a$ means that the foci are far apart; each focus is close to a vertex. Thus, the ellipse is elongated as shown in Figure 6.2.8(b).

□ **Applications of the Ellipse** Ellipses have a reflection property analogous to the one discussed in Section 6.1 for the parabola. It can be shown that if a light or sound source is placed at one focus of an ellipse, then all rays or waves will be reflected off the ellipse to the other focus. See FIGURE 6.2.9. For example, if a pool table is constructed in the form of an ellipse with a pocket at one focus, then any shot originating at the other focus will never miss the pocket. Similarly, if a ceiling is elliptical with two foci on (or near) the floor, but considerably distant from each other, then anyone whispering at one focus will be heard at the other. Some famous "whispering galleries" are the Statuary Hall at the Capitol in Washington, DC, the Mormon Tabernacle in Salt Lake City, and St. Paul's Cathedral in London.

Using his Law of Universal Gravitation, Isaac Newton was the first to prove Kepler's first law of planetary motion: The orbit of each planet about the Sun is an ellipse with the Sun at one focus.

## EXAMPLE 6    Eccentricity of Earth's Orbit

The perihelion distance of the Earth (the least distance between the Earth and the Sun) is approximately $9.16 \times 10^7$ miles, and its aphelion distance (the greatest distance between the Earth and the Sun) is approximately $9.46 \times 10^7$ miles. What is the eccentricity of Earth's orbit?

**Solution**  Let us assume that the orbit of the Earth is as shown in FIGURE 6.2.10. From the figure we see that

$$a - c = 9.16 \times 10^7$$
$$a + c = 9.46 \times 10^7.$$

Solving this system of equations gives $a = 9.31 \times 10^7$ and $c = 0.15 \times 10^7$. Thus the eccentricity $e = c/a$ is

$$e = \frac{0.15 \times 10^7}{9.31 \times 10^7} \approx 0.016.$$  ■

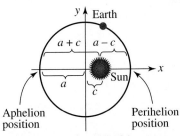

FIGURE 6.2.10 Graphical interpretation of data in Example 6

The orbits of seven of the planets have eccentricities less than 0.1 and, hence, the orbits are not far from circular. Mercury and Pluto are the exceptions. For example, the orbit of Pluto has the eccentricity 0.25. Many of the asteroids and comets have highly eccentric orbits. The orbit of the asteroid Hildago is one of the most eccentric, with $e = 0.66$. Another notable case is the orbit of Comet Halley. See Problem 43 in Exercises 6.2.

---

**6.2    Exercises**    Answers to selected odd-numbered problems begin on page ANS–18.

In Problems 1–20, find the center, foci, vertices, endpoints of the minor axis, and eccentricity of the given ellipse. Graph the ellipse.

**1.** $\dfrac{x^2}{25} + \dfrac{y^2}{9} = 1$

**2.** $\dfrac{x^2}{16} + \dfrac{y^2}{4} = 1$

**3.** $x^2 + \dfrac{y^2}{16} = 1$

**4.** $\dfrac{x^2}{4} + \dfrac{y^2}{10} = 1$

**5.** $9x^2 + 16y^2 = 144$

**6.** $2x^2 + y^2 = 4$

**7.** $9x^2 + 4y^2 = 36$

**8.** $x^2 + 4y^2 = 4$

**9.** $\dfrac{(x-1)^2}{49} + \dfrac{(y-3)^2}{36} = 1$

**10.** $\dfrac{(x+1)^2}{25} + \dfrac{(y-2)^2}{36} = 1$

**11.** $(x+5)^2 + \dfrac{(y+2)^2}{16} = 1$

**12.** $\dfrac{(x-3)^2}{64} + \dfrac{(y+4)^2}{81} = 1$

**13.** $4x^2 + (y + \frac{1}{2})^2 = 4$

**14.** $36(x+2)^2 + (y-4)^2 = 72$

**15.** $5(x-1)^2 + 3(y+2)^2 = 45$

**16.** $6(x-2)^2 + 8y^2 = 48$

**17.** $25x^2 + 9y^2 - 100x + 18y - 116 = 0$

**18.** $9x^2 + 5y^2 + 18x - 10y - 31 = 0$

**19.** $x^2 + 3y^2 + 18y + 18 = 0$

**20.** $12x^2 + 4y^2 - 24x - 4y + 1 = 0$

In Problems 21–40, find an equation of the ellipse that satisfies the given conditions.

**21.** Vertices $(\pm 5, 0)$, foci $(\pm 3, 0)$

**22.** Vertices $(\pm 9, 0)$, foci $(\pm 2, 0)$

**23.** Vertices $(0, \pm 3)$, foci $(0, \pm 1)$

**24.** Vertices $(0, \pm 7)$, foci $(0, \pm 3)$

**25.** Vertices $(0, \pm 3)$, endpoints of minor axis $(\pm 1, 0)$

**26.** Vertices $(\pm 4, 0)$, endpoints of minor axis $(0, \pm 2)$

**27.** Vertices $(-3, -3)$, $(5, -3)$, endpoints of minor axis $(1, -1)$, $(1, -5)$

**28.** Vertices $(1, -6)$, $(1, 2)$, endpoints of minor axis $(-2, -2)$, $(4, -2)$

**29.** One focus $(0, -2)$, center at origin, $b = 3$

**30.** One focus $(1, 0)$, center at origin, $a = 3$

**31.** Foci $(\pm \sqrt{2}, 0)$, length of minor axis 6

**32.** Foci $(0, \pm \sqrt{5})$, length of major axis 16

**33.** Foci $(0, \pm 3)$, passing through $(-1, 2\sqrt{2})$

**34.** Vertices $(\pm 5, 0)$, passing through $(\sqrt{5}, 4)$

**35.** Vertices $(\pm 4, 1)$, passing through $(2\sqrt{3}, 2)$

**36.** Center $(1, -1)$, one focus $(1, 1)$, $a = 5$

**37.** Center $(1, 3)$, one focus $(1, 0)$, one vertex $(1, -1)$

**38.** Center $(5, -7)$, length of vertical major axis 8, length of minor axis 6

**39.** Endpoints of minor axis $(0, 5)$, $(0, -1)$, one focus $(6, 2)$

**40.** Endpoints of major axis $(2, 4)$, $(13, 4)$, one focus $(4, 4)$

**41.** The orbit of the planet Mercury is an ellipse with the Sun at one focus. The length of the major axis of this orbit is 72 million miles and the length of the minor axis is 70.4 million miles. What is the least distance (perihelion) between Mercury and the Sun? What is the greatest distance (aphelion)?

**42.** What is the eccentricity of the orbit of Mercury in Problem 41?

**43.** The orbit of Comet Halley is an ellipse whose major axis is $3.34 \times 10^9$ miles long, and whose minor axis is $8.5 \times 10^8$ miles long. What is the eccentricity of the comet's orbit?

**44.** A satellite orbits the Earth in an elliptical path with the center of the Earth at one focus. It has a minimum altitude of 200 mi and a maximum altitude of 1000 mi above the surface of the Earth. If the radius of the Earth is 4000 mi, what is an equation of the satellite's orbit?

## Miscellaneous Applications

**45.** **Archway** A semielliptical archway has a vertical major axis. The base of the arch is 10 ft across and the highest part of the arch is 15 ft. Find the height of the arch above the point on the base of the arch 3 ft from the center.

**46.** **Gear Design** An elliptical gear rotates about its center and is always kept in mesh with a circular gear that is free to move horizontally. See **FIGURE 6.2.11**. If the origin of the $xy$-coordinate system is placed at the center of the ellipse, then the equation of the ellipse in its present position is $3x^2 + 9y^2 = 24$. The diameter of the circular gear equals the length of the minor axis of the elliptical gear. Given that the units are centimeters, how far does the center of the circular gear move horizontally during the rotation from one vertex of the elliptical gear to the next?

**47.** **Carpentry** A carpenter wishes to cut an elliptical top for a coffee table from a rectangular piece of wood that is 4 ft by 3 ft utilizing the entire length and width available. If the ellipse is to be drawn using the string and tack method illustrated in Figure 6.2.2, how long should the piece of string be and where should the tacks be placed?

**48.** **Park Design** The Ellipse is a park in Washington, DC. It is bounded by an elliptical path with a major axis of length 458 m and a minor axis of length 390 m. Find the distance between the foci of this ellipse.

**FIGURE 6.2.11** Elliptical and circular gears in Problem 46

49. **Whispering Gallery** Suppose that a room is constructed on a flat elliptical base by rotating a semiellipse 180° about its major axis. Then, by the reflection property of the ellipse, anything whispered at one focus will be distinctly heard at the other focus. If the height of the room is 16 ft and the length is 40 ft, find the location of the whispering and listening posts.

50. **Focal Width** The focal width of the ellipse is the length of a focal chord, that is, a line segment, perpendicular to the major axis, through a focus with endpoints on the ellipse. See **FIGURE 6.2.12**.
    (a) Find the focal width of the ellipse $x^2/9 + y^2/4 = 1$.
    (b) Show that, in general, the focal width of the ellipse $x^2/a^2 + y^2/b^2 = 1$ is $2b^2/a$.

51. Find an equation of the ellipse with foci $(0, 2)$ and $(8, 6)$ and fixed distance sum $2a = 12$. [*Hint*: Here the major axis is neither horizontal nor vertical; thus none of the standard forms from this section apply. Use the definition of the ellipse.]

52. Proceed as in Problem 51, and find an equation of the ellipse with foci $(-1, -3)$ and $(-5, 7)$ and fixed distance sum $2a = 20$.

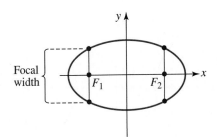

**FIGURE 6.2.12** Focal width in Problem 50

## For Discussion

53. The graph of the ellipse $x^2/4 + (y - 1)^2/9 = 1$ is shifted 4 units to the right. What are the center, foci, vertices, and endpoints of the minor axis for the shifted graph?

54. The graph of the ellipse $(x - 1)^2/9 + (y - 4)^2 = 1$ is shifted 5 units to the left and 3 units up. What are the center, foci, vertices, and endpoints of the minor axis for the shifted graph?

55. In engineering the eccentricity of an ellipse is often expressed only in terms of $a$ and $b$. Show that $e = \sqrt{1 - b^2/a^2}$.

## 6.3 The Hyperbola

☐ **Introduction** The definition of a hyperbola is basically the same as the definition of the ellipse with only one exception: the word *sum* is replaced by the word *difference*.

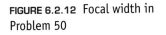

**HYPERBOLA**

A **hyperbola** is the set of points $P(x, y)$ in the plane such that the difference of the distances between $P$ and two fixed points $F_1$ and $F_2$ is constant. The fixed points $F_1$ and $F_2$ are called **foci** (plural for **focus**). The midpoint of the line segment joining points $F_1$ and $F_2$ is called the **center** of the hyperbola.

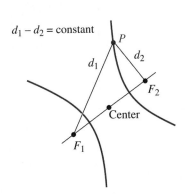

**FIGURE 6.3.1** A hyperbola

As shown in **FIGURE 6.3.1**, a hyperbola consists of two **branches**. If $P$ is a point on the hyperbola, then

$$|d_1 - d_2| = k, \qquad (1)$$

where $d_1 = d(F_1, P)$ and $d_2 = d(F_2, P)$.

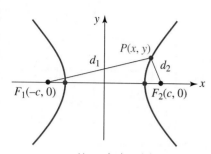

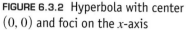

**FIGURE 6.3.2** Hyperbola with center $(0, 0)$ and foci on the $x$-axis

☐ **Hyperbola with Center $(0, 0)$** Proceeding as for the ellipse, we place the foci on the $x$-axis at $F_1(-c, 0)$ and $F_2(c, 0)$ as shown in **FIGURE 6.3.2** and choose the constant $k$ to be $2a$ for algebraic convenience. It follows from (1) that

$$d_1 - d_2 = \pm 2a. \tag{2}$$

As drawn in Figure 6.3.2, $P$ is on the right branch of the hyperbola and so $d_1 - d_2 = 2a > 0$. If $P$ is on the left branch then the difference is $-2a$. Writing (2) as

$$\sqrt{(x + c)^2 + y^2} - \sqrt{(x - c)^2 + y^2} = \pm 2a$$

or

$$\sqrt{(x + c)^2 + y^2} = \pm 2a + \sqrt{(x - c)^2 + y^2}$$

we square, simplify, and square again:

$$(x + c)^2 + y^2 = 4a^2 \pm 4a\sqrt{(x - c)^2 + y^2} + (x - c)^2 + y^2$$

$$\pm a\sqrt{(x - c)^2 + y^2} = cx - a^2$$

$$a^2[(x - c)^2 + y^2] = c^2x^2 - 2a^2cx + a^4$$

$$(c^2 - a^2)x^2 - a^2y^2 = a^2(c^2 - a^2). \tag{3}$$

From Figure 6.3.2, we see that the triangle inequality gives

$$d_1 < d_2 + 2c \qquad \text{and} \qquad d_2 < d_1 + 2c,$$

or

$$d_1 - d_2 < 2c \qquad \text{and} \qquad d_2 - d_1 < 2c.$$

Using $d_1 - d_2 = \pm 2a$, the last two inequalities imply that $2a < 2c$ or $a < c$. since $c > a > 0$, $c^2 - a^2$ is a positive constant. If we let $b^2 = c^2 - a^2$, (3) becomes $b^2x^2 - a^2y^2 = a^2b^2$ or, after dividing by $b^2$,

$$\frac{x^2}{a^2} - \frac{y^2}{b^2} = 1. \tag{4}$$

Equation (4) is called the **standard form** of the equation of a hyperbola centered at $(0, 0)$ with foci $(-c, 0)$ and $(c, 0)$, where $c$ is defined by $b^2 = c^2 - a^2$.

When the foci lie on the $y$-axis, a repetition of the foregoing algebra leads to

$$\frac{y^2}{a^2} - \frac{x^2}{b^2} = 1. \tag{5}$$

Note of Caution ▶

Equation (5) is the **standard form** of the equation of a hyperbola centered at $(0, 0)$ with foci $(0, -c)$ and $(0, c)$. Here again, $c > a$ and $b^2 = c^2 - a^2$.

For the hyperbola (unlike the ellipse), bear in mind that in (4) and (5) there is no relationship between the relative sizes of $a$ and $b$; rather, $a^2$ is always the denominator of the *positive term* and the intercepts *always* have $\pm a$ as a coordinate.

☐ **Transverse and Conjugate Axes** The line segment with endpoints on the hyperbola and lying on the line through the foci is called the **transverse axis**; its endpoints are called the **vertices** of the hyperbola. For the hyperbola described by equation (4), the transverse axis lies on the $x$-axis. Therefore, the coordinates of the vertices are the $x$-intercepts. Setting $y = 0$ gives $x^2/a^2 = 1$, or $x = \pm a$. Thus, as shown in **FIGURE 6.3.3** the vertices are $(-a, 0)$ and $(a, 0)$; the **length of the transverse axis** is $2a$. Notice that by setting $y = 0$ in (4), we get $-y^2/b^2 = 1$ or $y^2 = -b^2$, which has no real solutions. Hence the graph of any equation in that form has no $y$-intercepts. Nonetheless, the numbers $\pm b$ are important. The line segment through the center of the hyperbola perpendicular to the transverse axis and with endpoints $(0, -b)$ and $(0, b)$ is called the **conjugate**

**axis**. Similarly, the graph of an equation in standard form (5) has no $x$-intercepts. The conjugate axis for (5) is the line segment with endpoints $(-b, 0)$ and $(b, 0)$.

This information for equations (4) and (5) is summarized in Figure 6.3.3.

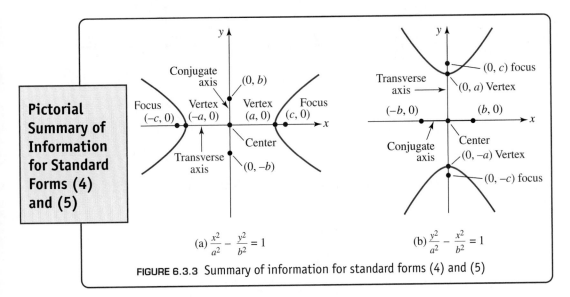

**Pictorial Summary of Information for Standard Forms (4) and (5)**

(a) $\dfrac{x^2}{a^2} - \dfrac{y^2}{b^2} = 1$   (b) $\dfrac{y^2}{a^2} - \dfrac{x^2}{b^2} = 1$

**FIGURE 6.3.3** Summary of information for standard forms (4) and (5)

☐ **Asymptotes** Every hyperbola possesses a pair of slant asymptotes that pass through its center. These asymptotes are indicative of end behavior, and as such are an invaluable aide in sketching the graph of a hyperbola. Solving (4) for $y$ in terms of $x$ gives

$$y = \pm \frac{b}{a} x \sqrt{1 - \frac{a^2}{x^2}}.$$

As $x \to -\infty$ or as $x \to \infty$, $a^2/x^2 \to 0$, and thus $\sqrt{1 - a^2/x^2} \to 1$. Therefore, for large values of $|x|$, points on the graph of the hyperbola are close to the points on the lines

$$y = \frac{b}{a} x \qquad \text{and} \qquad y = -\frac{b}{a} x. \tag{6}$$

By a similar analysis we find that the slant asymptotes for (5) are

$$y = \frac{a}{b} x \qquad \text{and} \qquad y = -\frac{a}{b} x. \tag{7}$$

Each pair of asymptotes intersect at the origin, which is the center of the hyperbola. Note, too, in **FIGURE 6.3.4(a)** that the asymptotes are simply the *extended diagonals* of a rectangle of width $2a$ (the length of the transverse axis) and height $2b$ (the length of the conjugate axis); in Figure 6.3.4(b) the asymptotes are the extended diagonals of a rectangle of width $2b$ and height $2a$.

We recommend that you *do not* memorize the equations in (6) and (7). There is an easy method for obtaining the asymptotes of a hyperbola. For example, since $y = \pm \dfrac{b}{a} x$ is equivalent to

$$\frac{x^2}{a^2} = \frac{y^2}{b^2}$$

the asymptotes of the hyperbola given in (4) are obtained from a single equation

$$\frac{x^2}{a^2} - \frac{y^2}{b^2} = 0. \tag{8}$$

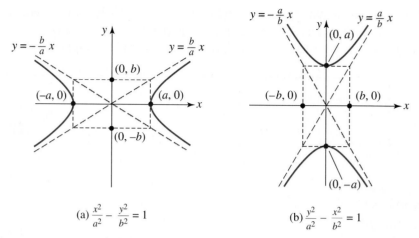

(a) $\dfrac{x^2}{a^2} - \dfrac{y^2}{b^2} = 1$  (b) $\dfrac{y^2}{a^2} - \dfrac{x^2}{b^2} = 1$

**FIGURE 6.3.4** Hyperbolas (4) and (5) with slant asymptotes

Note that (8) factors as the difference of two squares:

$$\left(\frac{x}{a} - \frac{y}{b}\right)\left(\frac{x}{a} + \frac{y}{b}\right) = 0.$$

This is a mnemonic, or memory device. It ▶ has no geometric significance.

Setting each factor equal to zero and solving for $y$ gives an equation of an asymptote. You do not even have to memorize (8) because it is simply the left-hand side of the standard form of the equation of a hyperbola given in (4). In like manner, to obtain the asymptotes for (5) just replace 1 by 0 in the standard form, factor $y^2/a^2 - x^2/b^2 = 0$, and solve for $y$.

### EXAMPLE 1  Hyperbola Centered at $(0, 0)$

Find the vertices, foci, and asymptotes of the hyperbola $9x^2 - 25y^2 = 225$. Graph.

   **Solution** We first put the equation into standard form by dividing the left-hand side by 225:

$$\frac{x^2}{25} - \frac{y^2}{9} = 1. \tag{9}$$

From this equation we see that $a^2 = 25$ and $b^2 = 9$, and so $a = 5$ and $b = 3$. Therefore the vertices are $(-5, 0)$ and $(5, 0)$. Since $b^2 = c^2 - a^2$ implies $c^2 = a^2 + b^2$, we have $c^2 = 34$, and so the foci are $(-\sqrt{34}, 0)$ and $(\sqrt{34}, 0)$. To find the slant asymptotes we use the standard form (9) with 1 replaced by 0:

$$\frac{x^2}{25} - \frac{y^2}{9} = 0 \quad \text{factors as} \quad \left(\frac{x}{5} - \frac{y}{3}\right)\left(\frac{x}{5} + \frac{y}{3}\right) = 0.$$

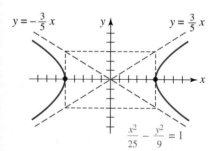

**FIGURE 6.3.5** Hyperbola in Example 1

Setting each factor equal to zero and solving for $y$ gives the asymptotes $y = \pm 3x/5$. We plot the vertices and graph the two lines through the origin. Both branches of the hyperbola must become arbitrarily close to the asymptotes as $x \to \pm\infty$. See **FIGURE 6.3.5**. ■

### EXAMPLE 2  Finding an Equation of a Hyperbola

Find an equation of the hyperbola with vertices $(0, -4,)$, $(0, 4)$ and asymptotes $y = -\frac{1}{2}x$, $y = \frac{1}{2}x$.

**Solution** The center of the hyperbola is $(0, 0)$. This is revealed by the fact that the asymptotes intersect at the origin. Moreover, the vertices are on the $y$-axis and are 4 units on either side of the origin. Thus the equation we seek is of form (5). From (7) or Figure 6.3.4(b), the asymptotes must be of the form $y = \pm\dfrac{a}{b}x$ so that $a/b = 1/2$. From the given vertices we identify $a = 4$, and so

$$\frac{4}{b} = \frac{1}{2} \quad \text{implies} \quad b = 8.$$

The equation of the hyperbola is then

$$\frac{y^2}{4^2} - \frac{x^2}{8^2} = 1 \quad \text{or} \quad \frac{y^2}{16} - \frac{x^2}{64} = 1. \qquad \blacksquare$$

☐ **Hyperbola with Center $(h, k)$** When the center of the hyperbola is $(h, k)$ the **standard form** analogues of equations (4) and (5) are, in turn,

$$\frac{(x - h)^2}{a^2} - \frac{(y - k)^2}{b^2} = 1 \qquad (10)$$

and

$$\frac{(y - h)^2}{a^2} - \frac{(x - k)^2}{b^2} = 1. \qquad (11)$$

As in (4) and (5), the numbers $a^2$, $b^2$, and $c^2$ are related by $b^2 = c^2 - a^2$.

You can locate vertices and foci using the fact that $a$ is the distance from the center to a vertex and $c$ is the distance from the center to a focus. The slant asymptotes for (10) can be obtained by factoring

$$\frac{(x - h)^2}{a^2} - \frac{(y - k)^2}{b^2} = 0$$

as

$$\left(\frac{x - h}{a} - \frac{y - k}{b}\right)\left(\frac{x - h}{a} + \frac{y - k}{b}\right) = 0.$$

Similarly, the asymptotes for (11) can be obtained from factoring $\dfrac{(y - h)^2}{a^2} - \dfrac{(x - k)^2}{b^2} = 0$, setting each factor equal to zero and solving for $y$ in terms of $x$. As a check on your work, remember that $(h, k)$ must be a point that lies on each asymptote.

**EXAMPLE 3**        **Hyperbola Centered at $(h, k)$**

Find the center, vertices, foci, and asymptotes of the hyperbola $4x^2 - y^2 - 8x - 4y - 4 = 0$. Graph.

**Solution** Before completing the square in $x$ and $y$, we factor 4 from the two $x$-terms and factor $-1$ from the two $y$-terms so that the leading coefficient in each expression is 1. Then we have

$$4(x^2 - 2x \quad) + (-1)(y^2 + 4y \quad) = 4$$
$$4(x^2 - 2x + 1) - (y^2 + 4y + 4) = 4 + 4 \cdot 1 + (-1) \cdot 4$$
$$4(x - 1)^2 - (y + 2)^2 = 4$$
$$\frac{(x - 1)^2}{1} - \frac{(y + 2)^2}{4} = 1.$$

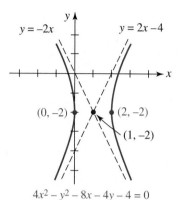

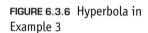

FIGURE 6.3.6 Hyperbola in Example 3

We see now that the center is $(1, -2)$. Since the term in the standard form involving $x$ has the positive coefficient, the transverse axis is horizontal along the line $y = -2$, and we identify $a = 1$ and $b = 2$. The vertices are 1 unit to the left and to the right of the center at $(0, -2)$ and $(2, -2)$, respectively. From $b^2 = c^2 - a^2$, we have

$$c^2 = a^2 + b^2 = 1 + 4 = 5,$$

and so $c = \sqrt{5}$. Hence the foci are $\sqrt{5}$ units to the left and the right of the center $(1, -2)$ at $(1 - \sqrt{5}, -2)$ and $(1 + \sqrt{5}, -2)$.

To find the asymptotes, we solve

$$\frac{(x-1)^2}{1} - \frac{(y+2)^2}{4} = 0 \quad \text{or} \quad \left(x - 1 - \frac{y+2}{2}\right)\left(x - 1 + \frac{y+2}{2}\right) = 0$$

for $y$. From $y + 2 = \pm 2(x - 1)$ we find that the asymptotes are $y = -2x$ and $y = 2x - 4$. Observe that by substituting $x = 1$, both equations give $y = -2$, which means that both lines pass through the center. We then locate the center, plot the vertices, and graph the asymptotes. As shown in FIGURE 6.3.6, the graph of the hyperbola passes through the vertices and becomes closer and closer to the asymptotes as $x \to \pm\infty$. ∎

### ▮ EXAMPLE 4    Finding an Equation of a Hyperbola

Find an equation of the hyperbola with center $(2, -3)$, passing through the point $(4, 1)$, and having one vertex $(2, 0)$.

**Solution** Since the distance from the center to one vertex is $a$, we have $a = 3$. From the location of the center and the vertex, it follows that the transverse axis is vertical and lies along the line $x = 2$. Therefore, the equation of the hyperbola must be of form (11):

$$\frac{(y + 3)^2}{3^2} - \frac{(x - 2)^2}{b^2} = 1, \tag{12}$$

where $b^2$ is yet to be determined. Since the point $(4, 1)$ is on the graph on the hyperbola, its coordinates must satisfy equation (12). From

$$\frac{(1 + 3)^2}{3^2} - \frac{(4 - 2)^2}{b^2} = 1$$

$$\frac{16}{9} - \frac{4}{b^2} = 1$$

$$\frac{7}{9} = \frac{4}{b^2}$$

we find $b^2 = \frac{36}{7}$. We conclude that the desired equation is

$$\frac{(y + 3)^2}{3^2} - \frac{(x - 2)^2}{\frac{36}{7}} = 1. \qquad ∎$$

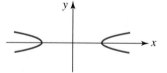

(a) $e$ close to 1

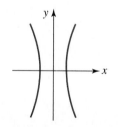

(b) $e$ much greater than 1

FIGURE 6.3.7 Effect of eccentricity on the shape of a hyperbola

☐ **Eccentricity** Like the ellipse, the equation that defines the **eccentricity** of a hyperbola is $e = c/a$. Except in this case the number $c$ is given by $c = \sqrt{a^2 + b^2}$. Since $0 < a < \sqrt{a^2 + b^2}$, the eccentricity of an ellipse satisfies $e > 1$. As with the ellipse, the magnitude of the eccentricity of a hyperbola is an indicator of its shape. FIGURE 6.3.7 shows examples of two extreme cases: $e \approx 1$ and $e$ much bigger than 1.

EXAMPLE 5

Find the eccentricity of the hyperbola $\dfrac{y^2}{2} - \dfrac{(x-1)^2}{36} = 1$.

**Solution** Identifying $a^2 = 2$ and $b^2 = 36$, we get $c^2 = 2 + 36 = 38$. Thus the eccentricity of the given hyperbola is

$$e = \frac{c}{a} = \frac{\sqrt{38}}{\sqrt{2}} \approx 4.4.$$

We conclude that the hyperbola is one whose branches open widely as in Figure 6.3.7(b).  ∎

☐ **Applications of the Hyperbola** The hyperbola has several important applications involving sounding techniques. In particular, several navigational systems utilize hyperbolas as follows. Two fixed radio transmitters at a known distance from each other transmit synchronized signals. The difference in reception times by a navigator determines the difference $2a$ of the distances from the navigator to the two transmitters. This information locates the navigator somewhere on the hyperbola with foci at the transmitters and fixed difference in distances from the foci equal to $2a$. By using two sets of signals obtained from a single master station paired with each of two second stations, the long-range navigation system LORAN locates a ship or plane at the intersection of two hyperbolas. See FIGURE 6.3.8.

The next example illustrates the use of a hyperbola in another situation involving sounding techniques.

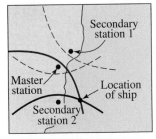

FIGURE 6.3.8 The idea behind LORAN

EXAMPLE 6

The sound of a dynamite blast is heard at different times by two observers at points $A$ and $B$. Knowing that the speed of sound is approximately 1100 ft/s or 335 m/s, it is determined that the blast occurred 1000 meters closer to point $A$ than to point $B$. If $A$ and $B$ are 2600 meters apart, show that the location of the blast lies on a branch of a hyperbola. Find an equation of the hyperbola.

**Solution** In FIGURE 6.3.9, we have placed the points $A$ and $B$ on the $x$-axis at $(1300, 0)$ and $(-1300, 0)$, respectively. If $P(x, y)$ denotes the location of the blast, then

$$d(P, B) - d(P, A) = 1000.$$

From the definition of the hyperbola on page 343 and the derivation following it, we see that this is the equation for the right branch of a hyperbola with fixed distance difference $2a = 1000$ and $c = 1300$. Thus the equation has the form

$$\frac{x^2}{a^2} - \frac{y^2}{b^2} = 1, \text{ where } x \geq 0,$$

or after solving for $x$,

$$x = a\sqrt{1 + \frac{y^2}{b^2}}.$$

With $a = 500$ and $c = 1300$, $b^2 = (1300)^2 - (500)^2 = (1200)^2$. Substituting in the foregoing equation gives

$$x = 500\sqrt{1 + \frac{y^2}{(1200)^2}} \quad \text{or} \quad x = \frac{5}{12}\sqrt{(1200)^2 + y^2}.  ∎$$

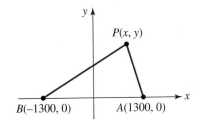

FIGURE 6.3.9 Graph for Example 6

To find the exact location of the blast in Example 6 we would need another observer hearing the blast at a third point $C$. Knowing the time between when this observer hears the blast and when the observer at $A$ hears the blast, we find a second hyperbola. The actual point of detonation is a point of intersection of the two hyperbolas.

There are many other applications of the hyperbola. As shown in FIGURE 6.3.10(a), a plane flying at a supersonic speed parallel to level ground leaves a hyperbolic sonic "footprint" on the ground. Like the parabola and ellipse, a hyperbola also possesses a reflecting property. The Cassegrain reflecting telescope shown in Figure 6.3.10(b) utilizes a convex hyperbolic secondary mirror to reflect a ray of light back through a hole to an eyepiece (or camera) behind the parabolic primary mirror. This telescope construction makes use of the fact that a beam of light directed along a line through one focus of a hyperbolic mirror will be reflected on a line through the other focus.

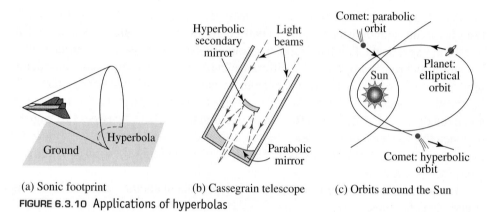

(a) Sonic footprint      (b) Cassegrain telescope      (c) Orbits around the Sun

FIGURE 6.3.10 Applications of hyperbolas

Orbits of objects in the universe can be parabolic, elliptic, or hyperbolic. When an object passes close to the Sun (or a planet), it is not necessarily captured by the gravitational field of the larger body. Under certain conditions, the object picks up a fractional amount of orbital energy of this much larger body and the resulting "slingshot-effect" orbit of the object as it passes the Sun is hyperbolic. See Figure 6.3.10(c).

**6.3** Exercises   Answers to selected odd-numbered problems begin on page ANS–19.

In Problems 1–20, find the center, foci, vertices, asymptotes, and eccentricity of the given hyperbola. Graph the hyperbola.

**1.** $\dfrac{x^2}{16} - \dfrac{y^2}{25} = 1$

**2.** $\dfrac{x^2}{4} - \dfrac{y^2}{4} = 1$

**3.** $\dfrac{y^2}{64} - \dfrac{x^2}{9} = 1$

**4.** $\dfrac{y^2}{6} - 4x^2 = 1$

**5.** $4x^2 - 16y^2 = 64$

**6.** $5x^2 - 5y^2 = 25$

**7.** $y^2 - 5x^2 = 20$

**8.** $9x^2 - 16y^2 + 144 = 0$

**9.** $\dfrac{(x-5)^2}{4} - \dfrac{(y+1)^2}{49} = 1$

**10.** $\dfrac{(x+2)^2}{10} - \dfrac{(y+4)^2}{25} = 1$

**11.** $\dfrac{(y-4)^2}{36} - x^2 = 1$

**12.** $\dfrac{(y-\frac{1}{4})^2}{4} - \dfrac{(x+3)^2}{9} = 1$

**13.** $25(x-3)^2 - 5(y-1)^2 = 125$

**14.** $10(x+1)^2 - 2(y-\frac{1}{2})^2 = 100$

**15.** $8(x + 4)^2 - 5(y - 7)^2 + 40 = 0$

**16.** $9(x - 1)^2 - 81(y - 2)^2 = 9$

**17.** $5x^2 - 6y^2 - 20x + 12y - 16 = 0$

**18.** $16x^2 - 25y^2 - 256x - 150y + 399 = 0$

**19.** $4x^2 - y^2 - 8x + 6y - 4 = 0$

**20.** $2x^2 - 9y^2 - 18x + 20y + 5 = 0$

In Problems 21–44, find an equation of the hyperbola that satisfies the given conditions.

**21.** Foci $(\pm 5, 0)$, $a = 3$

**22.** Foci $(\pm 10, 0)$, $b = 2$

**23.** Foci $(0, \pm 4)$, one vertex $(0, -2)$

**24.** Foci $(0, \pm 3)$, one vertex $(0, -\frac{3}{2})$

**25.** Foci $(\pm 4, 0)$, length of transverse axis 6

**26.** Foci $(0, \pm 7)$, length of transverse axis 10

**27.** Center $(0, 0)$, one vertex $(0, \frac{5}{2})$, one focus $(0, -3)$

**28.** Center $(0, 0)$, one vertex $(7, 0)$, one focus $(9, 0)$

**29.** Center $(0, 0)$, one vertex $(-2, 0)$, one focus $(-3, 0)$

**30.** Center $(0, 0)$, one vertex $(1, 0)$, one focus $(5, 0)$

**31.** Vertices $(0, \pm 8)$, asymptotes $y = \pm 2x$

**32.** Foci $(0, \pm 3)$, asymptotes $y = \pm \frac{3}{2}x$

**33.** Vertices $(\pm 2, 0)$, asymptotes $y = \pm \frac{4}{3}x$

**34.** Foci $(\pm 5, 0)$, asymptotes $y = \pm \frac{3}{5}x$

**35.** Center $(1, -3)$, one focus $(1, -6)$, one vertex $(1, -5)$

**36.** Center $(2, 3)$, one focus $(0, 3)$, one vertex $(3, 3)$

**37.** Foci $(-4, 2)$, $(2, 2)$, one vertex $(-3, 2)$

**38.** Vertices $(2, 5)$, $(2, -1)$, one focus $(2, 7)$

**39.** Vertices $(\pm 2, 0)$, passing through $(2\sqrt{3}, 4)$

**40.** Vertices $(0, \pm 3)$, passing through $(\frac{16}{5}, 5)$

**41.** Center $(-1, 3)$, one vertex $(-1, 4)$, passing through $(-5, 3 + \sqrt{5})$

**42.** Center $(3, -5)$, one vertex $(3, -2)$, passing through $(1, -1)$

**43.** Center $(2, 4)$, one vertex $(2, 5)$, one asymptote $2y - x - 6 = 0$

**44.** Eccentricity $\sqrt{10}$, endpoints of conjugate axis $(-5, 4)$, $(-5, 10)$

**45.** Three points are located at $A(-10, 16)$, $B(-2, 0)$, and $C(2, 0)$, where the units are kilometers. An artillery gun is known to lie on the line segment between $A$ and $C$, and using sounding techniques it is determined that the gun is 2 km closer to $B$ than to $C$. Find the point where the gun is located.

**46.** It can be shown that a ray of light emanating from one focus of a hyperbola will be reflected back along the line from the opposite focus. See **FIGURE 6.3.11**. A light ray from the left focus of the hyperbola $x^2/16 - y^2/20 = 1$ strikes the hyperbola at $(-6, -5)$. Find an equation of the reflected ray.

**47.** Find an equation of the hyperbola with foci $(0, -2)$ and $(8, 4)$ and fixed distance difference $2a = 8$. [*Hint:* See Problem 51 in Exercises 6.2.]

**48.** The **focal width** of a hyperbola is the length of a focal chord, that is, a line segment, perpendicular to the line containing the transverse axis and through a focus, with endpoints on the hyperbola. See **FIGURE 6.3.12**.

(a) Find the focal width of the hyperbola $x^2/4 - y^2/9 = 1$.

(b) Show that, in general, the focal width of the hyperbola $x^2/a^2 - y^2/b^2 = 1$ is $2b^2/a$.

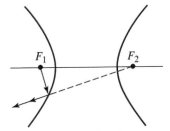

**FIGURE 6.3.11** Reflecting property in Problem 46

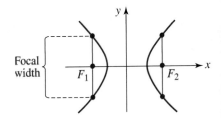

**FIGURE 6.3.12** Focal width in Problem 48

### For Discussion

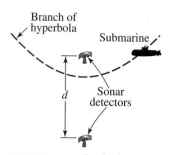

**FIGURE 6.3.13** Sonic detectors in Problem 49

**49.** Two sonar detectors are located at a distance $d$ from one another. Suppose that a sound (such as a sneeze aboard a submarine) is heard at the two detectors with a time delay $h$ between them. See **FIGURE 6.3.13**. Assume that sound travels in straight lines to the two detectors with speed $v$.

**(a)** Explain why $h$ cannot be larger than $d/v$.

**(b)** Explain why, for given values of $d$, $v$, and $h$, the source of the sound can be determined to lie on one branch of a hyperbola. [*Hint*: Where do you suppose that the foci might be?]

**(c)** Find an equation for the hyperbola in part (b), assuming that the detectors are at the points $(0, d/2)$ and $(0, -d/2)$. Express the answer in the standard form $y^2/a^2 - x^2/b^2 = 1$.

**50.** The hyperbolas

$$\frac{x^2}{a^2} - \frac{y^2}{b^2} = 1 \qquad \text{and} \qquad \frac{y^2}{b^2} - \frac{x^2}{a^2} = 1$$

are said to be **conjugates** of each other.

**(a)** Find the equation of the hyperbola that is conjugate to

$$\frac{x^2}{25} - \frac{y^2}{144} = 1.$$

**(b)** Discuss how the graphs of conjugate hyperbolas are related.

**51.** A **rectangular hyperbola** is one for which the asymptotes are perpendicular.

**(a)** Show that $y^2 - x^2 + 5y + 3x = 1$ is a rectangular hyperbola.

**(b)** Which of the hyperbolas given in Problems 1–20 are rectangular?

---

## 6.4 Polar Coordinates

☐ **Introduction** So far we have used the rectangular coordinate system to specify a point $P$ in the plane. We can regard this system as a grid of horizontal and vertical lines. The coordinates $(a, b)$ of a point $P$ are determined by the intersection of two lines: one line $x = a$ is perpendicular to the horizontal reference line called the $x$-axis, and the other $y = b$ is perpendicular to the vertical reference line called the $y$-axis. See **FIGURE 6.4.1(a)**. Another system for locating points in the plane is the **polar coordinate system**.

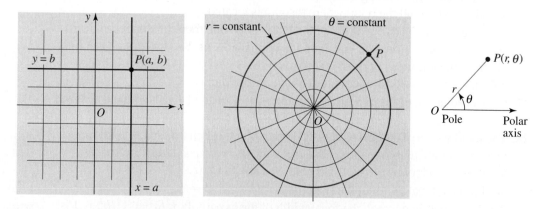

(a) Rectangular coordinate system  (b) Polar coordinate system  (c) Polar coordinates of $P$

**FIGURE 6.4.1** Comparison of rectangular and polar coordinates of a point $P$

□ **Polar Coordinates** To set up a **polar coordinate system**, we use a system of circles centered at a point $O$, called the **pole**, and straight lines or rays emanating from $O$. We take as a reference axis a horizontal half-line directed to the right of the pole and call it the **polar axis**. By specifying a directed (signed) distance $r$ from $O$ and an angle $\theta$ whose initial side is the polar axis, and whose terminal side is the ray $OP$, we label the point $P$ by $(r, \theta)$. We say that the ordered pair $(r, \theta)$ are the **polar coordinates** of $P$. See Figures 6.4.1(b) and 6.4.1(c).

Although the measure of the angle $\theta$ can be either in degrees or radians, in calculus radian measure is used almost exclusively. Consequently, we shall use only radian measure in this discussion.

In the polar coordinate system we adopt the following conventions.

## CONVENTIONS IN POLAR COORDINATES

(*i*) Angles $\theta > 0$ are measured counterclockwise from the polar axis, whereas angles $\theta < 0$ are measured clockwise.
(*ii*) To graph a point $(-r, \theta)$, where $-r < 0$, measure $|r|$ units along the ray $\theta + \pi$.
(*iii*) The coordinates of the pole $O$ are $(0, \theta)$, where $\theta$ is any angle.

### EXAMPLE 1     Plotting Polar Points

Plot the points whose polar coordinates are given.
(a) $(4, \pi/6)$    (b) $(2, -\pi/4)$    (c) $(-3, 3\pi/4)$

**Solution**

(a) Measure 4 units along the ray $\pi/6$ as shown in **FIGURE 6.4.2(a)**.
(b) Measure 2 units along the ray $-\pi/4$. See Figure 6.4.2(b).
(c) Measure 3 units along the ray $3\pi/4 + \pi = 7\pi/4$. Equivalently, we can measure 3 units along the ray $3\pi/4$ extended *backward* through the pole. Note carefully in Figure 6.4.2(c) that the point $(-3, 3\pi/4)$ is not in the same quadrant as the terminal side of the given angle.

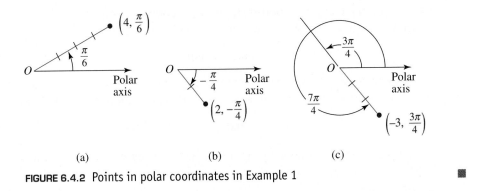

(a)             (b)             (c)

**FIGURE 6.4.2** Points in polar coordinates in Example 1

In contrast to the rectangular coordinate system, the description of a point in polar coordinates is not unique. This is an immediate consequence of the fact that

$$(r, \theta) \quad \text{and} \quad (r, \theta + 2n\pi), \ n \text{ an integer},$$

are equivalent. To compound the problem, negative values of $r$ can be used.

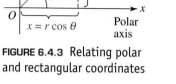

**FIGURE 6.4.3** Relating polar and rectangular coordinates

### ■ EXAMPLE 2      Equivalent Polar Points

The following coordinates are some alternative representations of the point $(2, \pi/6)$:

$$(2, 13\pi/6), \qquad (2, -11\pi/6), \qquad (-2, 7\pi/6), \qquad (-2, -5\pi/6). \quad ■$$

□ **Conversion of Polar Coordinates to Rectangular** By superimposing a rectangular coordinate system on a polar coordinate system, as shown in **FIGURE 6.4.3**, we can convert a polar description of a point to rectangular coordinates by using

$$x = r\cos\theta, \qquad y = r\sin\theta. \tag{1}$$

These conversion formulas hold true for any values of $r$ and $\theta$ in an equivalent polar representation of $(r, \theta)$.

### ■ EXAMPLE 3      Polar to Rectangular

Convert $(2, \pi/6)$ in polar coordinates to rectangular coordinates.
    **Solution** With $r = 2, \theta = \pi/6$, we have from (1),

$$x = 2\cos\frac{\pi}{6} = 2\left(\frac{\sqrt{3}}{2}\right) = \sqrt{3}$$

$$y = 2\sin\frac{\pi}{6} = 2\left(\frac{1}{2}\right) = 1.$$

Thus, $(2, \pi/6)$ is equivalent to $(\sqrt{3}, 1)$ in rectangular coordinates.    ■

□ **Conversion of Rectangular Coordinates to Polar** It should be evident from Figure 6.4.3 that $x, y, r$, and $\theta$ are also related by

$$r^2 = x^2 + y^2, \qquad \tan\theta = \frac{y}{x}. \tag{2}$$

The equations in (2) are used to convert the rectangular coordinates $(x, y)$ to the polar coordinates $(r, \theta)$.

### ■ EXAMPLE 4      Rectangular to Polar

Convert $(-1, 1)$ in rectangular coordinates to polar coordinates.
    **Solution** With $x = -1, y = 1$, we have from (2)

$$r^2 = 2 \qquad \text{and} \qquad \tan\theta = -1.$$

Now, $r^2 = 2$ or $r = \pm\sqrt{2}$, and two of many angles that satisfy $\tan\theta = -1$ are $3\pi/4$ and $7\pi/4$. From **FIGURE 6.4.4** we see that two polar representations for $(-1, 1)$ are

$$(\sqrt{2}, 3\pi/4) \qquad \text{and} \qquad (-\sqrt{2}, 7\pi/4). \quad ■$$

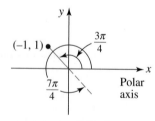

**FIGURE 6.4.4** Point in Example 4

    In Example 4, observe that we cannot pair just *any* angle $\theta$ and *any* value $r$ that satisfy (2); these solutions must also be consistent with (1). Because the points $(-\sqrt{2}, 3\pi/4)$ and $(\sqrt{2}, 7\pi/4)$ lie in the fourth quadrant, they are not polar representations of the second-quadrant point $(-1, 1)$.
    There are instances in calculus when a rectangular equation must be expressed as a polar equation $r = f(\theta)$. The next example shows how to do this using the conversion formulas in (1).

## EXAMPLE 5    Rectangular Equation to Polar Equation

Find a polar equation that has the same graph as the circle $x^2 + y^2 = 8x$.

**Solution** Substituting $x = r\cos\theta$, $y = r\sin\theta$, into the given equation we find

$$r^2\cos^2\theta + r^2\sin^2\theta = 8r\cos\theta$$

$$r^2(\cos^2\theta + \sin^2\theta) = 8r\cos\theta \qquad \leftarrow \cos^2\theta + \sin^2\theta = 1$$

$$r(r - 8\cos\theta) = 0.$$

The last equation implies that

$$r = 0 \quad \text{or} \quad r = 8\cos\theta.$$

since $r = 0$ determines only the pole $O$, we conclude that a polar equation of the circle is $r = 8\cos\theta$. Note that the circle $x^2 + y^2 = 8x$ passes through the origin since $x = 0$ and $y = 0$ satisfy the equation. Relative to the polar equation $r = 8\cos\theta$ of the circle, the origin or pole corresponds to the polar coordinates $(0, \pi/2)$. ∎

## EXAMPLE 6    Rectangular Equation to Polar Equation

Find a polar equation that has the same graph as the parabola $x^2 = 8(2 - y)$.

**Solution** We replace $x$ and $y$ in the given equation by $x = r\cos\theta$, $y = r\sin\theta$, and solve for $r$ in terms of $\theta$:

$$r^2\cos^2\theta = 8(2 - r\sin\theta)$$

$$r^2(1 - \sin^2\theta) = 16 - 8r\sin\theta$$

$$r^2 = r^2\sin^2\theta - 8r\sin\theta + 16 \qquad \leftarrow \text{right side is a perfect square}$$

$$r^2 = (r\sin\theta - 4)^2$$

$$r = \pm(r\sin\theta - 4).$$

Solving for $r$ gives two equations,

$$r = \frac{4}{1 + \sin\theta} \quad \text{or} \quad r = \frac{-4}{1 - \sin\theta}.$$

Now recall that by convention (*ii*), $(r, \theta)$ and $(-r, \theta + \pi)$ represent the same point. You should verify that if $(r, \theta)$ is replaced by $(-r, \theta + \pi)$ in the second of these two equations we obtain the first equation. In other words, the equations are equivalent and so we may simply take the polar equation of the parabola to be $r = 4/(1 + \sin\theta)$. ∎

## EXAMPLE 7    Polar Equation to Rectangular Equation

Find a rectangular equation that has the same graph as the polar equation $r^2 = 9\cos 2\theta$.

**Solution** First, we use the trigonometric identity for the cosine of a double angle:

$$r^2 = 9(\cos^2\theta - \sin^2\theta). \qquad \leftarrow \cos 2\theta = \cos^2\theta - \sin^2\theta$$

Then, from $r^2 = x^2 + y^2$, $\cos\theta = x/r$, $\sin\theta = y/r$, we have

$$x^2 + y^2 = 9\left(\frac{x^2}{x^2 + y^2} - \frac{y^2}{x^2 + y^2}\right) \quad \text{or} \quad (x^2 + y^2)^2 = 9(x^2 - y^2). \ ∎$$

The next section will be devoted to graphing polar equations.

In Problems 1–6, plot the point with the given polar coordinates.

**1.** $(3, \pi)$                                         **2.** $(2, -\pi/2)$
**3.** $(-\frac{1}{2}, \pi/2)$                               **4.** $(-1, \pi/6)$
**5.** $(-4, -\pi/6)$                              **6.** $(\frac{2}{3}, 7\pi/4)$

In Problems 7–12, find alternative polar coordinates that satisfy
**(a)** $r > 0, \theta < 0$                         **(b)** $r > 0, \theta > 2\pi$
**(c)** $r < 0, \theta > 0$                         **(d)** $r < 0, \theta < 0$
for each point with the given polar coordinates.

**7.** $(2, 3\pi/4)$                                **8.** $(5, \pi/2)$
**9.** $(4, \pi/3)$                               **10.** $(3, \pi/4)$
**11.** $(1, \pi/6)$                              **12.** $(3, 7\pi/6)$

In Problems 13–18, find the rectangular coordinates for each point with the given polar coordinates.

**13.** $(\frac{1}{2}, 2\pi/3)$                           **14.** $(-1, 7\pi/4)$
**15.** $(-6, -\pi/3)$                        **16.** $(\sqrt{2}, 11\pi/6)$
**17.** $(4, 5\pi/4)$                            **18.** $(-5, \pi/2)$

In Problems 19–24, find polar coordinates that satisfy
**(a)** $r > 0, -\pi < \theta \leq \pi$               **(b)** $r < 0, -\pi < \theta \leq \pi$
for each point with the given rectangular coordinates.

**19.** $(-2, -2)$                             **20.** $(0, -4)$
**21.** $(1, -\sqrt{3})$                          **22.** $(\sqrt{6}, \sqrt{2})$
**23.** $(7, 0)$                                 **24.** $(1, 2)$

In Problems 25–34, find a polar equation that has the same graph as the given rectangular equation.

**25.** $y = 5$                                   **26.** $x + 1 = 0$
**27.** $y = 7x$                                 **28.** $3x + 8y + 6 = 0$
**29.** $y^2 = -4x + 4$                        **30.** $x^2 - 12y - 36 = 0$
**31.** $x^2 + y^2 = 36$                      **32.** $x^2 - y^2 = 1$
**33.** $x^2 + y^2 + x = \sqrt{x^2 + y^2}$       **34.** $x^3 + y^3 - xy = 0$

## For Discussion

**35.** How would you express the distance $d$ between two points $(r_1, \theta_1)$ and $(r_2, \theta_2)$ in terms of their polar coordinates?

**36.** You know how to find a rectangular equation of a line through two points with rectangular coordinates. How would you find a polar equation of a line through two points with polar coordinates $(r_1, \theta_1)$ and $(r_2, \theta_2)$? Carry out your ideas by finding a polar equation of the line through $(3, 3\pi/4)$ and $(1, \pi/4)$. Find the polar coordinates of the $x$- and $y$-intercepts of the line.

# 6.5 Graphs of Polar Equations

☐ **Introduction** The graph of a polar equation $r = f(\theta)$ is the set of points $P$ with *at least* one set of polar coordinates that satisfies the equation. Since it is most likely that your classroom does not have a polar coordinate grid, to facilitate graphing and discussion of graphs of a polar equation $r = f(\theta)$ we will, as in the preceding section, superimpose a rectangular coordinate over the polar coordinate system.

We begin with some simple polar graphs.

### EXAMPLE 1    A Circle Centered at Origin

Graph $r = 3$.

**Solution** Since $\theta$ is not specified, the point $(3, \theta)$ lies on the graph of $r = 3$ for any value of $\theta$ and is 3 units from the origin. We see in **FIGURE 6.5.1** that the graph is the circle of radius 3 centered at the origin.

Alternatively, we know from (2) of Section 6.4 that $r = \pm\sqrt{x^2 + y^2}$ so that $r = 3$ yields the familiar rectangular equation $x^2 + y^2 = 3^2$ of a circle of radius 3 centered at the origin.  ■

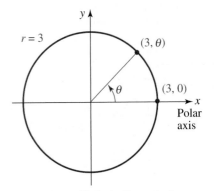

**FIGURE 6.5.1** Circle in Example 1

☐ **Circles Centered at the Origin** In general, if $a$ is any nonzero constant, the polar graph of

$$r = a \qquad (1)$$

is a circle of radius $|a|$ with center at the origin.

### EXAMPLE 2    A Ray Through the Origin

Graph $\theta = \pi/4$.

**Solution** Since $r$ is not specified, the point $(r, \pi/4)$ lies on the graph for any value of $r$. If $r > 0$, then this point lies on the half-line in the first quadrant; if $r < 0$, then the point lies on the half-line in the third quadrant. For $r = 0$, the point $(0, \pi/4)$ is the pole or origin. Therefore, the polar graph of $\theta = \pi/4$ is the line through the origin that makes an angle of $\pi/4$ with the polar axis or positive $x$-axis. See **FIGURE 6.5.2**.  ■

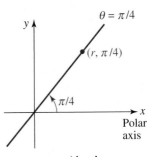

**FIGURE 6.5.2** Line in Example 2

☐ **Lines through the Origin** In general, if $\alpha$ is any nonzero real constant, the polar graph of

$$\theta = \alpha \qquad (2)$$

is a line through the origin that makes an angle of $\alpha$ radians with the polar axis.

### EXAMPLE 3    A Spiral

Graph $r = \theta$.

**Solution** As $\theta \geq 0$ increases, $r$ increases and the points $(r, \theta)$ wind around the pole in a counterclockwise manner. This is illustrated by the blue portion of the graph in **FIGURE 6.5.3**. The red portion of the graph is obtained by plotting points for $\theta < 0$.

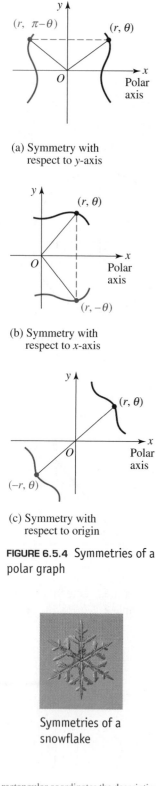

(a) Symmetry with respect to y-axis

(b) Symmetry with respect to x-axis

(c) Symmetry with respect to origin

**FIGURE 6.5.4** Symmetries of a polar graph

Symmetries of a snowflake

In rectangular coordinates the description of a point is unique. Hence in rectangular coordinates if a test for a particular type of symmetry fails, then we can definitely say that the graph does not possess that symmetry.

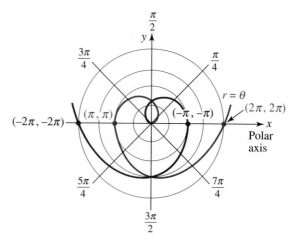

**FIGURE 6.5.3** Graph of equation in Example 3

☐ **Spirals** Many graphs in polar coordinates are given special names. The graph in Example 3 is a special case of

$$r = a\theta, \tag{3}$$

where $a$ is a constant. A graph of this equation is called a **spiral of Archimedes**. You are asked to graph other types of spiral curves in Problems 31 and 32 in Exercises 6.5.

In addition to basic point plotting, symmetry can often be utilized to graph a polar equation.

☐ **Symmetry** As shown in the figure, a polar graph can have three types of symmetry. A polar graph is **symmetric with respect to the y-axis** if whenever $(r, \theta)$ is a point on the graph, $(r, \pi - \theta)$ is also a point on the graph. A polar graph is **symmetric with respect to the x-axis** if whenever $(r, \theta)$ is a point on the graph, $(r, -\theta)$ is also a point on the graph. Finally, a polar graph is **symmetric with respect to the origin** if whenever $(r, \theta)$ is on the graph, $(-r, \theta)$ is also a point on the graph. **FIGURE 6.5.4** illustrates these three types of symmetries.

We have the following tests for symmetries.

### TESTS FOR SYMMETRY IN POLAR COORDINATES

The graph of a polar equation is:

 (i) **symmetric with respect to the y-axis** if replacing $(r, \theta)$ by $(r, \pi - \theta)$ results in the same equation;
 (ii) **symmetric with respect to the x-axis** if replacing $(r, \theta)$ by $(r, -\theta)$ results in the same equation;
 (iii) **symmetric with respect to the origin** if replacing $(r, \theta)$ by $(-r, \theta)$ results in the same equation.

Because the polar description of a point is not unique, the graph of a polar equation may still have a particular type of symmetry even though the test for it may fail. For example, if replacing $(r, \theta)$ by $(r, -\theta)$ fails to give the original polar equation, the graph of that equation may still possess symmetry with respect to the x-axis. Therefore, if one of the replacement tests in (i)–(iii) fails to give the same polar equation, the best we can say is "no conclusion."

**Graphing a Polar Equation**

Graph $r = 4 - 4\sin\theta$.

   **Solution** From the trigonometric sum formula for the sine function we have $\sin(\pi - \theta) = \sin\theta$. Thus by (*i*) the graph of $r = 3 - 3\sin\theta$ is symmetric with respect to the *y*-axis. By replacing, in turn, $\theta$ by $-\theta$ and $r$ by $-r$, we do not obtain equations that are the same as $r = 4 - 4\sin\theta$. Hence we can draw no conclusion about symmetry with respect to the *x*- or polar axis, and the origin or pole.

   Since the graph will be symmetric with respect to the *y*-axis, which corresponds to the line $\theta = \pi/2$, it suffices to make a table of *r* values using $-\pi/2 \le \theta \le \pi/2$. Plotting the points that correspond to the data in the following table and using symmetry, we obtain the graph given in **FIGURE 6.5.5**.

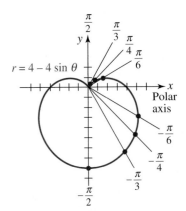

FIGURE 6.5.5 Graph of equation in Example 4

| $\theta$ | $-\pi/2$ | $-\pi/3$ | $-\pi/4$ | $-\pi/6$ | 0 | $\pi/6$ | $\pi/4$ | $\pi/3$ | $\pi/2$ |
|----------|----------|----------|----------|----------|---|---------|---------|---------|---------|
| $r$ | 8 | 7.5 | 6.8 | 6 | 4 | 2 | 1.2 | 0.5 | 0 |

☐ **Cardioids** The polar equation in Example 4 is a member of a family of equations that all have a "heart-shaped" graph that passes through the origin. A graph of any polar equation of the form

$$r = a \pm a\sin\theta \qquad \text{or} \qquad r = a \pm a\cos\theta \qquad (4)$$

is called a **cardioid**. The only difference in the graph of these four equations is their symmetry with respect to the *y*-axis $(r = a \pm a\sin\theta)$ or symmetry with respect to the *x*-axis $(r = a \pm a\cos\theta)$. See **FIGURE 6.5.6**.

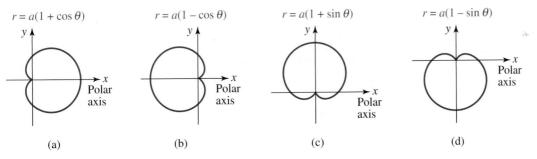

FIGURE 6.5.6 Cardioids

   By knowing the basic shape and orientation of a cardioid, you can obtain a quick and accurate graph by plotting the four points corresponding to $\theta = 0, \theta = \pi/2, \theta = \pi$, and $\theta = 3\pi/2$.

☐ **Limaçons** Cardioids are special cases of polar curves known as **limaçons**:

$$r = a \pm b\sin\theta \qquad \text{or} \qquad r = a \pm b\cos\theta. \qquad (5)$$

The shape of a limaçon depends on the relative magnitudes of *a* and *b*. Let us assume that $a > 0$ and $b > 0$. For $a/b < 1$, we get a **limaçon with an interior loop** as shown in **FIGURE 6.5.7(a)**. When $a = b$, or equivalently $a/b = 1$, we get a **cardioid**. For $1 < a/b < 2$, we get a **dimpled limaçon** as shown in Figure 6.5.7(b). For $a/b \ge 2$, the curve is called a **convex limaçon**. See Figure 6.5.7(c).

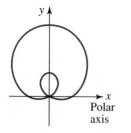

(a) Limaçon with interior loop

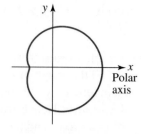

(b) Dimpled limaçon

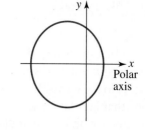

(c) Convex limaçon

**FIGURE 6.5.7** Three kinds of limaçons

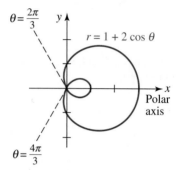

**FIGURE 6.5.8** Graph of equation in Example 6

### EXAMPLE 5     A Limaçon

The graph of $r = 3 - \sin\theta$ is a convex limaçon, since $a = 3, b = 1$, and $a/b = 3 > 2$. ∎

### EXAMPLE 6     A Limaçon

The graph of $r = 1 + 2\cos\theta$ is a limaçon with an interior loop, since $a = 1$, $b = 2$, and $a/b = \frac{1}{2} < 1$. For $\theta \geq 0$, notice in **FIGURE 6.5.8** that the limaçon starts at $\theta = 0$ or $(3, 0)$. The graph passes through the $y$-axis at $(1, \pi/2)$ and then enters the origin $(r = 0)$ for the first angle for which $r = 0$ or $1 + 2\cos\theta = 0$ or $\cos\theta = -\frac{1}{2}$. This implies that $\theta = 2\pi/3$. At $\theta = \pi$, the curve passes through $(-1, \pi)$. The remainder of the graph can then be completed using the fact that it is symmetric with respect to the $x$-axis. ∎

### EXAMPLE 7     A Rose Curve

Graph $r = 2\cos 2\theta$.

    **Solution** since

$$\cos(-2\theta) = \cos 2\theta \qquad \text{and} \qquad \cos 2(\pi - \theta) = \cos 2\theta$$

we conclude by (*i*) and (*ii*) of the tests for symmetry that the graph is symmetric with respect to both the $x$- and the $y$-axes. A moment of reflection should convince you that we need only consider $0 \leq \theta \leq \pi/2$. Using the data in the following table, we see that the dashed portion of the graph given in **FIGURE 6.5.9** is that completed by symmetry. The graph is called a **rose curve with four petals**.

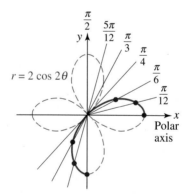

**FIGURE 6.5.9** Graph of equation in Example 7

| $\theta$ | 0 | $\pi/12$ | $\pi/6$ | $\pi/4$ | $\pi/3$ | $5\pi/12$ | $\pi/2$ |
|---|---|---|---|---|---|---|---|
| $r$ | 2 | 1.7 | 1 | 0 | $-1$ | $-1.7$ | 12 |

∎

☐  **Rose Curves**  In general, if $n$ is a positive integer, the graphs of

$$r = a\sin n\theta \qquad \text{or} \qquad r = a\cos n\theta, \quad n \geq 2 \tag{6}$$

are called **rose curves**, although as you can see in **FIGURE 6.5.10** the curve looks more like a daisy. When $n$ is odd, the number of **loops** or **petals** of the curve is $n$; if $n$ is even the curve has $2n$ petals. To graph a rose curve we can start by graphing one petal. To begin, we find an angle $\theta$ for which $r$ is a maximum. This gives the center line of the petal. We then find corresponding values of $\theta$ for which the rose curve enters the origin $(r = 0)$. To complete the graph we use the fact that the center lines of the petals are spaced $2\pi/n$ radians $(360/n$ degrees) apart if $n$ is odd, and $2\pi/2n = \pi/n$ radians $(180/n$ degrees) apart if $n$ is even. In Figure 6.5.10 we have drawn the graph of $r = a\sin 5\theta$, $a > 0$. The spacing between the center lines of the five petals is $2\pi/5$ radians $(72°)$.

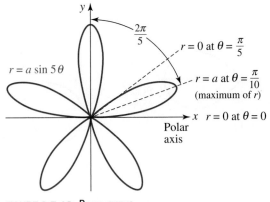

FIGURE 6.5.10 Rose curve

In Example 5 in Section 6.4 we saw that the polar equation $r = 8\cos\theta$ is equivalent to the rectangular equation $x^2 + y^2 = 8x$. By completing the square in $x$ in the rectangular equation, we recognize

$$(x - 4)^2 + y^2 = 16$$

as a circle of radius 4 centered at $(4, 0)$ on the $x$-axis. Polar equations such as $r = 8\cos\theta$ or $r = 8\sin\theta$ are circles and are also special cases of rose curves. See **FIGURE 6.5.11**.

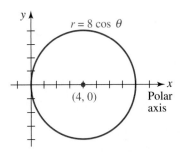

FIGURE 6.5.11 Graph of equation $r = 8\cos\theta$

☐ **Circles with Centers on an Axis** When $n = 1$ in (6) we get

$$r = a\sin\theta \qquad \text{or} \qquad r = a\cos\theta, \tag{7}$$

which are polar equations of circles passing through the origin with diameter $|a|$ with centers $(a/2, 0)$ on the $x$-axis $(r = a\cos\theta)$, or with centers $(0, a/2)$ on the $y$-axis $(r = a\sin\theta)$. **FIGURE 6.5.12** illustrates the graphs of the equations in (7) in the cases when $a > 0$ and $a < 0$.

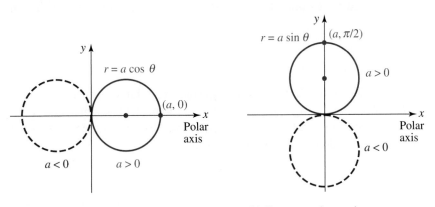

(a) Centers on the $x$-axis

(b) Centers on the $y$-axis

FIGURE 6.5.12 Circles through the origin with centers on an axis

☐ **Lemniscates** If $n$ is a positive integer, the graphs of

$$r^2 = a\cos 2\theta \qquad \text{or} \qquad r^2 = a\sin 2\theta \tag{8}$$

where $a > 0$, are called **lemniscates**. By (iii) of the tests for symmetry you can see the graphs of both of the equations in (1) are symmetric with respect to the origin. Moreover, by (ii) of the tests for symmetry the graph of $r^2 = a\cos 2\theta$ is symmetric with respect to the $x$-axis. **FIGURES 6.5.13(a)** and $6.5.13.$(b) show typical graphs of the equations $r^2 = a\cos 2\theta$ and $r^2 = a\sin 2\theta$, respectively.

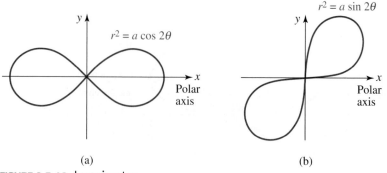

$r^2 = a \sin 2\theta$

$r^2 = a \cos 2\theta$

(a)　　　　　　　　　　(b)

**FIGURE 6.5.13** Lemniscates

☐ **Postscript—Points of Intersection** In rectangular coordinates we can find the points $(x, y)$ where the graphs of two functions $y = f(x)$ and $y = g(x)$ intersect by equating the $y$ values. The real solutions of the equation $f(x) = g(x)$ correspond to *all* the $x$-coordinates of the points where the graphs intersect. In contrast, problems may arise in polar coordinates when we try the same method to determine where the graphs of two polar equations $r = f(\theta)$ and $r = g(\theta)$ intersect.

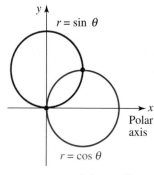

$r = \sin \theta$

$r = \cos \theta$

**FIGURE 6.5.14** Intersecting circles in Example 8

### ▮ EXAMPLE 8　　Intersecting Circles

**FIGURE 6.5.14** shows that the circles $r = \sin\theta$ and $r = \cos\theta$ have two points of intersection. By equating the $r$ values, the equation $\sin\theta = \cos\theta$ leads to $\theta = \pi/4$. Substituting this value into either equation yields $r = \sqrt{2}/2$. Thus we have found only a single polar point $(\sqrt{2}/2, \pi/4)$ where the graphs intersect. From the figure, it is apparent that the graphs also intersect at the origin. But the problem here is that the origin or pole is $(0, \pi/2)$ on the graph of $r = \cos\theta$ but is $(0, 0)$ on the graph of $r = \sin\theta$. This situation is analogous to the curves reaching the same point at different times.　■

Example 8 illustrates one of several frustrating difficulties of working in polar coordinates:

> *A point can be on the graph of a polar equation even though its coordinates do not satisfy the equation.*

You should verify that $(2, \pi/2)$ is an alternative polar description of the point $(-2, 3\pi/2)$. Moreover, verify that $(-2, 3\pi/2)$ is a point on the graph of $r = 1 + 3\sin\theta$ by showing that the coordinates satisfy the equation. However, note that the alternative coordinates $(2, \pi/2)$ do not satisfy the equation.

▶ This cannot happen in rectangular coordinates.

| **6.5** | Exercises | Answers to selected odd-numbered problems begin on page ANS–20. |

In Problems 1–30, identify by name the graph of the given polar equation. Then sketch the graph of the equation.

1. $r = 6$
2. $r = -1$
3. $\theta = \pi/3$
4. $\theta = 5\pi/6$
5. $r = 2\theta, \theta \le 0$
6. $r = 3\theta, \theta \ge 0$
7. $r = 1 + \cos\theta$
8. $r = 5 - 5\sin\theta$
9. $r = 2(1 + \sin\theta)$
10. $2r = 1 - \cos\theta$

**11.** $r = 1 - 2\cos\theta$

**12.** $r = 2 + 4\sin\theta$

**13.** $r = 4 - 3\sin\theta$

**14.** $r = 3 + 2\cos\theta$

**15.** $r = 4 + \cos\theta$

**16.** $r = 4 - 2\sin\theta$

**17.** $r = \sin 2\theta$

**18.** $r = 3\sin 4\theta$

**19.** $r = 3\cos 3\theta$

**20.** $r = 2\sin 3\theta$

**21.** $r = \cos 5\theta$

**22.** $r = 2\sin 9\theta$

**23.** $r = 6\cos\theta$

**24.** $r = -2\cos\theta$

**25.** $r = -3\sin\theta$

**26.** $r = 5\sin\theta$

**27.** $r^2 = 4\sin 2\theta$

**28.** $r^2 = 4\cos 2\theta$

**29.** $r^2 = -25\cos 2\theta$

**30.** $r^2 = -9\sin 2\theta$

In Problems 31 and 32, the graph of the given equation is a spiral. Sketch its graph.

**31.** $r = 2^\theta, \theta \geq 0$ (logarithmic)

**32.** $r\theta = \pi, \theta > 0$ (hyperbolic)

In Problems 33–36, find the points of intersection of the graphs of the given pair of polar equations.

**33.** $r = 2, \ r = 4\sin\theta$

**34.** $r = \sin\theta, \ r = \sin 2\theta$

**35.** $r = 1 - \cos\theta, \ r = 1 + \cos\theta$

**36.** $r = 3 - 3\cos\theta, \ r = 3\cos\theta$

## Calculator/Computer Problems

**37.** Use a graphing utility to obtain the graph of the **bifolium** $r = 4\sin\theta \cos^2\theta$ and the circle $r = \sin\theta$ on the same axes. Find all points of intersection of the graphs.

**38.** Use a graphing utility to verify that the cardioid $r = 1 + \cos\theta$ and the lemniscate $r^2 = 4\cos\theta$ intersect at four points. Find these points of intersection of the graphs.

## For Discussion

In Problems 39 and 40, suppose $r = f(\theta)$ is a polar equation. Graphically interpret the given property.

**39.** $f(-\theta) = f(\theta)$ (even function)

**40.** $f(-\theta) = -f(\theta)$ (odd function)

## 6.6 Conic Sections in Polar Coordinates

☐ **Introduction** In the first three sections of this chapter we derived equations for the parabola, ellipse, and hyperbola using the distance formula in rectangular coordinates. By using polar coordinates and the concept of eccentricity, we can now give one general definition of a conic section that encompasses all three curves.

> **CONIC SECTION**
>
> Let $L$ be a fixed line in the plane, and let $F$ be a point not on the line. A **conic section** is the set of points $P$ in the plane for which the distance from $P$ to $F$ divided by the distance from $P$ to $L$ is a constant.

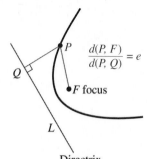

**FIGURE 6.6.1** Geometric interpretation of (1)

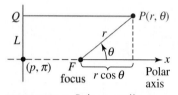

**FIGURE 6.6.2** Polar coordinate interpretation of (2)

The fixed line $L$ is called a **directrix**, and the point $F$ is a **focus**. The fixed constant is the **eccentricity** $e$ of the conic. As **FIGURE 6.6.1** shows, the point $P$ lies on the conic if and only if

$$\frac{d(P, F)}{d(P, Q)} = e, \tag{1}$$

where $Q$ denotes the foot of the perpendicular from $P$ to $L$. In (1), if

- $e = 1$, the conic is a **parabola**;
- $0 < e < 1$, the conic is an **ellipse**; and
- $e > 1$, the conic is a **hyperbola**.

☐ **Polar Equations of Conics** Equation (1) is readily interpreted using polar coordinates. Suppose $F$ is placed at the pole and $L$ is $P$ units ($p > 0$) to the left of $F$ perpendicular to the extended polar axis. We see from **FIGURE 6.6.2** that (1) written as $d(P, F) = e\,d(P, Q)$ is the same as

$$r = e(p + r\cos\theta) \qquad \text{or} \qquad r - er\cos\theta = ep. \tag{2}$$

Solving for $r$ yields,

$$r = \frac{ep}{1 - e\cos\theta}. \tag{3}$$

To see that (3) yields the familiar equations of the conics, let us superimpose a rectangular coordinate system on the polar coordinate system with origin at the pole and the positive $x$-axis coinciding with the polar axis. We then express the first equation in (2) in rectangular coordinates and simplify:

$$\pm\sqrt{x^2 + y^2} = ex + ep$$
$$x^2 + y^2 = e^2x^2 + 2e^2px + e^2p^2$$
$$(1 - e^2)x^2 - 2e^2px + y^2 = e^2p^2. \tag{4}$$

Choosing $e = 1$, (4) becomes

$$-2px + y^2 = p^2 \qquad \text{or} \qquad y^2 = 2p\left(x + \frac{p}{2}\right),$$

which is an equation in standard form of a parabola whose axis is the $x$-axis, vertex is at $(-p/2, 0)$ and, consistent with the placement of $F$, whose focus is at the origin.

It is a good exercise in algebra to show that (2) yields standard form equations of an ellipse in the case $0 < e < 1$ and a hyperbola in the case $e > 1$. See Problem 35 in Exercises 6.6. Thus, depending on the value of $e$, the polar equation (3) can have three possible graphs as shown in **FIGURE 6.6.3**.

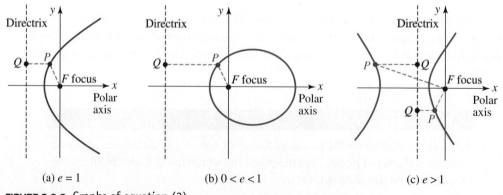

(a) $e = 1$      (b) $0 < e < 1$      (c) $e > 1$

**FIGURE 6.6.3** Graphs of equation (3)

CHAPTER 6 CONIC SECTIONS

If we had placed the focus $F$ to the *left* of the directrix in our derivation of the polar equation (3), then the equation $r = ep/(1 + e\cos\theta)$ would be obtained. When the directrix $L$ is chosen parallel to the polar axis (that is, horizontal), then the equation of the conic is found to be either $r = ep/(1 - e\sin\theta)$ or $r = ep/(1 + e\sin\theta)$. A summary of the preceding discussion is given next.

## POLAR EQUATIONS OF CONICS

Any polar equation of the form

$$r = \frac{ep}{1 \pm e\cos\theta} \qquad (5)$$

or

$$r = \frac{ep}{1 \pm e\sin\theta} \qquad (6)$$

is a conic section with focus at the origin and axis along a coordinate axis. The axis of the conic section is along the $x$-axis for equations of the form (5) and along the $y$-axis for equations of the form (6). The conic is a parabola if $e = 1$, an ellipse if $0 < e < 1$, and a hyperbola if $e > 1$.

## EXAMPLE 1        Identifying Conics

Identify each of the following conics:

**(a)** $r = \dfrac{2}{1 - 2\sin\theta}$     **(b)** $r = \dfrac{3}{4 + \cos\theta}$

**Solution**

**(a)** A term-by-term comparison of the given equation with the polar form $r = ep/(1 - e\sin\theta)$ enables us to make the identification $e = 2$. Hence the conic is a hyperbola.

**(b)** In order to identify the conic section, we divide the numerator and the denominator of the given equation by 4. This puts the equation into the form

$$r = \frac{\frac{3}{4}}{1 + \frac{1}{4}\cos\theta}.$$

Then by comparison with $r = ep/(1 + e\cos\theta)$ we see that $e = \frac{1}{4}$. Hence the conic is an ellipse. ∎

☐ **Graphs**  A rough graph of a conic defined by (5) or (6) can be obtained by knowing the orientation of its axis, finding the $x$- and $y$-intercepts, and finding the vertices. In the case of (5),

- the two vertices of the **ellipse** or a **hyperbola** occur at $\theta = 0$ and $\theta = \pi$; the vertex of a **parabola** can occur at only one of the values: $\theta = 0$ or $\theta = \pi$.

For (6),

- the two vertices of an **ellipse** or a **hyperbola** occur at $\theta = \pi/2$ and $\theta = 3\pi/2$; the vertex of a **parabola** can occur at only one of the values: $\theta = \pi/2$ or $\theta = 3\pi/2$.

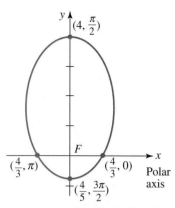

FIGURE 6.6.4 Graph of polar equation in Example 2

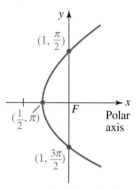

FIGURE 6.6.5 Graph of polar equation in Example 3

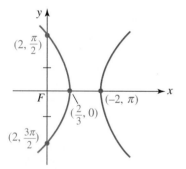

FIGURE 6.6.6 Graph of polar equation in Example 4

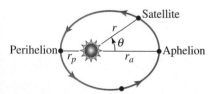

FIGURE 6.6.7 Orbit of satellite around the Sun

**EXAMPLE 2**        **Graphing a Conic**

Graph $r = \dfrac{4}{3 - 2\sin\theta}$.

**Solution** By writing the equation as $r = \dfrac{\frac{4}{3}}{1 - \frac{2}{3}\sin\theta}$, we see that the eccentricity is $e = \frac{2}{3}$ and so the conic is an ellipse. Moreover, because the equation is of the form given in (6), we know that the axis of the ellipse is vertical along the $y$-axis. Now in view of the discussion preceding this example, we obtain:

$$\begin{aligned} vertices: &\quad (4, \pi/2), \left(\tfrac{4}{5}, 3\pi/2\right) \\ x\text{-}intercepts: &\quad \left(\tfrac{4}{3}, 0\right), \left(\tfrac{4}{3}, \pi\right). \end{aligned}$$

The graph of the equation is given in **FIGURE 6.6.4**.  ■

**EXAMPLE 3**        **Graphing a Conic**

Graph $r = \dfrac{1}{1 - \cos\theta}$.

**Solution** Inspection of the equation shows that it is of the form given in (5) with $e = 1$. Hence the conic section is a parabola whose axis is horizontal along the $x$-axis. Since $r$ is undefined at $\theta = 0$, the vertex of the parabola occurs at $\theta = \pi$:

$$\begin{aligned} vertex: &\quad \left(\tfrac{1}{2}, \pi\right) \\ y\text{-}intercepts: &\quad (1, \pi/2), (1, 3\pi/2). \end{aligned}$$

The graph of the equation is given in **FIGURE 6.6.5**.  ■

**EXAMPLE 4**        **Graphing a Conic**

Graph $r = \dfrac{2}{1 + 2\cos\theta}$.

**Solution** From (5) we see that $e = 2$, and so the conic section is a hyperbola whose axis is horizontal along the $x$-axis. The vertices—the endpoints of the transverse axis of the hyperbola—occur at $\theta = 0$ and at $\theta = \pi$:

$$\begin{aligned} vertices: &\quad \left(\tfrac{2}{3}, 0\right), (-2, \pi) \\ y\text{-}intercepts: &\quad (2, \pi/2), (2, 3\pi/2). \end{aligned}$$

The graph of the equation is given in **FIGURE 6.6.6**.  ■

☐ **Applications** Equations of the type in (4) and (5) are well-suited to describe a closed orbit of satellite around the Sun (Earth or Moon) since such an orbit is an ellipse with the Sun (Earth or Moon) at one focus. Suppose that an equation of the orbit is given by $r = ep/(1 - e\cos\theta), 0 < e < 1$, and $r_p$ is the value of $r$ at perihelion (perigee or perilune) and $r_a$ is the value of $r$ at aphelion (apogee or apolune). These are the points in the orbit, occurring on the $x$-axis, at which the satellite is closest and farthest, respectively, from the Sun (Earth or Moon). See **FIGURE 6.6.7**. It is left as an exercise to show that the eccentricity $e$ of the orbit is related to $r_p$ and $r_a$ by

$$e = \frac{r_a - r_p}{r_a + r_p}. \tag{7}$$

## EXAMPLE 5     Finding a Polar Equation of an Orbit

Find a polar equation of the orbit of the planet Mercury around the Sun if $r_p = 2.85 \times 10^7$ miles and $r_a = 4.36 \times 10^7$ miles.

**Solution** From (7), the eccentricity of Mercury's orbit is

$$e = \frac{4.36 \times 10^7 - 2.85 \times 10^7}{4.36 \times 10^7 + 2.85 \times 10^7} = 0.21.$$

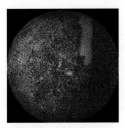

Mercury is the closest planet to the Sun

Hence
$$r = \frac{0.21p}{1 - 0.21\cos\theta}.$$

All we need do now is to solve for the quantity $0.21p$. To do this we use the fact that aphelion occurs at $\theta = 0$:

$$4.36 \times 10^7 = \frac{0.21p}{1 - 0.21}.$$

The last equation yields $0.21p = 3.44 \times 10^7$. Hence a polar equation of Mercury's orbit is

$$r = \frac{3.44 \times 10^7}{1 - 0.21\cos\theta}. \qquad ∎$$

---

## 6.6   Exercises    Answers to selected odd-numbered problems begin on page ANS–21.

In Problems 1–10, determine the eccentricity, identify the conic, and sketch its graph.

**1.** $r = \dfrac{2}{1 - \sin\theta}$          **2.** $r = \dfrac{2}{2 - \cos\theta}$

**3.** $r = \dfrac{16}{4 + \cos\theta}$          **4.** $r = \dfrac{5}{2 + 2\sin\theta}$

**5.** $r = \dfrac{4}{1 + 2\sin\theta}$          **6.** $r = \dfrac{-4}{\cos\theta - 1}$

**7.** $r = \dfrac{18}{3 - 6\cos\theta}$          **8.** $r = \dfrac{4\csc\theta}{3\csc\theta + 2}$

**9.** $r = \dfrac{6}{1 - \cos\theta}$          **10.** $r = \dfrac{2}{2 + 5\cos\theta}$

In Problems 11–14, determine the eccentricity $e$ of the given conic. Then convert the polar equation to a rectangular equation and verify that $e = c/a$.

**11.** $r = \dfrac{6}{1 + 2\sin\theta}$          **12.** $r = \dfrac{10}{2 - 3\cos\theta}$

**13.** $r = \dfrac{12}{3 - 2\cos\theta}$          **14.** $r = \dfrac{2\sqrt{3}}{\sqrt{3} + \sin\theta}$

In Problems 15–20, find a polar equation of the conic with focus at the origin that satisfies the given conditions.

**15.** $e = 1$, directrix $x = 3$

**16.** $e = \frac{3}{2}$, directrix $y = 2$

**17.** $e = \frac{2}{3}$, directrix $y = -2$

**18.** $e = \frac{1}{2}$, directrix $x = 4$

**19.** $e = 2$, directrix $x = 6$

**20.** $e = 1$, directrix $y = -2$

In Problems 21–26, find a polar equation of the parabola with focus at the origin and the given vertex.

**21.** $\left(\frac{3}{2}, 3\pi/2\right)$

**22.** $(2, \pi)$

**23.** $\left(\frac{1}{2}, \pi\right)$

**24.** $(2, 0)$

**25.** $\left(\frac{1}{4}, 3\pi/2\right)$

**26.** $\left(\frac{3}{2}, \pi/2\right)$

**27.** A communications satellite is 12,000 km above the Earth at its apogee. The eccentricity of its elliptical orbit is 0.2. Use (7) to find its perigee distance.

**28.** Find a polar equation $r = ep/(1 - e\cos\theta)$ of the orbit of the satellite in Problem 27.

**29.** Find a polar equation of the orbit of the Earth around the Sun if $r_p = 1.47 \times 10^8$ km and $r_a = 1.52 \times 10^8$ km.

**30. (a)** The eccentricity of the elliptical orbit of Comet Halley is 0.97 and the length of the major axis of its orbit is $3.34 \times 10^9$ mi. Find a polar equation of its orbit of the form $r = ep/(1 - e\cos\theta)$.

    **(b)** Use the equation in part (a) to obtain $r_p$ and $r_a$ for the orbit of Comet Halley.

Next visit of Comet Halley to the Solar System will be in 2061

## Calculator/Computer Problems

The orbital characteristics (eccentricity, perigee, and major axis) of a satellite near the Earth gradually degrade over time due to many small forces acting on the satellite other than the gravitational force of the Earth. These forces include atmospheric drag, the gravitational attractions of the Sun and the Moon, and magnetic forces. Approximately once a month tiny rockets are activated for a few seconds in order to "boost" the orbital characteristics back into the desired range. Rockets are turned on longer to create a major change in the orbit of a satellite. The most fuel-efficient way to move from an inner orbit to an outer orbit, called a **Hohmann transfer**, is to add velocity in the direction of flight at the time the satellite reaches perigee on the inner orbit, follow the Hohmann transfer ellipse halfway around to its apogee, and add velocity again to achieve the outer orbit. A similar process (subtracting velocity at apogee on the outer orbit and subtracting velocity at perigee on the Hohmann transfer orbit) moves a satellite from an outer orbit to an inner orbit.

In Problems 31–34, use a graphic calculator or computer to superimpose the graphs of the given three polar equations on the same coordinate axes. Print out your result and use a colored pencil to trace out the Hohmann transfer.

**31.** Inner orbit $r = \dfrac{24}{1 + 0.2\cos\theta}$, Hohmann transfer $r = \dfrac{32}{1 + 0.6\cos\theta}$,

    outer orbit $r = \dfrac{56}{1 + 0.3\cos\theta}$

**32.** Inner orbit $r = \dfrac{5.5}{1 + 0.1\cos\theta}$, Hohmann transfer $r = \dfrac{7.5}{1 + 0.5\cos\theta}$,

    outer orbit $r = \dfrac{13.5}{1 + 0.1\cos\theta}$

**33.** Inner orbit $r = 9$, Hohmann transfer $r = \dfrac{15.3}{1 + 0.7\cos\theta}$,

outer orbit $r = 51$

**34.** Inner orbit $r = \dfrac{73.5}{1 + 0.05\cos\theta}$, Hohmann transfer $r = \dfrac{77}{1 + 0.1\cos\theta}$,

outer orbit $r = \dfrac{84.7}{1 + 0.01\cos\theta}$

## For Discussion

**35.** Show that (2) yields standard form equations of an ellipse in the case $0 < e < 1$ and a hyperbola in the case $e > 1$.

**36.** The graph of polar equation

$$r = \frac{4}{3 + 3\cos(\theta - 5\pi/6)}$$

is a conic section. Graph the equation and find an equation of the axis of the conic.

## 6.7   Parametric Equations

$\displaystyle\int$ **Calculus PREVIEW**    ☐   Rectangular equations and polar equations are not the only— and often not the most convenient—ways of describing curves in the coordinate plane. In this section we will consider a different way of representing a curve that is important in the many applications of calculus. Let's consider one example. The motion of a particle along a curve, in contrast to a straight line, is called *curvilinear motion*. If it is assumed that a golf ball is hit off the ground, perfectly straight (no hook or slice), and that its path stays in a coordinate plane as shown in **FIGURE 6.7.1**, then with the help of physics and calculus it can be shown that its $x$- and $y$-coordinates at time $t$ are given by

$$x = (v_0\cos\theta_0)t, \qquad y = -\tfrac{1}{2}gt^2 + (v_0\sin\theta_0)t, \tag{1}$$

where $\theta_0$ is the launch angle, $v_0$ is its initial velocity, and $g = 32$ ft/s$^2$ is the acceleration due to gravity. These equations, which give the golf ball's position in the coordinate plane at time $t$, are said to be **parametric equations**. The third variable $t$ in (1) is called a **parameter** and restricted to some interval $0 \le t \le T$, where $t = 0$ gives the origin $(0, 0)$, and $t = T$ is the time the ball hits the ground.

In general, a curve in a coordinate plane can be *defined* in terms of parametric equations.

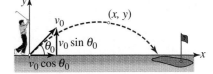

**FIGURE 6.7.1** Fore!

---

### PLANE CURVE

A **plane curve** is a set $C$ of ordered pairs $(f(t), g(t))$, where $f$ and $g$ are functions defined on a common interval $I$. The equations

$$x = f(t), y = g(t), \text{ for } t \text{ in } I,$$

are called **parametric equations** for $C$. The variable $t$ is called a **parameter**.

---

It is also common practice to refer to $x = f(t), y = g(t)$, for $t$ in $I$, as a **parameterization** for $C$.

The **graph** of a plane curve $C$ is the set of all points $(x, y)$ in the coordinate plane corresponding to the ordered pairs $(f(t), g(t))$. Hereafter, we will refer to a plane curve as a **curve** or as a **parameterized curve**.

### EXAMPLE 1          Graph of a Parametric Curve

Graph the curve $C$ that has the parametric equations

$$x = t^2, \qquad y = t^3, \qquad -1 \leq t \leq 2.$$

**Solution** As shown in the accompanying table, for any choice of $t$ in the interval $[-1, 2]$, we obtain a single ordered pair $(x, y)$. By connecting the points with a curve, we obtain the graph in **FIGURE 6.7.2**.

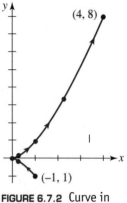

**FIGURE 6.7.2** Curve in Example 1

| $t$ | $-1$ | $-\frac{1}{2}$ | $0$ | $\frac{1}{2}$ | $1$ | $\frac{3}{2}$ | $2$ |
|---|---|---|---|---|---|---|---|
| $x$ | $1$ | $\frac{1}{4}$ | $0$ | $\frac{1}{4}$ | $1$ | $\frac{9}{4}$ | $4$ |
| $y$ | $-1$ | $-\frac{1}{8}$ | $0$ | $\frac{1}{8}$ | $1$ | $\frac{27}{8}$ | $8$ |

In Example 1, if we think in terms of motion and $t$ as time, then as $t$ increases from $-1$ to $2$, a point $P$ defined as $(t^2, t^3)$ starts from $(1, -1)$, advances up the lower branch to the origin $(0, 0)$, passes to the upper branch, and finally stops at $(4, 8)$. In general, as we plot points corresponding to *increasing values* of the parameter, the curve $C$ is traced out by $(f(t), g(t))$ in a certain *direction* indicated by the arrowheads on the curve in Figure 6.7.2. This direction is called the **orientation** of the curve $C$.

A parameter need have no relation to time. When the interval $I$ over which $f$ and $g$ in (1) are defined is a closed interval $[a, b]$, we say that $(f(a), g(a))$ is the **initial point** of the curve $C$ and that $(f(b), g(b))$ is its **terminal point**. In Example 1 the initial point is $(1, -1)$ and the terminal point is $(4, 8)$. If the terminal point is the same as the initial point, that is,

$$(f(a), g(a)) = (f(b), g(b))$$

then $C$ is a **closed curve**. If $C$ is closed but does not cross itself, then it is called a **simple closed curve**. In **FIGURE 6.7.3**, $A$ and $B$ represent the initial and terminal points, respectively.

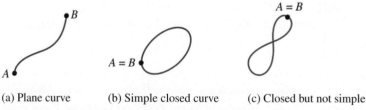

(a) Plane curve          (b) Simple closed curve          (c) Closed but not simple

**FIGURE 6.7.3** Some plane curves

The next example illustrates a simple closed curve.

| EXAMPLE 2 | **A Parameterization of a Circle** |

Find a parameterization for the circle $x^2 + y^2 = a^2$.

**Solution** The circle has center at the origin and radius $a$. If $t$ represents the central angle, that is, an angle with vertex at the origin and initial side coinciding with the positive $x$-axis, then as shown in FIGURE 6.7.4 the equations

$$x = a\cos t, \qquad y = a\sin t, \qquad 0 \le t \le 2\pi \qquad (2)$$

give every point $P$ on the circle. For example, at $t = \pi/2$ we get $x = 0$ and $y = a$; in other words, the initial point is $(0, a)$. The terminal point corresponds to $t = 2\pi$ and is also $(a, 0)$. Since the initial and terminal points are the same, this proves the obvious—that the curve $C$ defined by the parametric equations (2) is a closed curve. Note the orientation of $C$ in Figure 6.7.4; as $t$ increases from 0 to $2\pi$, the point $P$ traces out $C$ in a counterclockwise direction. ■

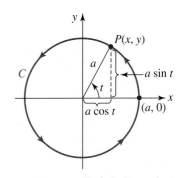

FIGURE 6.7.4 Circle in Example 2

In Example 2, the upper *semicircle* $x^2 + y^2 = a^2, 0 \le y \le a$, is defined parametrically by restricting the parameter interval to $[0, \pi]$,

$$x = a\cos t, \qquad y = a\sin t, \qquad 0 \le t \le \pi.$$

Observe that when $t = \pi$, the terminal point is now $(-a, 0)$. On the other hand, if we wish to describe *two* complete counterclockwise revolutions around the circle, we again modify the parameter interval by writing

$$x = a\cos t, \qquad y = a\sin t, \qquad 0 \le t \le 4\pi.$$

☐ **Eliminating the Parameter** Given a set of parametric equations, we sometimes desire to eliminate or clear the parameter to obtain a rectangular equation for the curve. To eliminate the parameter in (2), we simply square $x$ and $y$ and add the two equations:

$$x^2 + y^2 = a^2\cos^2 t + a^2\sin^2 t \qquad \text{implies} \qquad x^2 + y^2 = a^2$$

since $\sin^2 t + \cos^2 t = 1$.

| EXAMPLE 3 | **Eliminating the Parameter** |

**(a)** From the first equation in (2) we have $t = x/(v_0\cos\theta_0)$. Substituting this into the second equation then gives

$$y = -\frac{g}{2(v_0\cos\theta_0)^2}x^2 + (\tan\theta_0)x.$$

Since $v_0$, $\theta_0$, and $g$ are constants, the last equation has the form $y = ax^2 + bx$ and so the trajectory of any projectile launched at the angle $0 < \theta_0 < \pi/2$ is a parabolic arc.
**(b)** In Example 1, we can eliminate the parameter by solving the second equation for $t$ in terms of $y$ and then substituting in the first equation. We find

$$t = y^{1/3} \qquad \text{and so} \qquad x = (y^{1/3})^2 = y^{2/3}.$$

The curve shown in Figure 6.7.2 is only a portion of the graph of $x = y^{2/3}$. For $-1 \le t \le 2$, we have correspondingly $-1 \le y \le 8$. Thus, a rectangular equation for the curve in Example 2 is given by $x = y^{2/3}, -1 \le y \le 8$. ■

A curve *C* can have many different parameterizations.

A curve *C* can have more than one parameterization. For example, an alternative parameterization for the circle in Example 2 is

$$x = a\cos 2t, \qquad y = a\sin 2t, \qquad 0 \le t \le \pi.$$

Note that the parameter interval is now $[0, \pi]$. We see that as $t$ increases from 0 to $\pi$, the new angle $2t$ increases from 0 to $2\pi$.

### ■ EXAMPLE 4        Alternative Parameterizations

Consider the curve *C* that has the parametric equations $x = t$, $y = 2t^2$, $-\infty < t < \infty$. We can eliminate the parameter by using $t = x$ and substituting in $y = 2t^2$. This gives the rectangular equation $y = 2x^2$, which we recognize as a parabola. Moreover, since $-\infty < t < \infty$ is equivalent to $-\infty < x < \infty$, the point $(t, 2t^2)$ traces out the complete parabola $y = 2x^2$, $-\infty < x < \infty$.

    An alternative parameterization of *C* is given by $x = t^3/4$, $y = t^6/8$, $-\infty < t < \infty$. Using $t^3 = 4x$ and substituting in $y = t^6/8$ or $y = (t^3 \cdot t^3)/8$ gives $y = ((4x)^2/8) = 2x^2$. Moreover, $-\infty < t < \infty$ implies $-\infty < t^3 < \infty$ and so $-\infty < x < \infty$.   ■

We note in Example 4 that a point on *C* need not correspond to the same value of the parameter in each set of parametric equations for *C*. For example, $(1, 2)$ is obtained for $t = 1$ in $x = t$, $y = 2t^2$, but $t = \sqrt[3]{4}$ yields $(1, 2)$ in $x = t^3/4$, $y = t^6/8$.

### ■ EXAMPLE 5        Example 4 Revisited

Proceed with caution when eliminating the parameter.

One has to be careful when working with parametric equations. Eliminating the parameter in $x = t^2$, $y = 2t^4$, $-\infty < t < \infty$, would seem to yield the same parabola $y = 2x^2$ as in Example 4. However, this is *not* the case because for any value of $t$, $t^2 \ge 0$ and so $x \ge 0$. In other words, the last set of equations is a parametric representation of only the right-hand branch of the parabola, that is, $y = 2x^2$, $0 \le x < \infty$.   ■

### ■ EXAMPLE 6        Eliminating the Parameter

Consider the curve *C* defined parametrically by

$$x = \sin t, \qquad y = \cos 2t, \qquad 0 \le t \le \pi/2.$$

Eliminate the parameter and obtain a rectangular equation for *C*.

    **Solution** Using the double-angle formula $\cos 2t = \cos^2 t - \sin^2 t$, we can write

$$
\begin{aligned}
y &= \cos^2 t - \sin^2 t \\
&= (1 - \sin^2 t) - \sin^2 t \\
&= 1 - 2\sin^2 t \qquad \leftarrow \text{substitute } \sin t = x \\
&= 1 - 2x^2.
\end{aligned}
$$

Now the curve *C* described by the parametric equations does not consist of the complete parabola, that is, $y = 1 - 2x^2$, $-\infty < x < \infty$. See **FIGURE 6.7.5(a)**. For $0 \le t \le \pi/2$ we have $0 \le \sin t \le 1$ and $-1 \le \cos 2t \le 1$. This means that *C* is only that portion of the parabola for which the coordinates of a point $P(x, y)$ satisfy $0 \le x \le 1$ *and* $-1 \le y \le 1$. The curve *C*, along with its orientation, is shown in Figure 6.7.5(b). A rectangular equation for *C* is $y = 1 - 2x^2$ with the restricted domain $0 \le x \le 1$.   ■

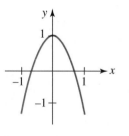

(a) $y = 1 - 2x^2$

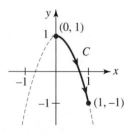

(b) $x = \sin t$, $y = \cos 2t$,
   $0 \le t \le \pi/2$

**FIGURE 6.7.5** Curve *C* in Example 6

☐ **Intercepts** We can get intercepts of a curve $C$ without finding its rectangular equation. For instance, in Example 6 we can find the $x$-intercept by finding the value of $t$ in the parameter interval for which $y = 0$. The equation $\cos 2t = 0$ yields $2t = \pi/2$ so that $t = \pi/4$. The corresponding point at which $C$ crosses the $x$-axis is $(\sqrt{2}, 0)$. Similarly, the $y$-intercept of $C$ is found by solving $x = 0$. From $\sin t = 0$ we immediately conclude $t = 0$ and so the $y$-intercept is $(0, 1)$.

☐ **Applications of Parametric Equations** Cycloidal curves were a popular topic of study by mathematicians in the seventeenth century. Suppose a point $P(x, y)$, marked on a circle of radius $a$, is at the origin when its diameter lies along the $y$-axis. As the circle rolls along the $x$-axis, the point $P$ traces out a curve $C$ that is called a **cycloid**. See FIGURE 6.7.6.*

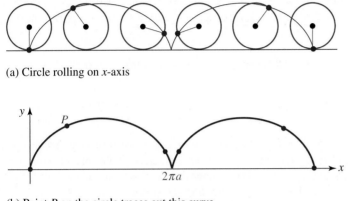

(a) Circle rolling on $x$-axis

(b) Point $P$ on the circle traces out this curve

**FIGURE 6.7.6** Cycloid

Two problems were extensively studied in the seventeenth century. Consider a flexible (frictionless) wire fixed at points $A$ and $B$ and a bead free to slide down the wire starting at $P$. See FIGURE 6.7.7. Is there a particular shape of the wire so that, regardless of where the bead starts, the time to slide down the wire to $B$ will be the same? Also, what would the shape of the wire be so that the bead slides from $P$ to $B$ in the shortest time? The so-called **tautochrone** (same time) and **brachistochrone** (least time) were shown to be an inverted half-arch of a cycloid.

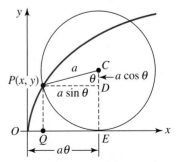

**FIGURE 6.7.7** Sliding bead

### ▮ EXAMPLE 7          Parameterization of a Cycloid

Find a parameterization for the cycloid shown in Figure 6.7.6(b).

**Solution** A circle of radius $a$ whose diameter initially lies along the $y$-axis rolls along the $x$-axis without slipping. We take as a parameter the angle $\theta$ (in radians) through which the circle has rotated. The point $P(x, y)$ starts at the origin, which corresponds to $\theta = 0$. As the circle rolls through an angle $\theta$, its distance from the origin is the arc $PE = \overline{OE} = a\theta$. From FIGURE 6.7.8 we then see that the $x$-coordinate of $P$ is

$$x = \overline{OE} - \overline{QE} = a\theta - a\sin\theta.$$

Now the $y$-coordinate of $P$ is seen to be

$$y = \overline{CE} - \overline{CD} = a - a\cos\theta.$$

**FIGURE 6.7.8** The angle $\theta$ is the parameter for the cycloid

---

*For an animation of the rolling circle go to http://mathworld.wolfram.com/Cycloid.html.

Hence parametric equations for the cycloid are

$$x = a\theta - a\sin\theta, \qquad y = a - a\cos\theta.$$

As shown in Figure 6.7.6(a), one arch of a cycloid is generated by one rotation of the circle and corresponds to the parameter interval $0 \le \theta \le 2\pi$. ∎

☐ **Parameterizations of Rectangular and Polar Curves** A curve $C$ described by a continuous function $y = f(x)$ can always be parameterized by letting $x = t$. Parametric equations for $C$ are then

$$x = t, \qquad y = f(t). \tag{3}$$

Also, it is sometimes convenient to use parametric equations to plot the graphs of polar equations. This can be done using the conversion formulas $x = r\cos\theta$, $y = r\sin\theta$. If $r = f(\theta)$, $\alpha \le \theta \le \beta$ describes a polar curve $C$, then a parameterization for $C$ is given by

$$x = f(\theta)\cos\theta, \qquad y = f(\theta)\sin\theta, \qquad \alpha \le \theta \le \beta. \tag{4}$$

## NOTES FROM THE CLASSROOM

In this section we have focused on **plane curves,** curves $C$ defined parametrically in two dimensions. In the study of multivariable calculus you will see curves and surfaces in three dimensions that are defined by means of parametric equations. For example, a **space curve** $C$ consists of a set of ordered triples $(f(t), g(t), h(t))$, where $f$, $g$, and $h$ are defined on a common interval. Parametric equations for $C$ are $x = f(t)$, $y = g(t)$, $z = h(t)$. For example, the **circular helix** such as shown in FIGURE 6.7.9 is a space curve whose parametric equations are

$$x = a\cos t, \quad y = a\cos t, \quad z = bt, \quad t \ge 0. \tag{5}$$

Surfaces in three dimensions can be represented by a set of parametric equations involving *two* parameters, $x = f(u, v)$, $y = g(u, v)$, $z = h(u, v)$. For example, the **circular helicoid** shown in FIGURE 6.7.10 arises from the study of minimal surfaces and is defined by the set of parametric equations similar to those in (5):

$$x = u\cos v, \quad y = u\sin v, \quad z = bv,$$

where $b$ is a constant. The circular helicoid has a circular helix as its boundary. You might recognize the helicoid as the model for the rotating curved blade in machinery such as post hole diggers, ice augers, and snow blowers.

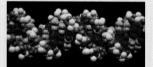

DNA is a double helix

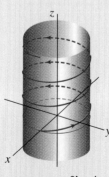

FIGURE 6.7.9 Circular helix

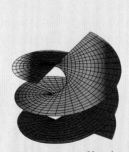

FIGURE 6.7.10 Circular helicoid

Helical antenna

CHAPTER 6 CONIC SECTIONS

In Problems 1 and 2, fill in the table for the given set of parametric equations. Find the x- and y-intercepts . Sketch the curve and indicate its orientation.

**1.** $x = t + 2, \ = 3 + \frac{1}{2}t, \ -\infty < t < \infty$

| t | −3 | −2 | −1 | 0 | 1 | 2 | 3 |
|---|---|---|---|---|---|---|---|
| x | | | | | | | |
| y | | | | | | | |

**2.** $x = 2t + 1, y = t^2 + t, \ -\infty < t < \infty$

| t | −3 | −2 | −1 | 0 | 1 | 2 | 3 |
|---|---|---|---|---|---|---|---|
| x | | | | | | | |
| y | | | | | | | |

In Problems 3–10, sketch the curve that has the given set of parametric equations.

**3.** $x = t - 1, y = 2t - 1, \quad -1 \le t \le 5$
**4.** $x = t^2 - 1, y = 3t, \quad -2 \le t \le 3$
**5.** $x = \sqrt{t}, y = 5 - t, \quad t \ge 0$
**6.** $x = t^3 + 1, y = t^2 - 1, \quad -2 \le t \le 2$
**7.** $x = 3\cos t, y = 5\sin t, \quad 0 \le t \le 2\pi$
**8.** $x = 3 + 2\sin t, y = 4 + \sin t, \quad -\pi/2 \le t \le \pi/2$
**9.** $x = e^t, y = e^{3t}, \quad 0 \le t \le \ln 2$
**10.** $x = -e^t, y = e^{-t}, \quad t \ge 0$

In Problems 11–18, eliminate the parameter from the given set of parametric equations and obtain a rectangular equation that has the same graph.

**11.** $x = t^2, y = t^4 + 3t^2 - 1$
**12.** $x = t^3 + t + 4, y = -2(t^3 + t)$
**13.** $x = \cos 2t, y = \sin t, \quad -\pi/2 \le t \le \pi/2$
**14.** $x = e^t, y = \ln t, \quad t > 0$
**15.** $x = t^3, y = 3\ln t, \quad t > 0$
**16.** $x = \tan t, y = \sec t, \quad -\pi/2 < t < \pi/2$
**17.** $x = 4\cos t, y = 2\sin t, \quad 0 \le t \le 2\pi$
**18.** $x = -1 + \cos t, y = 2 + \sin t, \quad 0 \le t \le 2\pi$

In Problems 19–24, graphically show the difference between the given curves.

**19.** $y = x$ and $x = \sin t, y = \sin t$
**20.** $y = x^2$ and $x = -\sqrt{t}, y = t$
**21.** $y = \frac{1}{4}x^2 - 1$ and $x = 2t, y = t^2 - 1, -1 \le t \le 2$
**22.** $y = -x^2$ and $x = e^t, y = -e^{2t}, t \ge 0$
**23.** $x^2 - y^2 = 1$ and $x = \cosh t, y = \sinh t \quad \leftarrow$ See (14) and (15) in Section 5.4.
**24.** $y = 2x - 2$ and $x = t^2 - 1, y = 2t^2 - 4$

In Problems 25–28 graphically show the difference between the given curves. Assume that $a > 0$ and $b > 0$.

**25.** $x = a\cos t, y = a\sin t, 0 \le t \le \pi$
$x = a\sin t, y = a\cos t, 0 \le t \le \pi$

**26.** $x = a\cos t, y = b\sin t, a > b, \pi \le t \le 2\pi$
$x = a\sin t, y = b\cos t, a > b, \pi \le t \le 2\pi$

**27.** $x = a\cos t, y = a\sin t, -\pi/2 \le t \le \pi/2$
$x = a\cos 2t, y = a\sin 2t, -\pi/2 \le t \le \pi/2$

**28.** $x = a\cos\dfrac{t}{2}, y = a\sin\dfrac{t}{2}, 0 \le t \le \pi$

$x = a\cos\left(-\dfrac{t}{2}\right), y = a\sin\left(-\dfrac{t}{2}\right), -\pi \le t \le 0$

In Problems 29 and 30, find the $x$- and $y$-intercepts of the given curves.

**29.** $x = t^2 - 2t, y = t + 1, -2 \le t < 4$
**30.** $x = t^2 + t, y = t^2 + t - 6, -5 \le t < 5$

**31.** Show that parametric equations for a line through $(x_1, y_1)$ and $(x_2, y_2)$ are

$$x = x_1 + (x_2 - x_1)t, \qquad y = y_1 + (y_2 - y_1)t, \qquad -\infty < t < \infty.$$

What do these equations represent when $0 \le t \le 1$?

**32. (a)** Use the result of Problem 31 to find parametric equations of the line through $(-2, 5)$ to $(4, 8)$.
   **(b)** Eliminate the parameter in part (a) to obtain a rectangular equation for the line.
   **(c)** Find parametric equations for the line segment with $(-2, 5)$ as the initial point and $(4, 8)$ as the terminal point.

**33.** A famous golfer can generate a club head speed of approximately 130 mi/h or $v_0 = 190$ ft/s. If the golf ball leaves the ground at an angle $\theta_0 = 45°$, use (1) to find parametric equations for the path of the ball. What are the coordinates of the ball at $t = 2$ s?

**34.** Use the parametric equations obtained in Problem 33 to determine
   **(a)** how long the golf ball is in the air,
   **(b)** its maximum height, and
   **(c)** the horizontal distance that the golf ball travels.

**35.** As shown in **FIGURE 6.7.11**, a piston is attached by means of a rod of length $L$ to a circular crank mechanism of radius $r$. Parameterize the coordinates of the point $P$ in terms of the angle $\phi$ shown in the figure.

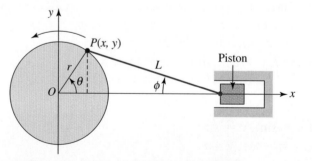

**FIGURE 6.7.11** Crank mechanism in Problem 35

**36.** Consider a circle of radius $a$, which is tangent to the $x$-axis at the origin. Let $B$ be a point on the horizontal line $y = 2a$ and let the line segment $OB$ cut the circle at point $A$. As shown in **FIGURE 6.7.12**, the projection of $AB$ on the vertical gives the line segment $BP$. Using the angle $\theta$ in the figure as a parameter, find parametric equations of the curve traced by the point $P$ as $A$ varies around the circle. The curve, more historically famous than useful, is called the **witch of Agnesi.***

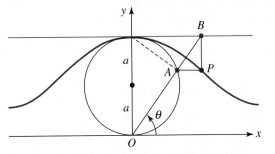

**FIGURE 6.7.12** Witch of Agnesi in Problem 36

## Calculator/Computer Problems

In Problems 37–42, use a graphing utility to obtain the graph of the given set of parametric equations.

**37.** $x = 4\sin 2t, y = 2\sin t, 0 \leq t \leq 2\pi$
**38.** $x = 6\cos 3t, y = 4\sin 2t, 0 \leq t \leq 2\pi$
**39.** $x = 6\sin 4t, y = 4\sin t, 0 \leq t \leq 2\pi$
**40.** $x = \cos t + t\sin t, y = \sin t - t\cos t, 0 \leq t \leq 3\pi$
**41.** $x = 4\cos t - \cos 4t, y = 4\sin t - \sin 4t, 0 \leq t \leq 2\pi$
**42.** $x = \cos^3 t, y = \sin^3 t, 0 \leq t \leq 2\pi$

In Problems 43–46, use (4) to parameterize the curve whose polar equation is given. Use a graphing utility to obtain the graph of the resulting set of parametric equations.

**43.** $r = 2\sin\dfrac{\theta}{2}, 0 \leq \theta \leq 4\pi$

**44.** $r = 2\sin\dfrac{\theta}{4}, 0 \leq \theta \leq 8\pi$

**45.** $r = 2\cos\dfrac{\theta}{5}, 0 \leq \theta \leq 6\pi$

**46.** $r = 2\cos\dfrac{3\theta}{2}, 0 \leq \theta \leq 6\pi$

| **CHAPTER 6** | Review Exercises | Answers to selected odd-numbered problems begin on page ANS-22. |

In Problems 1–20, fill in the blanks.

**1.** An equation in the standard form $y^2 = 4cx$ of a parabola with focus $(5, 0)$ is _____ .

**2.** An equation in the standard form $x^2 = 4cy$ of a parabola through $(2, 6)$ is _____ .

**3.** A rectangular equation of a parabola with focus $(1, -3)$ and directrix $y = -7$ is _____ .

---

*No, the curve has nothing to do with witches and goblins. This curve, called *versoria*, which is Latin for a kind of rope, was included in a text on analytic geometry written in 1748 by the Italian mathematician Maria Agnesi. A translator of the text confused *versoria* with the Italian word *versiera,* which means *female goblin.* In English, *female goblin* became a *witch.*

**4.** The directrix and vertex of a parabola are $x = -3$ and $(-1, -2)$, respectively. The focus of the parabola is _____.

**5.** The focus and directrix of a parabola are $(0, \frac{1}{4})$ and $y = -\frac{1}{4}$, respectively. The vertex of the parabola is _____.

**6.** The vertex and focus of the parabola $8(x + 4)^2 = y - 2$ are _____.

**7.** The eccentricity of a parabola is $e = $ _____.

**8.** The center and vertices of the ellipse $\dfrac{(x - 2)^2}{16} + \dfrac{(y + 5)^2}{4} = 1$ are _____.

**9.** The center and vertices of the hyperbola $y^2 - \dfrac{(x + 3)^2}{4} = 1$ are _____.

**10.** The oblique asymptotes of the hyperbola $y^2 - (x - 1)^2 = 1$ are _____.

**11.** The $y$-intercepts of the hyperbola $y^2 - (x - 1)^2 = 1$ are _____.

**12.** The eccentricity of the hyperbola $x^2 - y^2 = 1$ is _____.

**13.** If the graph of an ellipse is very elongated, then its eccentricity $e$ is close to _____. (Fill in with 0 or 1.)

**14.** The rectangular coordinates of the point with polar coordinates $(-\sqrt{2}, 5\pi/4)$ are _____.

**15.** Polar coordinates of the point with rectangular coordinates $(0, -10)$ are _____.

**16.** Approximate polar coordinates of the point with rectangular coordinates $(-1, 3)$ are _____.

**17.** On the graph of the polar equation $r = 4\cos\theta$, two pairs of coordinates of the pole or origin are _____.

**18.** The equations $x = t + 2$, $y = 3 + \frac{1}{2}t$, $-\infty < t < \infty$, are a parametric representation of a _____.

**19.** The point on the curve defined by the parametric equations $x = -4 + 2\cos t$, $y = 2 + \sin t$, $-\infty < t < \infty$, corresponding to $t = 5\pi/2$ is _____.

**20.** The $y$-intercepts of the curve defined by the parametric equations $x = t^2 - 4$, $y = t^3 - 3t$, $-\infty < t < \infty$ are _____.

In Problems 21–36, answer true or false.

**21.** Rectangular coordinates of a point in the plane are unique. _____

**22.** For an ellipse, the length of the major axis is always greater than the length of the minor axis. _____

**23.** The vertex and focus are both on the axis of symmetry of a parabola. _____

**24.** The asymptotes for $(x - h)^2/a^2 - (y - k)^2/b^2 = 1$ must pass through $(h, k)$. _____

**25.** An ellipse with eccentricity $e = 0.01$ is nearly circular. _____

**26.** The transverse axis of the hyperbola $x^2/9 - y^2/49 = 1$ is vertical. _____

**27.** The two hyperbolas $x^2 - y^2/25 = 1$ and $y^2/25 - x^2 = 1$ have the same pair of slant asymptotes. _____

**28.** The graph of the polar equation $r = 5\sec\theta$ is a line. _____

**29.** $(3, \pi/6)$ and $(-3, -5\pi/6)$ are polar coordinates of the same point. _____

**30.** The graph of the ellipse $r = 90/(15 - \sin\theta)$ is nearly circular. _____

**31.** The graph of the curve $x = t^2$, $y = t^4 + 1$ is the same as the graph of $y = x^2 + 1$. _____

**32.** If $P$ is a point on a parabola, then the perpendicular distance between $P$ and the directrix equals the distance between $P$ and the vertex. _____

**33.** The graph of $r = 2 + 4\sin\theta$ is a limaçon with an interior loop. _____

**34.** The graph of the rose curve $r = 5\sin 6\theta$ has 6 petals. _____

**35.** The polar graph $r^2 = 4\sin 2\theta$ is symmetric with respect to the origin. _____

**36.** The curve with parametric equations $x = 1 + \cos t, y = 1 + \sin t, 0 \le t \le 2\pi,$ is a circle of radius 1 centered at $(1, 1)$. _____

In Problems 37 and 38, find a rectangular equation that has the same graph as the given polar equation.

**37.** $r = \cos\theta + \sin\theta$

**38.** $r(\cos\theta + \sin\theta) = 1$

In Problems 39 and 40, find a polar equation that has the same graph as the given rectangular equation.

**39.** $x^2 + y^2 - 4y = 0$

**40.** $(x^2 + y^2 - 2x)^2 = 9(x^2 + y^2).$

**41.** Determine the rectangular coordinates of the vertices of the ellipse whose polar equation is $r = 2/(2 - \sin\theta)$.

**42.** Find a polar equation of the hyperbola with focus at the origin, vertices (in rectangular coordinates) $\left(0, -\frac{4}{3}\right)$ and $(0, -4)$, and eccentricity 2.

In Problems 43 and 44, find polar coordinates satisfying **(a)** $r > 0, -\pi < \theta \le \pi,$ and **(b)** $r < 0, -\pi < \theta \le \pi,$ for each point given in rectangular coordinates.

**43.** $(\sqrt{3}, -\sqrt{3})$

**44.** $\left(-\frac{1}{4}, \frac{1}{4}\right)$

In Problems 45–50, sketch the graph of the given conic section. Depending on the conic, find the center, vertex, vertices, endpoints of the minor axis, and asymptotes.

**45.** $(x - 1)^2 + 4(y - 1)^2 = 16$

**46.** $4y^2 - x^2 = 4$

**47.** $4x^2 + y^2 + 8x - 6y + 9 = 0$

**48.** $y^2 + 10y + 8x + 41 = 0$

**49.** $16y^2 - 9x^2 - 64y - 80 = 0$

**50.** $x^2 - 2x + 4y + 1 = 0$

In Problems 51–62, identify and sketch the graph of the given polar equation.

**51.** $r = 5$

**52.** $\theta = -\pi/3$

**53.** $r = 5\sin\theta$

**54.** $r = -2\cos\theta$

**55.** $r = 4 - 4\cos\theta$

**56.** $r = 1 + \sin\theta$

**57.** $r = 2 + \sin\theta$

**58.** $r = 1 - 2\cos\theta$

**59.** $r = \sin 3\theta$

**60.** $r = 3\sin 4\theta$

**61.** $r = \dfrac{8}{3 - 2\cos\theta}$

**62.** $r = \dfrac{1}{1 + \cos\theta}$

**63.** A satellite orbits the planet Neptune in an elliptical orbit with the center of the planet at one focus. If the length of the major axis of the orbit is $2 \times 10^9$ m and the length of the minor axis is $6 \times 10^8$ m, find the maximum distance between the satellite and the center of the planet.

**64.** A parabolic mirror has a depth of 7 cm at its center and the distance across the top of the mirror is 20 cm. Find the distance from the vertex to the focus.

Try to answer the following questions without referring back to the text.

In Problems 1–14, fill in the blank.

1. Completing the square in $x$ for $2x^2 + 6x + 5$ gives _____.
2. In the binomial expansion of $(1 - 2x)^3$ the coefficient of $x^2$ is _____.
3. In interval notation, the solution set of $\dfrac{x(x^2 - 9)}{x^2 - 25} \geq 0$ is _____.
4. If $a - 3$ is a negative number, then $|a - 3| =$ _____.
5. If $|5x| = 80$, then $x =$ _____.
6. If $(a, b)$ is a point in the third quadrant, then $(-a, b)$ is a point in the _____ quadrant.
7. The point $(1, 7)$ is on a graph in the Cartesian plane. Give the coordinates of another point on the graph if the graph is:
   **(a)** symmetric with respect to the $x$-axis. _____
   **(b)** symmetric with respect to the $y$-axis. _____
   **(c)** symmetric with respect to the origin. _____
8. The lines $6x + 2y = 1$ and $kx - 9y = 5$ are parallel if $k =$ _____. The lines are perpendicular if $k =$ _____.
9. The complete factorization of the function $f(x) = x^3 - 2x^2 - 6x$ is _____.
10. The only potential rational zeros of $f(x) = x^3 + 4x + 2$ are _____.
11. The phase shift of the graph of $y = 5\sin(4x + \pi)$ is _____.
12. If $f(x) = x^4 \arctan(x/2)$, then the exact value of $f(-2)$ is _____.
13. $5\ln 2 - \ln\frac{2}{3} = \ln$ _____.
14. The graph of $y = \ln(2x + 5)$ has the vertical asymptote $x =$ _____.

In Problems 15–32, answer true or false.

15. The absolute value of any real number $x$ is positive. _____
16. The inequality $|x| > -1$ has no solutions. _____
17. For any function $f$, if $f(a) = f(b)$, then $a = b$. _____
18. The graph of $y = f(x + c), c > 0$, is the graph of $y = f(x)$ shifted $c$ units to the right. _____
19. The points $(1, 3)$, $(3, 11)$, and $(5, 19)$ are collinear. _____
20. The function $f(x) = x^5 - 4x^3 + 2$ is an odd function. _____
21. $x + \frac{1}{4}$ is a factor of the function $f(x) = 64x^4 + 16x^3 + 48x^2 - 36x - 12$.
   _____
22. If $b^2 - 4ac < 0$, the graph of $f(x) = ax^2 + bx + c, a \neq 0$, does not cross the $x$-axis. _____
23. $f(x) = \dfrac{\sqrt{x}}{2x + 1}$ is a rational function. _____
24. If $f(x) = x^5 + 3x - 1$, then there exists a number $c$ in $[-1, 1]$ such that $f(c) = 0$. _____

**25.** The graph of the function $f(x) = \dfrac{1}{x-1} + \dfrac{1}{x-2}$ has no $x$-intercepts. _____

**26.** $x = 0$ is a vertical asymptote for the graph of the rational function

$f(x) = \dfrac{x^2 - 2x}{x}$. _____

**27.** The graph of $y = \cos(x/6)$ is the graph of $y = \cos x$ stretched horizontally. _____

**28.** $f(x) = \csc x$ is not defined at $x = \pi/2$. _____

**29.** The function $f(x) = e^{-4x^2}$ is not one-to-one. _____

**30.** The exponential function $f(x) = \left(\tfrac{3}{2}\right)^x$ increases on the interval $(-\infty, \infty)$.

_____

**31.** The domain of the function $f(x) = \ln x + \ln(x - 4)$ is $(4, \infty)$. _____

**32.** The solutions of the equation $\ln x^2 = \ln 3x$ are $x = 0$ and $x = 3$. _____

**33.** Match the given interval with the appropriate inequality.

(*i*) $[2, 4]$        (*ii*) $[2, 4)$

(*iii*) $(2, 4)$        (*iv*) $(2, 4]$

**(a)** $|x - 3| \leq 1$        **(b)** $1 < x - 1 \leq 3$

**(c)** $-2 < 2 - x \leq 0$        **(d)** $|x - 3| < 1$

**34.** Write the solution of the absolute-value inequality $|3x - 1| > 7$ using interval notation.

**35.** The answer to a problem given in the back of a mathematics text is $1 + \sqrt{3}$, but your answer is $2/(\sqrt{3} - 1)$. Are you correct?

**36.** In which quadrants in the Cartesian plane is the quotient $x/y$ negative?

**37.** Which one of the following equations best describes a circle that passes through the origin? The symbols $a, b, c, d$, and $e$ stand for different nonzero real constants.

**(a)** $ax^2 + by^2 + cx + dy + e = 0$      **(b)** $ax^2 + ay^2 + cx + dy + e = 0$

**(c)** $ax^2 + ay^2 + cx + dy = 0$      **(d)** $ax^2 + by^2 + cx + dy = 0$

**(e)** $ax^2 + ay^2 + e = 0$      **(f)** $ax^2 + ay^2 + cx + e = 0$

**38.** Match the given rational function $f$ with the most appropriate phrase.

(*i*) $f(x) = \dfrac{x^4}{x^2 - 2}$        (*ii*) $f(x) = \dfrac{x^2}{x^2 + 2}$

(*iii*) $f(x) = \dfrac{x^5}{x^2 + 2}$        (*iv*) $f(x) = \dfrac{x^3}{x^2 + 2}$

**(a)** slant asymptote        **(b)** no asymptotes

**(c)** horizontal asymptote        **(d)** vertical asymptote

**39.** What is the range of the rational function $f(x) = \dfrac{10}{x^2 + 1}$?

**40.** What is the domain of the function $f(x) = \dfrac{\sqrt{x + 2}}{x^2}$?

**41.** Find an equation of the line that passes through the origin and through the point of intersection of the graphs of $x + y = 1$ and $2x - y = 7$.

**42.** Find a quadratic function $f$ whose graph has the $y$-intercept $(0, -6)$ and the vertex of the graph is $(1, 4)$.

In calculus you are often required to rewrite a function either in a simpler form or in a form that is more helpful in solving the problem. In Problems 43–48, rewrite each function by following the given instruction. In calculus you would be expected to recognize what to do from the context of the actual problem.

**43.** $f(x) = \sqrt{x^6 + 4} - x^3$. Express $f$ as a quotient using rationalization and simplification.

**44.** $f(x) = \dfrac{5x^3 - 4x^2\sqrt{x} + 8}{\sqrt[3]{x}}$. Carry out the indicated division and express each term as a power of $x$.

**45.** $f(x) = \dfrac{7x^2 - 7x - 6}{x^3 - x^2}$. Decompose $f$ into partial fractions.

**46.** $f(x) = \dfrac{1}{1 + \sin x}$. Express $f$ in terms of $\sec x$ and $\tan x$.

**47.** $f(x) = e^{3\ln x}$. Express $f$ as a power of $x$.

**48.** $f(x) = |x^2 - 3x|$. Express $f$ without absolute value signs.

In calculus you are often required to find zeros of a function. In Problems 49 and 50, solve the equation $f(x) = 0$ by following the given instruction.

**49.** $f(x) = x^2\frac{1}{2}(4 - x^2)^{-1/2}(-2x) + 2x\sqrt{4 - x^2}$. Rewrite $f$ as a single expression without negative exponents.

**50.** $f(x) = 2\sin x \cos x - \sin x$. Find the zeros of $f$ on the interval $[-\pi, \pi]$.

In Problems 51 and 52, compute and simplify the difference quotient $\dfrac{f(x + h) - f(x)}{h}$ for the given function.

**51.** $f(x) = \dfrac{3x}{2x + 5}$

**52.** $f(x) = -x^3 + 10x^2$

**53.** Consider the trigonometric function $y = -8\sin(\pi x/3)$. What is the amplitude of the function? Give an interval over which one cycle of the graph is completed.

**54.** If $\tan\theta = \sqrt{5}$ and $\pi < \theta < 3\pi/2$, then what is the value of $\cos\theta$?

**55.** Suppose $f(x) = \sin x$ and $f(c) = 0.7$. What is the value of

$$2f(-c) + f(c + 2\pi) + f(c - 6\pi)?$$

**56.** Suppose $f(x) = \sin x$ and $g(x) = \ln x$. Solve $(f \circ g)(x) = 0$.

**57.** Find the $x$- and $y$-intercepts of the parabola whose equation is $(y + 4)^2 = 4(x + 1)$.

**58.** Find the center, foci, vertices, and endpoints of the minor axis of the ellipse whose equation is

$$x^2 + 2y^2 + 2x - 20y + 49 = 0.$$

**59.** The slant asymptotes of a hyperbola are $y = -5x + 2$ and $y = 5x - 8$. What is the center of the hyperbola?

**60.** From a point 220 ft from the base of a cell-phone antenna, a person measures a 30° angle of inclination from the ground to the top of the antenna. What is the angle of inclination to the top of the antenna if the person moves 100 ft closer to its base?

**61.** Iodine 131 is radioactive and is used in certain medical procedures. Assume that iodine 131 decays exponentially. If the half-life of I-131 is 8 days, then how much of a sample of 5 grams remains at the end of 15 days?

**62.** The polar coordinate equation $r = 3\cos 4\theta$ is a rose curve with eight petals. Find all radian-measure angles satisfying $0 \le \theta \le 2\pi$ for which $|r| = 3$.

# Answers to Selected Odd-Numbered Problems

**Exercises 1.1**  Page 9

**1.** $a + 2 > 0$

**3.** $a + b \geq 0$

**5.** $2b + 4 \geq 100$

**7.** $(-\infty, 0)$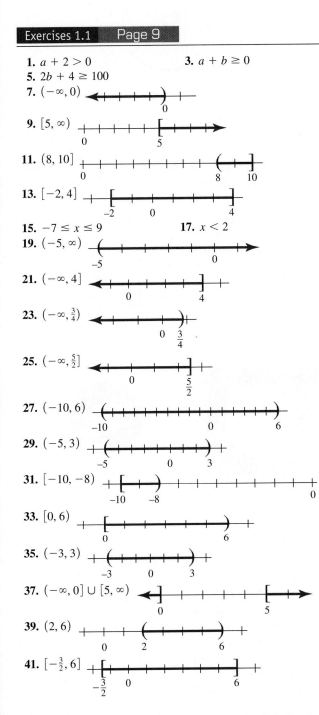

**9.** $[5, \infty)$

**11.** $(8, 10]$

**13.** $[-2, 4]$

**15.** $-7 \leq x \leq 9$

**17.** $x < 2$

**19.** $(-5, \infty)$

**21.** $(-\infty, 4]$

**23.** $\left(-\infty, \frac{3}{4}\right)$

**25.** $\left(-\infty, \frac{5}{2}\right]$

**27.** $(-10, 6)$

**29.** $(-5, 3)$

**31.** $[-10, -8)$

**33.** $[0, 6)$

**35.** $(-3, 3)$

**37.** $(-\infty, 0] \cup [5, \infty)$

**39.** $(2, 6)$

**41.** $\left[-\frac{3}{2}, 6\right]$

**43.** $(-\infty, -1) \cup (2, 4)$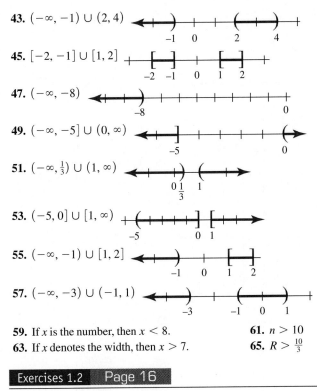

**45.** $[-2, -1] \cup [1, 2]$

**47.** $(-\infty, -8)$

**49.** $(-\infty, -5] \cup (0, \infty)$

**51.** $\left(-\infty, \frac{1}{3}\right) \cup (1, \infty)$

**53.** $(-5, 0] \cup [1, \infty)$

**55.** $(-\infty, -1) \cup [1, 2]$

**57.** $(-\infty, -3) \cup (-1, 1)$

**59.** If $x$ is the number, then $x < 8$.

**61.** $n > 10$

**63.** If $x$ denotes the width, then $x > 7$.

**65.** $R > \frac{10}{3}$

**Exercises 1.2**  Page 16

**1.** $4 - \pi$

**3.** $8 - \sqrt{63}$

**5.** 4

**7.** $-h$

**9.** $-x + 6$

**11.** 0

**13.** $-2x + 7$

**15.** 3

**17.** $2x - 2$

**19.** 4

**21.** 4; 5

**23.** 3; 0

**25.** $a = 2, b = 8$

**27.** $m = 4 + \pi, b = 4 + 2\pi$

**29.** $-\frac{1}{4}, \frac{3}{4}$

**31.** $-\frac{1}{2}, \frac{5}{6}$

**33.** $\frac{2}{3}, 2$

**35.** $\left(-\frac{4}{5}, \frac{4}{5}\right)$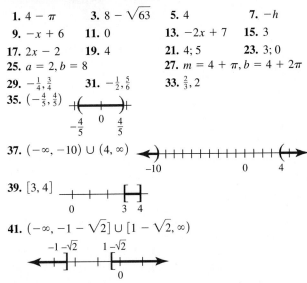

**37.** $(-\infty, -10) \cup (4, \infty)$

**39.** $[3, 4]$

**41.** $(-\infty, -1 - \sqrt{2}] \cup [1 - \sqrt{2}, \infty)$

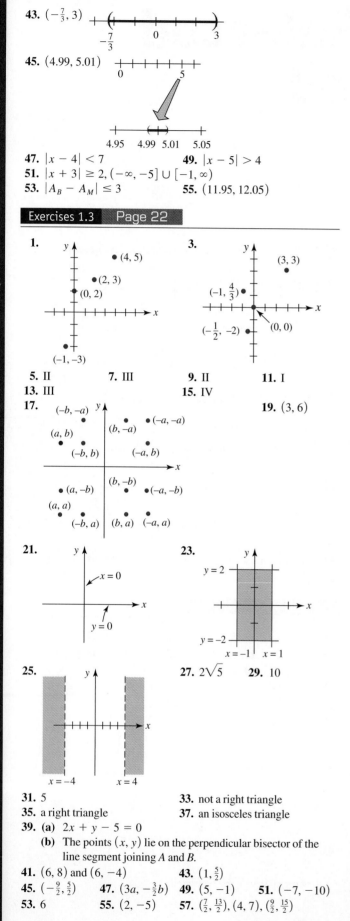

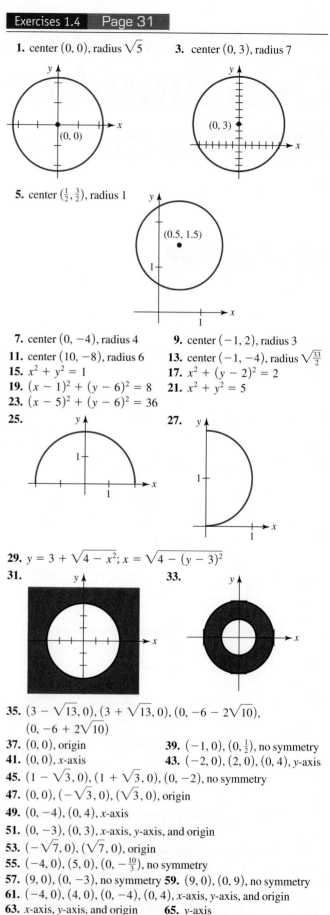

**Answers to Selected Odd-Numbered Problems, CHAPTER 1**

**43.** $\left(-\frac{7}{3}, 3\right)$

**45.** $(4.99, 5.01)$

**47.** $|x - 4| < 7$        **49.** $|x - 5| > 4$
**51.** $|x + 3| \geq 2, (-\infty, -5] \cup [-1, \infty)$
**53.** $|A_B - A_M| \leq 3$        **55.** $(11.95, 12.05)$

Exercises 1.3    Page 22

**1.**

**3.**

**5.** II        **7.** III        **9.** II        **11.** I
**13.** III        **15.** IV
**17.**        **19.** $(3, 6)$

**21.**        **23.**

**25.**        **27.** $2\sqrt{5}$        **29.** $10$

**31.** 5        **33.** not a right triangle
**35.** a right triangle        **37.** an isosceles triangle
**39. (a)** $2x + y - 5 = 0$
   **(b)** The points $(x, y)$ lie on the perpendicular bisector of the line segment joining $A$ and $B$.
**41.** $(6, 8)$ and $(6, -4)$        **43.** $\left(1, \frac{5}{2}\right)$
**45.** $\left(-\frac{9}{2}, \frac{5}{2}\right)$    **47.** $\left(3a, -\frac{3}{2}b\right)$    **49.** $(5, -1)$    **51.** $(-7, -10)$
**53.** 6        **55.** $(2, -5)$        **57.** $\left(\frac{7}{2}, \frac{13}{2}\right), (4, 7), \left(\frac{9}{2}, \frac{15}{2}\right)$

Exercises 1.4    Page 31

**1.** center $(0, 0)$, radius $\sqrt{5}$        **3.** center $(0, 3)$, radius 7

**5.** center $\left(\frac{1}{2}, \frac{3}{2}\right)$, radius 1

**7.** center $(0, -4)$, radius 4        **9.** center $(-1, 2)$, radius 3
**11.** center $(10, -8)$, radius 6    **13.** center $(-1, -4)$, radius $\sqrt{\frac{33}{2}}$
**15.** $x^2 + y^2 = 1$        **17.** $x^2 + (y - 2)^2 = 2$
**19.** $(x - 1)^2 + (y - 6)^2 = 8$    **21.** $x^2 + y^2 = 5$
**23.** $(x - 5)^2 + (y - 6)^2 = 36$
**25.**        **27.**

**29.** $y = 3 + \sqrt{4 - x^2}; x = \sqrt{4 - (y - 3)^2}$
**31.**        **33.**

**35.** $(3 - \sqrt{13}, 0), (3 + \sqrt{13}, 0), (0, -6 - 2\sqrt{10}),$
   $(0, -6 + 2\sqrt{10})$
**37.** $(0, 0)$, origin        **39.** $(-1, 0), (0, \frac{1}{2})$, no symmetry
**41.** $(0, 0)$, $x$-axis        **43.** $(-2, 0), (2, 0), (0, 4)$, $y$-axis
**45.** $(1 - \sqrt{3}, 0), (1 + \sqrt{3}, 0), (0, -2)$, no symmetry
**47.** $(0, 0), (-\sqrt{3}, 0), (\sqrt{3}, 0)$, origin
**49.** $(0, -4), (0, 4)$, $x$-axis
**51.** $(0, -3), (0, 3)$, $x$-axis, $y$-axis, and origin
**53.** $(-\sqrt{7}, 0), (\sqrt{7}, 0)$, origin
**55.** $(-4, 0), (5, 0), (0, -\frac{10}{3})$, no symmetry
**57.** $(9, 0), (0, -3)$, no symmetry **59.** $(9, 0), (0, 9)$, no symmetry
**61.** $(-4, 0), (4, 0), (0, -4), (0, 4)$, $x$-axis, $y$-axis, and origin
**63.** $x$-axis, $y$-axis, and origin        **65.** $y$-axis

**67.**

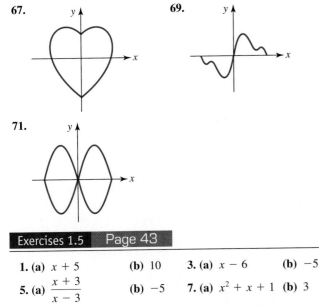

**69.**

**71.**

**7.** $-2x^2 + 3x, -8a^2 + 6a, -2a^4 + 3a^2, -50x^2 - 15x,$
$-8a^2 - 2a + 1, -2x^2 - 4xh - 2h^2 + 3x + 3h$

**9.** $-2, 2$     **11.** $[\frac{1}{2}, \infty)$

**13.** $(-\infty, 1)$     **15.** $\{x \mid x \neq 0, x \neq 3\}$

**17.** $\{x \mid x \neq 5\}$     **19.** $(-\infty, \infty)$

**21.** $[-5, 5]$     **23.** $(-\infty, 0] \cup [5, \infty)$

**25.** $(-2, 3]$     **27.** not a function

**29.** function     **31.** $[-4, 4], [0, 5]$

**33.** $[1, 9], [1, 6]$     **35.** $-\frac{6}{5}$

**37.** $2, 3$     **39.** $0, \frac{1}{3}, -9$

**41.** $-1, 1$     **43.** $(8, 0), (0, -4)$

**45.** $(\frac{3}{2}, 0), (\frac{5}{2}, 0), (0, 15)$     **47.** $(0, -\frac{1}{4})$

**49.** $(-2, 0), (2, 0), (0, 3)$

**51.** $f_1(x) = \sqrt{x + 5}, f_2(x) = -\sqrt{x + 5}; [-5, \infty)$

**53.** $0, -3.4, 0.3, 2, 3.8, 2.9; (0, 2)$

**55.** $3.6, 2, 3.3, 4.1, 2, -4.1; (-3.2, 0), (2.3, 0), (3.8, 0)$

**57. (a)** $2; 6; 120; 5040$     **(c)** $(n + 1)(n + 2)$

---

**Exercises 1.5**    **Page 43**

**1. (a)** $x + 5$    **(b)** $10$     **3. (a)** $x - 6$    **(b)** $-5$

**5. (a)** $\dfrac{x + 3}{x - 3}$    **(b)** $-5$     **7. (a)** $x^2 + x + 1$ **(b)** $3$

**9. (a)** $\dfrac{x^2 + x + 1}{x + 4}$   **(b)** $\frac{3}{5}$     **11. (a)** $\dfrac{x + 1}{x^2 - x + 1}$ **(b)** $0$

**13. (a)** $4 + h$    **(b)** $4$     **15. (a)** $4(x + 2)$   **(b)** $12$

**17. (a)** $3 + 3x + x^2$ **(b)** $3$     **19. (a)** $2h^2 + h - 4$ **(b)** $-4$

**21. (a)** $\dfrac{1}{x + 4}$    **(b)** $\frac{1}{6}$     **23. (a)** $\dfrac{1}{x + 10}$    **(b)** $\frac{1}{20}$

**25. (a)** $-\dfrac{4 + h}{4(2 + h)^2}$   **(b)** $-\frac{1}{4}$     **27. (a)** $\dfrac{1}{\sqrt{x + 3}}$    **(b)** $\frac{1}{6}$

**29. (a)** $\sqrt{7 + x} + \sqrt{7}$ **(b)** $2\sqrt{7}$   **31. (a)** $5 + \sqrt{t}$    **(b)** $10$

**33. (a)** $4(\sqrt{y^2 + y + 1} + \sqrt{y + 1})$   **(b)** $8$

**35.** $\dfrac{ax - 1}{ax}$

**37.** $2(2x - 3)^3(2x - 1)(9x + 10)$

**39.** $\dfrac{6x(2 - x)}{(-4x + 6)^{3/2}}$     **41.** $y' = \dfrac{x + y}{3y^2 - x}$

**43.** $y' = \dfrac{2x}{1 - 2y}$     **45.** $y' = \dfrac{x^2 - 2xy + 2y + y^2}{2x}$

---

**Chapter 1 Review Exercises**    **Page 46**

**1.** $x < 9$     **3.** II

**5.** $(2, -3)$     **7.** $(x + 2)^2 + (y + 5)^2 = 36$

**9.** $\sqrt{10}$     **11.** $(-\frac{5}{2}, 0), (\frac{5}{2}, 0), (0, -5)$

**13.** center $(8, 0)$, radius 8     **15.** $(-3, 4), (-3, -4)$

**17.** $x^2 + y^2 > 36$     **19.** $|x - \sqrt{2}| > 3$

**21.** false    **23.** true     **25.** false    **27.** true

**29.** true    **31.** true     **33.** true    **35.** $a^2 < ab$

**37.** $a < a + b$     **39.** 10

**41.** $\leq$     **43.** $-4 \leq x \leq 3$

**45.** $a = 4, b = 6$     **47.** $(-\infty, -3]$

**49.** $(4, 12)$     **51.** $(-\infty, -10) \cup (10, \infty)$

**53.** $(-\frac{1}{3}, 3)$     **55.** $[-1, \frac{5}{2}]$

**57.** $(-1, 0) \cup (1, \infty)$     **59.** $\frac{1}{2}$

**61.** 32

---

**Exercises 2.1**    **Page 55**

**1.** $24, 2, 8, 35$     **3.** $0, 1, 2, \sqrt{6}$

**5.** $-\frac{3}{2}, 0, \frac{3}{2}, \sqrt{2}$

---

**Exercises 2.2**    **Page 64**

**1.** even     **3.** neither even nor odd

**5.** odd     **7.** even     **9.** even     **11.** odd

**13.** neither even nor odd

**15. (a)**     **(b)**

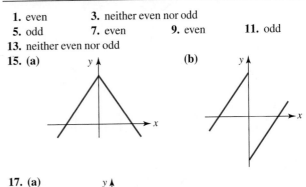

**17. (a)**

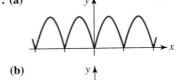

**(b)**

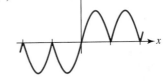

**19.** $f(2) = 4, f(-3) = 7$     **21.** $g(1) = 5, g(-4) = -8$

**23.** $(-2, 3), (3, -2)$     **25.** $(-8, 1), (-3, -4)$

**27.** $(-6, 2), (-1, -3)$     **29.** $(2, 1), (-3, -4)$

**31.** $(-2, 15), (3, -60)$

**33. (a)**     **(b)**

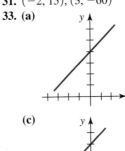

**(c)**     **(d)**

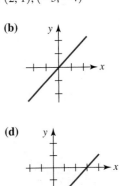

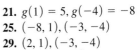

**(e)**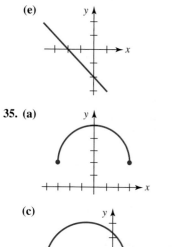

**(f)**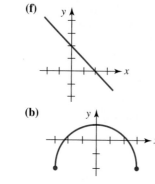

**35. (a)**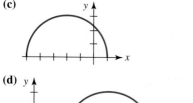

**(b)**

**(c)**

**(d)**

**(e)**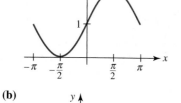

**(f)**

**37. (a)**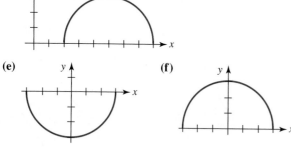

**(b)**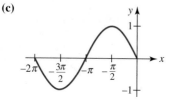

**(c)**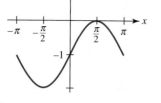

**(d)**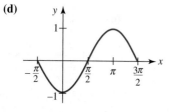

**(e)**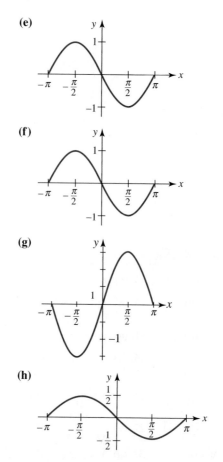

**(f)**

**(g)**

**(h)**

**39.** $y = (x - 1)^3 + 5$

**41.** $y = -(x + 7)^4$

**1.** $-\frac{7}{2}$;

**3.** 5;

**5.** $-1$;

**7.** $-\frac{5}{12}$

**9.** $\frac{3}{4}$; $(-4, 0)$, $(0, 3)$;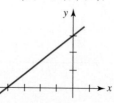

**11.** $\frac{2}{3}$; $\left(\frac{9}{2}, 0\right)$, $(0, -3)$;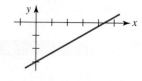

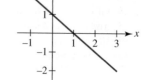
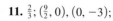
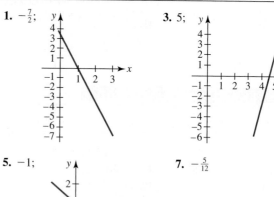

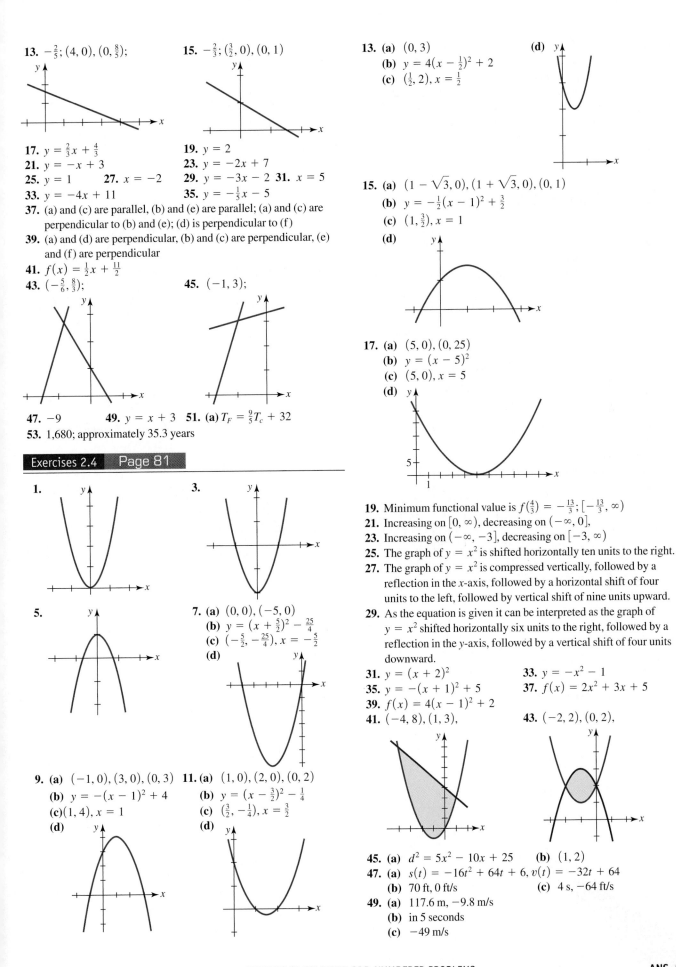

13. $-\frac{2}{5}$; $(4, 0)$, $(0, \frac{8}{5})$;

15. $-\frac{2}{3}$; $(\frac{3}{2}, 0)$, $(0, 1)$

17. $y = \frac{2}{3}x + \frac{4}{3}$
19. $y = 2$
21. $y = -x + 3$
23. $y = -2x + 7$
25. $y = 1$   27. $x = -2$
29. $y = -3x - 2$  31. $x = 5$
33. $y = -4x + 11$
35. $y = -\frac{1}{5}x - 5$
37. (a) and (c) are parallel, (b) and (e) are parallel; (a) and (c) are perpendicular to (b) and (e); (d) is perpendicular to (f)
39. (a) and (d) are perpendicular, (b) and (c) are perpendicular, (e) and (f) are perpendicular
41. $f(x) = \frac{1}{2}x + \frac{11}{2}$
43. $(-\frac{5}{6}, \frac{8}{3})$;
45. $(-1, 3)$;

47. $-9$   49. $y = x + 3$   51. (a) $T_F = \frac{9}{5}T_c + 32$
53. 1,680; approximately 35.3 years

**Exercises 2.4    Page 81**

1.

3.

5.

7. (a) $(0, 0)$, $(-5, 0)$
   (b) $y = (x + \frac{5}{2})^2 - \frac{25}{4}$
   (c) $(-\frac{5}{2}, -\frac{25}{4})$, $x = -\frac{5}{2}$
   (d)

9. (a) $(-1, 0)$, $(3, 0)$, $(0, 3)$
   (b) $y = -(x - 1)^2 + 4$
   (c) $(1, 4)$, $x = 1$
   (d)

11. (a) $(1, 0)$, $(2, 0)$, $(0, 2)$
    (b) $y = (x - \frac{3}{2})^2 - \frac{1}{4}$
    (c) $(\frac{3}{2}, -\frac{1}{4})$, $x = \frac{3}{2}$
    (d)

13. (a) $(0, 3)$
    (b) $y = 4(x - \frac{1}{2})^2 + 2$
    (c) $(\frac{1}{2}, 2)$, $x = \frac{1}{2}$
    (d)

15. (a) $(1 - \sqrt{3}, 0)$, $(1 + \sqrt{3}, 0)$, $(0, 1)$
    (b) $y = -\frac{1}{2}(x - 1)^2 + \frac{3}{2}$
    (c) $(1, \frac{3}{2})$, $x = 1$
    (d)

17. (a) $(5, 0)$, $(0, 25)$
    (b) $y = (x - 5)^2$
    (c) $(5, 0)$, $x = 5$
    (d)

19. Minimum functional value is $f(\frac{4}{3}) = -\frac{13}{3}$; $[-\frac{13}{3}, \infty)$
21. Increasing on $[0, \infty)$, decreasing on $(-\infty, 0]$,
23. Increasing on $(-\infty, -3]$, decreasing on $[-3, \infty)$
25. The graph of $y = x^2$ is shifted horizontally ten units to the right.
27. The graph of $y = x^2$ is compressed vertically, followed by a reflection in the $x$-axis, followed by a horizontal shift of four units to the left, followed by vertical shift of nine units upward.
29. As the equation is given it can be interpreted as the graph of $y = x^2$ shifted horizontally six units to the right, followed by a reflection in the $y$-axis, followed by a vertical shift of four units downward.
31. $y = (x + 2)^2$
33. $y = -x^2 - 1$
35. $y = -(x + 1)^2 + 5$
37. $f(x) = 2x^2 + 3x + 5$
39. $f(x) = 4(x - 1)^2 + 2$
41. $(-4, 8)$, $(1, 3)$,
43. $(-2, 2)$, $(0, 2)$,

45. (a) $d^2 = 5x^2 - 10x + 25$   (b) $(1, 2)$
47. (a) $s(t) = -16t^2 + 64t + 6$, $v(t) = -32t + 64$
    (b) 70 ft, 0 ft/s   (c) 4 s, $-64$ ft/s
49. (a) 117.6 m, $-9.8$ m/s
    (b) in 5 seconds
    (c) $-49$ m/s

**51. (a)** The graph of $R(D) = -kD^2 + kPD$ is a parabola with vertex at $-b/2a = (-kP)/(-2k) = P/2$. Since $k$ is positive, the graph opens downward, and so $R(D)$ is a maximum at this value. Since $R(D)$ measures the rate at which the disease spreads, we conclude that the disease spreads most rapidly when exactly one-half the population is infected.

**(b)** $3 \times 10^{-5}$  **(c)** approximately 48

**(d)** approximately 62, 79, 102, and 130

---

### Exercises 2.5    Page 88

**1.** $2, 4, -5$       **3.** $3, 0, 8, 2 + 2\sqrt{2}$

**5. (a)** 1    **(b)** 1    **(c)** 0    **(d)** 1

   **(e)** 1    **(f)** 0

**7. (a)** 3    **(b)** $-1, \sqrt{2}$    **(c)** $\sqrt[3]{-2}, 1$    **(d)** $\sqrt[3]{-3}, 0$

   **(e)** $\sqrt{3}$    **(f)** $-2$

**9.** $(0, 0)$, continuous,      **11.** $(0, 0)$, continuous,

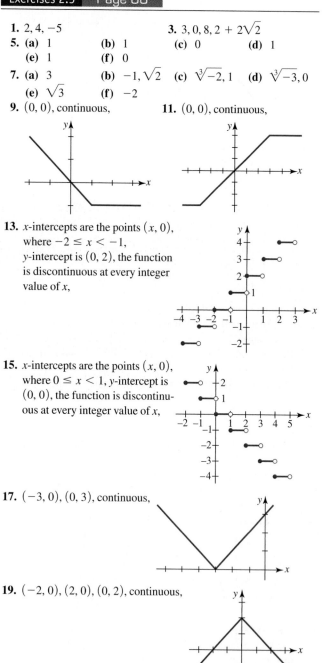

**13.** $x$-intercepts are the points $(x, 0)$, where $-2 \le x < -1$, $y$-intercept is $(0, 2)$, the function is discontinuous at every integer value of $x$,

**15.** $x$-intercepts are the points $(x, 0)$, where $0 \le x < 1$, $y$-intercept is $(0, 0)$, the function is discontinuous at every integer value of $x$,

**17.** $(-3, 0), (0, 3)$, continuous,

**19.** $(-2, 0), (2, 0), (0, 2)$, continuous,

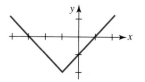

**23.** $(\frac{5}{3}, 0), (0, -5)$, continuous,

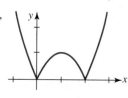

**25.** $(-1, 0), (1, 0), (0, 1)$, continuous,

**27.** $(0, 0), (2, 0)$, continuous,

**29.** $(-2, 0), (2, 0), (0, 2)$, continuous,

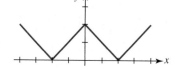

**31.** $(1, 0), (0, 1)$, continuous,     **33.** $(1, 0), (0, 1)$, continuous,

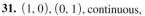
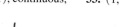
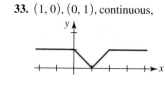

**35.** $\{-1, 1\}$

**37.** $f(x) = \begin{cases} x + 2, & x < 0 \\ -2x + 2, & 0 \le x < 2 \\ -2, & x \ge 2 \end{cases}$

**39.** $f(x) = \begin{cases} -x, & x < -3 \\ \sqrt{9 - x^2}, & -3 \le x < 3 \\ x, & x \ge 3 \end{cases}$

**41.**

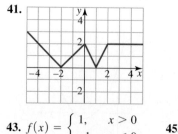

**43.** $f(x) = \begin{cases} 1, & x > 0 \\ -1, & x < 0 \end{cases}$    **45.** $k = 1$

**47.** $f(x) = [x] = \begin{cases} -2, & -3 < x \le -2 \\ -1, & -2 < x \le -1 \\ 0, & -1 < x \le 0 \\ 1, & 0 < x \le 1 \\ 2, & 1 < x \le 2 \\ 3, & 2 < x \le 3 \end{cases}$

Exercises 2.6    Page 95

**1.** $(f + g)(x) = 3x^2 - x + 1$, domain: $(-\infty, \infty)$
$(f - g)(x) = -x^2 + x + 1$, domain: $(-\infty, \infty)$
$(fg)(x) = 2x^4 - x^3 + 2x^2 - x$, domain: $(-\infty, \infty)$
$(f/g)(x) = (x^2 + 1)/(2x^2 - x)$,
domain: real numbers except $x = 0$ and $x = \frac{1}{2}$

**3.** $(f + g)(x) = x + \sqrt{x - 1}$, domain: $[1, \infty)$,
$(f - g)(x) = x - \sqrt{x - 1}$, domain: $[1, \infty)$,
$(fg)(x) = x\sqrt{x - 1}$, domain: $[1, \infty)$,
$(f/g)(x) = x/\sqrt{x - 1}$, domain: $(1, \infty)$

**5.** $(f + g)(x) = 3x^3 - 3x^2 + 3x + 1$, domain: $(-\infty, \infty)$,
$(f - g)(x) = 3x^3 - 5x^2 + 7x - 1$, domain: $(-\infty, \infty)$,
$(fg)(x) = 3x^5 - 10x^4 + 16x^3 - 14x^2 + 5x$,
domain: $(-\infty, \infty)$,
$(f/g)(x) = (3x^3 - 4x^2 + 5x)/(1 - x)^2$,
domain: real numbers except $x = 1$

**7.** $(f + g)(x) = \sqrt{x + 2} + \sqrt{5 - 5x}$, domain: $[-2, 1]$,
$(f - g)(x) = \sqrt{x + 2} - \sqrt{5 - 5x}$, domain: $[-2, 1]$,
$(fg)(x) = \sqrt{5(x + 2)(1 - x)}$, domain: $[-2, 1]$,
$\left(\dfrac{f}{g}\right)(x) = \sqrt{\dfrac{x + 2}{5 - 5x}}$, domain: $[-2, 1)$

**9.** $10, 8, -1, 2, 0$

**11.** $(f \circ g)(x) = x$, domain: $[1, \infty)$,
$(g \circ f)(x) = \sqrt{x^2} = |x|$, domain: $(-\infty, \infty)$

**13.** $(f \circ g)(x) = \dfrac{1}{2x^2 + 1}$, domain: $(-\infty, \infty)$,
$(g \circ f)(x) = \dfrac{4x^2 - 4x + 2}{4x^2 - 4x + 1}$,
domain: real numbers except $x = \frac{1}{2}$

**15.** $(f \circ g)(x) = x, (g \circ f)(x) = x$

**17.** $(f \circ g)(x) = \dfrac{x^3 + 1}{x}, (g \circ f)(x) = \dfrac{x^2}{x^3 + 1}$

**19.** $(f \circ g)(x) = x + 1 + \sqrt{x - 1}, (g \circ f)(x) = x + 1 + \sqrt{x}$

**21.** $(f \circ f)(x) = 4x + 18, \left(f \circ \dfrac{1}{f}\right)(x) = \dfrac{6x + 19}{x + 3}$

**23.** $(f \circ f)(x) = x^4, \left(f \circ \dfrac{1}{f}\right)(x) = \dfrac{1}{x^4}$

**25.** $(f \circ g \circ h)(x) = |x - 1|$    **27.** $(f \circ g \circ g)(x) = 54x^4 + 7$

**29.** $(f \circ f \circ f)(x) = 8x - 35$    **31.** $f(x) = x^5, g(x) = x^2 - 4x$

**33.** $f(x) = x^2 + 4\sqrt{x}, g(x) = x - 3$

**35.**

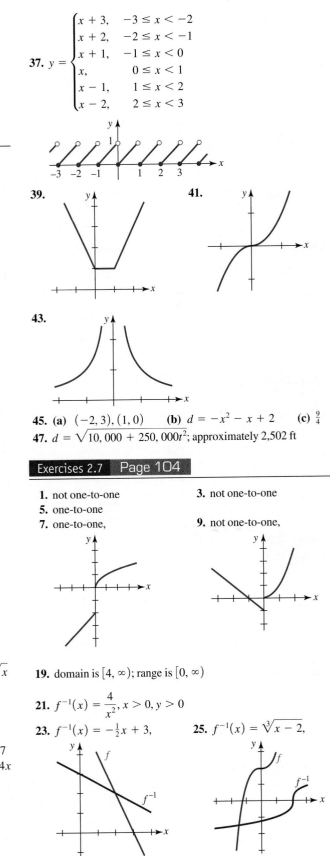

**37.** $y = \begin{cases} x + 3, & -3 \le x < -2 \\ x + 2, & -2 \le x < -1 \\ x + 1, & -1 \le x < 0 \\ x, & 0 \le x < 1 \\ x - 1, & 1 \le x < 2 \\ x - 2, & 2 \le x < 3 \end{cases}$

**39.**    **41.**

**43.**

**45.** (a) $(-2, 3), (1, 0)$    (b) $d = -x^2 - x + 2$    (c) $\frac{9}{4}$

**47.** $d = \sqrt{10,000 + 250,000t^2}$; approximately 2,502 ft

Exercises 2.7    Page 104

**1.** not one-to-one    **3.** not one-to-one
**5.** one-to-one
**7.** one-to-one,    **9.** not one-to-one,

**19.** domain is $[4, \infty)$; range is $[0, \infty)$

**21.** $f^{-1}(x) = \dfrac{4}{x^2}, x > 0, y > 0$

**23.** $f^{-1}(x) = -\frac{1}{2}x + 3$,    **25.** $f^{-1}(x) = \sqrt[3]{x - 2}$,

**27.** $f^{-1}(x) = (2 - x)^2, x \le 2,$ y

**29.** $f^{-1}(x) = \dfrac{x + 1}{2x}$, domain of $f^{-1}$ is the set of real numbers except $x = 0$, range of $f^{-1}$ is the set of real numbers except $y = \frac{1}{2}$, range of $f$ is the set of real numbers except $y = 0$

**31.** $f^{-1}(x) = \dfrac{3x}{2x - 7}$, domain of $f^{-1}$ is the set of real numbers except $x = \frac{7}{2}$, range of $f^{-1}$ is the set of real numbers except $y = \frac{3}{2}$, range of $f$ is the set of real numbers except $y = \frac{7}{2}$

**33.** $(20, 2)$          **35.** $(12, 9)$

**37.**

**39.**

**41.** $f^{-1}(x) = \frac{1}{2}\sqrt{x - 2}, x \ge 2$  y

**43.** $f^{-1}(x) = 2\sqrt{1 - x^2}, [0, 1]$  y

## Exercises 2.8    Page 111

**1.** $S(x) = x + \dfrac{50}{x}; (0, \infty)$    **3.** $S(x) = 3x^2 - 4x + 2; [0, 1]$

**5.** $A(x) = 100x - x^2; [0, 100]$   **7.** $A(x) = 2x - \frac{1}{2}x^2; [0, 4]$

**9.** $d(x) = \sqrt{2x^2 + 8}; (-\infty, \infty)$  **11.** $P(A) = 4\sqrt{A}; (0, \infty)$

**13.** $d(C) = C/\pi; (0, \infty)$       **15.** $A(h) = \dfrac{1}{\sqrt{3}}h^2; (0, \infty)$

**17.** $A(x) = \dfrac{1}{4\pi}x^2; (0, \infty)$      **19.** $s(h) = \dfrac{30h}{25 - h}; [0, 25)$

**21.** $S(w) = 3w^2 + \dfrac{1200}{w}; (0, \infty)$

**23.** $d(t) = 20\sqrt{13t^2 + 8t + 4}; (0, \infty)$

**25.** $V(h) = \begin{cases} 120h^2, & 0 \le h < 5 \\ 1200h - 3000, & 5 \le h \le 8 \end{cases}; [0, 8]$

**27.** $f(x) = x - x^2; (-\infty, \infty)$    **29.** $F(x) = 4x + \dfrac{8000}{x}; (0, \infty)$

**31.** $C(x) = 5x + \dfrac{512,000}{x}; (0, \infty)$

**33.** $A(x) = \frac{1}{2}xp - x^2; [0, \frac{1}{2}p]$

**35. (a)** $A(x) = x^2 + \dfrac{128,000}{x}; (0, \infty)$

  **(b)** $A(x) = 2x^2 + \dfrac{128,000}{x}; (0, \infty)$

**37.** $V(x) = 20x - 40x^2; [0, \frac{1}{2}]$

**39.** $A(x) = 40 + 4x + \dfrac{64}{x}; (0, \infty)$

**41.** $L(x) = x + \dfrac{8x}{\sqrt{x^2 - 64}}; (8, \infty)$

**43.** $L(x) = \dfrac{1}{4\pi}(L^2x - x^3); [0, L]$

**45.** $V(x) = 5x\sqrt{64 - x^2}; [0, 8]$

## Exercises 2.9    Page 122

**1. (a)** $6 + h$         **(b)** $6$        **(c)** $y = 6x - 15$
**3. (a)** $-1 + h$        **(b)** $-1$       **(c)** $y = -x - 1$
**5. (a)** $-23 - 12h - 2h^2$ **(b)** $-23$   **(c)** $y = -23x + 32$
**7. (a)** $\dfrac{1}{2(-1 + h)}$   **(b)** $-\frac{1}{2}$   **(c)** $y = -\frac{1}{2}x - 1$
**9. (a)** $\dfrac{1}{\sqrt{4 + h} + 2}$  **(b)** $\frac{1}{4}$  **(c)** $y = \frac{1}{4}x + 1$
**11. (a)** $0$                    **(b)** $f'(x) = 0$
**13. (a)** $-8x - 4h$             **(b)** $f'(x) = -8x$
**15. (a)** $6x + 3h - 1$          **(b)** $f'(x) = 6x - 1$
**17. (a)** $3x^2 + 3xh + h^2 + 5$ **(b)** $f'(x) = 3x^2 + 5$
**19. (a)** $\dfrac{1}{(4 - x)(4 - x - h)}$  **(b)** $f'(x) = \dfrac{1}{(4 - x)^2}$
**21. (a)** $\dfrac{-1}{(x - 1)(x + h - 1)}$  **(b)** $f'(x) = \dfrac{-1}{(x - 1)^2}$
**23. (a)** $1 - \dfrac{1}{x(x + h)}$  **(b)** $f'(x) = 1 - \dfrac{1}{x^2}$
**25. (a)** $\dfrac{2}{\sqrt{x + h} + \sqrt{x}}$  **(b)** $f'(x) = \dfrac{1}{\sqrt{x}}$
**27.** $(2, 17); 11; y = 11x - 5$
**29.** $(1, 2); 8; y = 8x - 6$    **31.** $(\frac{1}{2}, \frac{5}{2}); -3; y = -3x + 4$
**33. (a)** $6a + 3h$              **(b)** $f'(a) = 6a$
**35. (a)** $30a^2 + 30ah + 10h^2$ **(b)** $f'(a) = 30a^2$
**37. (a)** $\dfrac{-1}{a(a + h)}$  **(b)** $f'(a) = \dfrac{-1}{a^2}$
**39. (a)** $\dfrac{7}{\sqrt{7a + 7h} + \sqrt{7a}}$  **(b)** $f'(a) = \dfrac{1}{2}\sqrt{\dfrac{7}{a}}$

## Chapter 2 Review Exercises    Page 123

**1.** $-\frac{1}{3}$                  **3.** $(-\infty, 5)$
**5.** $x = 0, x = 2$                  **7.** $k = -\frac{6}{5}$
**9.** $m = -\frac{3}{2}$
**11.** $(1 - \sqrt{2}, 0), (1 + \sqrt{2}, 0), (0, -1)$
**13.** $f(x) = \frac{7}{4}(x + 2)^2$    **15.** $(10, 2)$
**17.** $(0, 5)$                       **19.** $a = -\frac{1}{4}$ and $a = 8$

**21.** true     **23.** false     **25.** true     **27.** true

**29.** true     **31.** true     **33.** true     **35.** false

**37.** false     **39.** true     **41.** $f(x) = x^2, g(x) = \dfrac{3x - 5}{x}$

**43.** (a) $y = (x + 3)^3 - 2$     (b) $y = x^3 - 7$

     (c) $y = (x - 1)^3$     (d) $y = -x^3 + 2$

     (e) $y = -x^3 - 2$     (f) $y = 3x^3 - 6$

**45.** domain: $(\pi/2, 3\pi/2)$, range: $(-\infty, \infty)$

**47.** $f(x) = \begin{cases} x + 1, & x < 0 \\ -x + 1, & 0 \le x < 1, \\ x - 1, & x \ge 1 \end{cases}$

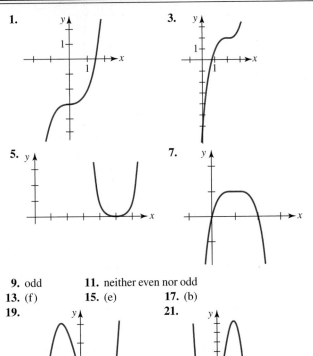

**49.** From the graph of $f$ it is seen that $f(x) > 0$ for all $x$. Thus the domain of $g$ is $(-\infty, \infty)$.

**51.** $f^{-1}(x) = -1 + \sqrt[3]{x}$     **53.** $A(h) = h^2(1 - \pi/4)$

**55.** $d(s) = \sqrt{3}s$

**57.** (a) $d(t) = 6t$     (b) $d(t) = \sqrt{90^2 + (90 - 6t)^2}$

**59.** $S(x) = 20x + \dfrac{5}{x}, x > 0$

**61.** $A(x) = 2x(1 - \pi x)$, where $x$ is the radius of the semicircle

**63.** $f'(x) = -6x + 16, y = 4x + 24$

**65.** $f'(x) = \dfrac{1}{x^3}, y = 8x - 6$

---

**Exercises 3.1**    **Page 137**

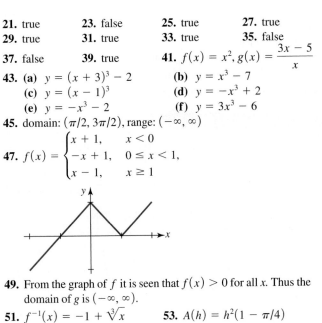

**9.** odd     **11.** neither even nor odd

**13.** (f)     **15.** (e)     **17.** (b)

**19.**      **21.**

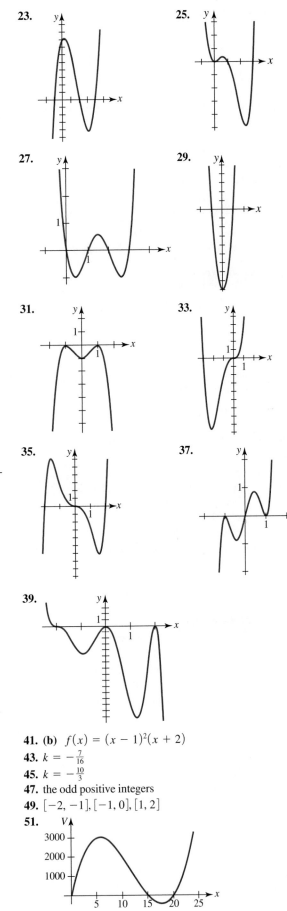

**41.** (b) $f(x) = (x - 1)^2(x + 2)$

**43.** $k = -\dfrac{7}{16}$

**45.** $k = -\dfrac{10}{3}$

**47.** the odd positive integers

**49.** $[-2, -1], [-1, 0], [1, 2]$

**51.**

## Exercises 3.2    Page 143

**1.** $f(x) = x^2 \cdot 8 + 4x - 7$

**3.** $f(x) = (x^2 + x - 1) \cdot (5x - 12) + 21x - 11$

**5.** $f(x) = (x + 2)^2 \cdot (2x - 4) + 5x + 21$

**7.** $f(x) = (3x^2 - x) \cdot (9x + 3) + 4x - 2$

**9.** $f(x) = (6x^2 + 4x + 1) \cdot (x^3 - 2) + 12x^2 + 8x + 2$

**11.** $r = 6$    **13.** $r = \frac{29}{8}$    **15.** $r = 76$    **17.** $f(2) = 2$

**19.** $f(-5) = -74$      **21.** $f(\frac{1}{2}) = \frac{303}{16}$

**23.** $q(x) = 2x + 3, r = 11$

**25.** $q(x) = x^2 - 4x + 12, r = -34$

**27.** $q(x) = x^3 + 2x^2 + 4x + 8, r = 32$

**29.** $q(x) = x^4 - 4x^3 + 16x^2 - 8x + 32, r = -132$

**31.** $q(x) = x^2 - 2x + \sqrt{3}, r = 0$

**33.** $f(-3) = 51$      **35.** $f(1) = 1$

**37.** $f(4) = 5369$   **39.** $k = -1$   **41.** $k = -\frac{1}{5}$    **43.** $k = -4$

## Exercises 3.3    Page 150

**1.** $f(x) = 4(x - \frac{1}{4})(x - 1)^2$    **3.** 5 is not a zero

**5.** $f(x) = 3(x + \frac{2}{3})(x - 2 + \sqrt{2})(x - 2 - \sqrt{2})$

**7.** $f(x) = 4(x + 3)(x - 5)(x - \frac{1}{2})(x + \frac{1}{2})$

**9.** $f(x) = 9(x - 1)(x + \frac{1}{3})^2(x + 8)$

**11.** $x - 5$ is not a factor

**13.** $f(x) = (x - 1)(x + \frac{1}{2} + \frac{1}{2}\sqrt{7}i)(x + \frac{1}{2} - \frac{1}{2}\sqrt{7}i)$

**15.** $x - \frac{1}{3}$ is not a factor

**17.** $f(x) = (x - 1)(x - 2)(x - 2i)(x + 2i)$

**19.** $f(x) = 2(x - 1)^2(x + 1)(x + \frac{3}{2})$

**21.** $f(x) = 3(x - \frac{5}{3})(x + 2i)(x - 2i)$

**23.** $f(x) = 5(x - \frac{2}{5})(x + 1 - i)(x + 1 + i)$

**25.** $f(x) = (x - 3)(x + 3)(x - 1 + 2i)(x - 1 - 2i)$

**27.** $f(x) = (x - 2)(x - 1)(x + 3)^2$
$\quad = x^4 + 3x^3 - 7x^2 - 15x + 18$

**29.** $f(x) = x^5 - 6x^4 + 10x^3$    **31.** $f(x) = x^2 - 2x + 37$

**33.** 0 is a simple zero, $\frac{5}{4}$ is a zero of multiplicity two, $\frac{1}{2}$ is a zero of multiplicity three

**35.** $-\frac{2}{3}$ is a zero of multiplicity two, $\frac{2}{3}$ is a zero of multiplicity two

**37.** $k = -36$      **39.** $f(x) = -\frac{1}{16}(x - 4)(x + 2)^2$

## Exercises 3.4    Page 157

**1.** $\frac{2}{5}$      **3.** 3

**5.** $\frac{1}{2}$ (multiplicity 2)      **7.** no rational zeros

**9.** $\frac{1}{3}, \frac{3}{2}$      **11.** 0, 1      **13.** $-3, 0, 2$      **15.** $\frac{3}{2}$

**17.** $-\frac{1}{5}$      **19.** $-\frac{1}{2}$ (multiplicity two), $\frac{1}{3}$ (multiplicity two)

**21.** $\frac{3}{8}, -\frac{1}{2} - \frac{1}{2}\sqrt{5}, -\frac{1}{2} + \frac{1}{2}\sqrt{5}$;
$\quad f(x) = (8x - 3)(x + \frac{1}{2} + \frac{1}{2}\sqrt{5})(x + \frac{1}{2} - \frac{1}{2}\sqrt{5})$

**23.** $\frac{4}{5}, \frac{5}{2}, -\sqrt{2}, \sqrt{2}$;
$\quad f(x) = (5x - 4)(2x - 5)(x + \sqrt{2})(x - \sqrt{2})$

**25.** $-4, -1, 1, -\sqrt{5}, \sqrt{5}$;
$\quad f(x) = (x + 4)(x + 1)(x - 1)(x + \sqrt{5})(x - \sqrt{5})$

**27.** $0, 1, 3, -1 - \sqrt{2}, -1 + \sqrt{2}$;
$\quad f(x) = x(x - 1)(x - 3)(x + 1 + \sqrt{2})(x + 1 - \sqrt{2})$

**29.** $-1, \frac{1}{4}$ (multiplicity 2);
$\quad f(x) = (x + 1)(4x - 1)^2(x^2 - 2x + 3)$

**31.** $-\frac{1}{2}$      **33.** $-\frac{3}{2}, 2, -2 - \sqrt{3}, -2 + \sqrt{3}$

**35.** 1 (mutliplicity 3)

**37.** $f(x) = 3x^4 - x^3 - 39x^2 + 49x - 12$

**39.** $\frac{3}{4}$      **41.** $f(x) = -\frac{1}{6}(x - 1)(x - 2)(x - 3)$

**43.** 3 inches or $\frac{1}{2}(7 - \sqrt{33}) \approx 0.63$ inches

## Exercises 3.5    Page 170

**1.**

| $x$ | 3.1 | 3.01 | 3.001 | 3.0001 | 3.00001 |
|-----|------|-------|--------|---------|----------|
| $f(x)$ | 62 | 602 | 6,002 | 60,002 | 600,002 |
| $x$ | 2.9 | 2.99 | 2.999 | 2.9999 | 2.99999 |
| $f(x)$ | $-58$ | $-598$ | $-5,998$ | $-59,998$ | $-599,998$ |

**3.** Asymptotes: $x = 2, y = 0$
Intercepts: $(0, -\frac{1}{2})$

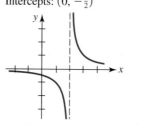

**5.** Asymptotes: $x = -1, y = 1$
Intercepts: $(0, 0)$

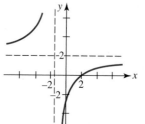

**7.** Asymptotes: $x = -\frac{2}{3}, y = 2$
Intercepts: $(\frac{9}{4}, 0), (10, -3)$

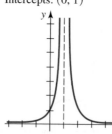

**9.** Asymptotes: $x = -1, y = -1$
Intercepts: $(1, 0), (0, 1)$

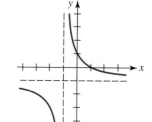

**11.** Asymptotes: $x = 1, y = 0$
Intercepts: $(0, 1)$

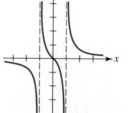

**13.** Asymptotes: $x = 0, y = 0$
Intercepts: none

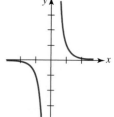

**15.** Asymptotes: $x = 1, x = -1, y = 0$
Intercepts: $(0, 0)$

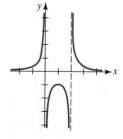

**17.** Asymptotes: $x = 0, x = 2, y = 0$
Intercepts: none

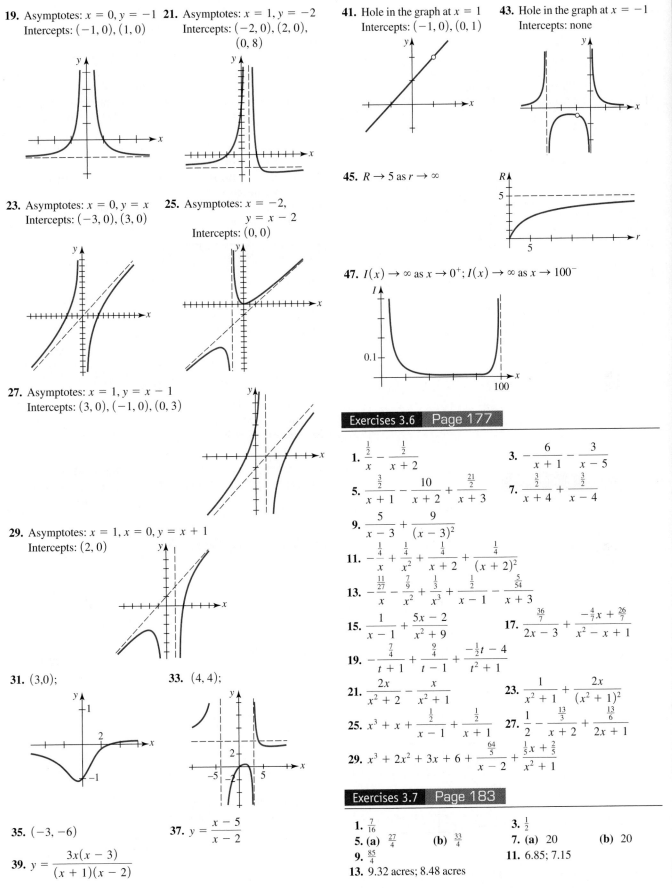

**19.** Asymptotes: $x = 0, y = -1$
Intercepts: $(-1, 0), (1, 0)$

**21.** Asymptotes: $x = 1, y = -2$
Intercepts: $(-2, 0), (2, 0),$
$(0, 8)$

**23.** Asymptotes: $x = 0, y = x$
Intercepts: $(-3, 0), (3, 0)$

**25.** Asymptotes: $x = -2,$
$y = x - 2$
Intercepts: $(0, 0)$

**27.** Asymptotes: $x = 1, y = x - 1$
Intercepts: $(3, 0), (-1, 0), (0, 3)$

**29.** Asymptotes: $x = 1, x = 0, y = x + 1$
Intercepts: $(2, 0)$

**31.** $(3, 0);$

**33.** $(4, 4);$

**35.** $(-3, -6)$

**37.** $y = \dfrac{x - 5}{x - 2}$

**39.** $y = \dfrac{3x(x - 3)}{(x + 1)(x - 2)}$

**41.** Hole in the graph at $x = 1$
Intercepts: $(-1, 0), (0, 1)$

**43.** Hole in the graph at $x = -1$
Intercepts: none

**45.** $R \to 5$ as $r \to \infty$

**47.** $I(x) \to \infty$ as $x \to 0^+$; $I(x) \to \infty$ as $x \to 100^-$

### Exercises 3.6  Page 177

**1.** $\dfrac{\frac{1}{2}}{x} - \dfrac{\frac{1}{2}}{x + 2}$

**3.** $-\dfrac{6}{x + 1} - \dfrac{3}{x - 5}$

**5.** $\dfrac{\frac{3}{2}}{x + 1} - \dfrac{10}{x + 2} + \dfrac{\frac{21}{2}}{x + 3}$

**7.** $\dfrac{\frac{3}{2}}{x + 4} + \dfrac{\frac{3}{2}}{x - 4}$

**9.** $\dfrac{5}{x - 3} + \dfrac{9}{(x - 3)^2}$

**11.** $-\dfrac{\frac{1}{4}}{x} + \dfrac{\frac{1}{4}}{x^2} + \dfrac{\frac{1}{4}}{x + 2} + \dfrac{\frac{1}{4}}{(x + 2)^2}$

**13.** $-\dfrac{\frac{11}{27}}{x} - \dfrac{\frac{7}{9}}{x^2} + \dfrac{\frac{1}{3}}{x^3} + \dfrac{\frac{1}{2}}{x - 1} - \dfrac{\frac{5}{54}}{x + 3}$

**15.** $\dfrac{1}{x - 1} + \dfrac{5x - 2}{x^2 + 9}$

**17.** $\dfrac{\frac{36}{7}}{2x - 3} + \dfrac{-\frac{4}{7}x + \frac{26}{7}}{x^2 - x + 1}$

**19.** $-\dfrac{\frac{7}{4}}{t + 1} + \dfrac{\frac{9}{4}}{t - 1} + \dfrac{-\frac{1}{2}t - 4}{t^2 + 1}$

**21.** $\dfrac{2x}{x^2 + 2} - \dfrac{x}{x^2 + 1}$

**23.** $\dfrac{1}{x^2 + 1} + \dfrac{2x}{(x^2 + 1)^2}$

**25.** $x^3 + x + \dfrac{\frac{1}{2}}{x - 1} + \dfrac{\frac{1}{2}}{x + 1}$

**27.** $\dfrac{1}{2} - \dfrac{\frac{13}{3}}{x + 2} + \dfrac{\frac{13}{6}}{2x + 1}$

**29.** $x^3 + 2x^2 + 3x + 6 + \dfrac{\frac{64}{5}}{x - 2} + \dfrac{\frac{1}{5}x + \frac{2}{5}}{x^2 + 1}$

### Exercises 3.7  Page 183

**1.** $\frac{7}{16}$

**3.** $\frac{1}{2}$

**5. (a)** $\frac{27}{4}$  **(b)** $\frac{33}{4}$

**7. (a)** 20  **(b)** 20

**9.** $\frac{85}{4}$

**11.** 6.85; 7.15

**13.** 9.32 acres; 8.48 acres

**Chapter 3 Review Exercises   Page 186**

**1.** $(1, 0); (0, 0), (5, 0)$          **3.** $f(x) = x^4$
**5.** $k = \frac{2}{3}$                          **7.** $x = 1, x = 4$
**9.** $y = -\frac{1}{2}$                        **11.** $n = 0, n = 1, n = 2$
**13.** true          **15.** true          **17.** true          **19.** true
**21.** true          **23.** false          **25.** false          **27.** false

**29.** $3x^3 - \frac{1}{2}x + 1 + \dfrac{-\frac{1}{2}x + 5}{2x^2 - 1}$

**31.** $7x^3 + 14x^2 + 22x + 53 + \dfrac{109}{x - 2}$

**33.** $r = f(-3) = -198$          **35.** $n$ an odd positive integer
**37.** $\pm 1, \pm 3, \pm 5, \pm 15, \pm\frac{1}{2}, \pm\frac{3}{2}, \pm\frac{5}{2}, \pm\frac{15}{2}, \pm\frac{1}{4}, \pm\frac{3}{4}, \pm\frac{5}{4}, \pm\frac{15}{4},$
      $\pm\frac{1}{8}, \pm\frac{3}{8}, \pm\frac{15}{8}$
**39.** $f(x) = (x - 2)(x - \frac{7}{2} + \frac{1}{2}\sqrt{3}i)(x - \frac{7}{2} - \frac{1}{2}\sqrt{3}i)$
**41.** $k = -\frac{21}{2}$                    **43.** $k = \frac{3}{2}$
**45.** $\dfrac{\frac{1}{3}}{x} + \dfrac{\frac{1}{4}}{x - 1} - \dfrac{\frac{7}{12}}{x + 3}$          **47.** $\dfrac{1}{x^2 + 4} - \dfrac{4}{(x^2 + 4)^2}$
**49.** $f(x) = 3x^2(x + 2)^2(x - 1)$          **51.** (f)
**53.** (d)          **55.** (h)          **57.** (c)          **59.** (b)
**61.** $y = 0, x = -4, x = 2, (-2, 0), (0, -\frac{1}{4}),$

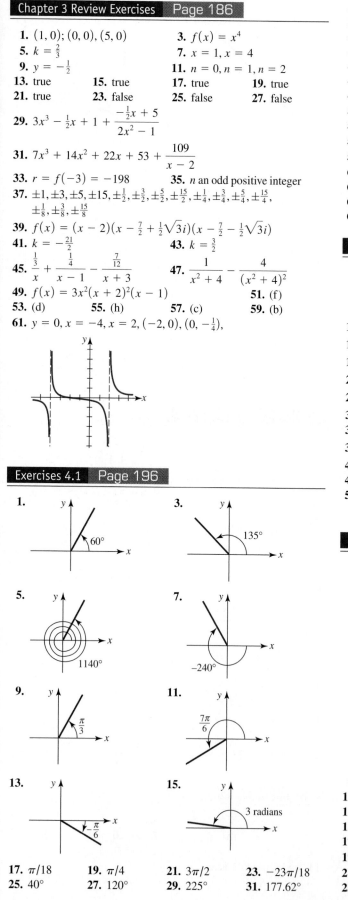

**Exercises 4.1   Page 196**

**1.**          **3.**

**5.**          **7.**

**9.**          **11.**

**13.**          **15.**

**17.** $\pi/18$          **19.** $\pi/4$          **21.** $3\pi/2$          **23.** $-23\pi/18$
**25.** $40°$          **27.** $120°$          **29.** $225°$          **31.** $177.62°$

**33.** $155°, -205°$          **35.** $110°, -250°$
**37.** $7\pi/4, -\pi/4$          **39.** $1.3\pi, -0.7\pi$
**41.** (a) $2\pi - 4 \approx 2.28$          (b) $-4$
**43.** (a) $41.75°$          (b) $131.75°$
**45.** (a) An obtuse angle does not have a complement.          (b) $81.6°$
**47.** (a) $\pi/4$          (b) $3\pi/4$
**49.** (a) An obtuse angle does not have a complement.          (b) $\pi/3$
**51.** (a) $216°, 1.2\pi$          (b) $-1845°, -10.25\pi$
**53.** $60°, \pi/3$
**55.** (a) 16 h          (b) 2 h          **57.** (a) 9          (b) 15
**59.** (a) 1.5          (b) $85.94°$
**63.** (a) $5.17°$          (b) $10.42°$          (c) $10.6547°$          (d) $143.1172°$
**65.** 1.15 statute miles
**67.** (a) $\frac{3}{26}$ radian          (b) $975$ cm$^2$
**69.** (a) $3\pi$ radians/s          (b) $180\pi$ cm/s

**Exercises 4.2   Page 204**

**1.** $\sqrt{21}/5$          **3.** $-\sqrt{5}/3$
**5.** $\pm 3\sqrt{5}/7$          **7.** $\pm 2\sqrt{6}/5 \approx \pm 0.98$
**9.** $\sin t = \pm 1/\sqrt{5}, \cos t = \pm 2/\sqrt{5}$
**11.** (a) $-1$          (b) 0          **13.** (a) 0          (b) 1
**15.** $\pi/3, \sqrt{3}/2, -\frac{1}{2}$          **17.** $\pi/4, -\sqrt{2}/2, -\sqrt{2}/2$
**19.** $\pi/6, -\frac{1}{2}, \sqrt{3}/2$          **21.** $\pi/4, -\sqrt{2}/2, \sqrt{2}/2$
**23.** $\pi/6, -\frac{1}{2}, -\sqrt{3}/2$          **25.** $\pi/3, \sqrt{3}/2, \frac{1}{2}$
**27.** $\sqrt{3}/2$          **29.** $\sqrt{2}/2$          **31.** $-1$
**33.** $\sin(t + 2\pi) = \sin t$ for $t = \pi$
**35.** $\sin(-t) = -\sin t$ for $t = 3 + \pi$
**37.** $\cos(-t) = \cos t$ for $t = 0.43$          **39.** $\sqrt{2}/2$
**41.** $-\sqrt{3}/2$     **43.** $\sqrt{3}/2$     **45.** $-\sqrt{3}/2$     **47.** $0, \pi$
**49.** $\pi/4, 7\pi/4$     **51.** $30°, 330°$     **53.** $225°, 315°$     **55.** 4.81 m
**57.** (a) $978.0309$ cm/s$^2$          (b) $983.21642$ cm/s$^2$
      (c) $980.61796$ cm/s$^2$

**Exercises 4.3   Page 213**

**1.**          **3.**

**5.**          **7.** $y = -3\sin x$

**9.** $y = 1 - 3\cos x$
**11.** $(n, 0)$, where $n$ is an integer
**13.** $((2n + 1)\pi, 0)$, where $n$ is an integer
**15.** $(\pi/4 + n\pi, 0)$, where $n$ is an integer
**17.** $(\pi/2, 0); (\pi/2 + 2n\pi, 0)$, where $n$ is an integer
**19.** $y = 3\sin 2x$
**21.** $y = \frac{1}{2}\cos \pi x$
**23.** $y = -\sin \pi x$

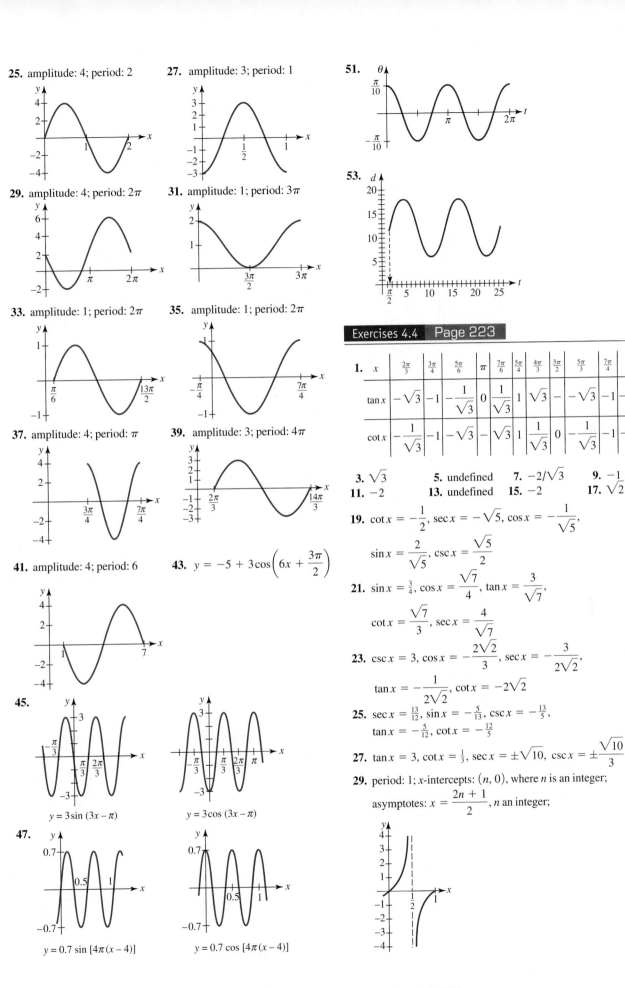

**25.** amplitude: 4; period: 2

**27.** amplitude: 3; period: 1

**29.** amplitude: 4; period: $2\pi$

**31.** amplitude: 1; period: $3\pi$

**33.** amplitude: 1; period: $2\pi$

**35.** amplitude: 1; period: $2\pi$

**37.** amplitude: 4; period: $\pi$

**39.** amplitude: 3; period: $4\pi$

**41.** amplitude: 4; period: 6

**43.** $y = -5 + 3\cos\left(6x + \dfrac{3\pi}{2}\right)$

**45.**

$y = 3\sin(3x - \pi)$

$y = 3\cos(3x - \pi)$

**47.**

$y = 0.7\sin[4\pi(x - 4)]$

$y = 0.7\cos[4\pi(x - 4)]$

**51.**

**53.**

**1.**

| $x$ | $\frac{2\pi}{3}$ | $\frac{3\pi}{4}$ | $\frac{5\pi}{6}$ | $\pi$ | $\frac{7\pi}{6}$ | $\frac{5\pi}{4}$ | $\frac{4\pi}{3}$ | $\frac{3\pi}{2}$ | $\frac{5\pi}{3}$ | $\frac{7\pi}{4}$ | $\frac{11\pi}{6}$ | $2\pi$ |
|---|---|---|---|---|---|---|---|---|---|---|---|---|
| $\tan x$ | $-\sqrt{3}$ | $-1$ | $-\frac{1}{\sqrt{3}}$ | $0$ | $\frac{1}{\sqrt{3}}$ | $1$ | $\sqrt{3}$ | $-$ | $-\sqrt{3}$ | $-1$ | $-\frac{1}{\sqrt{3}}$ | $0$ |
| $\cot x$ | $-\frac{1}{\sqrt{3}}$ | $-1$ | $-\sqrt{3}$ | $-$ | $\sqrt{3}$ | $1$ | $\frac{1}{\sqrt{3}}$ | $0$ | $-\frac{1}{\sqrt{3}}$ | $-1$ | $-\sqrt{3}$ | $-$ |

**3.** $\sqrt{3}$   **5.** undefined   **7.** $-2/\sqrt{3}$   **9.** $-1$

**11.** $-2$   **13.** undefined   **15.** $-2$   **17.** $\sqrt{2}$

**19.** $\cot x = -\dfrac{1}{2}$, $\sec x = -\sqrt{5}$, $\cos x = -\dfrac{1}{\sqrt{5}}$,

$\sin x = \dfrac{2}{\sqrt{5}}$, $\csc x = \dfrac{\sqrt{5}}{2}$

**21.** $\sin x = \frac{3}{4}$, $\cos x = \dfrac{\sqrt{7}}{4}$, $\tan x = \dfrac{3}{\sqrt{7}}$,

$\cot x = \dfrac{\sqrt{7}}{3}$, $\sec x = \dfrac{4}{\sqrt{7}}$

**23.** $\csc x = 3$, $\cos x = -\dfrac{2\sqrt{2}}{3}$, $\sec x = -\dfrac{3}{2\sqrt{2}}$,

$\tan x = -\dfrac{1}{2\sqrt{2}}$, $\cot x = -2\sqrt{2}$

**25.** $\sec x = \frac{13}{12}$, $\sin x = -\frac{5}{13}$, $\csc x = -\frac{13}{5}$,

$\tan x = -\frac{5}{12}$, $\cot x = -\frac{12}{5}$

**27.** $\tan x = 3$, $\cot x = \frac{1}{3}$, $\sec x = \pm\sqrt{10}$, $\csc x = \pm\dfrac{\sqrt{10}}{3}$

**29.** period: 1; $x$-intercepts: $(n, 0)$, where $n$ is an integer;

asymptotes: $x = \dfrac{2n + 1}{2}$, $n$ an integer;

**31.** period: $\pi/2$;

x-intercepts: $\left(\dfrac{2n+1}{4}\pi, 0\right)$,

where $n$ is an integer;

asymptotes: $x = n\pi/2$, $n$ an integer;

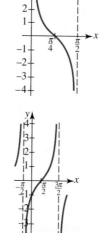

**33.** period: $2\pi$;

x-intercepts: $\left(\dfrac{\pi}{2} + 2n\pi, 0\right)$,

where $n$ is an integer;

asymptotes: $x = \dfrac{3\pi}{2} + 2n\pi$,

$n$ an integer;

**35.** period: 1;

x-intercepts: $\left(\frac{1}{4} + n, 0\right)$,

where $n$ is an integer;

asymptotes: $x = n$,

$n$ an integer;

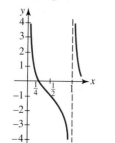

**37.** period: $2\pi$;

asymptotes: $x = \dfrac{2n+1}{2}\pi$,

$n$ an integer;

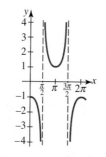

**39.** period: 2;

asymptotes: $x = n$,

$n$ an integer;

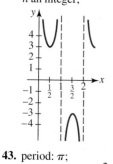

**41.** period: $2\pi/3$;

asymptotes: $x = \dfrac{n\pi}{3}$,

$n$ an integer;

**43.** period: $\pi$;

asymptotes: $x = \dfrac{2n-1}{4}\pi$,

$n$ an integer;

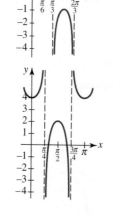

**45.** $\cot x = -\tan\left(x - \dfrac{\pi}{2}\right)$

**1.** $a\sin\theta$     **3.** $a\tan\theta$     **5.** $\tan\theta$     **7.** $\dfrac{\sqrt{7}}{7}\cos\theta$

**9.** $\dfrac{\sqrt{2}}{4}(1 + \sqrt{3})$     **11.** $\dfrac{\sqrt{2}}{4}(1 + \sqrt{3})$

**13.** $\dfrac{\sqrt{2}}{4}(1 + \sqrt{3})$     **15.** $2 + \sqrt{3}$

**17.** $\dfrac{\sqrt{2}}{4}(1 - \sqrt{3})$     **19.** $\dfrac{\sqrt{2}}{4}(\sqrt{3} - 1)$

**21.** $-\dfrac{\sqrt{2}}{4}(1 + \sqrt{3})$     **23.** $-2 + \sqrt{3}$

**25.** $\dfrac{\sqrt{2}}{4}(1 - \sqrt{3})$     **27.** $\dfrac{\sqrt{2}}{4}(\sqrt{3} + 1)$

**29.** $-\dfrac{\sqrt{2}}{4}(1 + \sqrt{3})$     **31.** $\sin 2\beta$

**33.** $\cos\dfrac{2\pi}{5}$

**35.** (a) $\frac{5}{9}$     (b) $-\dfrac{2\sqrt{14}}{9}$     (c) $-\dfrac{2\sqrt{14}}{5}$

**37.** (a) $\frac{3}{5}$     (b) $\frac{4}{5}$     (c) $\frac{4}{3}$

**39.** (a) $-\frac{119}{169}$     (b) $-\frac{120}{169}$     (c) $\frac{120}{119}$

**41.** $\dfrac{\sqrt{2+\sqrt{3}}}{2}$  **43.** $\dfrac{\sqrt{2+\sqrt{2}}}{2}$  **45.** $\dfrac{\sqrt{2-\sqrt{2}}}{2}$  **47.** $\dfrac{-2}{\sqrt{2-\sqrt{3}}}$

**49.** (a) $\dfrac{2\sqrt{13}}{13}$     (b) $\dfrac{3\sqrt{13}}{13}$     (c) $\frac{3}{2}$

**51.** (a) $-\sqrt{(5-\sqrt{5})/10}$     (b) $\sqrt{(5+\sqrt{5})/10}$

(c) $-(1 + \sqrt{5})/2$

**53.** (a) $\dfrac{\sqrt{30}}{6}$     (b) $\dfrac{\sqrt{6}}{6}$     (c) $\dfrac{\sqrt{5}}{5}$

**55.** (a) $-2(\sqrt{10} + 1)/9$     (b) $(\sqrt{5} - 4\sqrt{2})/9$

(c) $2(1 - \sqrt{10})/9$     (d) $(\sqrt{5} + 4\sqrt{2})/9$

**57.** $y = \sqrt{2}\sin(\pi x + 3\pi/4)$;

amplitude: $\sqrt{2}$; period: 2;

phase shift: $\frac{3}{4}$;

one cycle of the graph is

**59.** $y = 2\sin(2x + \pi/6)$;

amplitude: 2; period: $\pi$;

phase shift: $\pi/12$;

one cycle of the graph is

**61.** $y = \sin(x - 5\pi/4)$;

amplitude: 1; period: $2\pi$;

phase shift: $5\pi/4$;

one cycle of the graph is

**63.** $2\sqrt{2 + \sqrt{3}} \approx 3.86$

**1.** $x = \dfrac{\pi}{3} + 2n\pi$ or $x = \dfrac{2\pi}{3} + 2n\pi$, where $n$ is an integer

**3.** $x = \dfrac{\pi}{4} + 2n\pi$ or $x = \dfrac{7\pi}{4} + 2n\pi$, where $n$ is an integer

**5.** $x = \dfrac{5\pi}{6} + n\pi$, where $n$ is an integer

**7.** $x = \pi + 2n\pi = (2n + 1)\pi$, where $n$ is an integer

**9.** $x = n\pi$, where $n$ is an integer

**11.** $x = \dfrac{3\pi}{2} + 2n\pi$, where $n$ is an integer

**13.** $\theta = 60° + 360°n$ or $\theta = 120° + 360°n$, where $n$ is an integer

**15.** $\theta = 135° + 180°n$, where $n$ is an integer

**17.** $\theta = 120° + 360°n$ or $\theta = 240° + 360°n$, where $n$ is an integer

**19.** $x = n\pi$, where $n$ is an integer

**21.** no solutions

**23.** $\theta = 120° + 360°n$ or $\theta = 240° + 360°n$, where $n$ is an integer

**25.** $\theta = 90° + 180°n$ or $\theta = 135° + 180°n$, where $n$ is an integer

**27.** $x = \dfrac{\pi}{2} + n\pi$, where $n$ is an integer

**29.** $\theta = 10° + 120°n$ or $\theta = 50° + 120°n$, where $n$ is an integer

**31.** $x = \dfrac{\pi}{2} + 2n\pi$, where $n$ is an integer

**33.** $x = n\pi$, $x = \dfrac{2\pi}{3} + 2n\pi$, or $x = \dfrac{4\pi}{3} + 2n\pi$, where $n$ is an integer

**35.** $\theta = 30° + 360°n$, $\theta = 150° + 360°n$, or $\theta = 270° + 360°n$, where $n$ is an integer

**37.** $x = \dfrac{\pi}{2} + n\pi$, where $n$ is an integer

**39.** $x = n\pi$, where $n$ is an integer

**41.** $\theta = 90° + 180°n$, where $n$ is an integer

**43.** $x = \dfrac{\pi}{6} + 2n\pi$ or $x = \dfrac{5\pi}{6} + 2n\pi$, where $n$ is an integer

**45.** $\theta = 90° + 180°n$ or $\theta = 360°n$, where $n$ is an integer

**47.** $(\pi/3, 0), (2\pi/3, 0), (\pi, 0)$   **49.** $(\frac{2}{3}, 0), (\frac{10}{3}, 0), (\frac{14}{3}, 0)$

**51.** $(\pi, 0), (2\pi, 0), (3\pi, 0)$   **53.** $(\pi/3, 0), (\pi, 0), (5\pi/3, 0)$

**55.** The equation has infinitely many solutions.

**57.** The equation has infinitely many solutions.

**59.** 30° or 150°   **61.** 60°

**63.** $t = \frac{1}{120}(\frac{1}{6} + n)$, where $n$ is an integer.

**65.** (a) 36.93 million square kilometers
(b) $w = 31$ weeks   (c) August

**1.** 0   **3.** $\pi$   **5.** $\pi/3$   **7.** $-\pi/3$

**9.** $\pi/4$   **11.** $-\pi/6$   **13.** $-\pi/4$   **15.** $\frac{4}{5}$

**17.** $-\sqrt{5}/2$   **19.** $\sqrt{5}/5$   **21.** $\frac{5}{3}$   **23.** $\frac{1}{5}$

**25.** 1.2   **27.** $\pi/16$   **29.** 0   **31.** $\pi/4$

**33.** $\dfrac{x}{\sqrt{1 + x^2}}$   **35.** $\dfrac{x}{\sqrt{1 - x^2}}$   **37.** $\dfrac{\sqrt{1 - x^2}}{x}$   **39.** $\dfrac{\sqrt{1 + x^2}}{x}$

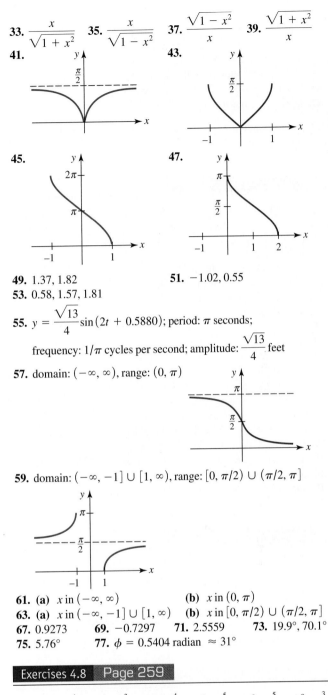

**41.**   **43.**

**45.**   **47.**

**49.** 1.37, 1.82   **51.** $-1.02, 0.55$

**53.** 0.58, 1.57, 1.81

**55.** $y = \dfrac{\sqrt{13}}{4}\sin(2t + 0.5880)$; period: $\pi$ seconds; frequency: $1/\pi$ cycles per second; amplitude: $\dfrac{\sqrt{13}}{4}$ feet

**57.** domain: $(-\infty, \infty)$, range: $(0, \pi)$

**59.** domain: $(-\infty, -1] \cup [1, \infty)$, range: $[0, \pi/2) \cup (\pi/2, \pi]$

**61.** (a) $x$ in $(-\infty, \infty)$   (b) $x$ in $(0, \pi)$

**63.** (a) $x$ in $(-\infty, -1] \cup [1, \infty)$   (b) $x$ in $[0, \pi/2) \cup (\pi/2, \pi]$

**67.** 0.9273   **69.** $-0.7297$   **71.** 2.5559   **73.** 19.9°, 70.1°

**75.** 5.76°   **77.** $\phi = 0.5404$ radian $\approx 31°$

**1.** $\sin\theta = \frac{4}{5}$, $\cos\theta = \frac{3}{5}$, $\tan\theta = \frac{4}{3}$, $\csc\theta = \frac{5}{4}$, $\sec\theta = \frac{5}{3}$, $\cot\theta = \frac{3}{4}$

**3.** $\sin\theta = 3\sqrt{10}/10$, $\cos\theta = \sqrt{10}/10$, $\tan\theta = 3$, $\csc\theta = \sqrt{10}/3$, $\sec\theta = \sqrt{10}$, $\cot\theta = \frac{1}{3}$

**5.** $\sin\theta = \frac{2}{5}$, $\cos\theta = \sqrt{21}/5$, $\tan\theta = 2\sqrt{21}/21$, $\csc\theta = \frac{5}{2}$, $\sec\theta = 5\sqrt{21}/21$, $\cot\theta = \sqrt{21}/2$

**7.** $\sin\theta = \frac{1}{3}$, $\cos\theta = 2\sqrt{2}/3$, $\tan\theta = \sqrt{2}/4$, $\csc\theta = 3$, $\sec\theta = 3\sqrt{2}/4$, $\cot\theta = 2\sqrt{2}$

**9.** $\sin\theta = y/\sqrt{x^2 + y^2}$, $\cos\theta = x/\sqrt{x^2 + y^2}$, $\tan\theta = y/x$, $\csc\theta = \sqrt{x^2 + y^2}/y$, $\sec\theta = \sqrt{x^2 + y^2}/x$, $\cot\theta = x/y$

**11.** $b = 2.04$, $c = 4.49$   **13.** $a = 11.71$, $c = 14.19$

**15.** $\alpha = 60°$, $\beta = 30°$, $a = 2.6$   **17.** $\alpha = 21.8°$, $\beta = 68.2°$, $c = 10.8$

**19.** $\alpha = 48.6°$, $\beta = 41.4°$, $c = 7.9$   **21.** $a = 8.5$, $c = 21.7$

**23.** 52.1 m    **25.** 66.4 ft
**27.** 409.7 ft    **29.** height: 15.5 ft; distance: 12.6 ft
**31.** 8.7°    **33.** 34,157 ft ≈ 6.5 mi
**35.** 20.2 ft    **37.** 6617 ft
**39.** Yes, since the altitude of the storm is 6.3 km.
**43.** 227,100 mi   **45.** $h = c/(\cot\alpha + \cot\beta)$
**47.** $h(\theta) = 1.25\tan\theta$

**49.** $\theta(x) = \arctan\left(\dfrac{1}{x}\right) - \arctan\left(\dfrac{1}{2x}\right)$, where $x$ is measured in

    meters

### Exercises 4.9   Page 272

**1.** $\gamma = 80°, a = 20.16, c = 20.16$
**3.** $\alpha = 92°, b = 3.01, c = 3.89$
**5.** $\alpha = 79.6°, \gamma = 28.4°, a = 12.41$
**7.** no solution
**9.** $\alpha = 24.46°, \beta = 140.54°, b = 12.28;$
    $\alpha = 155.54°, \beta = 9.46°, b = 3.18$
**11.** $\alpha = 37.59°, \beta = 77.41°, c = 7.43$
**13.** $\alpha = 52.62°, \beta = 83.33°, \gamma = 44.05°$
**15.** $\alpha = 25°, \beta = 57.67°, c = 7.04$
**17.** $\alpha = 76.45°, \beta = 57.1°, \gamma = 46.45°$
**19.** $\alpha = 27.66°, \beta = 40.54°, \gamma = 111.8°$
**21.** no solution
**23.** $\alpha = 36.87°, \beta = 53.13°, \gamma = 90°$
**25.** $\alpha = 45.58°, \gamma = 104.42°, c = 13.56;$
    $\alpha = 134.42°, \gamma = 15.58°, c = 3.76$
**27.** $\alpha = 26.38°, \beta = 36.34°, \gamma = 117.28°$
**29.** $\beta = 10.24°, \gamma = 147.76°, a = 6.32$
**31.** no solution   **33.** 15.80 ft    **35.** 9.1 m      **37.** 13 ft 5 in.
**39.** 35.94 nautical miles
**41. (a)** S 33.66° W      **(b)** S 2.82° E
**43.** $\alpha = 119.45°, \beta = 67.98°$

### Exercises 4.10   Page 279

**1.** $\frac{1}{2}$      **3.** $-1$      **5.** $-\sqrt{3}/2$    **7.** 5
**9.** 0      **11.** $-2$      **13.** 0      **15.** 1
**17.** 1      **19.** $y = x$

**21.** $y = \dfrac{1}{2} + \dfrac{\sqrt{3}}{2}\left(x - \dfrac{\pi}{6}\right)$      **23.** $f'(x) = -\sin x$

**25.** $f'(x) = 5\cos 5x$

### Chapter 4 Review Exercises   Page 282

**1.** 36°      **3.** $(-\sqrt{3}/2, 1/2)$
**5.** $\sqrt{3}$      **7.** $3/(2\sqrt{2})$    **9.** $(0, -2)$    **11.** $2/\sqrt{5}$
**13.** $6\sin(\pi x/2)$ **15.** $\sqrt{2}$      **17.** 10      **19.** $\pi/5$
**21.** false      **23.** true      **25.** true      **27.** true
**29.** true      **31.** true      **33.** true      **35.** false
**37.** false      **39.** false
**41.** $\pi/2, \pi$               **43.** $\pi/6, 5\pi/6, 7\pi/6, 11\pi/6$
**45.** $0, \pi/2, 2\pi$
**47.** $\gamma = 80°, a = 5.32, c = 10.48$
**49.** $a = 15.76, \beta = 99.44°, \gamma = 29.56°$
**51.** $2\pi/3$
**53.** $3/\sqrt{7}$      **55.** 0        **57.** $\frac{12}{13}$      **59.** $\sqrt{1 - x^2}$
**61.** $y = -\sin x; y = \cos(x + \pi/2)$
**63.** $y = 1 + \frac{1}{2}\sin(x + \pi/2); y = 1 + \frac{1}{2}\cos x$

**65.** 42.61 m    **67.** 118 ft      **69. (a)** 42.35°   **71.** 16,927.6 km
**73.** front: 18.88°; back: 35.12°
**75.** $V(\theta) = 160\sin 2\theta$          **77.** $L(\theta) = 3\csc\theta + 4\sec\theta$
**79.** $V(\theta) = 360 + 75\cot\theta$
**81.** $A(\phi) = 100\cos\phi + 50\sin 2\phi$

### Exercises 5.1   Page 295

**1.** $(0, 1)$, $y = 0$, decreasing      **3.** $(0, -1)$, $y = 0$, decreasing

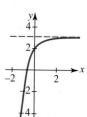

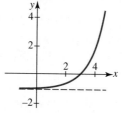

**5.** $(0, 2)$, $y = 0$, increasing      **7.** $(0, -4)$, $y = -5$, increasing

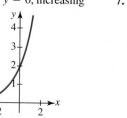

**9.** $(0, 2)$, $y = 3$, increasing      **11.** $(0, -1 + e^{-3})$, $y = -1$,
                                increasing

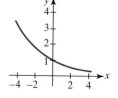

**13.** $f(x) = 6^x$             **15.** $f(x) = (e^{-2})^x = e^{-2x}$
**17.** $(5, \infty)$              **19.** $(2, 0), (0, -3)$
**21.** $(-10, 0), (0, 10)$      **23.** $(-2, 0), (-3, 0), (0, 0)$
**25.** $x > 4$              **27.** $x < 2$
**29.**

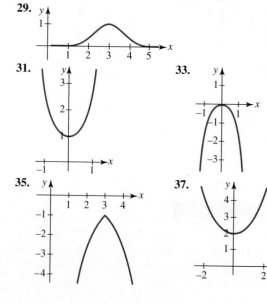

**31.**                           **33.**

**35.**                           **37.**

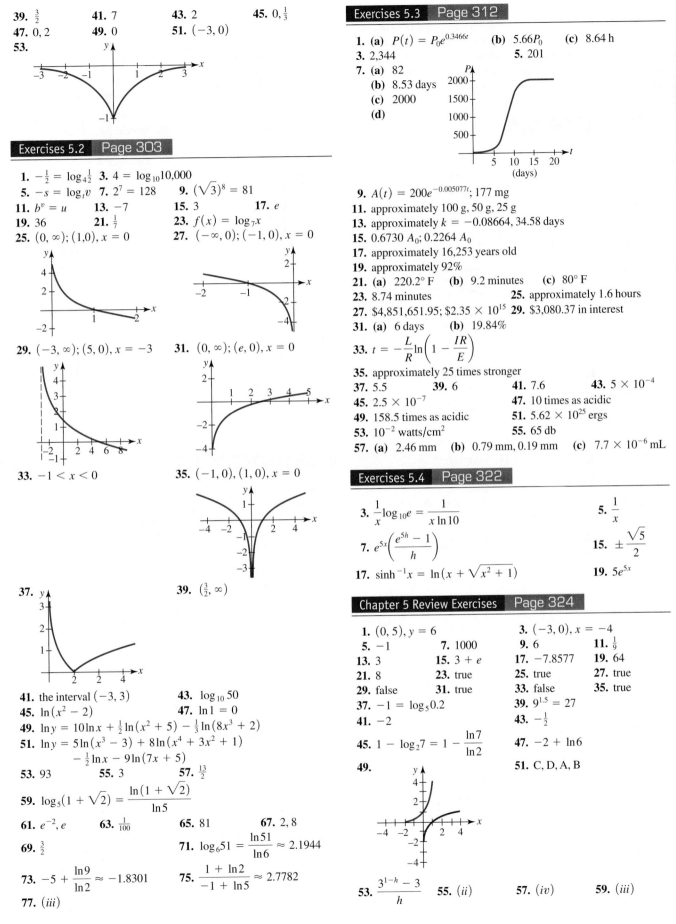

**39.** $\frac{3}{2}$     **41.** 7     **43.** 2     **45.** $0, \frac{1}{3}$

**47.** $0, 2$     **49.** 0     **51.** $(-3, 0)$

**53.**

Exercises 5.2    Page 303

**1.** $-\frac{1}{2} = \log_4 \frac{1}{2}$   **3.** $4 = \log_{10} 10{,}000$

**5.** $-s = \log_t v$   **7.** $2^7 = 128$    **9.** $(\sqrt{3})^8 = 81$

**11.** $b^v = u$    **13.** $-7$    **15.** 3    **17.** $e$

**19.** 36    **21.** $\frac{1}{7}$    **23.** $f(x) = \log_7 x$

**25.** $(0, \infty); (1, 0), x = 0$    **27.** $(-\infty, 0); (-1, 0), x = 0$

**29.** $(-3, \infty); (5, 0), x = -3$    **31.** $(0, \infty); (e, 0), x = 0$

**33.** $-1 < x < 0$    **35.** $(-1, 0), (1, 0), x = 0$

**37.**

**39.** $(\frac{3}{2}, \infty)$

**41.** the interval $(-3, 3)$    **43.** $\log_{10} 50$

**45.** $\ln(x^2 - 2)$    **47.** $\ln 1 = 0$

**49.** $\ln y = 10 \ln x + \frac{1}{2} \ln(x^2 + 5) - \frac{1}{3} \ln(8x^3 + 2)$

**51.** $\ln y = 5 \ln(x^3 - 3) + 8 \ln(x^4 + 3x^2 + 1)$
$- \frac{1}{2} \ln x - 9 \ln(7x + 5)$

**53.** 93    **55.** 3    **57.** $\frac{13}{2}$

**59.** $\log_5(1 + \sqrt{2}) = \dfrac{\ln(1 + \sqrt{2})}{\ln 5}$

**61.** $e^{-2}, e$    **63.** $\frac{1}{100}$    **65.** 81    **67.** $2, 8$

**69.** $\frac{3}{2}$    **71.** $\log_6 51 = \dfrac{\ln 51}{\ln 6} \approx 2.1944$

**73.** $-5 + \dfrac{\ln 9}{\ln 2} \approx -1.8301$    **75.** $\dfrac{1 + \ln 2}{-1 + \ln 5} \approx 2.7782$

**77.** $(iii)$

Exercises 5.3    Page 312

**1.** **(a)** $P(t) = P_0 e^{0.3466t}$    **(b)** $5.66 P_0$    **(c)** 8.64 h

**3.** 2,344      **5.** 201

**7.** **(a)** 82
**(b)** 8.53 days
**(c)** 2000
**(d)**

**9.** $A(t) = 200 e^{-0.005077t}$; 177 mg

**11.** approximately 100 g, 50 g, 25 g

**13.** approximately $k = -0.08664$, 34.58 days

**15.** $0.6730 A_0$; $0.2264 A_0$

**17.** approximately 16,253 years old

**19.** approximately 92%

**21.** **(a)** $220.2°$ F    **(b)** 9.2 minutes    **(c)** $80°$ F

**23.** 8.74 minutes      **25.** approximately 1.6 hours

**27.** \$4,851,651.95; $\$2.35 \times 10^{15}$   **29.** \$3,080.37 in interest

**31.** **(a)** 6 days    **(b)** 19.84%

**33.** $t = -\dfrac{L}{R} \ln\left(1 - \dfrac{IR}{E}\right)$

**35.** approximately 25 times stronger

**37.** 5.5    **39.** 6    **41.** 7.6    **43.** $5 \times 10^{-4}$

**45.** $2.5 \times 10^{-7}$      **47.** 10 times as acidic

**49.** 158.5 times as acidic    **51.** $5.62 \times 10^{25}$ ergs

**53.** $10^{-2}$ watts/cm$^2$      **55.** 65 db

**57.** **(a)** 2.46 mm    **(b)** 0.79 mm, 0.19 mm    **(c)** $7.7 \times 10^{-6}$ mL

Exercises 5.4    Page 322

**3.** $\dfrac{1}{x} \log_{10} e = \dfrac{1}{x \ln 10}$    **5.** $\dfrac{1}{x}$

**7.** $e^{5x}\left(\dfrac{e^{5h} - 1}{h}\right)$    **15.** $\pm\dfrac{\sqrt{5}}{2}$

**17.** $\sinh^{-1} x = \ln(x + \sqrt{x^2 + 1})$    **19.** $5e^{5x}$

Chapter 5 Review Exercises    Page 324

**1.** $(0, 5), y = 6$      **3.** $(-3, 0), x = -4$

**5.** $-1$    **7.** 1000    **9.** 6    **11.** $\frac{1}{9}$

**13.** 3    **15.** $3 + e$    **17.** $-7.8577$    **19.** 64

**21.** 8    **23.** true    **25.** true    **27.** true

**29.** false    **31.** true    **33.** false    **35.** true

**37.** $-1 = \log_5 0.2$      **39.** $9^{1.5} = 27$

**41.** $-2$      **43.** $-\frac{1}{2}$

**45.** $1 - \log_2 7 = 1 - \dfrac{\ln 7}{\ln 2}$    **47.** $-2 + \ln 6$

**49.**      **51.** C, D, A, B

**53.** $\dfrac{3^{1-h} - 3}{h}$    **55.** $(ii)$    **57.** $(iv)$    **59.** $(iii)$

**63.** $f(x) = 5e^{(-\frac{1}{6}\ln 5)x} = 5e^{-0.2682x}$ **65.** $f(x) = 5 + (\frac{1}{2})^x$

**67.** After doubling, it will take another 9 hours to double again. In other words, it will take a total of 18 hours for the population to grow to 4 times the initial population.

**69.** $0.0625A_0$ or $6\frac{1}{4}\%$ of the initial quantity **71.** 4.3%

**73.** $x = \frac{1}{c}[\ln b - \ln(\ln a - \ln y)]$

**Exercises 6.1** Page 334

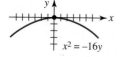

**1.** Vertex: $(0, 0)$
Focus: $(1, 0)$
Directrix: $x = -1$
Axis: $y = 0$

$y^2 = 4x$

**3.** Vertex: $(0, 0)$
Focus: $(-\frac{1}{3}, 0)$
Directrix: $x = \frac{1}{3}$
Axis: $y = 0$

$y^2 = -\frac{4}{3}x$

**5.** Vertex: $0, 0$
Focus: $(0, -4)$
Directrix: $y = 4$
Axis: $x = 0$

$x^2 = -16y$

**7.** Vertex: $(0, 0)$
Focus: $(0, 7)$
Directrix: $y = -7$
Axis: $x = 0$

$x^2 = 28y$

**9.** Vertex: $(0, 1)$
Focus: $(4, 1)$
Directrix: $x = -4$
Axis: $y = 1$

$(y - 1)^2 = 16x$

**11.** Vertex: $(-5, -1)$
Focus: $(-5, -2)$
Directrix: $y = 0$
Axis: $x = -5$

$(x + 5)^2 = -4(y + 1)$

**13.** Vertex: $(-5, -6)$
Focus: $(-4, -6)$
Directrix: $x = -6$
Axis: $y = -6$

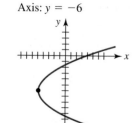

$(y + 6)^2 = 4(x + 5)$

**15.** Vertex: $(-\frac{5}{2}, -1)$
Focus: $(-\frac{5}{2}, -\frac{15}{16})$
Directrix: $y = -\frac{17}{16}$
Axis: $x = -\frac{5}{2}$

$(x + \frac{5}{2})^2 = \frac{1}{4}(y + 1)$

**17.** Vertex: $(3, 4)$
Focus: $(\frac{5}{2}, 4)$
Directrix: $x = \frac{7}{2}$
Axis: $y = 4$

**19.** Vertex: $(0, 0)$
Focus: $(0, \frac{1}{8})$
Directrix: $y = -\frac{1}{8}$
Axis: $x = 0$

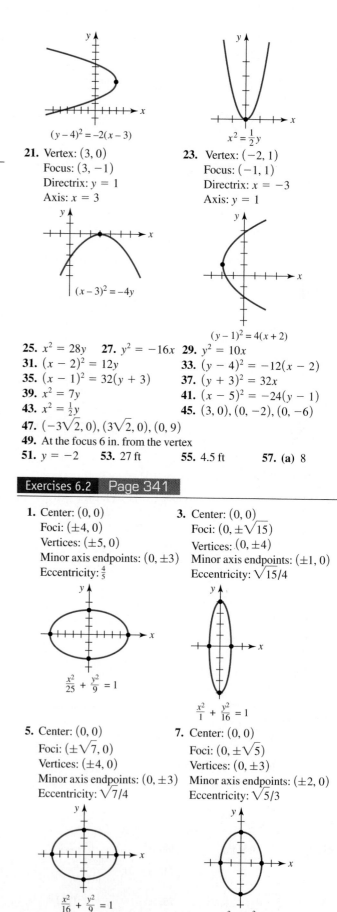

$(y - 4)^2 = -2(x - 3)$

$x^2 = \frac{1}{2}y$

**21.** Vertex: $(3, 0)$
Focus: $(3, -1)$
Directrix: $y = 1$
Axis: $x = 3$

$(x - 3)^2 = -4y$

**23.** Vertex: $(-2, 1)$
Focus: $(-1, 1)$
Directrix: $x = -3$
Axis: $y = 1$

$(y - 1)^2 = 4(x + 2)$

**25.** $x^2 = 28y$ **27.** $y^2 = -16x$ **29.** $y^2 = 10x$

**31.** $(x - 2)^2 = 12y$ **33.** $(y - 4)^2 = -12(x - 2)$

**35.** $(x - 1)^2 = 32(y + 3)$ **37.** $(y + 3)^2 = 32x$

**39.** $x^2 = 7y$ **41.** $(x - 5)^2 = -24(y - 1)$

**43.** $x^2 = \frac{1}{2}y$ **45.** $(3, 0), (0, -2), (0, -6)$

**47.** $(-3\sqrt{2}, 0), (3\sqrt{2}, 0), (0, 9)$

**49.** At the focus 6 in. from the vertex

**51.** $y = -2$ **53.** 27 ft **55.** 4.5 ft **57. (a)** 8

**Exercises 6.2** Page 341

**1.** Center: $(0, 0)$
Foci: $(\pm 4, 0)$
Vertices: $(\pm 5, 0)$
Minor axis endpoints: $(0, \pm 3)$
Eccentricity: $\frac{4}{5}$

$\frac{x^2}{25} + \frac{y^2}{9} = 1$

**3.** Center: $(0, 0)$
Foci: $(0, \pm\sqrt{15})$
Vertices: $(0, \pm 4)$
Minor axis endpoints: $(\pm 1, 0)$
Eccentricity: $\sqrt{15}/4$

$\frac{x^2}{1} + \frac{y^2}{16} = 1$

**5.** Center: $(0, 0)$
Foci: $(\pm\sqrt{7}, 0)$
Vertices: $(\pm 4, 0)$
Minor axis endpoints: $(0, \pm 3)$
Eccentricity: $\sqrt{7}/4$

$\frac{x^2}{16} + \frac{y^2}{9} = 1$

**7.** Center: $(0, 0)$
Foci: $(0, \pm\sqrt{5})$
Vertices: $(0, \pm 3)$
Minor axis endpoints: $(\pm 2, 0)$
Eccentricity: $\sqrt{5}/3$

$\frac{x^2}{4} + \frac{y^2}{9} = 1$

**9.** Center: $(1, 3)$
Foci: $(1 \pm \sqrt{13}, 3)$
Vertices: $(-6, 3), (8, 3)$
Minor axis endpoints: $(1, -3), (1, 9)$
Eccentricity: $\sqrt{13}/7$

$$\frac{(x-1)^2}{16} + \frac{(y-3)^2}{36} = 1$$

**11.** Center: $(-5, -2)$
Foci: $(-5, -2 \pm \sqrt{15})$
Vertices: $(-5, -6), (-5, 2)$
Minor axis endpoints: $(-6, -2)$,
$(-4, -2)$
Eccentricity: $\sqrt{15}/4$

$$\frac{(x+5)^2}{1} + \frac{(y+1)^2}{16} = 1$$

**13.** Center: $(0, -\frac{1}{2})$
Foci: $(0, -\frac{1}{2} \pm \sqrt{3})$
Vertices: $(0, -\frac{5}{2}), (0, \frac{3}{2})$
Minor axis endpoints: $(-1, -\frac{1}{2}), (1, -\frac{1}{2})$
Eccentricity: $\sqrt{3}/2$

$$\frac{x^2}{1} + \frac{(y+\frac{1}{2})^2}{4} = 1$$

**15.** Center: $(1, -2)$
Foci: $(1, -2 \pm \sqrt{6})$
Vertices: $(1, -2 \pm \sqrt{15})$
Minor axis endpoints: $(-2, -2), (4, -2)$
Eccentricity: $\sqrt{\frac{2}{5}}$

$$\frac{(x-1)^2}{9} + \frac{(y+2)^2}{15} = 1$$

**17.** Center: $(2, -1)$
Foci: $(2, -5), (2, 3)$
Vertices: $(2, -6), (2, 4)$
Minor axis endpoints: $(-1, -1), (5, -1)$
Eccentricity: $\frac{4}{5}$

$$\frac{(x-2)^2}{9} + \frac{(y+1)^2}{25} = 1$$

**19.** Center: $(0, -3)$
Foci: $(\pm\sqrt{6}, -3)$
Vertices: $(-3, -3), (3, -3)$
Minor axis endpoints: $(0, -3 \pm \sqrt{3})$
Eccentricity: $\sqrt{6}/3$

$$\frac{x^2}{9} + \frac{(y+3)^2}{3} = 1$$

**21.** $\dfrac{x^2}{25} + \dfrac{y^2}{16} = 1$

**23.** $\dfrac{x^2}{8} + \dfrac{y^2}{9} = 1$

**25.** $\dfrac{x^2}{1} + \dfrac{y^2}{9} = 1$

**27.** $\dfrac{(x-1)^2}{16} + \dfrac{(y+3)^2}{4} = 1$

**29.** $\dfrac{x^2}{9} + \dfrac{y^2}{13} = 1$

**31.** $\dfrac{x^2}{11} + \dfrac{y^2}{9} = 1$

**33.** $\dfrac{x^2}{3} + \dfrac{y^2}{12} = 1$

**35.** $\dfrac{x^2}{16} + \dfrac{(y-1)^2}{4} = 1$

**37.** $\dfrac{(x-1)^2}{7} + \dfrac{(y-3)^2}{16} = 1$

**39.** $\dfrac{x^2}{45} + \dfrac{(y-2)^2}{9} = 1$

**41.** greatest distance is 43.5 millions miles; least distance is 28.5 million miles

**43.** approximately 0.97    **45.** 12 ft

**47.** The piece of string should be 4 ft long. The tacks should be placed $\sqrt{7}/2$ ft from the center of the rectangle on the major axis of the ellipse.

**49.** on the major axis, 12 ft to either side from the center of the room

**51.** $5x^2 - 4xy + 8y^2 - 24x - 48y = 0$

**Exercises 6.3    Page 350**

**1.** Center: $(0, 0)$
Foci: $(\pm\sqrt{41}, 0)$
Vertices: $(\pm 4, 0)$
Asymptotes: $y = \pm\frac{5}{4}x$
Eccentricity: $\sqrt{41}/4$

$$\frac{x^2}{16} - \frac{y^2}{25} = 1$$

**3.** Center: $(0, 0)$
Foci: $(0, \pm\sqrt{73})$
Vertices: $(0, \pm 8)$
Asymptotes: $y = \pm\frac{8}{3}x$
Eccentricity: $\sqrt{73}/8$

$$\frac{y^2}{64} - \frac{x^2}{9} = 1$$

**5.** Center: $(0, 0)$
Foci: $(\pm 2\sqrt{5}, 0)$
Vertices: $(\pm 4, 0)$
Asymptotes: $y = \pm\frac{1}{2}x$
Eccentricity: $\sqrt{5}/2$

$$\frac{x^2}{16} - \frac{y^2}{4} = 1$$

**7.** Center: $(0, 0)$
Foci: $(0, \pm 2\sqrt{6})$
Vertices: $(0, \pm 2\sqrt{5})$
Asymptotes: $y = \pm\sqrt{5}x$
Eccentricity: $\sqrt{\frac{6}{5}}$

$$\frac{y^2}{20} - \frac{x^2}{4} = 1$$

**9.** Center: $(5, -1)$
Foci: $(5 \pm \sqrt{53}, -1)$
Vertices: $(3, -1), (7, -1)$
Asymptotes: $y = -1 \pm \frac{7}{2}(x - 5)$
Eccentricity: $\sqrt{53}/2$

$$\frac{(x-5)^2}{4} - \frac{(y+1)^2}{49} = 1$$

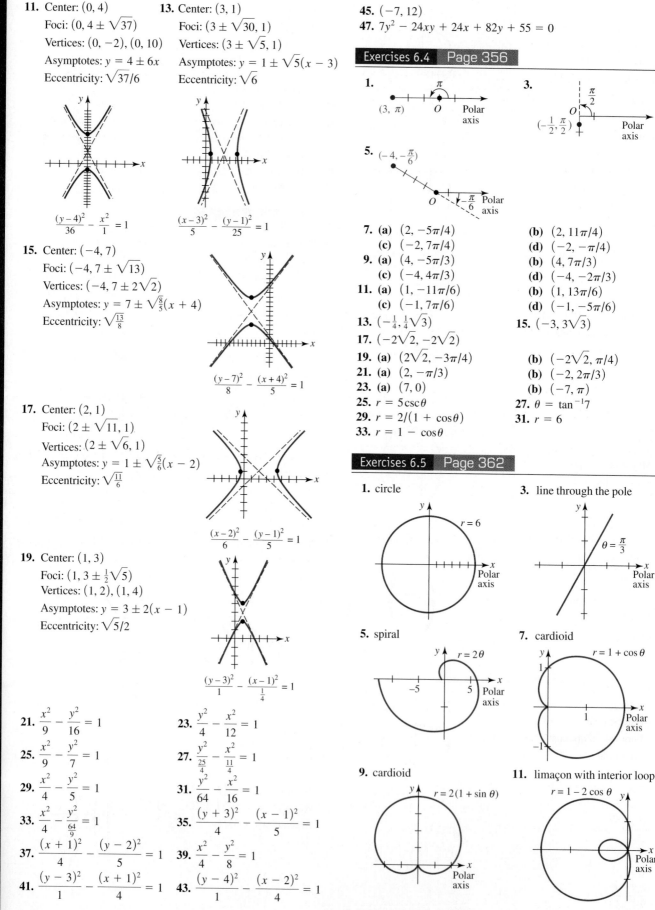

**11.** Center: $(0, 4)$
Foci: $(0, 4 \pm \sqrt{37})$
Vertices: $(0, -2), (0, 10)$
Asymptotes: $y = 4 \pm 6x$
Eccentricity: $\sqrt{37}/6$

$$\frac{(y-4)^2}{36} - \frac{x^2}{1} = 1$$

**13.** Center: $(3, 1)$
Foci: $(3 \pm \sqrt{30}, 1)$
Vertices: $(3 \pm \sqrt{5}, 1)$
Asymptotes: $y = 1 \pm \sqrt{5}(x - 3)$
Eccentricity: $\sqrt{6}$

$$\frac{(x-3)^2}{5} - \frac{(y-1)^2}{25} = 1$$

**15.** Center: $(-4, 7)$
Foci: $(-4, 7 \pm \sqrt{13})$
Vertices: $(-4, 7 \pm 2\sqrt{2})$
Asymptotes: $y = 7 \pm \sqrt{\frac{8}{5}}(x + 4)$
Eccentricity: $\sqrt{\frac{13}{8}}$

$$\frac{(y-7)^2}{8} - \frac{(x+4)^2}{5} = 1$$

**17.** Center: $(2, 1)$
Foci: $(2 \pm \sqrt{11}, 1)$
Vertices: $(2 \pm \sqrt{6}, 1)$
Asymptotes: $y = 1 \pm \sqrt{\frac{5}{6}}(x - 2)$
Eccentricity: $\sqrt{\frac{11}{6}}$

$$\frac{(x-2)^2}{6} - \frac{(y-1)^2}{5} = 1$$

**19.** Center: $(1, 3)$
Foci: $(1, 3 \pm \frac{1}{2}\sqrt{5})$
Vertices: $(1, 2), (1, 4)$
Asymptotes: $y = 3 \pm 2(x - 1)$
Eccentricity: $\sqrt{5}/2$

$$\frac{(y-3)^2}{1} - \frac{(x-1)^2}{\frac{1}{4}} = 1$$

**21.** $\dfrac{x^2}{9} - \dfrac{y^2}{16} = 1$

**23.** $\dfrac{y^2}{4} - \dfrac{x^2}{12} = 1$

**25.** $\dfrac{x^2}{9} - \dfrac{y^2}{7} = 1$

**27.** $\dfrac{y^2}{\frac{25}{4}} - \dfrac{x^2}{\frac{11}{4}} = 1$

**29.** $\dfrac{x^2}{4} - \dfrac{y^2}{5} = 1$

**31.** $\dfrac{y^2}{64} - \dfrac{x^2}{16} = 1$

**33.** $\dfrac{x^2}{4} - \dfrac{y^2}{\frac{64}{9}} = 1$

**35.** $\dfrac{(y+3)^2}{4} - \dfrac{(x-1)^2}{5} = 1$

**37.** $\dfrac{(x+1)^2}{4} - \dfrac{(y-2)^2}{5} = 1$

**39.** $\dfrac{x^2}{4} - \dfrac{y^2}{8} = 1$

**41.** $\dfrac{(y-3)^2}{1} - \dfrac{(x+1)^2}{4} = 1$

**43.** $\dfrac{(y-4)^2}{1} - \dfrac{(x-2)^2}{4} = 1$

**45.** $(-7, 12)$
**47.** $7y^2 - 24xy + 24x + 82y + 55 = 0$

**Exercises 6.4   Page 356**

**1.** $(3, \pi)$   $O$   Polar axis   ($\pi$)

**3.** $\left(-\frac{1}{2}, \frac{\pi}{2}\right)$   $O$   $\frac{\pi}{2}$   Polar axis

**5.** $\left(-4, -\frac{\pi}{6}\right)$   $O$   $-\frac{\pi}{6}$   Polar axis

**7. (a)** $(2, -5\pi/4)$   **(b)** $(2, 11\pi/4)$
**(c)** $(-2, 7\pi/4)$   **(d)** $(-2, -\pi/4)$
**9. (a)** $(4, -5\pi/3)$   **(b)** $(4, 7\pi/3)$
**(c)** $(-4, 4\pi/3)$   **(d)** $(-4, -2\pi/3)$
**11. (a)** $(1, -11\pi/6)$   **(b)** $(1, 13\pi/6)$
**(c)** $(-1, 7\pi/6)$   **(d)** $(-1, -5\pi/6)$
**13.** $\left(-\frac{1}{4}, \frac{1}{4}\sqrt{3}\right)$   **15.** $(-3, 3\sqrt{3})$
**17.** $(-2\sqrt{2}, -2\sqrt{2})$
**19. (a)** $(2\sqrt{2}, -3\pi/4)$   **(b)** $(-2\sqrt{2}, \pi/4)$
**21. (a)** $(2, -\pi/3)$   **(b)** $(-2, 2\pi/3)$
**23. (a)** $(7, 0)$   **(b)** $(-7, \pi)$
**25.** $r = 5\csc\theta$   **27.** $\theta = \tan^{-1}7$
**29.** $r = 2/(1 + \cos\theta)$   **31.** $r = 6$
**33.** $r = 1 - \cos\theta$

**Exercises 6.5   Page 362**

**1.** circle

$r = 6$

**3.** line through the pole

$\theta = \frac{\pi}{3}$

**5.** spiral

$r = 2\theta$

**7.** cardioid

$r = 1 + \cos\theta$

**9.** cardioid

$r = 2(1 + \sin\theta)$

**11.** limaçon with interior loop

$r = 1 - 2\cos\theta$

**13.** dimpled limaçon

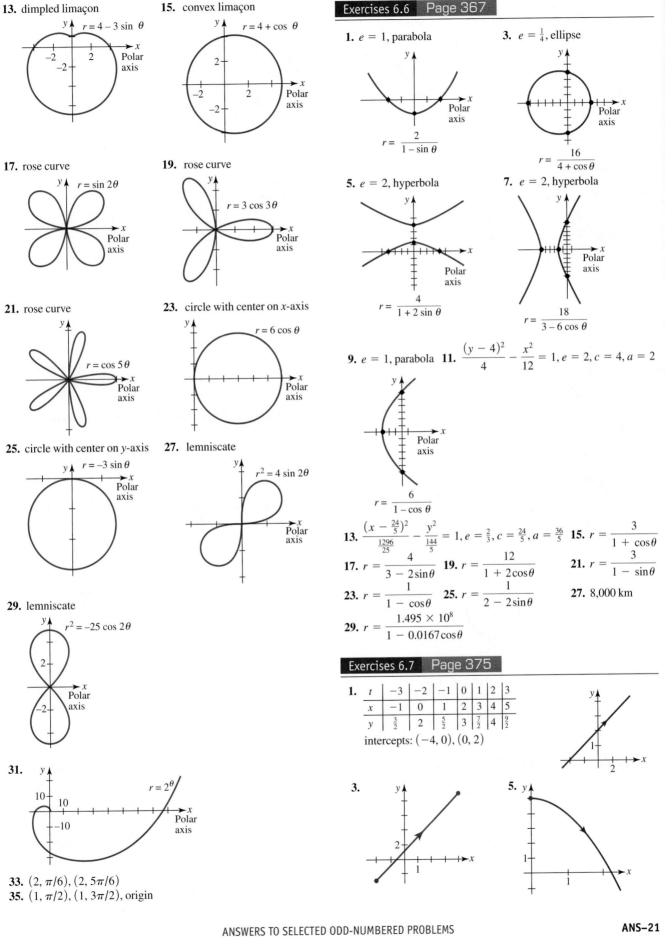

$r = 4 - 3\sin\theta$

**15.** convex limaçon

$r = 4 + \cos\theta$

**17.** rose curve

$r = \sin 2\theta$

**19.** rose curve

$r = 3\cos 3\theta$

**21.** rose curve

$r = \cos 5\theta$

**23.** circle with center on $x$-axis

$r = 6\cos\theta$

**25.** circle with center on $y$-axis

$r = -3\sin\theta$

**27.** lemniscate

$r^2 = 4\sin 2\theta$

**29.** lemniscate

$r^2 = -25\cos 2\theta$

**31.**

$r = 2\theta$

**33.** $(2, \pi/6), (2, 5\pi/6)$

**35.** $(1, \pi/2), (1, 3\pi/2),$ origin

**1.** $e = 1$, parabola

$r = \dfrac{2}{1 - \sin\theta}$

**3.** $e = \frac{1}{4}$, ellipse

$r = \dfrac{16}{4 + \cos\theta}$

**5.** $e = 2$, hyperbola

$r = \dfrac{4}{1 + 2\sin\theta}$

**7.** $e = 2$, hyperbola

$r = \dfrac{18}{3 - 6\cos\theta}$

**9.** $e = 1$, parabola  **11.** $\dfrac{(y-4)^2}{4} - \dfrac{x^2}{12} = 1, e = 2, c = 4, a = 2$

$r = \dfrac{6}{1 - \cos\theta}$

**13.** $\dfrac{(x - \frac{24}{5})^2}{\frac{1296}{25}} - \dfrac{y^2}{\frac{144}{5}} = 1, e = \frac{2}{3}, c = \frac{24}{5}, a = \frac{36}{5}$  **15.** $r = \dfrac{3}{1 + \cos\theta}$

**17.** $r = \dfrac{4}{3 - 2\sin\theta}$  **19.** $r = \dfrac{12}{1 + 2\cos\theta}$  **21.** $r = \dfrac{3}{1 - \sin\theta}$

**23.** $r = \dfrac{1}{1 - \cos\theta}$  **25.** $r = \dfrac{1}{2 - 2\sin\theta}$  **27.** 8,000 km

**29.** $r = \dfrac{1.495 \times 10^8}{1 - 0.0167\cos\theta}$

**1.**

| $t$ | $-3$ | $-2$ | $-1$ | $0$ | $1$ | $2$ | $3$ |
|---|---|---|---|---|---|---|---|
| $x$ | $-1$ | $0$ | $1$ | $2$ | $3$ | $4$ | $5$ |
| $y$ | $\frac{3}{2}$ | $2$ | $\frac{5}{2}$ | $3$ | $\frac{7}{2}$ | $4$ | $\frac{9}{2}$ |

intercepts: $(-4, 0), (0, 2)$

**3.**

**5.**

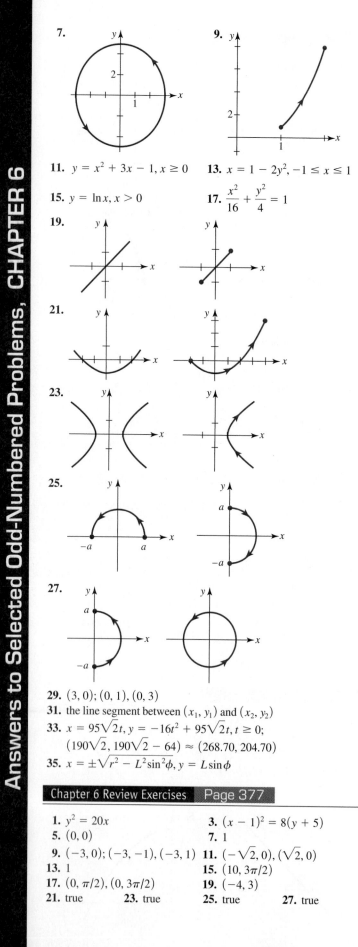

**7.**

**9.**

**11.** $y = x^2 + 3x - 1, x \geq 0$     **13.** $x = 1 - 2y^2, -1 \leq x \leq 1$

**15.** $y = \ln x, x > 0$     **17.** $\dfrac{x^2}{16} + \dfrac{y^2}{4} = 1$

**19.**

**21.**

**23.**

**25.**

**27.**

**29.** $(3, 0); (0, 1), (0, 3)$
**31.** the line segment between $(x_1, y_1)$ and $(x_2, y_2)$
**33.** $x = 95\sqrt{2}t, y = -16t^2 + 95\sqrt{2}t, t \geq 0;$
$(190\sqrt{2}, 190\sqrt{2} - 64) \approx (268.70, 204.70)$
**35.** $x = \pm\sqrt{r^2 - L^2\sin^2\phi}, y = L\sin\phi$

| Chapter 6 Review Exercises | Page 377 |

**1.** $y^2 = 20x$     **3.** $(x - 1)^2 = 8(y + 5)$
**5.** $(0, 0)$     **7.** 1
**9.** $(-3, 0); (-3, -1), (-3, 1)$     **11.** $(-\sqrt{2}, 0), (\sqrt{2}, 0)$
**13.** 1     **15.** $(10, 3\pi/2)$
**17.** $(0, \pi/2), (0, 3\pi/2)$     **19.** $(-4, 3)$
**21.** true     **23.** true     **25.** true     **27.** true

**29.** true     **31.** false     **33.** true     **35.** true
**37.** $(x - \tfrac{1}{2})^2 + (y - \tfrac{1}{2})^2 = \tfrac{1}{2}$     **39.** $r = 4\sin\theta$
**41.** $(0, 2), (0, -\tfrac{2}{3})$
**43.** (a) $(\sqrt{6}, -\pi/4)$     (b) $(-\sqrt{6}, 3\pi/4)$
**45.** Center $(1, 1)$,
Vertices $(-3, 1), (5, 1)$,
Endpoints of minor axis
$(1, 3), (1, -1)$

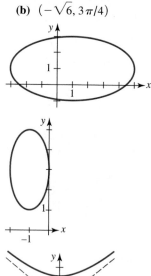

**47.** Center $(-1, 3)$,
Vertices $(-1, 5), (-1, 1)$,
Endpoints of minor axis
$(-2, 3), (0, 3)$

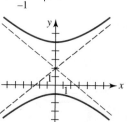

**49.** Center $(0, 2)$,
Vertices $(0, -1), (0, 5)$,
Asymptotes $y = 2 - \tfrac{3}{4}x$,
$y = 2 + \tfrac{3}{4}x$

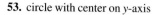

**51.** circle of radius 5 centered at the origin     **53.** circle with center on $y$-axis

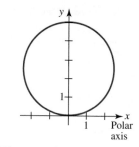

**55.** cardioid     **57.** convex limaçon

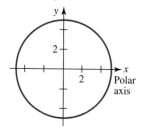
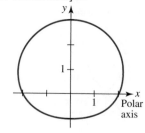

**59.** rose curve     **61.** ellipse

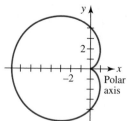
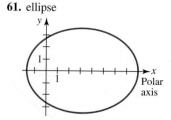

**63.** $1.95 \times 10^9$ m

# Index

Index

# D

# E

Index

Index

Index

# Credits

## Photo Credits

**Page iii (Calculus equations)**, © Digital Vision/Getty Images; **Page iii (Neon sign)**, © Paul A. Souders/ Corbis; **Page iv (Golden Gate Bridge)**, © Bill Draven/ShutterStock, Inc.; **Page iv (Violin)**, © GeoM/ ShutterStock, Inc.; **Page iv (St. Louis Arch)**, © Jose Gil/ShutterStock, Inc.; **Page v (Saturn)**, © AbleStock.

**Chapter 1**: **Page 2**, © Digital Vision/Getty Images; **Page 9**, © LWA-Dann Tardif/Corbis; **Page 11**, © Blend Images/Jupiterimages; **Page 42**, © Thom Lang/Corbis.

**Chapter 2**: **Page 48**, © Paul A. Souders/Corbis; **Page 54**, Courtesy of Texas Instruments; **Page 62**, © Paul A. Souders/Corbis; **Page 70**, © Andrew G. Davis/ShutterStock, Inc.; **Page 83**, © Grapes - Michaud/Photo Researchers, Inc.; **Page 95**, Courtesy of NASA; **Page 111**, © Paul Edmondson/Corbis; **Page 115**, © haveseen/ShutterStock, Inc.

**Chapter 3**: **Page 128**, © Bill Draven/ShutterStock, Inc.; **Page 153**, Printed with permission by Norway Post, Philatelic Service; **Page 169**, © Comstock Images/Alamy Images; **Page 177**, © Corbis; **Page 183**, © Tony Stewart/Corbis.

**Chapter 4**: **Page 190**, © GeoM/ShutterStock, Inc.; **Page 205**, © Pétur Ásgeirsson/ShutterStock, Inc.; **Page 232**, © G. Schuster/zefa/Corbis; **Page 254**, Courtesy of the National Park Service; **Page 263 (Mt. Rushmore)**, © Paul Fries/ShutterStock, Inc.; **Page 263 (sundial)**, © Peter Elvidge/ ShutterStock, Inc.; **Page 264**, © Mary Lane/ShutterStock, Inc.; **Page 271**, © Gabe Palmer/Corbis.

**Chapter 5**: **Page 288**, © Jose Gil/ShutterStock, Inc.; **Page 303**, © Wolfgang Flamisch/zefa/Corbis; **Page 306**, © Phototake/Alamy Images; **Page 307 (Thomas R. Malthus)**, © National Library of Medicine; **Page 307 (Marie Curie)**, © Mary Evans Picture Library/Alamy Images; **Page 308 (ibuprofen)**, © Jones and Bartlett Publishers. Photographed by Kimberly Potvin; **Page 308 (Willard Libby)**, © Time Inc./Time Life Pictures/Getty Images; **Page 308 (Psalms scroll)**, Courtesy of the Israel Antiquities Authority; **Page 309**, © Johanna Goodyear/ ShutterStock, Inc.; **Page 311**, © AP Photos; **Page 314 (cave drawing)**, © Robert Harding Picture Library Ltd./Alamy Images; **Page 314 (Shroud of Turin)**, © Targa/age footstock; **Page 315 (iceman)**, Photo courtesy of the Photo Archives South Tyrol Museum of Archaeology - www.iceman.it; **Page 315 (thermometer)**, © Ron Hilton/ShutterStock, Inc.; **Page 316**, © Thomas Weißenfels/ShutterStock, Inc.; **Page 317**, Courtesy of USGS; **Page 318**, © Neal Preston/Corbis; **Page 322**, © Jose Gil/ShutterStock, Inc.

**Chapter 6**: **Page 328**, © AbleStock; **Page 329 (Hypatia)**, Photo courtesy of Wikipedia; **Page 329 (solar system)**, Courtesy of NASA/JPL; **Page 334 (searchlight)**, © Popperfoto/Alamy Images; **Page 334 (satellite dishes)**, © Soundsnaps/ShutterStock, Inc.; **Page 334 (Brooklyn Bridge)**, © where-@tiscali.it/ShutterStock, Inc.; **Page 334 (basketball game)**, © Corbis; **Page 340**, © Brand X Pictures/Alamy Images; **Page 358**, © Kenneth Libbrecht/Visuals Unlimited; **Page 367**, Courtesy of Mariner 10, Astrogeology Team, and USGS; **Page 368**, Courtesy of the Observatories of the Carnegie Institution of Washington; **Page 374 (equations)**, © Corbis; **Page 374 (double helix)**, Courtesy of Robert Guy/National Cancer Institute.

# Review of Trigonometry

## Unit Circle Definition of Sine and Cosine

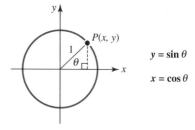

$y = \sin\theta$

$x = \cos\theta$

## Other Trigonometric Functions

$$\tan\theta = \frac{y}{x} = \frac{\sin\theta}{\cos\theta}, \qquad \cot\theta = \frac{x}{y} = \frac{\cos\theta}{\sin\theta}$$

$$\sec\theta = \frac{1}{x} = \frac{1}{\cos\theta}, \qquad \csc\theta = \frac{1}{y} = \frac{1}{\sin\theta}$$

## Right Triangle Definition of Sine and Cosine

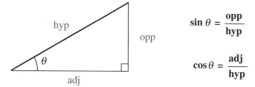

$$\sin\theta = \frac{\text{opp}}{\text{hyp}}$$

$$\cos\theta = \frac{\text{adj}}{\text{hyp}}$$

## Other Trigonometric Functions

$$\tan\theta = \frac{\text{opp}}{\text{adj}}, \qquad \cot\theta = \frac{\text{adj}}{\text{opp}}$$

$$\sec\theta = \frac{\text{hyp}}{\text{adj}}, \qquad \csc\theta = \frac{\text{hyp}}{\text{opp}}$$

## Signs of Sine and Cosine

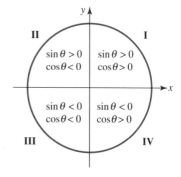

## Values of Sine and Cosine for Special Angles

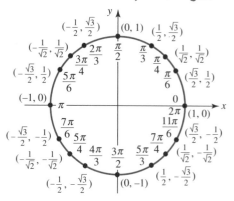

## Bounds for Sine and Cosine

$$-1 \le \sin\theta \le 1 \quad \text{and} \quad -1 \le \cos\theta \le 1$$

## Periodicity of Trigonometric Functions

$$\sin(\theta + 2\pi) = \sin\theta, \qquad \cos(\theta + 2\pi) = \cos\theta$$

$$\sec(\theta + 2\pi) = \sec\theta, \qquad \csc(\theta + 2\pi) = \csc\theta$$

$$\tan(\theta + \pi) = \tan\theta, \qquad \cot(\theta + \pi) = \cot\theta$$

## Cofunction Identities

$$\sin\left(\frac{\pi}{2} - \theta\right) = \cos\theta$$

$$\cos\left(\frac{\pi}{2} - \theta\right) = \sin\theta$$

$$\tan\left(\frac{\pi}{2} - \theta\right) = \cot\theta$$

## Pythagorean Identities

$$\sin^2\theta + \cos^2\theta = 1$$

$$1 + \tan^2\theta = \sec^2\theta$$

$$1 + \cot^2\theta = \csc^2\theta$$

## Even/Odd Identities

| Even | Odd |
|---|---|
| $\cos(-\theta) = \cos\theta$ | $\sin(-\theta) = -\sin\theta$ |
| $\sec(-\theta) = \sec\theta$ | $\csc(-\theta) = -\csc\theta$ |
| | $\tan(-\theta) = -\tan\theta$ |
| | $\cot(-\theta) = -\cot\theta$ |